Contemporary Developments in Mathematical Psychology

VOLUME II

Contemporary Developments
in Mathematical Psychology

VOLUME II

MEASUREMENT, PSYCHOPHYSICS, AND NEURAL INFORMATION PROCESSING

EDITED BY

DAVID H. KRANTZ
THE UNIVERSITY OF MICHIGAN

R. DUNCAN LUCE
UNIVERSITY OF CALIFORNIA, IRVINE

RICHARD C. ATKINSON
STANFORD UNIVERSITY

PATRICK SUPPES
STANFORD UNIVERSITY

W. H. FREEMAN AND COMPANY
San Francisco

Library of Congress Cataloging in Publication Data

Main entry under title:
Contemporary developments in mathematical psychology.

Includes bibliographies.
CONTENTS: v. 1. Learning, memory, and thinking.
v. 2. Measurement, psychophysics, and neural information processing.
1. Human behavior—Mathematical models. 2. Psychometrics. I. Krantz, David H., 1938– ed.
[DNLM: 1. Psychometrics. BF39 C761]
BF39.C59 150′.1′84 73-21887
ISBN 0-7167-0849.3 (v. 2)

Printed in the United States of America.

9 8 7 6 5 4 3 2 1

Contents

Introduction

The two volumes of this symposium are an attempt by the editors to define the concept of progress in mathematical psychology. The definition is ostensive, by means of a series of exemplars; this preface attempts to define the same concept briefly by means of attributes and to note and to account for possible discrepancies between the two definitions.

A necessary feature of progress in science is its cumulative nature. We interpret this quite strictly: no experiment or theory, however clever, sophisticated, or compelling, fits our concept of progress unless it forms part of a growing chain or lattice of development. Thus, our criteria for excellence in mathematical psychology need not be the same as those for progress or development. It is considerably easier to judge excellence; that is what we usually do when refereeing articles. Excellent work may not lead to progress, for any of a variety of reasons: it may fall outside the mainstream of development, or it may reveal clearly why a particular approach is a blind alley, or it may be ahead of its time. On the other hand, it is quite possible that work that contains glaring flaws may nevertheless provide a key element on which all later work in the same field will depend.

It is easy to point to excellent work in mathematical psychology, past and present. But in retrospect, cumulative progress is less easy to find. Psychology is full of beginnings, and in the past two decades mathematical psychology has contributed no small part of these. One need not be apologetic about the sinking of many pillars that support little superstructure. Historically, cumulative progress is the exception rather than the rule in the search for knowledge. Nevertheless, without apologizing for the past, we do need to ask ourselves whether we can do better in the coming decades. Is it possible that the lack of cumulative progress is partly due to that goal being subordinated to others, such as originality or technical mastery? If so, then that goal needs to be formulated more directly, and seeking it needs to be encouraged.

Our conception, in this project, was to identify a series of areas in which mathematical methods (broadly defined) have played an integral role in the recent cumulative development of psychological knowledge. Our goal is to influence the scientific judgments of mathematical psychologists, both in respect to choice of problem area and in respect to standards of accomplishment. We are not trying to influence the level of technical excellence, nor to emphasize creativity, but rather to direct attention to the criterion of cumulative substantive progress. We wrote to potential authors as follows:

> The purpose of this project is to survey research areas in which mathematical methods have led, in recent years, to advances in psychological knowledge that seem relatively permanent. The majority of the contributions will exposit research in which the use of formal models and/or quantitative methods has produced substantive knowledge that is known, or ought to be known, by all workers in the given field, regardless of their mathematical orientation. Other contributions—though a minority—will exposit advances in mathematical methodology that seem to have wide applicability.
>
> The volume should constitute a set of paradigms for Mathematical Psychology, in two senses. First, it may influence mathematically oriented graduate students and fellow scientists to enter an area in which real progress has been made; their work will be likely to extend the progress. Second, it should set a standard of quality of achievement, influencing thereby the goals of psychologists in all fields of research.

Naturally, perturbations have occurred along the way. We were undoubtedly wrong in a good many of our decisions, both for inclusion and for exclusion of particular areas. Moreover, some chapters did not turn out exactly as we expected, and three or four that we invited and very much wanted to include were not forthcoming. Since these remarks, though perfectly truthful, sound like a hedge against criticism of our final product, it may be well for us to discuss more fully a few specific inclusions and exclusions.

Perhaps the most striking exclusion is the entire area of preferential choice. There is no lack whatever of technically excellent papers in this area, but they give no sense of any real cumulation of knowledge. What are the established laws of preferential choice behavior? (Since three of the editors have worked in this area, our attitude may reflect some measure of our own frustration.)

Perhaps the most striking inclusion is the strong emphasis on sensory mechanisms in this volume. There is a chapter on vision and one on hearing (another chapter on vision was invited but was not written), and sensory mechanisms comprise the background and part of the substance of the three chapters on psychophysics as well as the chapter on brain function. The sensory emphasis may seem disproportionate when compared with the set of publications in mathematical psychology, but we think it is a natural consequence of our criterion of cumulative progress. Nowhere in psychology is there as much cumulative knowledge, much of it in quantitative form, as in the sensory area; so it is natural that new theoretical contributions should continue to build on the established paradigms.

Originally, we planned to include six chapters on methodological advances that we felt are so important that they are bound to have lasting value in psychology. They were: one chapter on the use of normative models, one on stochastic processes, and four on measurement and scaling. The chapter on normative models was not written; the one on stochastic processes turned out to be largely substantive (though we feel that perhaps its greatest contribution is in illustrating the advantage, and even the necessity, of translating vague verbal theories into precise models that have definite predictions); and one author, produced instead of a methodological piece, a new and integrative psychophysical theory. The latter is not in the least an exemplar of our original concept (though it could become one in time), but it is an interesting nonexemplar. In the end, we no longer tried to separate sharply the methodological from the substantive contributions. The chapter on covariance models is perhaps the most obviously methodological one, but even in it the examples illustrate important substantive developments in test theory.

During the summer of 1972, some of the authors and the editors met in Ann Arbor for three days presenting and discussing drafts of these chapters. On some issues there was surprising convergence—for example, no one seemed to take a reductionist position on the nature of truth in mathematical theorizing. On others, there was broad disagreement—for example, the future of learning theory, and the very criteria of cumulative progress. The areas of memory and thinking are beset with conflicting approaches; one critique is offered in the comments by an invited discussant.

The preparation of these two volumes and the 1972 summer conference were supported by funds from the Mathematical Social Science Board, which supervises a National Science Foundation grant administered through the Center for Advanced Study in the Behavioral Sciences. The editors wish to acknowledge the administrative contributions of the University of Michigan and of the Institute for Mathematical Studies in the Social Sciences, Stanford University. More specifically, we wish to mention the contributions of Amy Horowitz in the processing of manuscripts and organization of the summer conference and of Lillian O'Toole for copy editing and following up problems in the final manuscripts.

April 1974 *The Editors*

Contemporary Developments in Mathematical Psychology

VOLUME II

Analyzing Psychological Data by Structural Analysis of Covariance Matrices

Karl G. Jöreskog
UNIVERSITY INSTITUTE OF STATISTICS
UPPSALA, SWEDEN

1. INTRODUCTION

Quantitative studies in psychology, and in other behavioral sciences as well, usually involve several measurements that correlate with each other in various ways. Structural analysis of covariance matrices is a general method for analyzing such measurements in order to detect and assess latent sources of variation and covariation in the observed measurements. When successfully applied this technique yields latent variables that account for the intercorrelations between the observed variables and that contain all the essential information about the linear interrelationships among these variables.

It is convenient to distinguish between exploratory and confirmatory studies. In confirmatory studies the experimenter has obtained such knowledge about the variables measured that he is in a position to formulate a hypothesis that specifies the latent variables on which the observed variables depend. Such a hypothesis may arise because of a specified theory, a given experimental design, known experimental conditions, or as a result of previous studies based on extensive data. In exploratory analysis, on the other hand, no such knowledge is available, and the main object is to find a simple, but meaningful, interpretation of the experimental results.

In practice, the above distinction is not always clear cut. Many investigations are to some extent both exploratory and confirmatory, since they involve some variables of known and other variables of unknown composition. The former should be chosen with great care so that as much information as possible about the latter may be extracted. It is highly desirable that a hypothesis suggested by mainly exploratory procedures should subsequently be confirmed, or disproved, by obtaining new data and subjecting them to more rigorous statistical techniques.

In this chapter I describe a general method for analysis of covariance structures and give many examples of areas and problems in the behavioral sciences where this method is useful. Although the method may be used for exploratory analysis, it is most useful for confirmatory analysis and I shall mainly deal with this situation. In Section 2, the model is treated as a purely formal model without substantive interpretations, and the problems of estimation and testing of this model are briefly reviewed. Examples of particular models, as special cases of the general model, are presented in Sections 3 through 6 of the chapter.

2. THE GENERAL MODEL

Definition of the General Model

I shall describe a general method for analyzing data according to a general model for covariance structures. The model assumes that the population variance-covariance matrix $\boldsymbol{\Sigma}(p \times p) = (\sigma_{ij})$ of a set of variables has the form

$$\boldsymbol{\Sigma} = \mathbf{B}(\boldsymbol{\Lambda}\boldsymbol{\Phi}\boldsymbol{\Lambda}' + \boldsymbol{\Psi}^2)\mathbf{B}' + \boldsymbol{\Theta}^2, \tag{1}$$

where $\mathbf{B}(p \times q) = (\beta_{ik})$, $\boldsymbol{\Lambda}(q \times r) = (\lambda_{km})$, the symmetric matrix $\boldsymbol{\Phi}(r \times r) = (\varphi_{mn})$, and the diagonal matrices $\boldsymbol{\Psi}(q \times q) = (\delta_{kl}\psi_k)$ and $\boldsymbol{\Theta}(p \times p) = (\delta_{ij}\theta_i)$ are parameter matrices (δ_{ij} denotes Kronecker's delta, which is one if $i = j$ and zero otherwise). It is assumed that the mean vector of the variables is unconstrained so that the information about the covariance structure is provided by the usual sample variance-covariance matrix $\mathbf{S}(p \times p) = (s_{ij})$, which may be taken to be a correlation matrix if the model is scale free and if the units of measurements in the variables are arbitrary or irrelevant.

The covariance structure of Equation 1 arises when the observed variables $\mathbf{x}(p \times 1)$ are of the form

$$\mathbf{x} = \boldsymbol{\mu} + \mathbf{B}\boldsymbol{\Lambda}\boldsymbol{\xi} + \mathbf{B}\boldsymbol{\zeta} + \mathbf{e}, \tag{2}$$

where $\boldsymbol{\xi}(r \times 1)$, $\boldsymbol{\zeta}(q \times 1)$, and $\mathbf{e}(p \times 1)$ are uncorrelated random latent vectors, in general unobserved, with zero mean vectors and dispersion matrices $\boldsymbol{\Phi}$, $\boldsymbol{\Psi}^2$, and $\boldsymbol{\Theta}^2$, respectively, and where $\boldsymbol{\mu}$ is the mean vector of $\mathbf{x}$.

In any application of the model of Equation 1, the number of variables p is given by the data, and q and r are given by the particular application that the investigator has in mind. In any such application any parameter in $\mathbf{B}$, $\boldsymbol{\Lambda}$, $\boldsymbol{\Phi}$, $\boldsymbol{\Psi}$, or $\boldsymbol{\Theta}$ may be known a priori and one or more subsets of the remaining parameters may have identical but unknown values. Thus, parameters are of three kinds: (a) *fixed parameters* that have been assigned given values, (b) *constrained parameters* that are unknown, but equal to one or more other parameters, and (c) *free parameters* that are unknown and not constrained to be equal to any other parameter. The advantage of such an approach is the great generality and flexibility obtained by the various specifications that may be imposed. Thus the general model contains a wide range of specific models.

In the general model (Eq. 1), it should be noted that if $\mathbf{B}$ is replaced by $\mathbf{B}\mathbf{T}_1^{-1}$, $\boldsymbol{\Lambda}$ by $\mathbf{T}_1\boldsymbol{\Lambda}\mathbf{T}_2^{-1}$, $\boldsymbol{\Phi}$ by $\mathbf{T}_2\boldsymbol{\Phi}\mathbf{T}_2'$, and $\boldsymbol{\Psi}^2$ by $\mathbf{T}_1\boldsymbol{\Psi}^2\mathbf{T}_1'$ while $\boldsymbol{\Theta}$ is left unchanged, then $\boldsymbol{\Sigma}$ is unaffected. This holds for all nonsingular matrices $\mathbf{T}_1(q \times q)$ and $\mathbf{T}_2(r \times r)$ such that $\mathbf{T}_1\boldsymbol{\Psi}^2\mathbf{T}_1'$ is diagonal. Hence in order to obtain a unique set of parameters and a corresponding unique set of estimates, some restrictions must be imposed. This may be done by defining certain fixed and constrained parameters. In some cases these restrictions are given in a natural way by the particular application intended. In other cases they can be chosen in an almost arbitrary way. To make sure that all indeterminacies have been eliminated, one should verify that the only transformations $\mathbf{T}_1$ and $\mathbf{T}_2$ that preserve the specifications about fixed and constrained parameters are identity matrices.

Estimation and Testing of the Model

To determine the estimates of the unknown parameters, two different methods of fitting the model to the observed data may be used. One is the *generalized least squares method* (GLS) that minimizes

$$G = \mathrm{tr}(\mathbf{I} - \mathbf{S}^{-1}\boldsymbol{\Sigma})^2; \tag{3}$$

the other is the maximum-likelihood method (ML) that minimizes

$$M = \log|\boldsymbol{\Sigma}| + \mathrm{tr}(\mathbf{S}\boldsymbol{\Sigma}^{-1}) - \log|\mathbf{S}| - p. \tag{4}$$

Both G and M are regarded as functions of the independent elements of $\mathbf{B}$, $\boldsymbol{\Lambda}$, $\boldsymbol{\Phi}$, $\boldsymbol{\Psi}$, $\boldsymbol{\Theta}$. For derivation and justification of the GLS method, see Jöreskog and Goldberger (1972). The function M is a transform of the likelihood function obtained under the assumption that the observed variables have a multivariate normal distribution (see, e.g., Jöreskog, 1969). In a large sample of size N, $(N - 1)$ times the minimum value of G or M may be used as a χ^2 to test the goodness of fit of the model and for both methods approximate standard errors may be obtained for each estimated parameter by computing the inverse of the information matrix. The χ^2 test is a test of the specified model against the most general alternative that $\boldsymbol{\Sigma}$ is any positive definitive matrix.

Suppose H_0 represents one model under given specifications of fixed, free, and constrained parameters. Then it is possible, in large samples, to test the model H_0 against any more general model H_1, by estimating each of them separately and comparing their χ^2 goodness-of-fit values. The difference in χ^2 is asymptotically a χ^2 with degrees of freedom equal to the corresponding difference in degrees of freedom. In many situations, it is possible to set up a sequence of hypotheses such that each one is a special case of the preceding and to test these hypotheses sequentially.

The values of χ^2 should be interpreted very cautiously. In most empirical work many of the hypotheses may not be realistic. If a sufficiently large sample were obtained, the test statistic would, no doubt, indicate that any such hypothesis is statistically untenable. The hypothesis should rather be that Equation 1 represents a reasonable approximation to the population variance-covariance matrix. From this point of view the statistical problem is not one of testing a given hypothesis (which a priori may be considered false), but rather one of fitting various models with different numbers of parameters and of deciding when to stop fitting. In other words, the problem is to extract as much information as possible out of a sample of given size without going so far that the result is affected to a large extent by 'noise.' It is reasonable and likely that more information can be extracted from a large sample than from a small sample. In such a problem the differences between χ^2 values matter rather than the χ^2 values themselves. In an exploratory study, if a value of χ^2 is obtained, which is large compared to the number of degrees of freedom, the fit may be examined by an inspection of the residuals, i.e., the discrepancies between observed and reproduced values. Often the results of an analysis, an inspection of residuals or other considerations will suggest ways to relax the model somewhat by introducing more parameters. The new model usually yields a smaller χ^2. A large drop in χ^2, compared to the difference in degrees of freedom, indicates that the changes made in the model represent a real improvement. On the other hand, a drop in χ^2 close to the difference in number of degrees of freedom indicates that the improvement in fit is obtained by 'capitalizing on chance,' and the added parameters may not have real significance and meaning.

The function G or M is minimized numerically with respect to the independent parameters using a modification of the iterative method of Fletcher and Powell (1963) (see Gruvaeus & Jöreskog, 1970). The minimization method makes use of the first-order derivatives and large sample approximations to the elements of the matrix of second-order derivatives. The essential formulas and the basic algorithm are given in Jöreskog (1970a), and a computer program ACOVS is described in Jöreskog, Gruvaeus and van Thillo (1970).

In the following sections, several examples are given of models that are useful in the behavioral sciences, and some of these models are illustrated by means of real data. All analyses presented are based on the ML method and all χ^2 values have been obtained as $(N - 1)$ times the minimum value of M. Sometimes a probability level P for a χ^2 value is given that refers to the probability of obtaining a χ^2 larger than that actually obtained, given that the hypothesized model holds.

In presenting some models, it is sometimes convenient to use a path diagram in which the observed variables are enclosed in squares and latent variables in circles. Residuals (errors in equations) and errors of measurement are not enclosed. A one-way arrow between two variables indicates a direct causal influence of one variable on another, whereas a two-way arrow indicates correlation or covariation between two variables not dependent on other variables in the model.

3. MODELS FOR SETS OF CONGENERIC TESTS

Test Theory Models

Most measurements employed in the behavioral sciences contain sizable errors of measurements and any adequate theory or model must take this fact into account. Of particular importance is the study of congeneric measurements, i.e., those measurements that are assumed to measure the same thing. Jöreskog (1971) considered several models for sets of congeneric tests and some material in this section is drawn from that paper.

Classical test theory (Lord & Novick, 1968) assumes that a test score x is the sum of a true score τ and an error score e, where e and τ are uncorrelated. A set of test scores $x_1, \ldots, x_p$ with true scores $\tau_1, \ldots, \tau_p$ is said to be congeneric if every pair of true scores τ_i and τ_j have unit correlation. Such a set of test scores can be represented as

$$\mathbf{x} = \boldsymbol{\mu} + \boldsymbol{\beta}\tau + \mathbf{e},$$

where $\mathbf{x}' = (x_1, \ldots, x_p)$, $\boldsymbol{\beta}' = (\beta_1, \ldots, \beta_p)$ is a vector of regression coefficients, $\mathbf{e}' = (e_1, \ldots, e_p)$ is the vector of error scores, $\boldsymbol{\mu}$ is the mean vector of $\mathbf{x}$, and τ is a true score, scaled to zero mean and unit variance for convenience. The elements of $\mathbf{x}$, $\mathbf{e}$, and τ are regarded as random variables for a population of examinees. Let $\theta_1^2, \ldots, \theta_p^2$ be the variances of $e_1, \ldots, e_p$, respectively, i.e., the error variances. The corresponding true score variances are $\beta_1^2, \ldots, \beta_p^2$. One important problem is that of estimating these quantities. The variance-covariance matrix of $\mathbf{x}$ is

$$\boldsymbol{\Sigma} = \boldsymbol{\beta\beta}' + \boldsymbol{\Theta}^2, \tag{5}$$

where $\boldsymbol{\Theta} = \text{diag}(\theta_1, \ldots, \theta_p)$. This is a special case of Equation 1 obtained by specifying $q = r = 1$, $\mathbf{B} = \boldsymbol{\beta}$, $\boldsymbol{\Lambda} = \boldsymbol{\Phi} = 1$, and $\boldsymbol{\Psi} = 0$. The congeneric test model with four tests is illustrated in Figure 1.

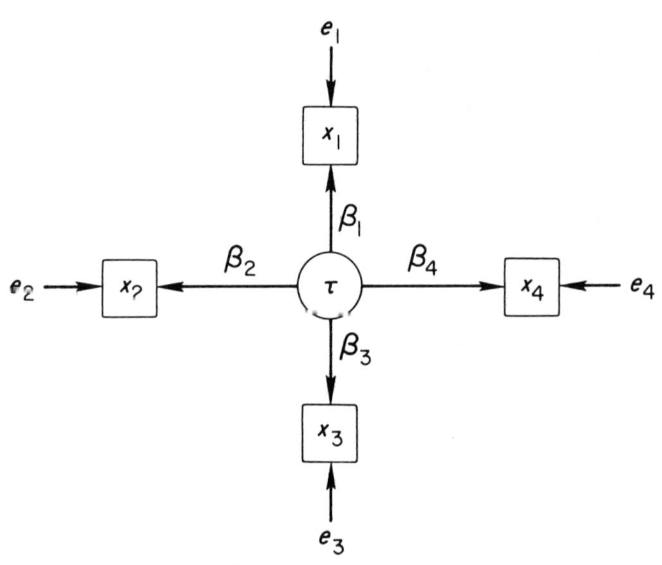

FIGURE 1.
The congeneric test model with four tests.

Parallel tests and tau-equivalent tests, in the sense of Lord and Novick (1968), are special cases of congeneric tests. Parallel tests have equal true score variances and equal error variances, i.e.,

$$\beta_1^2 = \cdots = \beta_p^2, \theta_1^2 = \cdots = \theta_p^2.$$

Tau-equivalent tests have equal true score variances, but possibly different error variances. These two models are obtained from Equation 1 by specification of equality of the corresponding set of parameters.

Parallel and tau-equivalent tests are homogenous in the sense that all covariances between pairs of test scores are equal. Scores on such tests are directly comparable, i.e., they represent measurements on the same scale. For tests composed of binary items this can hold only if the tests have the same number of items and are administrated under the same time limits. Congeneric tests, on the other hand, need not satisfy such strong restrictions. They need not even be tests consisting of items but can consist of ratings, for example, or even measurements produced by different measuring instruments.

Recently Kristof (1971) developed a model for tests that differ only in length. This model assumes that there is a 'length' parameter β_i associated with each test score x_i in such a way that the true score variance is proportional to β_i^4 and that the error variance is proportional to β_i^2. It can be shown that the covariance structure for this model is of the form

$$\boldsymbol{\Sigma} = \mathbf{D}_\beta(\boldsymbol{\beta}\boldsymbol{\beta}' + \psi^2\mathbf{I})\mathbf{D}_\beta,$$

where $\mathbf{D}_\beta = \text{diag}(\beta_1, \beta_2, \ldots, \beta_p)$, and $\boldsymbol{\beta}' = (\beta_1, \beta_2, \ldots, \beta_p)$. This is a special case of Equation 1, obtained by specifying $q = p$, $r = 1$, $\mathbf{B} = \mathbf{D}_\beta$, $\boldsymbol{\Lambda} = \boldsymbol{\beta}$, $\boldsymbol{\Phi} = 1$, $\boldsymbol{\Psi}^2 = \psi^2\mathbf{I}$, and $\boldsymbol{\Theta} = \mathbf{0}$. It should be noted that this model specifies equality constraints between the diagonal elements of $\mathbf{B}$ and the elements of the column vector $\boldsymbol{\Lambda}$, and also the equality of all the diagonal elements of $\boldsymbol{\Psi}$. The model has $p + 1$ independent parameters and is less restrictive than the parallel model, but is more restrictive than the congeneric model. A summary of the various test theory models and their number of parameters is given in Table 1. In the table, $\mathbf{j}$ denotes a column vector with all elements equal to one.

TABLE 1
Various test theory models

Model	Covariance Structure	No. of Parameters
Parallel	$\boldsymbol{\Sigma} = \beta^2\mathbf{jj}' + \theta^2\mathbf{I}$	2
Tau-equivalent	$\boldsymbol{\Sigma} = \beta^2\mathbf{jj}' + \boldsymbol{\theta}^2$	$p + 1$
Variable-length	$\boldsymbol{\Sigma} = \mathbf{D}_\beta(\boldsymbol{\beta}\boldsymbol{\beta}' + \psi^2\mathbf{I})\mathbf{D}_\beta$	$p + 1$
Congeneric	$\boldsymbol{\Sigma} = \boldsymbol{\beta}\boldsymbol{\beta}' + \boldsymbol{\theta}^2$	$2p$

As an illustration, consider the following variance-covariance matrix $\mathbf{S}$ taken from Kristof (1971):

$$\mathbf{S} = \begin{bmatrix} 54.85 & & \\ 60.21 & 99.24 & \\ 48.42 & 67.00 & 63.81 \end{bmatrix}.$$

This is based on candidates ($N = 900$) who took the January, 1969, administration of the Scholastic Aptitude Test (SAT). The first test, Verbal Omnibus, was administered in 30 minutes, and the second test, Reading Comprehension, in 45 minutes. These two tests contained 40 and 50 items, respectively. The third test is an additional section of the SAT not normally administered.

The following maximum-likelihood estimates were obtained with the program ACOVS: $\hat{\beta}_1 = 2.58$, $\hat{\beta}_2 = 3.03$, $\hat{\beta}_3 = 2.69$, and $\hat{\psi} = 1.60$. The goodness-of-fit test yielded $\chi^2 = 4.93$ with 2 degrees of freedom and a probability level of $P \doteq 0.09$.

A Statistical Model for Several Sets of Congeneric Test Scores

The previous model generalizes immediately to several sets of congeneric test scores. If there are q sets of such tests, with $m_1, m_2, \ldots, m_q$ tests, respectively, we write $\mathbf{x}' = (\mathbf{x}_1', \mathbf{x}_2', \ldots, \mathbf{x}_q')$, where $\mathbf{x}_g'$, $g = 1, 2, \ldots, q$ is the vector of observed scores for the gth set. Associated with the vector $\mathbf{x}_g$ is a true score τ_g and vectors $\boldsymbol{\mu}_g$ and $\boldsymbol{\beta}_g$ defined as in the previous section, so that

$$\mathbf{x}_g = \boldsymbol{\mu}_g + \boldsymbol{\beta}_g \tau_g + \mathbf{e}_g.$$

As before we may, without loss of generality, assume that τ_g is scaled to zero mean and unit variance. If the different true scores $\tau_1, \tau_2, \ldots, \tau_q$ are all mutually uncorrelated, then each set of tests can be analyzed separately as in the previous section. However, in most cases these true scores correlate with each other, and an overall analysis of the entire set of tests must be made. Let $p = m_1 + m_2 + \cdots + m_q$ be the total number of tests. Then $\mathbf{x}$ is of order p. Let $\boldsymbol{\mu}$ be the mean vector of $\mathbf{x}$, and let $\mathbf{e}$ be the vector of error scores. Furthermore, let

$$\boldsymbol{\tau}' = (\tau_1, \tau_2, \ldots, \tau_q)$$

and let $\mathbf{B}$ be the matrix of order $p \times q$, partitioned as

$$\mathbf{B} = \begin{bmatrix} \boldsymbol{\beta}_1 & 0 & \cdots & 0 \\ 0 & \boldsymbol{\beta}_2 & \cdots & 0 \\ \cdot & \cdot & \cdots & \cdot \\ \cdot & \cdot & \cdots & \cdot \\ 0 & 0 & \cdots & \boldsymbol{\beta}_q \end{bmatrix}. \tag{6}$$

Then $\mathbf{x}$ is represented as

$$\mathbf{x} = \boldsymbol{\mu} + \mathbf{B}\boldsymbol{\tau} + \mathbf{e}.$$

Let $\boldsymbol{\Gamma}$ be the correlation matrix of $\boldsymbol{\tau}$. Then the variance-covariance matrix $\boldsymbol{\Sigma}$ of $\mathbf{x}$ is

$$\boldsymbol{\Sigma} = \mathbf{B}\boldsymbol{\Gamma}\mathbf{B}' + \boldsymbol{\Theta}^2, \tag{7}$$

where $\boldsymbol{\Theta}^2$ is a diagonal matrix of order p containing the error variances. One such model is illustrated in Figure 2.

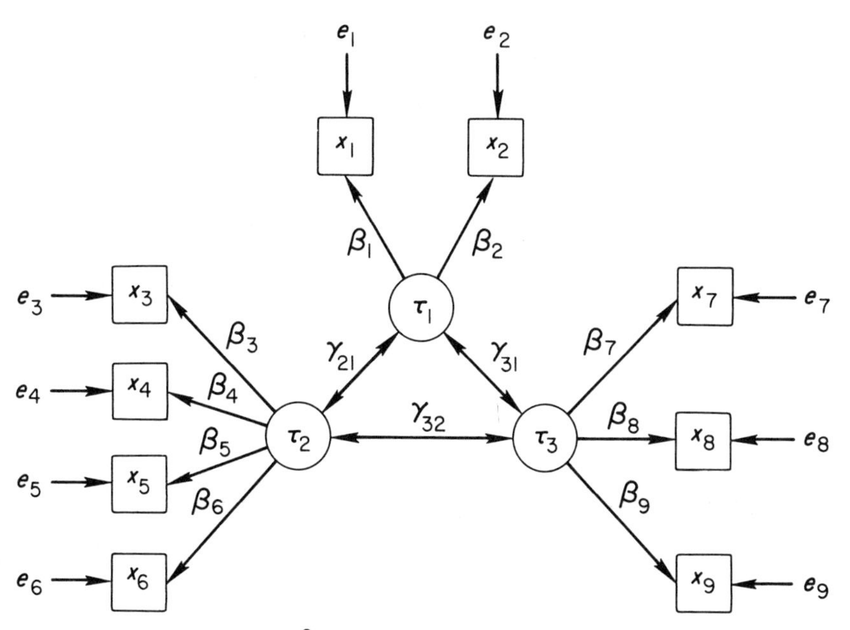

FIGURE 2.
A model with three sets of congeneric tests.

Testing the Hypothesis that the Correlation Coefficient Corrected for Attenuation Is Unity

The correlation coefficient corrected for attenuation between two tests x and y is the correlation between their true scores. If, on the basis of a sample of examinees, the corrected coefficient is near unity, the experimenter concludes that the two tests measure the same trait. Lord (1957) and McNemar (1958) have developed two different tests of the hypothesis that the population correlation coefficient corrected for attenuation $\rho_{\tau_x\tau_y}$ is equal to one. Lord's test is based on the large sample χ^2 approximation to the likelihood ratio statistic. McNemar's test is based on analysis of variance assuming homogeneity of error variances. Both tests employ two parallel forms x_1 and x_2 of x and y_1 and y_2 of y. McNemar's test, in addition, assumes that x and y are equally reliable. I formulate these two tests in terms of the model of Equation 7 and show how the hypothesis can also be tested under less restrictive assumptions.

Let x_1, x_2, y_1, and y_2 be four tests with zero means and let them satisfy the following model:

$$\begin{pmatrix} x_1 \\ x_2 \\ y_1 \\ y_2 \end{pmatrix} = \begin{pmatrix} \beta_1 & 0 \\ \beta_2 & 0 \\ 0 & \beta_3 \\ 0 & \beta_4 \end{pmatrix} \begin{pmatrix} \tau_x \\ \tau_y \end{pmatrix} + \begin{pmatrix} e_1 \\ e_2 \\ e_3 \\ e_4 \end{pmatrix},$$

with variance-covariance matrix

$$\boldsymbol{\Sigma} = \begin{pmatrix} \beta_1 & 0 \\ \beta_2 & 0 \\ 0 & \beta_3 \\ 0 & \beta_4 \end{pmatrix} \begin{pmatrix} 1 & \rho \\ \rho & 1 \end{pmatrix} \begin{pmatrix} \beta_1 & \beta_2 & 0 & 0 \\ 0 & 0 & \beta_3 & \beta_4 \end{pmatrix} + \begin{bmatrix} \theta_1^2 & 0 & 0 & 0 \\ 0 & \theta_2^2 & 0 & 0 \\ 0 & 0 & \theta_3^2 & 0 \\ 0 & 0 & 0 & \theta_4^2 \end{bmatrix}. \tag{8}$$

This model is shown in Figure 3.

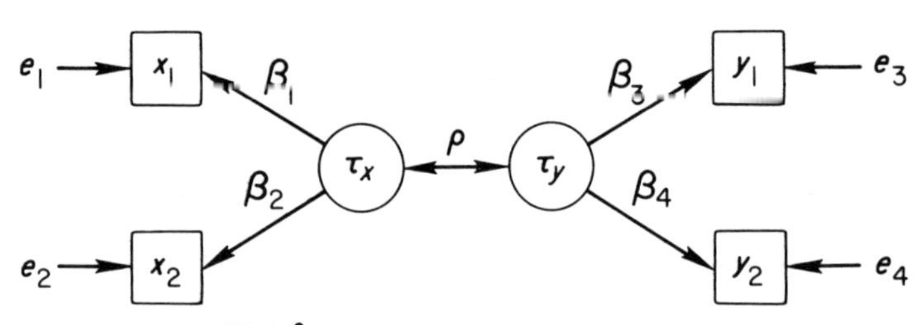

FIGURE 3.
A model with two sets of two congeneric tests.

McNemar's test is: Given $\beta_1 = \beta_2$, $\beta_3 = \beta_4$, $\theta_1^2 = \theta_2^2$, $\theta_3^2 = \theta_4^2$, $\beta_1^2/(\beta_1^2 + \theta_1^2) = \beta_3^2/(\beta_3^2 + \theta_3^2)$, test the hypothesis that $\rho = 1$. Lord's test is: Given that $\beta_1 = \beta_2$, $\beta_3 = \beta_4$, $\theta_1^2 = \theta_2^2$, $\theta_3^2 = \theta_4^2$, test the hypothesis that $\rho = 1$. McNemar's procedure requires a common metric for all four tests, whereas Lord's requires a common metric for x_1 and x_2 and also for y_1 and y_2. The conditions for these two tests are unnecessarily restrictive. It is possible to test the hypothesis assuming only that x_1 and x_2, as well as y_1 and y_2, are congeneric. This amounts to testing $\rho = 1$ under the model of Equation 8.

To illustrate the above ideas, I use some data from Lord (1957). The first two tests x_1 and x_2 are 15-item vocabulary tests administered under liberal time limits. The last two tests y_1 and y_2 are highly speeded 75-item vocabulary tests. The variance-covariance matrix is given in Table 2. I analyze these data under four different hypotheses:

H$_1$: $\beta_1 = \beta_2$, $\beta_3 = \beta_4$, $\theta_1^2 = \theta_2^2$, $\theta_3^2 = \theta_4^2$, $\rho = 1$;

H$_2$: $\beta_1 = \beta_2$, $\beta_3 = \beta_4$, $\theta_1^2 = \theta_2^2$, $\theta_3^2 = \theta_4^2$;

H$_3$: $\rho = 1$;

H$_4$: $\boldsymbol{\Sigma}$ is of the form of Equation 8 with $\beta_1, \beta_2, \beta_3, \beta_4, \theta_1, \theta_2, \theta_3, \theta_4$, and ρ unconstrained.

TABLE 2
Lord's vocabulary test data

Variance-Covariance Matrix, $N = 649$

	x_1	x_2	y_1	y_2
x_1	86.3979			
x_2	57.7751	86.2632		
y_1	56.8651	59.3177	97.2850	
y_2	58.8986	59.6683	73.8201	97.8192

Summary of Analyses

Hypothesis	No. Par.	χ^2	d.f.	P
H_1	4	37.33	6	0.00
H_2	5	1.93	5	0.86
H_3	8	36.21	2	0.00
H_4	9	0.70	1	0.70

Tests of Hypotheses

	Parallel	Congeneric	
$\rho = 1$	$\chi_6^2 = 37.33$	$\chi_2^2 = 36.21$	$\chi_4^2 = 1.12$
$\rho \neq 1$	$\chi_5^2 = 1.93$	$\chi_1^2 = 0.70$	$\chi_4^2 = 1.23$
	$\chi_1^2 = 35.40$	$\chi_1^2 = 35.51$	

The results are shown under Summary of Analyses. Each hypothesis is tested against the general alternative that $\boldsymbol{\Sigma}$ is unconstrained. To consider various hypotheses that can be tested, I recorded the four χ^2 values in a 2-×-2 table under Tests of Hypotheses. Lord's test is equivalent to testing H_1 against H_2, which gives $\chi^2 = 35.40$ with 1 degree of freedom. An alternative test is H_3 against H_4, which gives $\chi^2 = 35.51$ with 1 degree of freedom. Thus, regardless of whether we treat the two pairs of tests as parallel or congeneric, the hypothesis $\rho = 1$ is rejected. There is strong evidence that the unspeeded and speeded tests do not measure the same trait. I shall return to this question in the next section. The hypothesis of parallelism of the two pairs of tests can also be tested by means of the Tests of Hypotheses, which give $\chi^2 = 1.12$ or $\chi^2 = 1.23$ with 4 degrees of freedom, depending on whether we assume $\rho = 1$ or $\rho \neq 1$. Thus we cannot reject the hypothesis that the two pairs of tests are parallel. It appears that H_2 is the most reasonable of the four hypotheses. The maximum-likelihood estimate of ρ under H_2 is $\hat{\rho} = 0.899$ with a standard error of 0.019. An approximate 95-percent confidence interval for ρ is $0.86 < \rho < 0.94$.

Analysis of Speeded and Unspeeded Tests

Table 3 shows correlations and standard deviations for 15 tests. The three types of tests, vocabulary, intersections, and arithmetic reasoning, each has two levels of unspeeded or very little speeded tests (L) and three levels of highly speeded tests (S). These tests, together with several others, were analyzed by Lord (1956) with the objective being to isolate and to identify speed factors and their relationships to academic grades. Lord lists among others the following questions that his study was designed to answer: "Is speed on cognitive tests a unitary trait? Or are there different kinds of speed for different kinds of tasks? If so, how highly correlated are these kinds of speed? How highly correlated are speed and level on the same task? [p. 31]." Lord used maximum-likelihood factor analysis and oblique rotations, and verbal-speed and spatial-speed factors were clearly identified, but no arithmetic-reasoning speed factor was found. The smaller battery in Table 3 will be used here to answer the same kinds of questions by means of a series of confirmatory analyses.

I begin with the hypothesis that the five tests of each type are all congeneric. In principle, one could start by examining the more restrictive hypotheses of parallelism and tau-equivalence, but in view of the highly different time limits used for the unspeeded and the speeded tests, these hypotheses seem a priori implausible. If all tests of each kind are congeneric, it would mean that if errors of measurement were not present, speeded and unspeeded tests would measure the same thing. This, of course, is not very likely, a fact

TABLE 3
Lord's speed test data

$N = 649$

Type of Test		Standard Deviations	Correlations														
Vocabulary	L	0.930	1.00														
	L	0.929	0.67	1.00													
	S	0.986	0.62	0.65	1.00												
	S	0.961	0.69	0.70	0.77	1.00											
	S	0.989	0.64	0.65	0.76	0.85	1.00										
Intersections	L	0.955	0.15	0.18	0.16	0.12	0.09	1.00									
	L	0.967	0.06	0.14	0.14	0.06	0.07	0.72	1.00								
	S	0.980	0.12	0.17	0.19	0.13	0.14	0.70	0.71	1.00							
	S	0.983	0.09	0.13	0.17	0.11	0.10	0.70	0.74	0.80	1.00						
	S	0.988	0.08	0.12	0.16	0.10	0.11	0.70	0.75	0.79	0.83	1.00					
Arith. Reasoning	L	0.905	0.30	0.32	0.28	0.29	0.27	0.34	0.30	0.32	0.31	0.33	1.00				
	L	0.967	0.25	0.31	0.29	0.27	0.29	0.36	0.30	0.30	0.29	0.31	0.54	1.00			
	S	0.991	0.28	0.26	0.36	0.35	0.38	0.29	0.28	0.31	0.29	0.36	0.55	0.54	1.00		
	S	0.985	0.27	0.29	0.33	0.35	0.35	0.34	0.28	0.35	0.34	0.37	0.53	0.55	0.63	1.00	
	S	0.983	0.28	0.30	0.38	0.36	0.39	0.26	0.23	0.27	0.25	0.31	0.51	0.53	0.64	0.61	1.00

TABLE 4
Lord's speed test data. Summary of analyses

	Hypothesis	No. Par.	d.f.	χ^2	P
1	Three sets of congeneric tests	33	87	264.35	0.000
2	Four factors	42	78	140.50	0.000
3	Six sets of parallel tests	27	93	210.17	0.000
4	Six sets of tau-equivalent tests	36	84	138.72	0.000
5	Six sets of congeneric tests	45	75	120.57	0.001
6	Six factors	45	75	108.37	0.007

that is confirmed by the goodness-of-fit statistic in Table 4, the value of χ^2 being 264.35 with 87 degrees of freedom. This is an indication that the speeded tests measure something that the unspeeded tests do not. We therefore consider the hypothesis that there is a unitary factor associated with speed on which only the speeded tests are loaded. This leads to a matrix **B** with four columns, where the first three columns have nonzero loadings in the same position as the previous analysis and where the fourth column has nonzero loadings for the nine speeded tests. An analysis under this model yields a χ^2 of 140.50 with 78 degrees of freedom. Since this χ^2 is still large, we are led to consider the hypothesis that different kinds of speed are associated with different kinds of tests. We therefore consider the hypothesis that the two unspeeded and the three speeded tests of each kind are congeneric, i.e., we regard the tests as being six sets of congeneric tests. However, since tests in each set are equally speeded or unspeeded, we first examine the hypotheses of parallelism and tau-equivalence. It is seen in Table 4 that the hypothesis of parallelism has a very poor fit and that, when the restriction of equality of error variances is relaxed, the value of χ^2 drops significantly. However, the value 138.72 of χ^2 is still large compared to the degrees of freedom 84, so the restriction of equality of true score variances is also relaxed and the previously mentioned hypothesis of six sets of congeneric tests is examined. This gives $\chi^2 = 120.57$, a drop of 18.15 from the previous χ^2 with a corresponding drop of 9 in the degrees of freedom. Since this decrease in χ^2 is not very large compared to the drop in the degrees of freedom, there is an indication that the hypothesis of tau-equivalence does not fit much worse than the hypothesis of congenerism and that the true score variances are fairly equal. That this is indeed so can be seen in Table 5, which gives the solution under the hypothesis of congenerism. In fact, the variation between factor loadings in each column of $\hat{\mathbf{B}}$ is, in most cases, of the same size as the standard errors of estimate. An inspection of the correlations in the matrix $\hat{\boldsymbol{\Gamma}}$ reveals that true scores on speeded and unspeeded tests of the same kind are very highly correlated, whereas true scores on speeded tests of different kinds correlate

TABLE 5
Lord's speed test data. Results of analysis 5

i	Type of Test		$\hat{\beta}_{i1}$	$\hat{\beta}_{i2}$	$\hat{\beta}_{i3}$	$\hat{\beta}_{i4}$	$\hat{\beta}_{i5}$	$\hat{\beta}_{i6}$	$\hat{\theta}_i$
1	Vocabulary	L	0.75	0.*	0.*	0.*	0.*	0.*	0.55
2		L	0.77	0.*	0.*	0.*	0.*	0.*	0.52
3		S	0.*	0.*	0.*	0.82	0.*	0.*	0.54
4		S	0.*	0.*	0.*	0.90	0.*	0.*	0.33
5		S	0.*	0.*	0.*	0.90	0.*	0.*	0.42
6	Intersections	L	0.*	0.80	0.*	0.*	0.*	0.*	0.53
7		L	0.*	0.84	0.*	0.*	0.*	0.*	0.48
8		S	0.*	0.*	0.*	0.*	0.86	0.*	0.48
9		S	0.*	0.*	0.*	0.*	0.89	0.*	0.41
10		S	0.*	0.*	0.*	0.*	0.90	0.*	0.40
11	Reasoning	L	0.*	0.*	0.66	0.*	0.*	0.*	0.62
12		L	0.*	0.*	0.71	0.*	0.*	0.*	0.65
13		S	0.*	0.*	0.*	0.*	0.*	0.80	0.58
14		S	0.*	0.*	0.*	0.*	0.*	0.77	0.61
15		S	0.*	0.*	0.*	0.*	0.*	0.77	0.62

$\hat{\gamma}_{kl}$

k	l					
	1	2	3	4	5	6
1	1.*					
2	0.19	1.*				
3	0.49	0.51	1.*			
4	0.90	0.12	0.42	1.*		
5	0.16	0.93	0.47	0.15	1.*	
6	0.43	0.41	0.93	0.50	0.45	1.*

Note: Asterisks denote parameter values specified by hypothesis.

only moderately; the same holds for true scores on unspeeded tests of different kinds. This suggests that speeded and unspeeded tests of the same kind have a large common element that we may interpret as 'power' and that we should try to separate this power factor from the speed factor. If the hypothesis of congenerism had yielded a reasonably good fit, it would be natural to do this by a second-order analysis, i.e., by a factor analysis of $\widehat{\boldsymbol{\Gamma}}$ into one power factor and one speed factor. However, since $\chi^2 = 120.57$ with 75 degrees of freedom does not represent a good fit, it is more appropriate to separate speed from power at the first-order level by relaxing the hypothesis of congenerism, which is done by adding loadings for the three speeded tests on the first three factors. The resulting solution, given in Table 6, represents three power factors and three speed factors. The value of χ^2 for this solution

TABLE 6
Lord's speed test data. Results of analysis 6

i	Type of Test		$\hat{\beta}_{i1}$	$\hat{\beta}_{i2}$	$\hat{\beta}_{i3}$	$\hat{\beta}_{i4}$	$\hat{\beta}_{i5}$	$\hat{\beta}_{i6}$	$\hat{\theta}_i$
1	Vocabulary	L	0.75	0.*	0.*	0.*	0.*	0.*	0.55
2		L	0.77	0.*	0.*	0.*	0.*	0.*	0.51
3		S	0.76	0.*	0.*	0.31	0.*	0.*	0.54
4		S	0.82	0.*	0.*	0.37	0.*	0.*	0.35
5		S	0.78	0.*	0.*	0.48	0.*	0.*	0.38
6	Intersections	L	0.*	0.80	0.*	0.*	0.*	0.*	0.53
7		L	0.*	0.84	0.*	0.*	0.*	0.*	0.48
8		S	0.*	0.82	0.*	0.*	0.75	0.*	0.49
9		S	0.*	0.84	0.*	0.*	0.32	0.*	0.41
10		S	0.*	0.84	0.*	0.*	0.34	0.*	0.40
11	Reasoning	L	0.*	0.*	0.67	0.*	0.*	0.*	0.61
12		L	0.*	0.*	0.72	0.*	0.*	0.*	0.65
13		S	0.*	0.*	0.73	0.*	0.*	0.37	0.57
14		S	0.*	0.*	0.72	0.*	0.*	0.28	0.61
15		S	0.*	0.*	0.69	0.*	0.*	0.34	0.61

$\hat{\gamma}_{kl}$

k	l = 1	2	3	4	5	6
1	1.*					
2	0.17	1.*				
3	0.47	0.48	1.*			
4	0.*	0.*	0.*	1.*		
5	0.*	0.*	0.*	0.26	1.*	
6	0.*	0.*	0.*	0.70	0.39	1.*

Note: Asterisks denote parameter values specified by hypothesis.

is 108.37, which is 12.20 below that of Table 5. It should be noted, however, that the degrees of freedom for the two solutions are the same, due to the fact that the same number of parameters has been estimated in both solutions. Thus it appears that the solution of Table 6 is a better one than that of Table 5, both from the point of view of goodness of fit and of interpretation of the data. Table 6 clearly shows that three power factors and three speed factors can be isolated from these data. Speed factors are defined to be uncorrelated with the power factors. For each test, speed is defined to be the part that remains after the power factor has been eliminated. The three power factors correlate moderately and so do the three speed factors. If factor loadings and standard deviations in Table 6 are squared, one gets a decomposition of the unit variance of each test into components directly associated with power, speed, and error, respectively.

Analysis of Multitrait-Multimethod Data

A particular instance when sets of congeneric tests are employed is in multitrait-multimethod studies, where each of a number of traits is measured with a number of different methods or measuring instruments (see, e.g., Campbell & Fiske, 1959). One objective may be to find the best method of measuring each trait. In particular, one would like to get estimates of the trait, method, and error variance involved in each measure. A second objective is to study the internal relationships between the measures employed, in particular between the traits and between the methods.

Data from multitrait-multimethod studies are usually summarized in a correlation matrix giving correlations for all pairs of trait-method combinations. If there are m methods and n traits, this correlation matrix is of order $mn \times mn$. In analyzing such a correlation matrix, it seems natural to begin with the hypothesis that all methods are equivalent in measuring each trait, in the sense that scores obtained for a given trait with the different methods are congeneric. This hypothesis implies that all variation and covariation in the multitrait-multimethod matrix is due to trait factors only and may be tested by using a factor matrix $\mathbf{B}$ of order $mn \times n$ with one column for each trait. If the measurements are arranged with methods within traits, $\mathbf{B}$ is of the form of Equation 6. If, on the other hand, measurements are arranged with traits within methods, $\mathbf{B}$ has the form

$$\mathbf{B} = \begin{bmatrix} \boldsymbol{\Delta}_1 \\ \boldsymbol{\Delta}_2 \\ \vdots \\ \boldsymbol{\Delta}_m \end{bmatrix}, \tag{9}$$

where each $\boldsymbol{\Delta}_i$ is a diagonal matrix of order $n \times n$. In both cases, the model is given by Equation 7, where $\boldsymbol{\Gamma}$ is the correlation matrix for the trait factors and $\boldsymbol{\Theta}^2$ is the diagonal matrix of error variances. If this model fits the data, the interrelationships between the trait factors may be analyzed further by a factoring of $\boldsymbol{\Gamma}$ as described in Section 4. However, if the hypothesis of equivalent methods does not fit the data, this is an indication that method factors are present. It then seems best to postulate the existence of one method factor for each method. This leads to a factor matrix $\mathbf{B}$ of order $mn \times (m + n)$ of the form (with traits within methods)

$$\mathbf{B} = \begin{bmatrix} \boldsymbol{\Delta}_1 & \boldsymbol{\beta}_1 & \mathbf{0} & \cdots & \mathbf{0} \\ \boldsymbol{\Delta}_2 & \mathbf{0} & \boldsymbol{\beta}_2 & \cdots & \mathbf{0} \\ \vdots & \vdots & \vdots & \cdots & \vdots \\ \boldsymbol{\Delta}_m & \mathbf{0} & \mathbf{0} & \cdots & \boldsymbol{\beta}_m \end{bmatrix}, \tag{10}$$

where the $\boldsymbol{\Delta}$s are as before and each $\boldsymbol{\beta}_i$ is a column vector of order n. The correlation matrix $\boldsymbol{\Gamma}$ of the factors may be specified as

$$\boldsymbol{\Gamma} = \begin{pmatrix} \boldsymbol{\Gamma}_1 & \mathbf{0} \\ \mathbf{0} & \boldsymbol{\Gamma}_2 \end{pmatrix}, \tag{11}$$

where $\boldsymbol{\Gamma}_1$ is the correlation matrix for the trait factors and $\boldsymbol{\Gamma}_2$ is the correlation matrix for the method factors. One such model is illustrated in Figure 4. In this model it is assumed that trait factors and method factors are uncorrelated. This defines each method factor to be independent of the particular traits that the method is used to measure. However, the trait factors may also be allowed to correlate with the method factors, in which case $\boldsymbol{\Gamma}$ is defined as a full correlation matrix. Substituting Equations 10 and 11 into Equation 7 gives the variance-covariance matrix $\boldsymbol{\Sigma}$ under this model. An analysis of data under this model yields estimates of $\mathbf{B}$, $\boldsymbol{\Gamma}$, and $\boldsymbol{\Theta}$. If the two factor loadings in each row of $\mathbf{B}$ and the corresponding element of $\boldsymbol{\Theta}$ are squared, one obtains a partition of the total variance of each measurement into components due to traits, methods, and error, respectively. If the fit of the model is good, and there are many traits and/or methods, one may analyze the interrelationships in $\boldsymbol{\Gamma}_1$ and $\boldsymbol{\Gamma}_2$ further in a way similar to that of Section 4.

In analyzing data in accordance with the above model it sometimes happens that one or more correlations in $\hat{\boldsymbol{\Gamma}}_2$ are close to unity or else that $\hat{\boldsymbol{\Gamma}}_2$ is not Gramian. This means that two or more method factors are collinear and have to be combined into one factor.

Analysis of Assessment Ratings of Clinical Psychologists

Clinical psychology students rated themselves and three teammates on each of several traits or characteristics. The median of the three teammate ratings

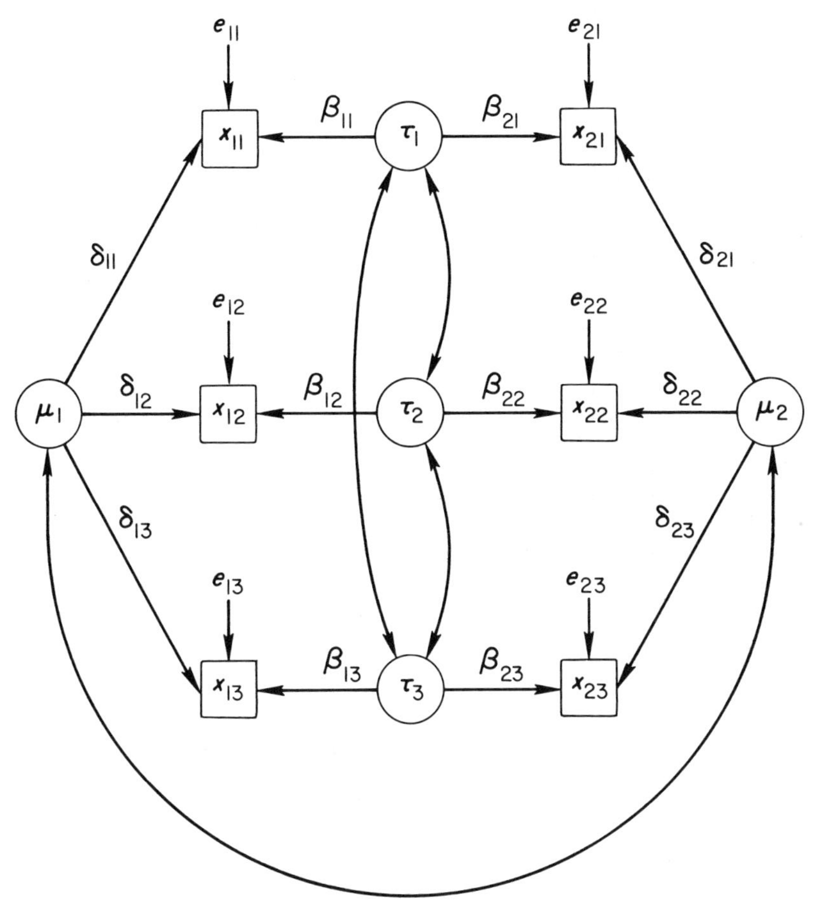

FIGURE 4.
A model for three traits measured by two methods.

was used as the score for this method of measurement. An assessment staff also rated the students, the ratings of the three members of the staff being pooled for a score on a given subject. The three methods of measurement thus are staff ratings, teammate ratings, and self-ratings. The study was conducted by Kelly and Fiske (1951) and full descriptions of the traits, methods, and data are contained in their article. On the basis of validity, Campbell and Fiske (1959) selected five traits—Assertive, Cheerful, Serious, Unshakable Poise, and Broad Interests—and published the 15 × 15 correlation matrix obtained from measurements on each of the five traits by each of the three methods. This correlation matrix is reproduced in Table 7 and is analyzed as outlined in the previous section.

TABLE 7
Ratings from assessment study of clinical psychologists

		N = 124														
		Staff Ratings					Teammate Ratings					Self-Ratings				
		A_1	B_1	C_1	D_1	E_1	A_2	B_2	C_2	D_2	E_2	A_3	B_3	C_3	D_3	E_3
Staff Ratings																
Assertive	A_1															
Cheerful	B_1	0.37														
Serious	C_1	−0.24	−0.14													
Unshakable Poise	D_1	0.25	0.46	0.08												
Broad Interests	E_1	0.35	0.19	0.09	0.31											
Teammate Ratings																
Assertive	A_2	0.71	0.35	−0.18	0.26	0.41										
Cheerful	B_2	0.39	0.53	−0.15	0.38	0.29	0.37									
Serious	C_2	−0.27	−0.31	0.43	−0.06	0.03	−0.15	−0.19								
Unshakable Poise	D_2	0.03	−0.05	0.03	0.20	0.07	0.11	0.23	0.19							
Broad Interests	E_2	0.19	0.05	0.04	0.29	0.47	0.33	0.22	0.19	0.29						
Self-Ratings																
Assertive	A_3	0.48	0.31	−0.22	0.19	0.12	0.46	0.36	−0.15	0.12	0.23					
Cheerful	B_3	0.17	0.42	−0.10	0.10	−0.03	0.09	0.24	−0.25	−0.11	−0.03	0.23				
Serious	C_3	−0.04	−0.13	0.22	--0.13	−0.05	−0.04	−0.11	0.31	0.06	0.06	−0.05	−0.12			
Unshakable Poise	D_3	0.13	0.27	−0.03	0.22	−0.04	0.10	0.15	0.00	0.14	−0.03	0.16	0.26	0.11		
Broad Interests	E_3	0.37	0.15	−0.22	0.09	0.26	0.27	0.12	−0.07	0.05	0.35	0.21	0.15	0.17	0.31	

The hypothesis that the three measurements of each trait are congeneric yields a χ^2 of 140.46 with 80 degrees of freedom. Since this does not represent a good fit, three method factors are added to the five trait factors, which yields a χ^2 of 57.91 with 62 degrees of freedom and represents an acceptable fit. However, the maximum-likelihood solution reveals that the two method factors for staff ratings and self-ratings correlate unity. These two factors are therefore combined into one. The final solution is shown in Table 8, which has a χ^2 of 61.51 with 64 degrees of freedom. Variance components of each measurement are given in Table 9.

It is interesting to see how Tables 8 and 9 can be interpreted directly in terms of what method best measures a given trait. The trait Assertiveness is best measured by staff ratings, but is also measured well by teammate ratings. When self-ratings are used, a less reliable measurement is obtained. Whatever method is used to measure Assertiveness, a very small method variance is involved. Thus it appears that all three methods produce valid measurements of Assertiveness. Also the trait Cheerfulness is validly measured by all three methods but the staff ratings are the most reliable measurements, though less reliable than staff ratings of Assertiveness. It should also be noted that Cheerfulness has fairly high correlations with Assertiveness and Unshakable Poise and is therefore likely to be confused with these traits. The trait Seriousness is best measured by teammate ratings, but cannot be measured as reliably as the other traits. The trait Unshakable Poise can be measured by staff ratings, but not with the other two methods. Broad Interests is reliably measured by self-ratings, but only with a large method variance.

4. FACTOR ANALYSIS MODELS

Factor analysis is a widely used technique, especially among psychologists and other behavioral scientists. The basic idea is that for a given set of response variates $x_1, \ldots, x_p$ one wants to find a set of underlying or latent factors $f_1, \ldots, f_k$, fewer in number than the observed variates, that will account for the intercorrelations of the response variates, in the sense that when the factors are partialled out from the observed variates no correlation remains between them. This leads to the model

$$\mathbf{x} = \boldsymbol{\mu} + \boldsymbol{\Lambda}\mathbf{f} + \mathbf{z}, \tag{12}$$

where $E(\mathbf{x}) = \boldsymbol{\mu}$, $E(\mathbf{f}) = \mathbf{0}$ and $E(\mathbf{z}) = \mathbf{0}$, $\mathbf{z}$ being uncorrelated with $\mathbf{f}$. Let $\boldsymbol{\Phi} = E(\mathbf{ff}')$, which may be taken as a correlation matrix and $\boldsymbol{\Psi}^2 = E(\mathbf{zz}')$, which is diagonal. Then the variance-covariance matrix $\boldsymbol{\Sigma}$ of $\mathbf{x}$ becomes

$$\boldsymbol{\Sigma} = \boldsymbol{\Lambda\Phi\Lambda}' + \boldsymbol{\Psi}^2. \tag{13}$$

TABLE 8
Ratings from assessment study of clinical psychologists.
Maximum-likelihood solution

		Factor Matrix							
		A	B	C	D	E	1–3	2	θ
Staff Ratings									
Assertive	A_1	0.87	0.*	0.*	0.*	0.*	0.11	0.*	0.49
Cheerful	B_1	0.*	0.84	0.*	0.*	0.*	0.02	0.*	0.55
Serious	C_1	0.*	0.*	0.57	0.*	0.*	−0.30	0.*	0.76
Unshakable Poise	D_1	0.*	0.*	0.*	0.78	0.*	−0.25	0.*	0.56
Broad Interests	E_1	0.*	0.*	0.*	0.*	0.69	−0.34	0.*	0.63
Teammate Ratings									
Assertive	A_2	0.83	0.*	0.*	0.*	0.*	0.*	0.16	0.54
Cheerful	B_2	0.*	0.70	0.*	0.*	0.*	0.*	0.29	0.68
Serious	C_2	0.*	0.*	0.72	0.*	0.*	0.*	0.32	0.59
Unshakable Poise	D_2	0.*	0.*	0.*	0.21	0.*	0.*	0.53	0.82
Broad Interests	E_2	0.*	0.*	0.*	0.*	0.60	0.*	0.44	0.65
Self-Ratings									
Assertive	A_3	0.55	0.*	0.*	0.*	0.*	0.11	0.*	0.83
Cheerful	B_3	0.*	0.45	0.*	0.*	0.*	0.22	0.*	0.87
Serious	C_3	0.*	0.*	0.43	0.*	0.*	0.23	0.*	0.88
Unshakable Poise	D_3	0.*	0.*	0.*	0.43	0.*	0.38	0.*	0.82
Broad Interests	E_3	0.*	0.*	0.*	0.*	0.70	0.62	0.*	0.41

	Factor Intercorrelations						
	A	B	C	D	E	1–3	2
A	1.*						
B	0.56	1.*					
C	−0.37	−0.44	1.*				
D	0.38	0.66	−0.08	1.*			
E	0.55	0.29	−0.02	0.43	1.*		
1–3	0.*	0.*	0.*	0.*	0.*	1.*	
2	0.*	0.*	0.*	0.*	0.*	−0.21	1.*

Note: Asterisks denote parameter values specified by hypothesis.

TABLE 9
Ratings from assessment study of clinical psychologists

		Variance Components		
		Trait	Method	Error
Staff Ratings				
Assertive	A_1	0.76	0.01	0.24
Cheerful	B_1	0.70	0.00	0.30
Serious	C_1	0.33	0.09	0.58
Unshakable Poise	D_1	0.61	0.06	0.32
Broad Interests	E_1	0.48	0.11	0.40
Teammate Ratings				
Assertive	A_2	0.69	0.03	0.29
Cheerful	B_2	0.48	0.09	0.46
Serious	C_2	0.52	0.10	0.35
Unshakable Poise	D_2	0.05	0.28	0.68
Broad Interests	E_2	0.36	0.19	0.43
Self-Ratings				
Assertive	A_3	0.30	0.01	0.69
Cheerful	B_3	0.21	0.05	0.75
Serious	C_3	0.18	0.05	0.77
Unshakable Poise	D_3	0.18	0.14	0.68
Broad Interests	E_3	0.49	0.39	0.17

If $(p - k)^2 < p + k$, this relationship can be tested statistically, unlike Equation 12, which involves hypothetical variates and cannot be verified directly. Equation 13 may be obtained from the general model (Eq. 1) by specifying $\mathbf{B} = \mathbf{I}$ and $\boldsymbol{\Theta} = \mathbf{0}$.

When $k > 1$ there is an indeterminacy in Equation 14 arising from the fact that a nonsingular linear transformation of $\mathbf{f}$ changes $\boldsymbol{\Lambda}$ and in general also $\boldsymbol{\Phi}$ but leaves $\boldsymbol{\Sigma}$ unchanged. The usual way to eliminate this indeterminacy in exploratory factor analysis (see, for example, Lawley & Maxwell, 1963; Jöreskog, 1967; Jöreskog & Lawley, 1968) is to choose $\boldsymbol{\Phi} = \mathbf{I}$ and $\boldsymbol{\Lambda}'\boldsymbol{\Psi}^{-1}\boldsymbol{\Lambda}$ to be diagonal and to estimate the parameters in $\boldsymbol{\Lambda}$ and $\boldsymbol{\Psi}$ subject to these conditions. This leads to an arbitrary set of factors that may then be subjected to a rotation or a linear transformation to another set of factors that can be given a more meaningful interpretation.

In terms of the general model (Eq. 1), the indeterminacy in Equation 13 may be eliminated by assigning zero values, or any other values, to k^2 elements in $\boldsymbol{\Lambda}$ and/or $\boldsymbol{\Phi}$ in such a way that the assigned values will be destroyed by all nonsingular transformations of the factors except the identity transformation. There may be an advantage in eliminating the indeterminacy this way, in that, if the fixed parameters are chosen in a reasonable way, the resulting solution will be directly interpretable and the subsequent rotation of factors may be avoided.

Specification of parameters a priori may also be used in a confirmatory factor analysis, where the experimenter has already obtained a certain amount of knowledge about the variates measured and is in a position to formulate a hypothesis that specifies the factors on which the variates depend. Such a hypothesis may be specified by assigning values to some parameters in $\boldsymbol{\Lambda}$, $\boldsymbol{\Phi}$, and $\boldsymbol{\Psi}$; (see, e.g., Jöreskog & Lawley, 1968; Jöreskog, 1969). If the number of fixed parameters in $\boldsymbol{\Lambda}$ and $\boldsymbol{\Phi}$ exceeds k^2, the hypothesis represents a restriction of the common factor space, and a solution obtained under such a hypothesis cannot be obtained by a rotation of an arbitrary solution such as that obtained in an exploratory analysis.

The model of Equation 5 is formally equivalent to a factor analytic model with one common factor, and the model of Equation 7 is equivalent to a factor analytic model with q correlated nonoverlapping factors. In the latter case the factors are the true scores $\boldsymbol{\tau}' = (\tau_1, \ldots, \tau_q)$ of the tests. These true scores may themselves satisfy a factor analytic model, i.e.,

$$\boldsymbol{\tau} = \boldsymbol{\Lambda}\mathbf{f} + \mathbf{s},$$

where $\mathbf{f}$ is a vector of order k of common true score factors, $\mathbf{s}$ is a vector of order q of specific true score factors, and $\boldsymbol{\Lambda}$ is a matrix of order $q \times k$ of factor loadings. Let $\boldsymbol{\Phi}$ be the variance-covariance matrix of $\mathbf{f}$, and let $\boldsymbol{\Psi}^2$ be a diagonal matrix whose diagonal elements are the variances of the specific true score factors $\mathbf{s}$. Then $\boldsymbol{\Gamma}$, the variance-covariance matrix of $\boldsymbol{\tau}$, becomes

$$\boldsymbol{\Gamma} = \boldsymbol{\Lambda\Phi\Lambda}' + \boldsymbol{\Psi}^2. \tag{14}$$

Substituting Equation 14 into Equation 7 gives $\boldsymbol{\Sigma}$ as

$$\boldsymbol{\Sigma} = \mathbf{B}(\boldsymbol{\Lambda\Phi\Lambda}' + \boldsymbol{\Psi}^2)\mathbf{B}' + \boldsymbol{\Theta}^2. \tag{15}$$

This model is a special case of Equation 1 by specifying zero values in $\mathbf{B}$ as before. To define $\boldsymbol{\Lambda}$ and $\boldsymbol{\Phi}$ uniquely, we must impose k^2 independent conditions on these to eliminate the indeterminacy due to rotation. The model of Equation 15 is a special case of the second-order factor analytic model. An example of a second-order factor analysis model is given in Figure 5.

5. VARIANCE AND COVARIANCE COMPONENTS

Estimation of Variance Components

Several authors (Bock, 1960; Bock & Bargmann, 1966; Wiley, Schmidt, & Bramble, 1973) have considered covariance structure analysis as an approach

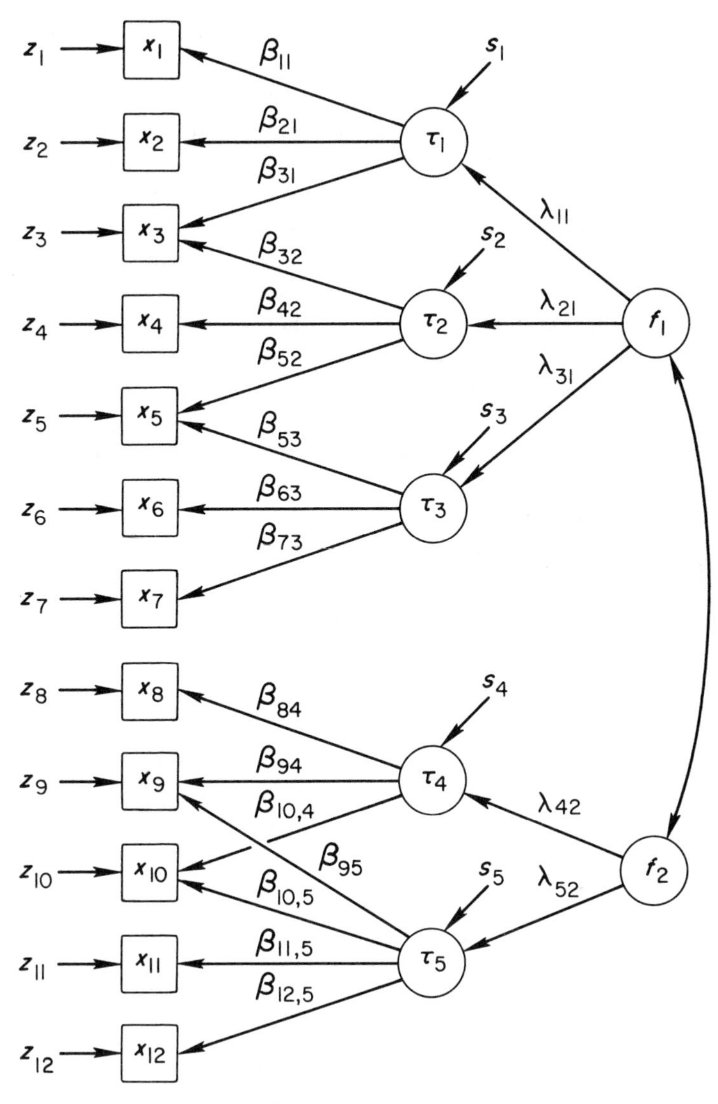

FIGURE 5.
A second-order factor analysis model.

to study differences in test performances when the tests have been constructed by assigning items or subtests according to objective features of content or format to subclasses of a factorial or hierarchical classification.

The idea of analyzing a design of test items or subtests is due to Burt (1947), who applied analysis of variance techniques to determine the effects on the test scores of the different classifications. Bock (1960) suggested that the scores of N subjects on a set of tests classified in 2^n factorial design may be viewed as data from an $N \times 2^n$ experimental design, where the subjects represent a random mode of classification and the tests represent n fixed modes of classification. Bock pointed out that conventional mixed-model analysis of variance gives useful information about the psychometric properties of the tests. In particular, the presence of nonzero variance components for the random mode of classification and for the interaction of the random and fixed modes of classification provides information about the number of dimensions in which the tests are able to discriminate among subjects. The relative size of these components measures the power of the tests to discriminate among subjects along the respective dimensions.

The multitrait-multimethod matrix of Campbell and Fiske (1959) is an example of a factorial design of tests. A more complex design is Guilford's (1956) structure of intellect. This design is based on a cross-classification of test items, not all of which may exist or be employed in any one study. Thus the classification scheme may be incomplete.

Consider an experimental design that has one random way of classification $\nu = 1, 2, \ldots, N$, one fixed way of classification $i = 1, 2, 3$ and another fixed way of classification $j = 1, 2, 3$ for $i = 1, 2$ and $j = 1, 2$ for $i = 3$. One model that may be considered is

$$x_{\nu ij} = \mu_{ij} + a_\nu + b_{\nu i} + c_{\nu j} + e_{\nu ij}, \tag{16}$$

where μ_{ij} is the mean of $x_{\nu ij}$ and where a_ν, $b_{\nu i}$, $c_{\nu j}$, and $e_{\nu ij}$ are uncorrelated random variables with zero means and variances σ_a^2, $\sigma_{b_i}^2$, $\sigma_{c_j}^2$, and $\sigma_{e_{ij}}^2$, respectively. Writing $\mathbf{x}'_\nu = (x_{\nu 11}, x_{\nu 12}, x_{\nu 13}, x_{\nu 21}, x_{\nu 22}, x_{\nu 23}, x_{\nu 31}, x_{\nu 32})$, $\mathbf{u}'_\nu = (a_\nu, b_{\nu 1}, b_{\nu 2}, b_{\nu 3}, c_{\nu 1}, c_{\nu 2}, c_{\nu 3})$ and

$$\mathbf{A} = \begin{bmatrix} 1 & 1 & 0 & 0 & 1 & 0 & 0 \\ 1 & 1 & 0 & 0 & 0 & 1 & 0 \\ 1 & 1 & 0 & 0 & 0 & 0 & 1 \\ 1 & 0 & 1 & 0 & 1 & 0 & 0 \\ 1 & 0 & 1 & 0 & 0 & 1 & 0 \\ 1 & 0 & 1 & 0 & 0 & 0 & 1 \\ 1 & 0 & 0 & 1 & 1 & 0 & 0 \\ 1 & 0 & 0 & 1 & 0 & 1 & 0 \end{bmatrix},$$

we may write Equation 16 as

$$\mathbf{x}_\nu = \boldsymbol{\mu} + \mathbf{A}\mathbf{u}_\nu + \mathbf{e}_\nu,$$

where $\boldsymbol{\mu}$ is the mean vector and $\mathbf{e}_\nu$ is a random error vector both of the same form as $\mathbf{x}_\nu$. The variance-covariance matrix of $\mathbf{x}_\nu$ is

$$\boldsymbol{\Sigma} = \mathbf{A}\boldsymbol{\Phi}\mathbf{A}' + \boldsymbol{\Psi}^2, \tag{17}$$

where $\boldsymbol{\Phi}$ is a diagonal matrix whose diagonal elements are σ_a^2, $\sigma_{b_1}^2$, $\sigma_{b_2}^2$, $\sigma_{b_3}^2$, $\sigma_{c_1}^2$, $\sigma_{c_2}^2$, and $\sigma_{c_3}^2$, and $\boldsymbol{\Psi}^2$ is a diagonal matrix whose elements are the $\sigma_{e_{ij}}^2$. In terms of the general model (Eq. 1), this model may be represented by choosing $\mathbf{B} = \mathbf{I}$, $\boldsymbol{\Lambda} = \mathbf{A}$, and $\boldsymbol{\Theta} = \mathbf{0}$. Matrices $\boldsymbol{\Phi}$ and $\boldsymbol{\Psi}^2$ are as defined in Equation 17. However, in this case the matrix $\boldsymbol{\Lambda}$ has rank 5, and only 5 linearly independent combinations of the components of $\mathbf{u}_\nu$ are estimable (see, e.g., Graybill, 1961, pp. 228–229). In conventional mixed-model analysis of variance one usually makes the assumptions that $\sigma_{b_i}^2 = \sigma_b^2$ for all $i = 1, 2, 3$, $\sigma_{c_j}^2 = \sigma_c^2$ for all $j = 1, 2, 3$, and $\sigma_{e_{ij}}^2 = \sigma_e^2$ for all i and j, but all these assumptions are not necessary.

In general, if $\mathbf{A}$ is of order $p \times r$ and of rank k, one may choose k independent linear functions, each one linearly dependent on the rows of $\mathbf{A}$ and estimate the mean vector and variance-covariance matrix of these functions. It is customary to choose linear combinations that are mutually uncorrelated, but this is not necessary in the analysis by our method. Let $\mathbf{L}$ be the matrix of coefficients of the chosen linear functions, and let $\mathbf{K}$ be any matrix such that $\mathbf{A} = \mathbf{KL}$. For example, $\mathbf{K}$ may be obtained from

$$\mathbf{K} = \mathbf{AL}'(\mathbf{LL}')^{-1}. \tag{18}$$

The model may then be reparameterized to full rank by defining $\mathbf{u}^* = \mathbf{Lu}$. We then have $\mathbf{x} = \mathbf{Au} + \mathbf{e} = \mathbf{KLu} + \mathbf{e} = \mathbf{Ku}^* + \mathbf{e}$. The variance-covariance matrix of $\mathbf{x}$ is represented as

$$\boldsymbol{\Sigma} = \mathbf{K}\boldsymbol{\Phi}^*\mathbf{K}' + \boldsymbol{\Psi}^2, \tag{19}$$

where $\boldsymbol{\Phi}^*$ is the variance-covariance matrix of $\mathbf{u}^*$ and $\boldsymbol{\Psi}^2$ is as before. The general method of analysis yields estimates of $\boldsymbol{\Psi}^2$ and $\boldsymbol{\Phi}^*$. The last matrix may be taken to be diagonal if desired.

A 2 × 2 Factorial Design with Replications

As an illustration, consider an example from Bock (1960). This example employs four subtests of an experimental form of the Language Modalities Survey (1959) to be used with aphasic subjects. The tests are classified according as the stimulus is a word or a picture and the response is oral or graphic:

		Response	
		Oral ($j = 1$)	Graphic ($j = 2$)
Stimulus	Picture ($i = 1$)	11	12
	Word ($i = 2$)	21	22

Let us assume that two tau-equivalent forms (replications) of each subtest are available. A suitable model is then

$$x_{\nu ijk} = \tau_{\nu ij} + e_{\nu ijk}, \tag{20}$$

where $x_{\nu ijk}$ is the score of person ν on the kth replication of the test classified as ij, and $\tau_{\nu ij}$ and $e_{\nu ijk}$ are the true and error scores, respectively. The true score $\tau_{\nu ij}$ is represented as

$$\tau_{\nu ij} = \mu_{ij} + a_\nu + b_{\nu i} + c_{\nu j} + d_{\nu ij}, \tag{21}$$

where μ_{ij} is the mean of $\tau_{\nu ij}$ in the population of individuals, a_ν is a component of score specific to individual ν and general to all tests, $b_{\nu i}$ and $c_{\nu j}$ are components of score specific to the performance of individual ν on tests of the 'Stimulus' and 'Response' classification, respectively, and $d_{\nu ij}$ is a component of score for individual ν specific to each test. The components a_ν, $b_{\nu i}$, $c_{\nu j}$, and $d_{\nu ij}$ are assumed to be uncorrelated random variables with zero mean and variances σ_a^2, σ_b^2, σ_c^2, and σ_d^2, respectively.

Equations 20 and 21 may be written in matrix form as

$$\begin{pmatrix} x_{\nu 111} \\ x_{\nu 112} \\ x_{\nu 121} \\ x_{\nu 122} \\ x_{\nu 211} \\ x_{\nu 212} \\ x_{\nu 221} \\ x_{\nu 222} \end{pmatrix} = \begin{bmatrix} 1 & 0 & 0 & 0 \\ 1 & 0 & 0 & 0 \\ 0 & 1 & 0 & 0 \\ 0 & 1 & 0 & 0 \\ 0 & 0 & 1 & 0 \\ 0 & 0 & 1 & 0 \\ 0 & 0 & 0 & 1 \\ 0 & 0 & 0 & 1 \end{bmatrix} \begin{pmatrix} \tau_{\nu 11} \\ \tau_{\nu 12} \\ \tau_{\nu 21} \\ \tau_{\nu 22} \end{pmatrix} + \begin{pmatrix} e_{\nu 111} \\ e_{\nu 112} \\ e_{\nu 121} \\ e_{\nu 122} \\ e_{\nu 211} \\ e_{\nu 212} \\ e_{\nu 221} \\ e_{\nu 222} \end{pmatrix},$$

$$\begin{pmatrix} \tau_{\nu 11} \\ \tau_{\nu 12} \\ \tau_{\nu 21} \\ \tau_{\nu 22} \end{pmatrix} = \begin{pmatrix} \mu_{11} \\ \mu_{12} \\ \mu_{21} \\ \mu_{22} \end{pmatrix} + \begin{bmatrix} 1 & 1 & 0 & 1 & 0 & 1 & 0 & 0 & 0 \\ 1 & 1 & 0 & 0 & 1 & 0 & 1 & 0 & 0 \\ 1 & 0 & 1 & 1 & 0 & 0 & 0 & 1 & 0 \\ 1 & 0 & 1 & 0 & 1 & 0 & 0 & 0 & 1 \end{bmatrix} \begin{pmatrix} a_\nu \\ b_{\nu 1} \\ b_{\nu 2} \\ c_{\nu 1} \\ c_{\nu 2} \\ d_{\nu 11} \\ d_{\nu 12} \\ d_{\nu 21} \\ d_{\nu 22} \end{pmatrix},$$

or

$$\mathbf{x}_\nu = \mathbf{B}\boldsymbol{\tau}_\nu + \mathbf{e}_\nu, \tag{22}$$

$$\boldsymbol{\tau}_\nu = \boldsymbol{\mu} + \mathbf{A}\mathbf{u}_\nu. \tag{23}$$

The variance-covariance matrix of $\mathbf{x}_\nu$ is

$$\boldsymbol{\Sigma} = \mathbf{B}\mathbf{A}\boldsymbol{\Phi}\mathbf{A}'\mathbf{B}' + \boldsymbol{\Theta}^2, \tag{24}$$

where

$$\boldsymbol{\Phi} = \operatorname{diag}(\sigma_a^2, \sigma_b^2, \sigma_b^2, \sigma_c^2, \sigma_c^2, \sigma_d^2, \sigma_d^2, \sigma_d^2, \sigma_d^2)$$

and $\boldsymbol{\Theta}^2$ is the diagonal matrix of error variances. The variance components in $\boldsymbol{\Phi}$ and the error variances in $\boldsymbol{\Theta}^2$ may be estimated by specifying the model as in Equation 24, by imposing the constraints on $\boldsymbol{\Phi}$ just described and by using the sample variance-covariance matrix **S** of all the eight test forms. Bock (1960), assuming the replicate forms to be parallel and the error variances for different tests to be homogenous, estimated the error variances separately and the variance components from the dispersion matrix of the means of the replicate scores. Our approach uses less restrictive assumptions and gives more efficient estimates of the variance components.

In Equation 24, the matrix **A** is of order 4×9 and of rank 4. However, reparameterization is not necessary if one is willing to assume that $\sigma^2_{b_{\nu i}} = \sigma^2_b$, $\sigma^2_{c_{\nu j}} = \sigma^2_c$, and $\sigma^2_{d_{\nu ij}} = \sigma^2_d$ for all i and j, since these constraints can be introduced directly in $\boldsymbol{\Phi}$. An alternative approach is to choose four independent linear functions of the components in $\boldsymbol{\tau}$ and to estimate the variance-covariance matrix of these. For example, one may choose the following functions:

$$\begin{aligned}
u^*_{\nu 1} &= 2a_\nu + b_{\nu 1} + b_{\nu 2} + c_{\nu 1} + c_{\nu 2} + \tfrac{1}{2}(d_{\nu 11} + d_{\nu 12} + d_{\nu 21} + d_{\nu 22}),\\
u^*_{\nu 2} &= b_{\nu 1} - b_{\nu 2} + \tfrac{1}{2}(d_{\nu 11} + d_{\nu 12} - d_{\nu 21} - d_{\nu 22}),\\
u^*_{\nu 3} &= c_{\nu 1} - c_{\nu 2} + \tfrac{1}{2}(d_{\nu 11} - d_{\nu 12} + d_{\nu 21} - d_{\nu 22}),\\
u^*_{\nu 4} &= \tfrac{1}{2}(d_{\nu 11} - d_{\nu 12} - d_{\nu 21} + d_{\nu 22});
\end{aligned}$$

i.e., we choose **L** as

$$\mathbf{L} = \begin{bmatrix} 2 & 1 & 1 & 1 & 1 & \tfrac{1}{2} & \tfrac{1}{2} & \tfrac{1}{2} & \tfrac{1}{2} \\ 0 & 1 & -1 & 0 & 0 & \tfrac{1}{2} & \tfrac{1}{2} & -\tfrac{1}{2} & -\tfrac{1}{2} \\ 0 & 0 & 0 & 1 & -1 & \tfrac{1}{2} & -\tfrac{1}{2} & \tfrac{1}{2} & -\tfrac{1}{2} \\ 0 & 0 & 0 & 0 & 0 & \tfrac{1}{2} & -\tfrac{1}{2} & -\tfrac{1}{2} & \tfrac{1}{2} \end{bmatrix}.$$

Then from Equation 18, **K** becomes

$$\mathbf{K} = \tfrac{1}{2}\begin{bmatrix} 1 & 1 & 1 & 1 \\ 1 & 1 & -1 & -1 \\ 1 & -1 & 1 & -1 \\ 1 & -1 & -1 & 1 \end{bmatrix},$$

Equation 23 becomes

$$\boldsymbol{\tau}_\nu = \boldsymbol{\mu} + \mathbf{K}\mathbf{u}^*_\nu$$

and the variance-covariance matrix of **x** is

$$\boldsymbol{\Sigma} = \mathbf{B}\mathbf{K}\boldsymbol{\Phi}^*\mathbf{K}'\mathbf{B}' + \boldsymbol{\Theta}^2,$$

where $\boldsymbol{\Phi}^*$ is the variance-covariance matrix of $u^*_{\nu 1}$, $u^*_{\nu 2}$, $u^*_{\nu 3}$, and $u^*_{\nu 4}$.

The estimated variances in $\hat{\boldsymbol{\Phi}}^*$ give useful information for the interpretation of individual differences. If there are no effects due to test classification and specificity, i.e., if $\sigma_b^2 = \sigma_c^2 = \sigma_d^2 = 0$, $u_{\nu 1}$ estimates separately the level of general ability of individual ν with respect to the tests used. In the absence of specific effects, $u_{\nu 2}$ and $u_{\nu 3}$ estimate separately contrasts in abilities defined by the 'Stimulus' and 'Response' dichotomies of the test classification. Finally, $u_{\nu 4}$ estimates separately a function of the specific effects, namely, the interaction effects

$$\tfrac{1}{2}(d_{\nu 11} - d_{\nu 12}) - \tfrac{1}{2}(d_{\nu 21} - d_{\nu 22}).$$

A positive $u_{\nu 4}$ indicates that specific effects cause test scores in the first categories of the 'Stimulus' classification to differ more than those in the second. Covariances between the various us are also useful. For example, the covariance between $u_{\nu 1}$ and $u_{\nu 2}$ gives an indication of the extent to which $\sigma_{b_1}^2 = \sigma_{b_2}^2$.

The classical formulation of the mixed model with the assumptions of homogenous error variances and uncorrelated latent random variables is not realistic in most applications in the behavioral sciences. Bock, Dicken, and van Pelt (1969) demonstrated heuristically the inadequacy of the specification of uncorrelated latent variables, i.e., the inadequacy of $\boldsymbol{\Phi}$ being specified as a diagonal matrix, which is the case considered by Bock (1960) and by Bock and Bargmann (1966). As already pointed out, in our method of covariance structure analysis, the assumption that $\boldsymbol{\Phi}$ be diagonal is not necessary. If the model provides information enough, so that all the variances and covariances of the latent variables are identified, these may also be estimated, and the assumption of zero covariances may be examined empirically.

A General Class of Components of Covariance Models

Wiley, Schmidt, and Bramble (1973) suggested the study of a general class of components of covariance models. This class of models is a special case of Equation 1, namely, when $\mathbf{B}$ is diagonal, $\boldsymbol{\Lambda}$ is known a priori, $\boldsymbol{\Phi}$ is symmetric and positive definite, and $\boldsymbol{\Psi}$ or $\boldsymbol{\Theta}$ are either zero or diagonal. The covariance matrix $\boldsymbol{\Sigma}$ will then be of the form

$$\boldsymbol{\Sigma} = \boldsymbol{\Delta}\mathbf{A}\boldsymbol{\Phi}\mathbf{A}'\boldsymbol{\Delta} + \boldsymbol{\Theta}^2 \quad \text{or} \quad \boldsymbol{\Sigma} = \boldsymbol{\Delta}(\mathbf{A}\boldsymbol{\Phi}\mathbf{A}' + \boldsymbol{\Psi}^2)\boldsymbol{\Delta}. \qquad (25a, 25b)$$

The matrix $\mathbf{A}(p \times k)$ is assumed to be known and gives the coefficient of the linear functions connecting the manifest and latent variables, $\boldsymbol{\Delta}$ is a $p \times p$ diagonal matrix of unknown scale factors, $\boldsymbol{\Phi}$ is the $k \times k$ symmetric and positive definite covariance matrix of the latent variables and $\boldsymbol{\Psi}^2$ and $\boldsymbol{\Theta}^2$ are $p \times p$ diagonal matrices of error variances.

Within this class of models eight different special cases are of interest. These are generated by the combination of the following set of conditions:

$$\text{on } \boldsymbol{\Delta}: \quad \begin{Bmatrix} \boldsymbol{\Delta} = \mathbf{I} \\ \boldsymbol{\Delta} \neq \mathbf{I} \end{Bmatrix};$$

$$\text{on } \boldsymbol{\Phi}: \quad \left\{ \begin{matrix} \boldsymbol{\Phi} \text{ is diagonal} \\ \boldsymbol{\Phi} \text{ is not diagonal} \end{matrix} \right\};$$

$$\text{on } \boldsymbol{\Psi}^2 \text{ or } \boldsymbol{\Theta}^2: \quad \left\{ \begin{matrix} \boldsymbol{\Psi}^2 \text{ or } \boldsymbol{\Theta}^2 = \sigma^2\mathbf{I} \\ \boldsymbol{\Psi}^2 \text{ or } \boldsymbol{\Theta}^2 \text{ general diagonal} \end{matrix} \right\}.$$

The classical formulation of the mixed model and its generalizations assume that $\boldsymbol{\Delta} = \mathbf{I}$. This is appropriate if the observed variables are in the same metric as, for example, when the observed variables represent physical measurements, time to criterion measures, reaction times, or items similarly scaled such as semantic differential responses. However, if the observed variables are measured in different metrics, then the classical model will not fit. In such cases the inclusion of $\boldsymbol{\Delta}$ in the model as a general diagonal matrix of scaling factors will provide a useful alternative specification. It should be pointed out that the elements of $\boldsymbol{\Delta}$ do not have to be related to the variances of the variables.

The classical components of the variance model assume that $\boldsymbol{\Phi}$ is diagonal. However, as has already been pointed out, there are no substantive reasons for assuming this.

The two conditions on $\boldsymbol{\Psi}^2$ or $\boldsymbol{\Theta}^2$ correspond to homogeneous and heterogeneous error variances. If the variables are in the same metric and if the measurement situation is sufficiently similar from variable to variable, then it seems reasonable to hypothesize that the variances of the errors of measurement ought to be homogeneous, i.e., in Equation 25a we take $\boldsymbol{\Delta} = \mathbf{I}$ and $\boldsymbol{\Theta}^2 = \sigma^2\mathbf{I}$.

If, on the other hand, the scale of measurement is the same, but the measurement situation from variable to variable is different enough so as to generate different kinds of error structures, then the variances of the errors of measurement might differ systematically from variable to variable. For this situation it seems best to take $\boldsymbol{\Delta} = \mathbf{I}$ but to leave $\boldsymbol{\Theta}^2$ free in Equation 25a. This situation could arise where the variable time to task completion is measured under different treatment conditions on the same individuals. Suppose that under the first set of conditions, time to completion takes on the average of 1 minute while in the second treatment group the average time required is 10 minutes. It is reasonable to hypothesize that some recording procedures will vary in their degree of accuracy, since in the first treatment condition the observations will most likely be more accurate than under the second set of conditions. It is also possible that individuals are inherently more variable under some conditions than others. This will result in different error variances despite the constant metric of time.

If the manifest variables are in different metrics, then clearly the error variances in the observed metric will most likely be heterogeneous. One useful hypothesis to test in this context is that the standard deviations of the errors of measurement are proportional to the rescaling factors. This corresponds to taking $\boldsymbol{\Psi}^2 = \sigma^2\mathbf{I}$ in Equation 25b. When both $\boldsymbol{\Delta}$ and $\boldsymbol{\Psi}^2$ are free, Equations 25a and 25b are equivalent.

Analysis of 2^3 Factorial Design

This example is taken from Wiley, Schmidt, and Bramble (1973). The original data are from a study by Miller and Lutz (1966) and consist of the scores of 51 education students on a test designed to assess teachers' judgments about the effects of situation and instruction factors on the facilitation of pupil learning. The items used were designed according to three factors that were hypothesized to influence classroom learning situations and teaching practices. The three factors and their levels are given as follows (see Miller & Lutz, 1966).

> Grade Level (G). The levels of this factor, the first grade (G_1) and the sixth grade (G_2), were chosen to represent extremes of the elementary grades. In this way it was possible to maximize the opportunity for observing any differences in teachers' judgments that might occur as a result of variations due to grade level.
>
> Teacher Approach (T). The teacher-centered approach (T_1) and the pupil-centered approach (T_2) were distinguished as levels of this factor on the basis of the locus of described activity control, and the direction of described pupils' attention. In the case of the teacher-centered approach, the locus and direction were oriented to the teacher; in the case of the pupil-centered approach, the locus and direction were oriented to the pupil.
>
> Teaching Method (M). Level one of this factor was drill (M_1) which was used strictly to refer to rote learning activities; discovery (M_2) was used to refer to an approach in which the teacher attempts to develop pupil understanding through procedures aimed at stimulating insight without recourse to rote memorization or rigid learning routines.

The eight subtest scores are each based on eight items. The eight subtests conform to a 2^3 factorial arrangement, which is given by

$$\boldsymbol{A} = \begin{bmatrix} 1 & 1 & 1 & 1 \\ 1 & 1 & 1 & -1 \\ 1 & 1 & -1 & 1 \\ 1 & 1 & -1 & -1 \\ 1 & -1 & 1 & 1 \\ 1 & -1 & 1 & -1 \\ 1 & -1 & -1 & 1 \\ 1 & -1 & -1 & -1 \end{bmatrix}.$$

The sample covariance matrix $\mathbf{S}$ for these data is given in Table 10.

For this data the use of $\boldsymbol{\Delta} = \boldsymbol{I}$ was considered appropriate, because the same scale (1-7) was employed for each item; the numbers of items in the subtests were also equal.

The data were analyzed under each of the four remaining cases in the general class of models considered in the previous section. Note also the χ^2 values and their degrees of freedom as shown in Table 10. In the margin of this table, χ^2 values are given for testing the hypotheses of uncorrelated components and of homogeneous error variances. It appears that these hypotheses were both rejected by the tests. The only model that seems reasonable is the one

TABLE 10
Miller-Lutz data

Sample Covariance Matrix

$$\mathbf{S} = \begin{bmatrix} 18.74 & & & & & & & \\ 9.28 & 18.80 & & & \text{(symmetric)} & & & \\ 15.51 & 7.32 & 21.93 & & & & & \\ 3.98 & 15.27 & 4.10 & 26.62 & & & & \\ 15.94 & 4.58 & 13.79 & -2.63 & 19.82 & & & \\ 7.15 & 13.63 & 3.86 & 15.33 & 3.65 & 16.81 & & \\ 11.69 & 6.05 & 10.18 & 1.13 & 13.55 & 5.72 & 16.58 & \\ 2.49 & 12.35 & 0.03 & 16.93 & -0.86 & 14.33 & 2.99 & 18.26 \end{bmatrix}$$

χ^2 Values for Testing the Fit of Four Models

	$\boldsymbol{\Theta}^2 = \sigma^2\mathbf{I}$	$\boldsymbol{\Theta}^2 \neq \sigma^2\mathbf{I}$	
$\boldsymbol{\Phi}$ diagonal	$\chi^2_{31} = 68.25$	$\chi^2_{24} = 46.16$	$\chi^2_7 = 22.09$
$\boldsymbol{\Phi}$ not diagonal	$\chi^2_{25} = 51.00$	$\chi^2_{18} = 25.98$	$\chi^2_7 = 25.02$
	$\chi^2_6 = 17.25$	$\chi^2_6 = 20.18$	

Maximum-Likelihood Estimates of $\boldsymbol{\Phi}$ and $\boldsymbol{\Theta}^2$
(Standard Errors in Parentheses)

$$\hat{\boldsymbol{\Phi}} = \begin{bmatrix} 9.16\ (1.95) & & & \\ 0.75\ (0.48) & 0.70\ (0.34) & & \\ 0.63\ (0.43) & -0.05\ (0.33) & 0.43\ (0.91) & \\ -0.62\ (1.10) & -0.51\ (0.81) & 1.13\ (0.51) & 5.21\ (1.58) \end{bmatrix}$$

$$\hat{\boldsymbol{\Theta}}^2 = \text{diag}\ [1.52(0.83),\ 4.95(1.41),\ 8.25(1.88),\ 5.58(1.60),\ 1.95(0.96),\ 5.76(1.21),\ 2.52(0.92)]$$

Note: Miller-Lutz data taken from Wiley, Schmidt, and Bramble (1972).

that assumes both correlated components and heterogenous error variances. The maximum-likelihood estimates of the variances and covariances of the components and the error variances, together with their standard errors, are also given in Table 10.

The relative magnitudes of the estimated variance components for the latent variables indicate the major sources of variation in the performance of the subjects. The estimate of the first variance component ($\hat{\varphi}_{11} = 9.16$) is the largest as would be expected, since this component reflects the variation due to individual differences between the subjects. The estimated values of the other components indicate that another major source of variation in the responses of the subjects is due to the different type of teaching method specified in the content of the item (i.e., $\hat{\varphi}_{44} = 5.21$). Apparently the contrast between the drill and the discovery methods of instruction caused the responses of the education students to vary considerably. Variation contributed by grade level was intermediate in magnitude ($\hat{\varphi}_{22} = 0.70$). The estimate of the variance component for the teacher-approach factor and its large standard error (i.e., $\hat{\varphi}_{33} = 0.43$, S.E. $(\hat{\varphi}_{33}) = 0.91$) indicate that this was not an important source of variation in the performance of the subjects. One of the estimated covariances was relatively large—that between the teaching-method factor and the teacher-approach factor—indicating these latent variables to be highly correlated. This would indicate that the responses of the education students to the different types of teaching methods specified in the items were related to their responses to the teacher-approach factor found in the items.

6. SIMPLEX AND CIRCUMPLEX MODELS

Various Types of Simplex Models

Since the fundamental paper of Guttman (1954) on the simplex structure for correlations between ordered tests, many investigators have found data displaying the typical simplex structure. Guttman gave several examples of this structure. His Table 5 is reproduced here as Table 11. In this example all the tests involve verbal ability and are ordered according to increasing complexity.

The typical property in a simplex correlation structure, such as that in Table 11, is that the correlations decrease as one moves away from the main diagonal. Such data will not usually fit a factor analysis model with one common factor for the following reasons. Let the tests be ordered so that the factor loadings decrease. Then if the factor model holds, the correlations in the first row decrease as one moves away from the diagonal, but the correlations in the last row *increase* as one moves away from the diagonal. Also the

TABLE 11
Intercorrelations of six verbal-ability tests for 1046 Bucknell College sophomores

Test	Spelling	Punctuation	Grammar	Vocabulary	Literature	Foreign Literature
	A	C	B	D	E	H
A	—	0.621	0.564	0.476	0.394	0.389
C	0.621	—	0.742	0.503	0.461	0.411
B	0.564	0.742	—	0.577	0.472	0.429
D	0.476	0.503	0.577	—	0.688	0.548
E	0.394	0.461	0.472	0.688	—	0.639
H	0.389	0.411	0.429	0.548	0.639	—
Total	2.444	2.738	2.784	2.792	2.654	2.416

correlations just below the diagonal *decrease markedly* as one moves down, which does not hold in Table 11.

Jöreskog (1970b) considered several statistical models for such simplex structures. Following Anderson (1960), he formulated these models in terms of the well-known Wiener and Markov stochastic processes. A distinction was made between a perfect simplex and a quasi-simplex. A perfect simplex is reasonable only if the measurement errors in the test scores are negligible. A quasi-simplex on the other hand allows for sizable errors of measurement.

Table 12 shows six types of simplex models and some of their character-

TABLE 12
Six types of simplex models

Model	Covariance Structure	Scale Dependence	No. of Parameters
Markov Simplex	$\boldsymbol{\Sigma} = \mathbf{D}_\alpha \mathbf{T} \mathbf{D}_{\delta*} \mathbf{T}' \mathbf{D}_\alpha$	Scale Free	$2p - 1$
Wiener Simplex	$\boldsymbol{\Sigma} = \mathbf{T} \mathbf{D}_{\delta*} \mathbf{T}'$	Scale Dependent	p
Quasi-Wiener Simplex	$\boldsymbol{\Sigma} = \mathbf{T} \mathbf{D}_{\delta*} \mathbf{T}' + \boldsymbol{\Theta}^2$	Scale Dependent	$2p - 1$
Quasi-Wiener Simplex with Equal Error Variances	$\boldsymbol{\Sigma} = \mathbf{T} \mathbf{D}_{\delta*} \mathbf{T}' + \theta^2 \mathbf{I}$	Scale Dependent	$p + 1$
Quasi-Markov Simplex	$\boldsymbol{\Sigma} = \mathbf{D}_\alpha \mathbf{T} \mathbf{D}_{\delta*} \mathbf{T}' \mathbf{D}_\alpha + \boldsymbol{\Theta}^2$	Scale Free	$3p - 3$
Restricted Quasi-Markov Simplex	$\boldsymbol{\Sigma} = \mathbf{D}_\alpha (\mathbf{T} \mathbf{D}_{\delta*} \mathbf{T}' + \psi^2 \mathbf{I}) \mathbf{D}_\alpha$	Scale Free	$2p$

istics. The Wiener simplexes are scale-dependent models and are appropriate only when the units of measurement are the same for all tests. The Markov simplexes, on the other hand, are scale-free models and can therefore be used with data like those in Table 11, where the units of measurements are arbitrary.

The Markov Simplex

Consider a real stochastic process $X(t)$ and arbitrary scale points $t_1 < t_2 < \cdots < t_p$. The Markov simplex is defined by

$$\begin{aligned} E[X(t_i)] &= \mu_i, \\ \mathrm{Var}[X(t_i)] &= \sigma_i^2, \\ \mathrm{Corr}[X(t_i), X(t_j)] &= \rho^{|t_i - t_j|}, \end{aligned}$$

where $0 < \rho < 1$.

Consider the following transformation from $t' = (t_1, t_2, \ldots, t_p)$ to $t^{*\prime} = (t_1^*, t_2^*, \ldots, t_p^*)$, where

$$\left.\begin{aligned} t_1^* &= t_1 \\ t_2^* &= t_2 - t_1 \\ t_3^* &= t_3 - t_2 \\ &\vdots \\ t_p^* &= t_p - t_{p-1} \end{aligned}\right\}.$$

This transformation is one-to-one and the ts may be obtained from the t^*s as follows

$$\left.\begin{aligned} t_1 &= t_1^* \\ t_2 &= t_1^* + t_2^* \\ t_3 &= t_1^* + t_2^* + t_3^* \\ &\vdots \\ t_p &= t_1^* + t_2^* + \cdots + t_p^* \end{aligned}\right\}.$$

In matrix form these transformations are

$$\mathbf{t}^* = \mathbf{T}^{-1}\mathbf{t},$$

$$\mathbf{t} = \mathbf{T}\mathbf{t}^*,$$

where

$$\mathbf{T} = \begin{bmatrix} 1 & 0 & 0 & \cdots & 0 \\ 1 & 1 & 0 & \cdots & 0 \\ 1 & 1 & 1 & \cdots & 0 \\ \vdots & \vdots & \vdots & & \vdots \\ 1 & 1 & 1 & & 1 \end{bmatrix}; \quad \mathbf{T}^{-1} = \begin{bmatrix} 1 & 0 & 0 & \cdots & 0 & 0 \\ -1 & 1 & 0 & \cdots & 0 & 0 \\ 0 & -1 & 1 & \cdots & 0 & 0 \\ \vdots & \vdots & \vdots & & \vdots & \vdots \\ 0 & 0 & 0 & \cdots & -1 & 1 \end{bmatrix}. \qquad (26a, 26b)$$

In terms of $t_2^*, t_3^*, \ldots, t_p^*$, the correlation matrix of $x' = [X(t_1), X(t_2), \ldots, X(t_p)]$ is

$$\mathbf{P} = \begin{bmatrix} 1 & & & \text{symmetric} \\ \rho^{t_2^*} & 1 & & \\ \rho^{t_2^*+t_3^*} & \rho^{t_3^*} & & \\ \vdots & \vdots & & \\ \rho^{t_2^*+\cdots+t_p^*} & \rho^{t_3^*+\cdots+t_p^*} & \cdots & 1 \end{bmatrix}. \tag{27}$$

It is seen that there are only $p - 1$ independent correlations, namely, those just below (or above) the main diagonal and the other correlations are products of these. For example,

$$\rho_{ji} = \prod_{k=i}^{k=j} \rho^{t_k^*} = \prod_{k=i}^{k=j} \rho_{k+1,k}, \quad i < j.$$

It is also seen that the correlations fall off as one moves away from the main diagonal. Thus, the model of a Markov simplex may be consistent with the simplex pattern referred to in the previous section. The variance-covariance matrix of $\mathbf{x}' = [X(t_1), X(t_2), \ldots, X(t_p)]$ is

$$\boldsymbol{\Sigma} = \mathbf{D}_\sigma \mathbf{P} \mathbf{D}_\sigma, \tag{28}$$

where $\mathbf{P}$ is defined by Equation 27 and $\mathbf{D}_\sigma = \text{diag}\,(\sigma_1, \sigma_2, \ldots, \sigma_p)$.

The values $t_2^*, t_3^*, \ldots, t_p^*$ are determined only up to a multiplicative constant, and since only differences between scale points enter into $t_2^*, t_3^*, \ldots, t_p^*$, it is evident that the origin and the unit of the scale are arbitrary. One may, for example, choose the origin at $t_1 = 0$ and define the unit so that $t_2 = 1$.

The Wiener Simplex

The Wiener process $X(s)$, with scale points $s_1 < s_2 \cdots < s_{p-1} < s_p$, is defined by

$$E[X(s_i)] = \mu_i,$$
$$\text{Var}[X(s_i)] = s_i,$$
$$\text{Cov}[X(s_i), X(s_j)] = s_i, \quad i < j.$$

The correlation between $X(s_i)$ and $X(s_j)$ is

$$\rho[X(s_i), X(s_j)] = \frac{s_i}{\sqrt{s_i s_j}} = \sqrt{\frac{s_i}{s_j}}, \quad i < j,$$

so that the correlation matrix of $\mathbf{x}' = [X(s_1), X(s_2), \ldots, X(s_p)]$ is of the form

$$\mathbf{P} = \begin{bmatrix} 1 & & & \\ \sqrt{\dfrac{s_1}{s_2}} & 1 & & \\ \sqrt{\dfrac{s_1}{s_3}} & \sqrt{\dfrac{s_2}{s_3}} & 1 & \\ \vdots & \vdots & & \\ \sqrt{\dfrac{s_1}{s_p}} & \sqrt{\dfrac{s_2}{s_p}} & \cdots & 1 \end{bmatrix}. \tag{29}$$

It is seen that **P** in Equation 29 has the same properties as the **P** in Equation 27. In fact, the correlation structures of the Markov simplex and the Wiener simplex are equivalent, for if $i < j$, then

$$\sqrt{\frac{s_i}{s_j}} = e^{-1/2(\log s_j - \log s_i)} = e^{-\alpha(t_j - t_i)} = \rho^{t_j - t_i},$$

with $\rho = e^{-\alpha}$, $\alpha t_i = \frac{1}{2} \log s_i$, and $\alpha > 0$ arbitrary. Thus, the scale for the Markov simplex is just a logarithmic transformation of the scale for the Wiener simplex. In the Markov simplex, correlations correspond to distances between scale points, whereas in the Wiener simplex, correlations correspond to square roots of ratios between scale points.

Although the correlation matrix **P** for the Markov simplex and the Wiener simplex are identical, it does not hold for the corresponding dispersion matrices. Writing $\mathbf{s}^* = \mathbf{T}^{-1}\mathbf{s}$, the dispersion matrix $\boldsymbol{\Sigma}$ for the Wiener simplex has the form

$$\boldsymbol{\Sigma} = \begin{bmatrix} s_1 & & & \\ s_1 & s_2 & & \\ \vdots & \vdots & \cdots & \\ s_1 & s_2 & \cdots & s_p \end{bmatrix} = \begin{bmatrix} s_1^* & & & \\ s_1^* & s_1^* + s_2^* & & \\ \vdots & \vdots & \cdots & \\ s_1^* & s_1^* + s_2^* & \cdots & s_1^* + s_2^* + \cdots + s_p^* \end{bmatrix} \tag{30}$$
$$= \mathbf{T}\mathbf{D}_{s^*}\mathbf{T}',$$

where **T** is defined by Equation 26 and $\mathbf{D}_{s^*} = \operatorname{diag}(s_1^*, s_2^*, \ldots, s_p^*)$. It is seen that $\boldsymbol{\Sigma}$ has p independent parameters, whereas the $\boldsymbol{\Sigma}$ for the Markov simplex has $2p - 1$ independent parameters. Thus the Wiener simplex is a more restricted model than the Markov simplex. Since **P** in Equations 27 and 29 are the same, namely, $\mathbf{P} = \mathbf{D}_s^{-1/2}\mathbf{T}\mathbf{D}_{s^*}\mathbf{T}'\mathbf{D}_s^{-1/2}$, this may be substituted into Equation 28 to give the variance-covariance matrix for the Markov simplex as

$$\boldsymbol{\Sigma} = \mathbf{D}_\sigma \mathbf{D}_s^{-1/2}\mathbf{T}\mathbf{D}_{s^*}\mathbf{T}'\mathbf{D}_s^{-1/2}\mathbf{D}_\sigma,$$

which, with $\mathbf{D}_\alpha = \mathbf{D}_\sigma \mathbf{D}_s^{-1/2} = \operatorname{diag}(\alpha_1, \alpha_2, \ldots, \alpha_p)$, becomes

$$\boldsymbol{\Sigma} = \mathbf{D}_\alpha \mathbf{T}\mathbf{D}_{s^*}\mathbf{T}'\mathbf{D}_\alpha. \tag{31}$$

It is clear that one α may be fixed at unity, so that there are $2p - 1$ independent parameters. Equations 30 and 31 are in the form of Equation 1. Although simpler methods are available (see, e.g., Jöreskog, 1970b), the Wiener and Markov simplexes may be estimated by means of the ACOVS program, which also provides for the testing of the goodness of fit of the models.

The Wiener simplex is most suitable when all variates $x_1, x_2, \ldots, x_p$ are measured in the same units. If they are not measured in the same units, the results will depend on the various units of measurement used. In contrast to the Markov simplex, the Wiener simplex is scale dependent. Analyses of $\mathbf{S}$ and $\mathbf{DSD}$, where $\mathbf{D}$ is an arbitrary diagonal matrix of scale factors, do not yield results that are properly related. Whereas the correlation matrix $\mathbf{R}$ may be used to estimate a Markov simplex, the dispersion matrix $\mathbf{S}$ must be used to estimate a Wiener simplex.

The Quasi-Wiener Simplex

In a perfect simplex it is assumed that the observed variates are infallible, containing no errors of measurement, and that the simplex structure holds for the observed variates themselves. In a quasi-simplex it is assumed that the observed variates contain errors of measurement and that the simplex structure holds for the true variates. Since most measurements in the behavioral sciences contain sizable errors, quasi-simplex models are more realistic for such data.

The Markov model and the Wiener model of the preceding subsections are both perfect simplexes. While the estimation of these models is a simple matter, this is not so for most quasi-simplexes. Here iterative techniques have to be used to obtain the estimates.

When the observed variates $x_i = X(s_i)$ contain errors of measurement, we write

$$x_i = \tau_i + e_i, \quad i = 1, 2, \ldots, p,$$

where τ_i is the true measurement and e_i is the error. About the errors we assume that $E(e_i) = 0$ and $E(e_i e_j) = 0$ for $i \neq j$ and that $\mathrm{Cov}(e_i, \tau_i) = 0$. Let $\boldsymbol{T}$ be the dispersion matrix of $\tau' = (\tau_1, \tau_2, \ldots, \tau_p)$ and let $\boldsymbol{\Theta}^2$ be the diagonal dispersion matrix of $\mathbf{e}' = (e_1, e_2, \ldots, e_p)$. Then the dispersion matrix $\boldsymbol{\Sigma}$ of the observed variates is

$$\boldsymbol{\Sigma} = \boldsymbol{T} + \boldsymbol{\Theta}^2. \tag{32}$$

We shall first consider the case where $\boldsymbol{T}$ has the Wiener simplex structure $\mathbf{T}\mathbf{D}_{s^*}\mathbf{T}'$ and then the case where $\boldsymbol{T}$ has the Markov simplex structure $\mathbf{D}_\alpha\mathbf{P}\mathbf{D}_\sigma = \mathbf{D}_\alpha\mathbf{T}\mathbf{D}_{s^*}\mathbf{T}'\mathbf{D}_\alpha$.

The dispersion matrix of the quasi-Wiener simplex is

$$\boldsymbol{\Sigma} = \begin{bmatrix} s_1 + \theta_1^2 & & & \\ s_1 & s_2 + \theta_2^2 & & \\ \vdots & \vdots & \cdots & \\ s_1 & s_2 & \cdots & s_p + \theta_p^2 \end{bmatrix} = \mathbf{TD}_{s^*}\mathbf{T}' + \boldsymbol{\Theta}^2. \tag{33}$$

It is seen that s_p and θ_p are only involved in σ_{pp} so that they are not separately identified; only their sum is identified. The other $2p - 1$ parameters in $\boldsymbol{\Sigma}$ are independently identified.

Equation 33 is in the form of Equation 1 and may therefore be estimated using the ACOVS program. There are many different ways in which this may be done. The simplest is probably to choose $\mathbf{B} = \mathbf{T}$, $\boldsymbol{\Lambda} = \mathbf{0}$, $\boldsymbol{\Phi} = \mathbf{0}$, $\boldsymbol{\Psi}^2 = \mathbf{D}_{s^*}$, and the last element of $\boldsymbol{\Theta}$ equal to zero. By specifying the model in this way, the likelihood function may be maximized with respect to $\boldsymbol{\Psi}$ and $\boldsymbol{\Theta}$, and as a consequence, the maximum-likelihood estimates $\hat{\mathbf{D}}_{s^*} = \hat{\boldsymbol{\Psi}}^2$ and $\hat{\boldsymbol{\Theta}}^2$ will be nonnegative. Standard errors may be obtained for each estimated parameter.

The Quasi-Wiener Simplex with Equal Error Variances

Since the quasi-Wiener simplex is applied to data where the units of measurements are the same for all variates, it is sometimes useful to consider a more restrictive model where the error variances are assumed to be equal. The variance-covariance matrix $\boldsymbol{\Sigma}$ for this model is

$$\boldsymbol{\Sigma} = \mathbf{TD}_{s^*}\mathbf{T}' + \theta^2\mathbf{I}, \tag{34}$$

where $\mathbf{D}_{s^*}$ is as before and θ^2 is the common error variance. It has $p + 1$ independent parameters.

The maximum-likelihood estimates of this model can be obtained as before with the ACOVS program by specifying the equality of all the elements of $\boldsymbol{\Theta}$. An illustrative example is given in Table 13. The variance-covariance matrix of a proficiency measure in six trials is shown. The data were obtained in a study of a two-hand coordination task that was conducted by Bilodeau (1957). The task requires the subject to move a pin around a clover-shaped runway by the coordinated turning of two control handles. The subjects were 152 basic airmen, and the trials were 60 seconds long with 30-second rest intervals between trials. The data have previously been analyzed by Bock and Bargmann (1966).

The data from the six trials were analyzed under the three models listed under Summary of Analyses. It is seen that a perfect Wiener simplex does not fit the data, whereas the quasi-Wiener simplex with equal error variances has a very good fit. In general, when such a good fit has been obtained there is seldom reason to relax the model further by introducing more parameters. However, for the sake of completeness we have also analyzed the data with-

TABLE 13
Bilodeau's data

Variance-Covariance Matrix of Proficiency Measures in Six Trials
$N = 152$

Trial						
1	521					
2	477	576				
3	484	536	601			
4	510	575	593	755		
5	523	580	598	718	797	
6	528	584	613	722	751	802

Summary of Analyses

	Model	χ^2	d.f.	P
1	Perfect Wiener Simplex	119.05	15	0.00
2	Quasi-Wiener Simplex with Equal Error Variances	9.39	14	0.81
3	Quasi-Wiener Simplex	8.36	10	0.59

Solution for Model 2

i	s_i^*	s.e.(s_i^*)	
1	482.6	58.7	
2	54.6	14.6	$\hat{\theta} = 6.73$
3	16.0	10.2	$\hat{\theta}^2 = 45.3$
4	81.4	14.9	s.e.$(\theta) = 0.35$
5	21.6	9.6	
6	1.61	10.3	

out the assumption of equal error variances. The drop in χ^2 is 1.03 with a drop in degrees of freedom equal to 4. Thus the hypothesis of equal error variances cannot be rejected. The maximum-likelihood solution for the model with equal error variances is given with standard errors for each parameter.

The Quasi-Markov Simplex

In the preceding example the data were obtained from learning trials where there is an a priori given ordering of the variables (trials) and where the unit of measurement in each variable is the same. For the data of Table 11, neither of these conditions is necessarily true, and it is therefore not appropriate to analyze the variance-covariance matrix according to the quasi-Wiener simplex model. In this case a scale-free model such as the quasi-Markov simplex is needed. The variance-covariance matrix of a quasi-Markov simplex is

$$\boldsymbol{\Sigma} = \mathbf{D}_\alpha \mathbf{T} \mathbf{D}_{s^*} \mathbf{T}' \mathbf{D}_\alpha + \boldsymbol{\Theta}^2. \tag{35}$$

In a four-variate case, $\boldsymbol{\Sigma}$ is explicitly given by

$$\Sigma = \begin{bmatrix} \alpha_1^2 s_1^* + \theta_1^2 & & & \\ \alpha_1\alpha_2 s_1^* & \alpha_2^2(s_1^* + s_2^*) + \theta_2^2 & & \\ \alpha_1\alpha_3 s_1^* & \alpha_2\alpha_3(s_1^* + s_2^*) & \alpha_3^2(s_1^* + s_2^* + s_3^*) + \theta^2 & \\ \alpha_1\alpha_4 s_1^* & \alpha_2\alpha_4(s_1^* + s_2^*) & \alpha_3\alpha_4(s_1^* + s_2^* + s_3^*) & \\ & & & \alpha_4^2(s_1^* + s_2^* + s_3^* + s_4^*) + \theta_4^2 \end{bmatrix}.$$

As pointed out by Anderson (1960; see also Jöreskog, 1970b), there are three kinds of indeterminacies in this model in that arbitrary numbers may be added to the two first and the last diagonal element of $\boldsymbol{\Theta}^2$, the effect of which may be counterbalanced by certain changes in $\mathbf{D}_\alpha$ and $\mathbf{D}_{s^*}$. To eliminate the indeterminacies, we must impose three independent conditions. The most convenient conditions seem to be to set s_1^*, s_2^*, and s_p^* at unity. It is clear that in this model one cannot determine the distance between s_1 and s_2 and the distance between s_{p-1} and s_p. Only distances between points $s_2 < s_3 < \cdots < s_{p-1}$ are estimable.

For the quasi-Markov simplex the maximum-likelihood estimates may be computed using the program ACOVS. To do so one specifies $\mathbf{B}$ as the diagonal matrix $\mathbf{D}_\alpha$, $\boldsymbol{\Lambda}$ the fixed matrix $\mathbf{T}$, $\boldsymbol{\Phi}$ the diagonal matrix $\mathbf{D}_{s^*}$ with $\varphi_{11} = 1$, $\varphi_{22} = 1$ and $\varphi_{pp} = 1$ and $\boldsymbol{\Theta}^2$ free. In the resulting solution the estimated error variances will be nonnegative. In some cases it may happen that one or more of the estimates $\hat{\varphi}_{33}, \hat{\varphi}_{44}, \ldots, \hat{\varphi}_{p-1,p-1}$ are negative. This is usually an indication that the variables are in the wrong order or that the model is otherwise wrong. The computer program also gives a large sample χ^2 for testing the goodness of fit of the model.

The maximum-likelihood method for the quasi-Markov simplex is scale free in the following sense. If, on the basis of a sample dispersion matrix $\mathbf{S}$, the estimates $\hat{\mathbf{D}}_\alpha$, $\hat{\mathbf{D}}_{s^*}$, and $\hat{\boldsymbol{\Delta}}$ are obtained with $s_1^* = 1$, $s_2^* = 1$, and $s_p^* = 1$, then if $\mathbf{DSD}$ is used instead of $\mathbf{S}$, where $\mathbf{D}$ is a diagonal matrix of arbitrary scale factors, the estimates will be $\mathbf{D}\hat{\mathbf{D}}_\alpha$, $\hat{\mathbf{D}}_{s^*}$, and $\hat{\mathbf{D}}^2\boldsymbol{\Theta}^2$. Hence one is free to use the correlation matrix $\mathbf{R}$ instead of $\mathbf{S}$. Whichever scaling is used, the estimate of $\mathbf{D}_{s^*}$ is the same. The maximum-likelihood method may be used when the units of measurement differ from one variate to another, and when they are arbitrary or irrelevant.

A Restricted Quasi-Markov Simplex

Equation 35 for the quasi-Markov simplex may also be written

$$\boldsymbol{\Sigma} = \mathbf{D}_\alpha(\mathbf{T}\mathbf{D}_{s^*}\mathbf{T}' + \mathbf{D}_\alpha^{-1}\boldsymbol{\Theta}^2\mathbf{D}_\alpha^{-1})\mathbf{D}_\alpha.$$

If one makes the assumption that $\mathbf{D}_\alpha^{-1}\boldsymbol{\Theta}^2\mathbf{D}_\alpha^{-1}$ is a scalar matrix, one obtains a certain restricted quasi-Markov simplex. The variance-covariance matrix for this model thus is

$$\boldsymbol{\Sigma} = \mathbf{D}_\alpha(\mathbf{T}\mathbf{D}_{s^*}\mathbf{T}' + \psi^2\mathbf{I})\mathbf{D}_\alpha, \tag{36}$$

where ψ^2 is a scalar. This may also be viewed as a scaling of the quasi-Wiener simplex with equal error variances. The advantage of the model of Equation 36 compared with that of Equation 35 is that the three indeterminacies with their interpretational difficulties are eliminated. The $\boldsymbol{\Sigma}$ in Equation 36 has $2p$ independent parameters. One element in $\mathbf{D}_\alpha$ or $\mathbf{D}_{s^*}$ or ψ may be fixed. The most meaningful choice is $s_1^* = 1$, which fixes the unit of the s scale or equivalently the zero point of the t scale.

The model of Equation 36 may also be estimated by the ACOVS method. One chooses $\mathbf{B}$ as the diagonal matrix $\mathbf{D}_\alpha$, $\boldsymbol{\Lambda}$ as $\mathbf{T}$, $\boldsymbol{\Phi}$ as the diagonal matrix $\mathbf{D}$ with $\varphi_{11} = 1$, $\boldsymbol{\Psi}$ of the form $\psi\mathbf{I}$ and $\boldsymbol{\Theta} = \mathbf{0}$. When the estimates $\hat{\mathbf{D}}_\alpha$, $\hat{\mathbf{D}}_{s^*}$, and $\hat{\psi}$ have been obtained, one can interpret $\hat{\mathbf{D}}_\alpha\mathbf{T}\hat{\mathbf{D}}_{s^*}\mathbf{T}\,\hat{\mathbf{D}}_\alpha$ as an estimate of the true variance-covariance matrix $\mathbf{T}$ and $\hat{\psi}^2\hat{\mathbf{D}}_\alpha^2$ as an estimate of the matrix of error variances. A test of goodness of fit of the model may also be obtained by the ACOVS method.

The model of Equation 36 is also scale free. The estimate $\hat{\mathbf{D}}_{s^*}$ with $s_1^* = 1$ is invariant under scale changes in the original variables.

Consider again the data of Table 11. This represents six verbal ability tests and the sample size is very large ($N = 1046$). Some results of analyses are shown in Table 14. It is seen that although the value 43.81 of χ^2 is significant, the fit appears to be very good as judged by the small residuals. Only two residuals 0.056 and 0.090 are larger than 0.030 in absolute value. It is clear that the fit is acceptable for most practical purposes. The estimated scale values on the t scale are given in the last column under Solution for Model 3.

Circumplex Models

Simplex models, as considered in the previous sections, are models for tests that may be conceived of as having a linear ordering. The circumplex is another model considered by Guttman (1954) and this yields a circular instead of a linear ordering. The circular order has no beginning and no end, but there is still a law of neighboring that holds.

TABLE 14
Guttman's simplex data

Summary of Analyses

	Model	No. Par.	χ^2	d.f.	P
1	Perfect Markov Simplex	11	202.64	10	0.000
2	Restricted Quasi-Markov Simplex	12	70.36	9	0.000
3	Quasi-Markov Simplex	15	43.81	6	0.000

Solution for Model 3

i	$\hat{\alpha}_i$	$\hat{s}_i^*$	$\hat{\theta}_i^2$	$\hat{s}_i$	$\hat{t}_i = \log \hat{s}_i$
1	0.989	1.000*	0.022	1.000	0.00
2	0.628	1.000*	0.212	2.000	0.69
3	0.586	0.290	0.212	2.290	0.83
4	0.425	2.065	0.216	4.355	1.47
5	0.370	1.357	0.218	5.712	1.74
6	0.302	1.000*	0.386	6.712	1.90

Residuals: $\mathbf{S} - \boldsymbol{\Sigma}$

Spelling	0.000					
Punctuation	−0.000	−0.000				
Grammar	−0.016	0.005	−0.000			
Vocabulary	0.056	−0.030	0.007	−0.000		
Literature	0.028	−0.004	−0.025	0.004	−0.000	
Foreign literature	0.090	0.031	0.023	−0.011	−0.000	0.000

The circumplex model suggested by Guttman (1954) is a circular moving average process. Let $\xi_1, \xi_2, \ldots, \xi_p$ be uncorrelated, random latent variables. Then the perfect circumplex of order m with p variables is defined by

$$x_i = \xi_i + \xi_{i+1} + \cdots + \xi_{i+m-1},$$

where $x_{p+i} = x_i$. In matrix form we may write this as $\mathbf{x} = \mathbf{C}\boldsymbol{\xi}$, where $\mathbf{C}$ is a matrix of order $p \times p$ with zeros and ones. In the case of $p = 6$ and $m = 3$,

$$\mathbf{C} = \begin{bmatrix} 1 & 1 & 1 & 0 & 0 & 0 \\ 0 & 1 & 1 & 1 & 0 & 0 \\ 0 & 0 & 1 & 1 & 1 & 0 \\ 0 & 0 & 0 & 1 & 1 & 1 \\ 1 & 0 & 0 & 0 & 1 & 1 \\ 1 & 1 & 0 & 0 & 0 & 1 \end{bmatrix}$$

Let $\varphi_1, \varphi_2, \ldots, \varphi_p$ be the variances of $\xi_1, \xi_2, \ldots, \xi_p$, respectively. Then the variance-covariance matrix of $\mathbf{x}$ is

$$\boldsymbol{\Sigma} = \mathbf{C}\mathbf{D}_\varphi\mathbf{C}', \tag{37}$$

where $\mathbf{D}_\varphi = \text{diag}(\varphi_1, \varphi_2, \ldots, \varphi_p)$. The variance of x_i is $\sum_{k=i}^{i+m-1} \varphi_k$, the covariance between x_i and x_j, for $i < j$, is $\sum_{k=j}^{i+m-1} \varphi_k$ for $j = i + 1, i + 2, \ldots, i + m - 1$ and 0 otherwise, and the correlation between x_i and x_j, $i < j$, is

$$\rho_{ij} = \frac{\sum_{k=j}^{i+m-1} \varphi_k}{\sqrt{\left(\sum_{k=i}^{i+m-1} \varphi_k\right)\left(\sum_{k=j}^{j+m-1} \varphi_k\right)}}.$$

Here φ_{p+k} should be interpreted as φ_k.

For a given i, the correlations ρ_{ij} decrease as j increases beyond i, reach a minimum, and then increase as j approaches $p + i$. It is convenient to think of $1, 2, \ldots, p$ as points on a circle. Then for adjacent points the correlation tends to be high; for points far apart, the correlation tends to be low or 0. If $m < p/2$, then $\rho_{ij} = 0$ if $j - i > m$ (modulus p). Since zero correlations are not expected in practice, we assume that m is chosen to be greater than or equal to $p/2$.

Guttman gives several examples of sets of tests showing this property of the correlations. His Table 19 is reproduced here as Table 15. Note that in each row the correlations fall away from the main diagonal, reach a minimum, and then increase. Guttman argues that the tests form a circular order. In

TABLE 15
Intercorrelations among tests of six different kinds of abilities for 710 Chicago school children

	Association	Incomplete Words	Multi-plication	Dot Patterns	ABC	Directions
Test	6	32	37	17	1	12
6	1.	0.446	0.321	0.213	0.234	0.442
32	0.446	1.	0.388	0.313	0.208	0.330
37	0.321	0.388	1.	0.396	0.325	0.328
17	0.213	0.313	0.396	1.	0.352	0.247
1	0.234	0.208	0.325	0.352	1.	0.347
12	0.442	0.330	0.328	0.247	0.347	1.

the example the Association test is about equally related to the Incomplete Words test and the Directions test. It is this circular ordering that is the significant feature of the correlation matrix.

The perfect circumplex is too restrictive a model. First, it cannot account for random measurement error in the test scores, and second, since the model is not scale free, it cannot be used to analyze correlations such as those in Table 15. However, these difficulties are easily remedied by considering the quasi-circumplex

$$\mathbf{x} = \mathbf{D}_\alpha \mathbf{C} \boldsymbol{\xi} + \mathbf{e}$$

with dispersion matrix

$$\boldsymbol{\Sigma} = \mathbf{D}_\alpha \mathbf{C} \mathbf{D}_\varphi \mathbf{C}' \mathbf{D}_\alpha + \boldsymbol{\Theta}^2, \tag{38}$$

where $\mathbf{e}$ is the vector of error scores with variances in the diagonal matrix $\boldsymbol{\Theta}^2$, and $\mathbf{D}_\alpha$ is a diagonal matrix of scale factors. One element in $\mathbf{D}_\alpha$ or $\mathbf{D}_\varphi$ must be fixed at unity. It seems most natural to fix $\varphi_1 = 1$. Then the model is scale free and may be estimated by the ACOVS program. The results of an analysis of the data in Table 15 are given in Table 16.

TABLE 16
Quasi-circumplex solution for the data of Table 15 ($m = 4$)

i	$\hat{\alpha}_i$	$\hat{\varphi}_i$	$\hat{\theta}_i$
1	0.268	1.000	0.625
2	0.219	2.012	0.709
3	0.210	3.133	0.714
4	0.235	2.129	0.711
5	0.220	2.793	0.758
6	0.254	2.767	0.639

Note: $x^2 = 16.47$ with 4 d.f.

7. PATH ANALYSIS MODELS

Introduction

Path analysis, due to Wright (1918), is a technique sometimes used to assess the direct causal contribution of one variable to another in a nonexperimental situation. The problem in general is that of estimating the parameters of a set of linear structural equations representing the cause and effect relationships hypothesized by the investigator. Recently, several models have been studied that involve hypothetical constructs, i.e., latent variables that, while not directly observed, have operational implications for relationships among observable variables (see, e.g., Werts & Linn, 1970; Hauser & Goldberger, 1971). In some models, the observed variables appear only as effects (indicators) of the hypothetical constructs, while in others, the observed variables appear as causes (components) or as both causes and effects of latent variables. We give one simple example of each kind of model to indicate how such models may be handled within the framework of covariance structure analysis.

A Model with Correlated Measurement Errors

Consider the model discussed by Costner (1969) shown in Figure 6. Note that the errors δ_3 and ϵ_3 are assumed to be correlated as might be the case, for

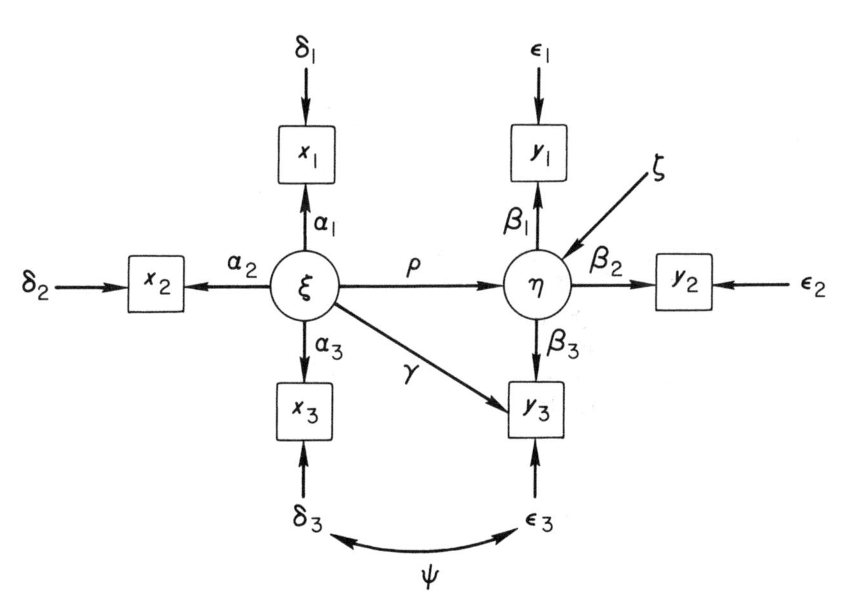

FIGURE 6.
A model with correlated measurement errors.

example, if x_3 and y_3 are scores from the same measuring instrument used at two different occasions. In algebraic form the model may be written, ignoring the means of the observed variables,

$$\begin{pmatrix} x_1 \\ x_2 \\ x_3 \\ y_1 \\ y_2 \\ y_3 \end{pmatrix} = \begin{bmatrix} \alpha_1 & 0 & 1 & 0 & 0 & 0 & 0 & 0 \\ \alpha_2 & 0 & 0 & 1 & 0 & 0 & 0 & 0 \\ \alpha_3 & 0 & 0 & 0 & 1 & 0 & 0 & 0 \\ \gamma & \beta_1 & 0 & 0 & 0 & 1 & 0 & 0 \\ 0 & \beta_2 & 0 & 0 & 0 & 0 & 1 & 0 \\ 0 & \beta_3 & 0 & 0 & 0 & 0 & 0 & 1 \end{bmatrix} \begin{pmatrix} \xi \\ \eta \\ \delta_1 \\ \delta_2 \\ \delta_3 \\ \epsilon_1 \\ \epsilon_2 \\ \epsilon_3 \end{pmatrix}. \tag{39}$$

Let $\boldsymbol{\Lambda}$ be the matrix in Equation 39 and $\boldsymbol{\Phi}$ the variance-covariance matrix of the vector on the right side. Then $\boldsymbol{\Phi}$ is of the form

$$\boldsymbol{\Phi} = \begin{bmatrix} 1 & & & & & & & \\ \rho & 1 & & & & & & \\ 0 & 0 & \theta_1^2 & & & & & \\ 0 & 0 & 0 & \theta_2^2 & & & & \\ 0 & 0 & 0 & 0 & \theta_3^2 & & & \\ 0 & 0 & 0 & 0 & 0 & \theta_4^2 & & \\ 0 & 0 & 0 & 0 & 0 & 0 & \theta_5^2 & \\ 0 & 0 & 0 & 0 & \varphi & 0 & 0 & \theta_6^2 \end{bmatrix},$$

where ρ is the correlation between the latent variables ξ and η, φ the covariance between δ_3 and ϵ_3, and $\theta_1^2, \theta_2^2, \ldots, \theta_6^2$ the variances of the errors $\delta_1, \delta_2, \delta_3, \epsilon_1, \epsilon_2, \epsilon_3$. The variance-covariance matrix of the observed variables is

$$\boldsymbol{\Sigma} = \boldsymbol{\Lambda\Phi\Lambda}'. \tag{40}$$

Note that in this example $\boldsymbol{\Lambda}$ has more columns than rows and includes also the error part of the model. This representation is necessary since the covariance matrix of the errors is not diagonal.

This model has 15 parameters to be estimated and the covariance matrix in Equation 40 has six degrees of freedom. The investigator may be interested in testing the specific hypothesis $\gamma = 0$, i.e., that ξ affects y_3 only via η. This may be done in large samples, assuming that the rest of the model holds, by a χ^2 test with one degree of freedom.

A Model with Multiple Causes and Multiple Indicators of a Single Latent Variable

Consider the model discussed by Hauser and Goldberger (1971) shown in Figure 7. This model involves a single hypothetical variable ξ, which appears as both cause and effect variable. The equations are

$$\xi = \boldsymbol{\alpha}'\mathbf{x} + v,$$

$$\mathbf{y} = \boldsymbol{\beta}\xi + \mathbf{u}.$$

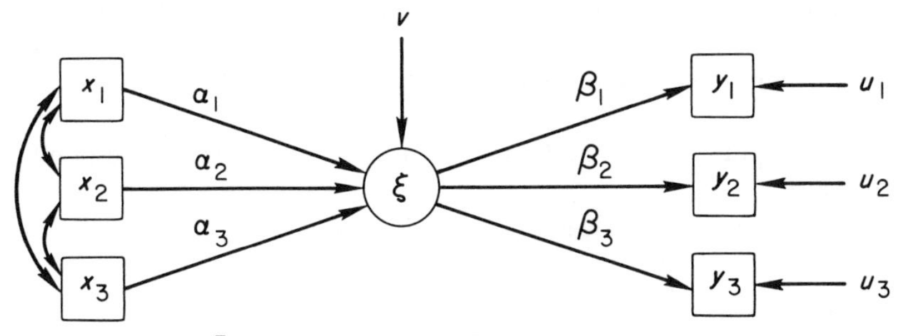

FIGURE 7.
A model with multiple causes and indicators of a latent variable.

The case where the residuals u_1, u_2, and u_3 are mutually correlated and $v = 0$ was considered by Hauser and Goldberger (1971). The case shown in Figure 7, where u_1, u_2, and u_3 are mutually uncorrelated, will be considered in detail in a forthcoming paper by Jöreskog and Goldberger (1973). In this case the structure of the variance-covariance matrix of the observed variables is

$$\begin{aligned} \boldsymbol{\Sigma}_{yy} &= \boldsymbol{\beta}\boldsymbol{\alpha}'\boldsymbol{\Sigma}_{xx}\boldsymbol{\alpha}\boldsymbol{\beta}' + \boldsymbol{\beta}\boldsymbol{\beta}' + \boldsymbol{\Theta}^2, \\ \boldsymbol{\Sigma}_{yx} &= \boldsymbol{\beta}\boldsymbol{\alpha}'\boldsymbol{\Sigma}_{xx}, \\ \boldsymbol{\Sigma}_{xx} & \quad \text{unconstrained.} \end{aligned}$$

The residual v may be scaled to unit variance, as assumed here. Alternatively, the latent variable ξ may be scaled to unit variance or one of the αs fixed at some nonzero value. It is readily verified that this model may be represented in terms of Equation 1 by specifying

$$\mathbf{B} = \begin{bmatrix} \beta_1 & 0 & 0 & 0 \\ \beta_2 & 0 & 0 & 0 \\ \beta_3 & 0 & 0 & 0 \\ 0 & 1 & 0 & 0 \\ 0 & 0 & 1 & 0 \\ 0 & 0 & 0 & 1 \end{bmatrix}, \quad \boldsymbol{\Lambda} = \begin{bmatrix} \alpha_1 & \alpha_2 & \alpha_3 \\ 1 & 0 & 0 \\ 0 & 1 & 0 \\ 0 & 0 & 1 \end{bmatrix}, \quad \boldsymbol{\Phi} = \boldsymbol{\Sigma}_{xx},$$

$$\boldsymbol{\Psi} = \text{diag}\,(1, 0, 0, 0), \; \boldsymbol{\Theta} = \text{diag}\,(\sigma_{u_1}, \sigma_{u_2}, \sigma_{u_3}, 0, 0, 0).$$

The model has 15 independent parameters and six degrees of freedom.

Models for Longitudinal Data

Many studies in the behavioral sciences involve longitudinal data, i.e., measurements that have been repeated on the same individuals over time or under different conditions. Jöreskog (1970c) developed a general model for analyzing multitest-multioccasion correlation matrices and Werts, Jöreskog, and Linn (1973) discussed the implications of this model for the study of growth. Here we confine ourselves to a path analysis formulation of the quasi-Markov simplex.

A path diagram of the model is shown in Figure 8. The corresponding equations are

$$x_i = \alpha_i \tau_i + \epsilon_i, \quad i = 1, 2, \ldots, p, \tag{41}$$

$$\tau_i = \beta_i \tau_{i-1} + \zeta_i, \quad i = 2, 3, \ldots, p, \tag{42}$$

where it is assumed, without loss of generality, that $\tau_1, \tau_2, \ldots, \tau_p$ are scaled to unit variance. With this scaling, α_i is the covariance between x_i and τ_i, and β_i is the correlation between τ_i and τ_{i-1}; the covariance between any two variates is the product of the path coefficients of the lines going from one variate to the other. For example, $\text{cov}(x_1, \tau_3) = \alpha_1\beta_2\beta_3$ and generally $\text{cov}(x_i, x_j) = \alpha_i\beta_{i+1}, \beta_{i+2} \ldots \beta_j\alpha_j$ for $i < j$.

When $p = 4$, the variance-covariance matrix is

$$\Sigma = \begin{bmatrix} \alpha_1^2 + \theta_1^2 & & & \\ \alpha_2\alpha_1\beta_2 & \alpha_2^2 + \theta_2^2 & & \\ \alpha_3\alpha_1\beta_2\beta_3 & \alpha_3\alpha_2\beta_3 & \alpha_3^2 + \theta_3^2 & \\ \alpha_4\alpha_1\beta_2\beta_3\beta_4 & \alpha_4\alpha_2\beta_3\beta_4 & \alpha_4\alpha_3\beta_4 & \alpha_4^2 + \theta_4^2 \end{bmatrix}.$$

If α_1 is replaced by $a\alpha_1$ and β_2 by (β_2/a), the off-diagonal elements in the first row and column are unchanged; if θ_1^2 is properly adjusted the first-diagonal element is also unchanged. Thus, α_1 and β_2 are not identified, but their product is, since

$$(\alpha_1\beta_2)^2 = \frac{\sigma_{21}\sigma_{13}}{\sigma_{23}} = \frac{\sigma_{21}\sigma_{14}}{\sigma_{24}}.$$

The same kind of argument can be used to show that α_4 and β_4 are not identified, but that $\alpha_4\beta_4$ is. The parameters α_2, α_3, and β_3 are identified since

$$\alpha_2^2 = \frac{\sigma_{12}\sigma_{23}}{\sigma_{13}} = \frac{\sigma_{12}\sigma_{24}}{\sigma_{14}},$$

$$\alpha_3^2 = \frac{\sigma_{13}\sigma_{34}}{\sigma_{14}} = \frac{\sigma_{23}\sigma_{34}}{\sigma_{24}},$$

$$\beta_3^2 = \frac{\sigma_{32}\sigma_{24}}{\sigma_{34}\alpha_2^2} = \frac{\sigma_{13}\sigma_{32}}{\sigma_{12}\alpha_3^2}.$$

Since α_2 and α_3 are identified, θ_2 and θ_3 are also identified.

The above reasoning generalizes to an arbitrary number of variables, p, in an obvious way. For the 'inner' variables the parameters $\alpha_2, \alpha_3, \ldots, \alpha_{p-1}$, $\beta_3, \beta_4, \ldots, \beta_{p-1}, \theta_2, \theta_3, \ldots, \theta_{p-1}$ are all identified, but for the 'outer' variates only the products $\alpha_1\beta_2$ and $\alpha_p\beta_p$ are identified. The indeterminacy in α_1 and β_2

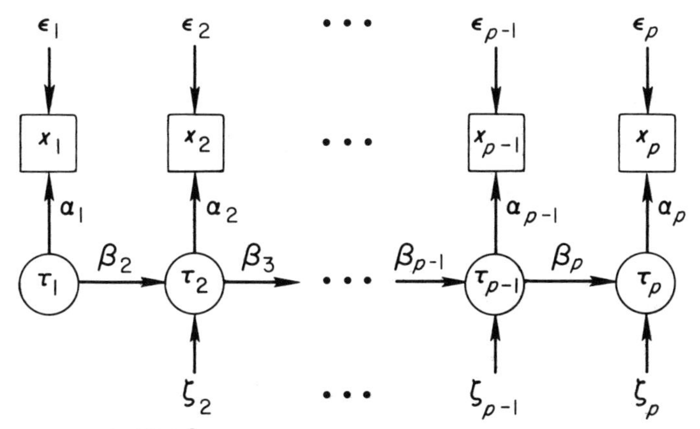

FIGURE 8.
Path analysis diagram of the quasi-Markov simplex.

and in α_p and β_p may be eliminated by choosing $\beta_2 = \beta_p = 1$ and Figure 8 may, for estimation purposes, be replaced by Figure 9. The condition $\beta_2 = \beta_p = 1$ is assumed in what follows.

The special cases $p = 3$ and $p = 4$ may now be examined more closely and shown to be equivalent to certain factor analysis models. In the case $p = 3$, Equation 41 may be written

$$\begin{pmatrix} x_1 \\ x_2 \\ x_3 \end{pmatrix} = \begin{pmatrix} \alpha_1 \\ \alpha_2 \\ \alpha_3 \end{pmatrix} \tau_2 + \begin{pmatrix} \epsilon_1 \\ \epsilon_2 \\ \epsilon_3 \end{pmatrix},$$

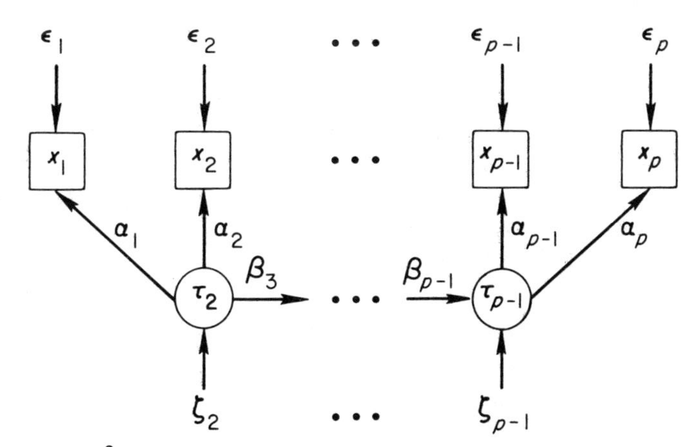

FIGURE 9.
Path analysis diagram of the quasi-Markov simplex after elimination of indeterminacies.

which is formally equivalent to a factor analysis model with one common factor. This model, however, has zero degrees of freedom, and if the data are consistent with the model the fit will be perfect. In the case $p = 4$, the equation may be written

$$\begin{pmatrix} x_1 \\ x_2 \\ x_3 \\ x_4 \end{pmatrix} = \begin{pmatrix} \alpha_1 & 0 \\ \alpha_2 & 0 \\ 0 & \alpha_3 \\ 0 & \alpha_4 \end{pmatrix} \begin{pmatrix} \tau_2 \\ \tau_3 \end{pmatrix} + \begin{pmatrix} \epsilon_1 \\ \epsilon_2 \\ \epsilon_3 \\ \epsilon_4 \end{pmatrix},$$

with β_3 equal to the correlation between τ_2 and τ_3. The corresponding $\boldsymbol{\Sigma}$ is

$$\Sigma = \begin{bmatrix} \alpha_1 & 0 \\ \alpha_2 & 0 \\ 0 & \alpha_3 \\ 0 & \alpha_4 \end{bmatrix} \begin{bmatrix} 1 & \beta_3 \\ \beta_3 & 1 \end{bmatrix} \begin{bmatrix} \alpha_1 & \alpha_2 & 0 & 0 \\ 0 & 0 & \alpha_3 & \alpha_4 \end{bmatrix} + \begin{bmatrix} \theta_1^2 & 0 & 0 & 0 \\ 0 & \theta_2^2 & 0 & 0 \\ 0 & 0 & \theta_3^2 & 0 \\ 0 & 0 & 0 & \theta_4^2 \end{bmatrix}. \quad (43)$$

$$= \boldsymbol{\Lambda\Phi\Lambda}' + \boldsymbol{\Theta}^2.$$

This model is equivalent to a factor analysis model with two nonoverlapping oblique factors. The model has one degree of freedom. Given this model, one can test each of the hypotheses $\beta_3 = 1$, $\alpha_2 = \alpha_3$, and $\theta_2^2 = \theta_3^2$ using a large sample χ^2 with one degree of freedom. Assumptions of this kind have traditionally been thought necessary in growth studies, but a test of their validity has not been available earlier (see, e.g., Harris, 1963, Chs. 1-3).

A Growth Data Example

To illustrate the model (Eq. 43), we use some data from a longitudinal growth study conducted at Educational Testing Service (Anderson & Maier, 1963; Hilton, 1969). A nationwide sample of fifth graders was tested in 1961 and again in 1963, 1965, and 1967 as seventh, ninth, and eleventh graders, respectively. Table 17 shows the variance-covariance matrix of the scores on the quantitative part of SCAT (Scholastic Aptitude Test) at the four different occasions for a random sample of 799 girls from all the girls that took all tests at all occasions. The tests have been scaled so that the unit of measurement is approximately the same at all occasions.

The matrix in Table 17 has been analyzed under five different hypotheses based on the model of Equation 43:

(1) Equation 43;
(2) Equation 43 with $\beta_3 = 1$;
(3) Equation 43 with $\alpha_2 = \alpha_3$;
(4) Equation 43 with $\theta_2 = \theta_3$;
(5) Equation 43 with $\alpha_2^* = \alpha_3^*$.

TABLE 17
SCATQ growth data

	Variance-Covariance Matrix $N = 799$			
Q_5	67.951			
Q_7	71.071	141.578		
Q_9	85.966	134.655	249.748	
Q_{11}	97.153	151.068	218.757	300.669

Summary of Analysis

Hypothesis	χ^2	d.f.	P
1	0.092	1	0.761
2	50.020	2	0.000
3	65.543	2	0.000
4	22.289	2	0.000
5	0.100	2	0.951

Solution for Model 1

$\hat{\alpha}_i$		$\hat{\psi}_i^2$
6.75		22.39
10.53	$\hat{\beta}_3 = 0.92$	30.71
13.95		55.28
15.69		54.60

In model 5 it is assumed that the standardized regression coefficients, i.e., the reliabilities, are equal. This condition may be specified by letting the true scores τ_2 and τ_3 be in the same metric as x_1 and x_2; the resulting regression coefficients α_2 and α_3 will then be standardized. The values of χ^2 for each hypothesis are given under Summary of Analyses. Each hypothesis has been tested against the general alternative that Σ is unconstrained. It is seen that models 1 and 5 have good fits, but the other models have very poor fits. Testing hypothesis 5 against hypothesis 1, one obtains $\chi^2 = 0.100 - 0.092 = 0.008$ with one degree of freedom. Thus, the hypothesis of equal reliabilities cannot be rejected. It should be noted that this hypothesis is equivalent to the assertion that the error variances are in a fixed proportion of the total variances. The maximum-likelihood solution under hypothesis 1 is given under Solution for Model 1.

REFERENCES

Anderson, S. B., & Maier, M. H. 34,000 pupils and how they grow. *Journal of Teacher Education*, 1963, **14,** 212–216.

Anderson, T. W. Some stochastic process models for intelligence test scores. In K. J. Arrow, S. Karlin, and P. Suppes (Eds.), *Mathematical methods in the social sciences*, 1959. Stanford, California: Stanford University Press, 1960.

Bilodeau, E. A. The relationship between a relatively complex motor skill and its components. *American Journal of Psychology*, 1957, **70,** 49–55.

Bock, R. D. Components of variance analysis as a structural and discriminal analysis for psychological tests. *British Journal of Statistical Psychology*, 1960, **13,** 151–163.

Bock, R. D., & Bargmann, R. E. Analysis of covariance structures. *Psychometrika*, 1966, **31,** 507–534.

Bock, R. D., Dicken, D., & van Pelt, J. Methodological implications of content-acquiescence correlation in the MMPI. *Psychological Bulletin*, 1969, **71,** 127–139.

Burt, C. Factor analysis and analysis of variance. *British Journal of Psychology*, Statistical Section, 1947, **1,** 3–26.

Campbell, D. T., & Fiske, D. W. Convergent and discriminant validation by the multitrait-multimethod matrix. *Psychological Bulletin*, 1959, **56,** 81–105.

Costner, H. L. Theory, deduction and rules of correspondence. *American Journal of Sociology*, 1969, **75,** 245–263.

Fletcher, R., & Powell, M. J. D. A rapidly convergent descent method for minimization. *The Computer Journal*, 1963, **6,** 163–168.

Graybill, F. A. *An introduction to linear statistical models*. Vol. 1. New York: McGraw-Hill, 1961.

Gruvaeus, G. T., & Jöreskog, K. G. *A computer program for minimizing a function of several variables*. Research Bulletin 70–14. Princeton, N.J.: Educational Testing Service, 1970.

Guilford, J. P. The structure of intellect. *Psychological Bulletin*, 1956, **53,** 267–293.

Guttman, L. A new approach to factor analysis: The Radex. In P. F. Lazarsfeld (Ed.), *Mathematical thinking in the social sciences*. New York: Columbia University Press, 1954.

Harris, C. W. (Ed.) *Problems in measuring change*. Madison: University of Wisconsin Press, 1963.

Hauser, R. M., & Goldberger, A. S. The treatment of unobservable variables in path analysis. In H. L. Costner (Ed.), *Sociological methodology*. London: Jossey-Bass, 1971.

Hilton, T. L. *Growth study annotated bibliography*. Progress Report 69–11. Princeton, N.J.: Educational Testing Service, 1969.

Jöreskog, K. G. Some contributions to maximum likelihood factor analysis. *Psychometrika*, 1967, **32,** 443–482.

Jöreskog, K. G. A general approach to confirmatory maximum likelihood factor analysis. *Psychometrika*, 1969, **34,** 183–202.

Jöreskog, K. G. A general method for analysis of covariance structures. *Biometrika*, 1970, **57,** 239–251. (a)

Jöreskog, K. G. Estimation and testing of simplex models. *British Journal of Mathematical and Statistical Psychology*, 1970, **23,** 121–145. (b)

Jöreskog, K. G. Factoring the multitest-multioccasion correlation matrix. In C. E. Lunneborg (Ed.), *Current problems and techniques in multivariate psychology*. (Proceedings of a Conference Honoring Professor Paul Horst.) Seattle: University of Washington, 1970. (c)

Jöreskog, K. G. Statistical analysis of sets of congeneric tests. *Psychometrika*, 1971, **36,** 109–133.

Jöreskog, K. G., & Goldberger, A. S. Factor analysis by generalized least squares. *Psychometrika*, 1972, **37,** 243–260.

Jöreskog, K. G., & Goldberger, A. S. *Estimation of a model with multiple indicators and multiple causes of a single latent variable*. Research Report 73-14. Uppsala: Dept. of Statistics, University of Uppsala, 1973. Also issued as Report 7328. Madison: Social Systems Res. Inst. Workshop Series, University of Wisconsin, 1973.

Jöreskog, K. G., Gruvaeus, G. T., & van Thillo, M. *ACOVS—A general computer program for analysis of covariance structures*. Research Bulletin 70–15. Princeton, N.J.: Educational Testing Service, 1970.

Jöreskog, K. G., & Lawley, D. N. New methods in maximum likelihood factor analysis. *British Journal of Mathematical and Statistical Psychology*, 1968, **21,** 85–96.

Kelley, E. L., & Fiske, D. W. *The prediction of performance in clinical psychology*. Ann Arbor: University of Michigan Press, 1951.

Kristof, W. On the theory of a set of tests which differ only in length. *Psychometrika*, 1971, **36,** 207–225.

Language modalities survey: A test for aphasia. Speech Clinic, University of Chicago and Psychometric Laboratory, University of North Carolina, Chapel Hill, 1959.

Lawley, D. N., & Maxwell, A. E. *Factor analysis as a statistical method*. London: Butterworths, 1963.

Lord, F. M. A study of speed factors in tests and academic grades. *Psychometrika*, 1956, **21,** 31–50.

Lord, F. M. A significance test for the hypothesis that two variables measure the same trait except for errors of measurement. *Psychometrika*, 1957, **22,** 207–220.

Lord, F. M., & Novick, M. R. *Statistical theories of mental test scores* (with contributions by A. Birnbaum). Reading, Massachusetts: Addison-Wesley, 1968.

McNemar, Q. Attenuation and interaction. *Psychometrika*, 1958, **23,** 259–265.

Miller, D. M., & Lutz, M. V. Item design for an inventory of teaching practices and learning situations. *Journal of Educational Measurement*, 1966, **3,** 53–61.

Werts, C. E., Jöreskog, K. G., & Linn, R. L. A multitrait-multimethod model for studying growth. *Educational and Psychological Measurement*, 1973, **33,** in press.

Werts, C. E., & Linn, R. L. Path analysis: Psychological examples. *Psychological Bulletin*, 1970, **67,** 193–212.

Wiley, D. E., Schmidt, W. H., & Bramble, W. J. Studies of a class of covariance structure models. *Journal of the American Statistical Association*, 1973, **68,** 317–323.

Wright, S. On the nature of size factors. *Genetics*, 1918, **3,** 367–374.

Models and Methods for Three-way Multidimensional Scaling

J. Douglas Carroll *Myron Wish*
BELL LABORATORIES
MURRAY HILL, NEW JERSEY

Multidimensional scaling (MDS) of proximities was originally developed to handle the case of an ordinary *two-way* matrix of proximities data (see, for example, Torgerson, 1958; Shepard, 1962a, 1962b; Ekman, 1963; Coombs, 1964; Kruskal, 1964a, 1964b; McGee, 1966; Young & Torgerson, 1967; Guttman, 1968; Roskam, 1968). Such a matrix is typically square and symmetric, although there are cases, such as off-diagonal conditional proximity matrices or confusion matrices, in which one or both of these conditions are violated. (We use the term 'proximities' to denote measures of similarity, or closeness, as well as dissimilarity.) We shall call MDS, as applied to a single such proximities matrix, *two-way* MDS.

The typical input and output for a two-way MDS analysis are shown schematically in Figure 1. (See Shepard, 1972a, and Wish, 1972, for a discussion of data appropriate for MDS.) The input is an $n \times n$ symmetric matrix of proximities, while the output is an $n \times r$ matrix of coordinates of n points in r dimensions. The *model* underlying two-way MDS (in the typical case in which the Euclidean metric is assumed) is summarized in Equations 1 and 2.

$$F(\partial_{jk}) \cong d_{jk}, \tag{1}$$

$$d_{jk} \equiv \sqrt{\sum_t (x_{jt} - x_{kt})^2}, \tag{2}$$

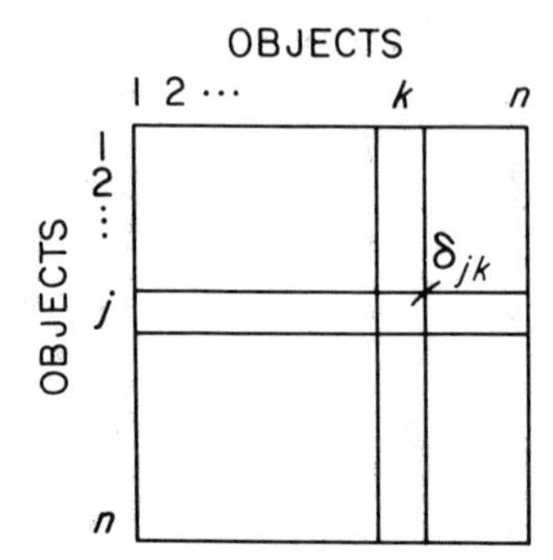

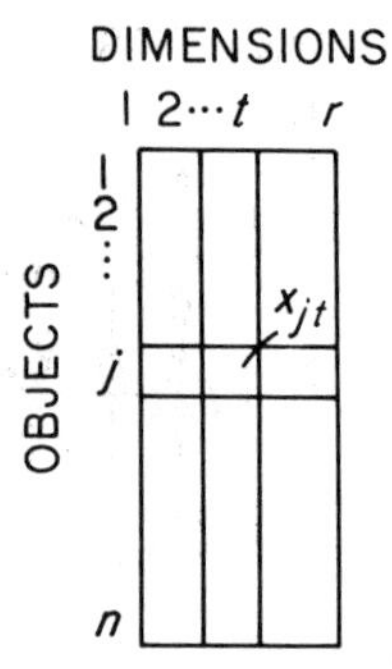

A. Input for two-way MDS B. Output from two-way MDS

FIGURE 1.
The typical input and output for two-way MDS are shown schematically in A and B, respectively. The input is a single $n \times n$ square symmetric matrix (or half matrix) of proximities (similarities or dissimilarities) data. $\partial_{jk} \equiv$ proximity of stimuli, or other objects, j and k. Symmetry implies $\partial_{kj} = \partial_{jk}$, for all j, k, so only a half matrix is required in most instances. Diagonals may be present or absent. The MDS output is a single $n \times r$ matrix of coordinates of the n stimuli in r dimensions ($x_{jt} \equiv$ coordinate of jth stimulus on the tth dimension).

where ∂_{jk} is the proximity of stimuli (objects) j and k, d_{jk} is Euclidean distance (as defined in Eq. 2), and x_{jt} is the projection of stimulus j on the tth coordinate axis. (Here and elsewhere $\cong$ will mean 'equal, except for unspecified error terms.' It will also have the more specific meaning in some contexts that least squares estimates of certain parameters are sought.)

In metric MDS, F is usually assumed to be a linear or polynomial function (Kruskal & Carmone, 1969; Kruskal, Young, & Seery, 1973). (Chang & Shepard, 1966, have discussed a method of metric MDS involving exponential functions.) In what is called the nonmetric case (Shepard, 1962a, 1962b; Kruskal, 1964a, 1964b; Guttman, 1968; Young & Torgerson, 1967), F is a nondecreasing or nonincreasing function, depending on whether δ is a dissimilarity or a similarity measure. Unlike the metric case, the solutions for nonmetric MDS are invariant under any monotonic transformations of the data.

Many applications have demonstrated the usefulness of both metric and nonmetric two-way MDS (Ekman, 1954; Messick, 1956, 1961; Abelson & Sermat, 1962; Shepard, 1963; Coombs, 1964; Wish, 1964; Gregson, 1965; Torgerson, 1965; Kruskal & Hart, 1966; Levelt, van de Geer, & Plomp, 1966; Künnapas, 1967; Wish, 1967; Behrman & Brown, 1968; Rosenberg, Nelson, & Vivekananthan, 1968; Yoshida, 1968; Bennett, 1969; Green & Carmone, 1970; Shepard & Chipman, 1970; Fillenbaum & Rapoport, 1971; Indow, 1973).

Many situations arise in the social and behavioral sciences (and elsewhere) in which several proximities matrices for the same stimuli (or other 'objects') are available. One would generally like to account for all of these data matrices in a single comprehensive analysis, based on an appropriate and psychologically (or otherwise) plausible model. The most typical situation arises when proximities matrices are available for each of a number of different individuals. In fact, the INDSCAL (*IN*dividual *D*ifferences *SCAL*ing) method (Carroll & Chang, 1970a) derives its name from this situation. However, the different matrices may be associated with other types of *data sources*, such as different experimental conditions, tasks, or occasions. We use the term 'three-way MDS' rather than the more restrictive term 'individual differences scaling' to characterize these models and methods. The latter general term should not be confused with the more specific name, INDSCAL, which refers to one particular three-way model and method.

Because INDSCAL is probably the most widely used three-way MDS method and is based on such a simple psychological model, we will use it to introduce the basic concepts of three-way MDS. After describing and illustrating INDSCAL, we discuss other three-way models and methods.

THE INDSCAL MODEL

Basic Concepts of INDSCAL

The input and output for an INDSCAL analysis are schematized in Figure 2. The input (for INDSCAL or *any* three-way method) consists (in the 'typical' case) of a three-way array that can be thought of as comprising a set of m (≥ 2) $n \times n$ symmetric proximities matrices. The output for INDSCAL (but not, in general, for other three-way methods) consists of *two* matrices. One is called the *'group' stimulus matrix*, defining coordinates of the n stimuli on each of r dimensions (x_{jt} = value of jth stimulus point on dimension t) in the 'group stimulus space.' The other, called the *subject matrix*, defines *weights* of dimensions for subjects (w_{it} = weight of dimension t for subject i). The 'space' defined by this matrix is called the 'subject' space. Of course, 'stimuli' may be entities other than stimuli, while 'subjects' may be data sources other than different subjects.

The model for INDSCAL is summarized in the equations

$$F_i[\partial_{jk}^{(i)}] \cong d_{jk}^{(i)}, \qquad (3)$$

$$d_{jk}^{(i)} = \sqrt{\sum_{t=1}^{r} w_{it}(x_{jt} - x_{kt})^2}. \qquad (4)$$

As before, the Fs will generally be considered to be either linear in the metric case, or monotonic in the nonmetric case. It is important to note, however,

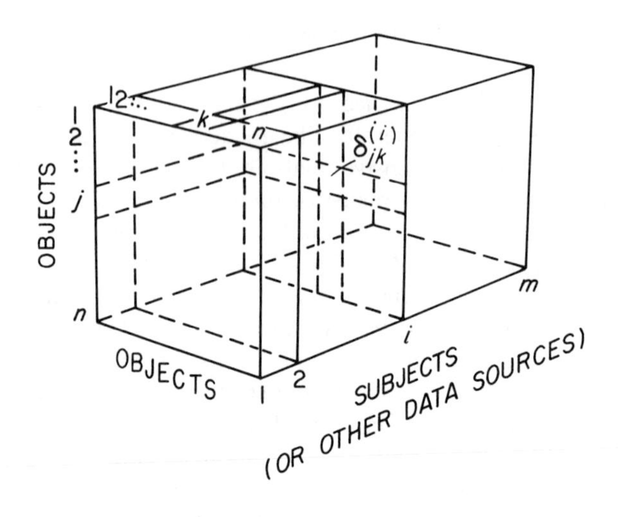

A. Input for three-way MDS

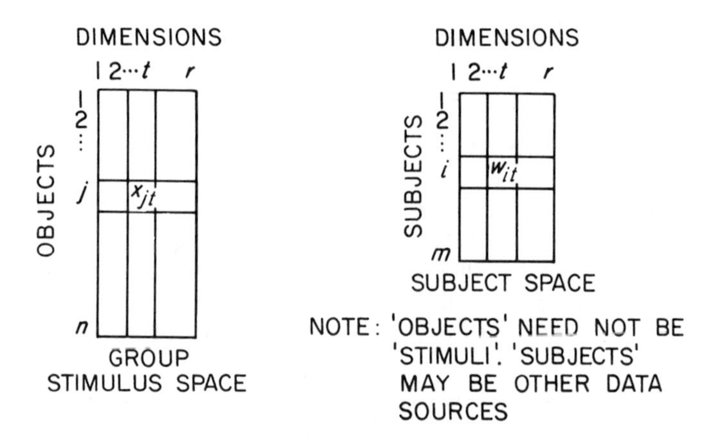

B. Output from three-way MDS

FIGURE 2.
Schematic diagrams of typical input (A) and output (B) for three-way MDS. The input is a set of $m(\geq 2) n \times n$ square symmetric data matrices (or half matrices), one for each of m subjects or other data sources. $\partial_{jk}^{(i)}$ is the proximity of objects j and k for subject i. Symmetry implies $\partial_{kj}^{(i)} = \partial_{jk}^{(i)}$ for all i, j, k. The output is comprised of two matrices, an $n \times r$ matrix of coordinates of the n objects on r dimensions (group stimulus space) and an $m \times r$ matrix of weights of m subjects on the same r dimensions (subject space). Both of these can be plotted graphically. A 'private space' for each subject can also be constructed (as illustrated in Fig. 3) by applying (square roots of) the subject weights to stimulus (object) dimensions (according to Eq. 5).

that a different F_i is assumed, in general, for each separate individual. Essentially, then, INDSCAL generalizes two-way MDS by substituting a *weighted* Euclidean metric for the ordinary (unweighted) Euclidean metric. Different patterns of weights are allowed for each individual or other data source.

A more geometric interpretation of the INDSCAL model is provided by Equation 5. Recall that x_{jt} is the tth coordinate of the jth stimulus in what we call the 'group stimulus space.' We may furthermore conceive of each individual (or other data source) as having what we may call a 'private perceptual space' whose general coordinate we will designate as $y_{jt}^{(i)}$. The $y^{(i)}$s, however, are derived from the xs by the very simple relation expressed in Equation 5.

$$y_{jt}^{(i)} = w_{it}^{1/2} x_{jt}. \tag{5}$$

The distances for 'individual' i are simply ordinary Euclidean distances computed in his own private stimulus space, as in Equation 6.

$$d_{jk}^{(i)} = \sqrt{\sum_{t}^{r} (y_{jt}^{(i)} - y_{kt}^{(i)})^2}. \tag{6}$$

It can easily be seen that by substituting Equation 5 into Equation 6 we get Equation 4. Thus Equations 5 and 6, which together provide an alternate interpretation of the 'weighted' generalization of the ordinary Euclidean metric defined in Equation 4, express the INDSCAL model in terms of a particularly simple class of transformations of the common space followed by computation of the ordinary Euclidean metric. The class of transformations can be described algebraically as linear transformations, with the transformation matrix constrained to being diagonal (sometimes called a strain transformation). More simply, it amounts to simply rescaling each dimension (by the square root of that particular subject's weight for that dimension). Geometrically we can think of this as differentially stretching or shrinking each dimension by a factor proportional to the square root of the weight (only the relative sizes of the weights are important here, however).

Hypothetical Example Illustrating the INDSCAL Model

This more geometric interpretation of the model is illustrated in Figure 3. The group stimulus space in the upper left shows nine stimuli, A through I, in a lattice configuration; the subject space in the upper right shows the weights, or perceptual saliences, of the dimensions for nine hypothetical subjects. These weights can be thought of as stretching factors that are applied to the dimensions of the group stimulus space.

We may think of the effect of these differential weights as producing, for each subject, a 'private perceptual space' by rescaling (stretching and con-

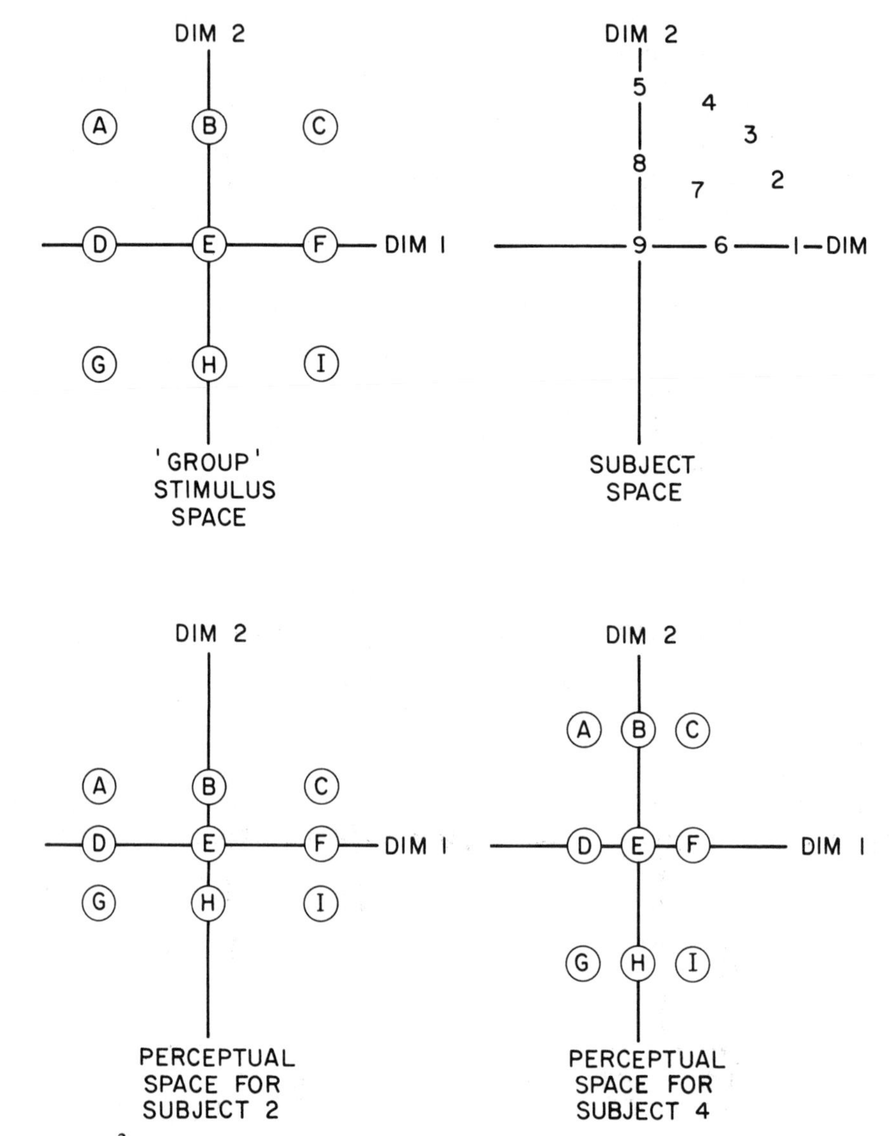

FIGURE 3.
Hypothetical example illustrating the INDSCAL model. Weights from the *subject space* are applied to the dimensions of the '*group*' *stimulus space* to produce *private perceptual spaces* for subjects 2 and 4. (While technically it is the square roots of the weights that are applied, the first power weights have been used in this illustration to accentuate the effects of differential weighting.)

tracting) the dimensions of the group stimulus space. In the illustration, for example, subject 3 has equal weights for the two dimensions; his private perceptual space would therefore look exactly the same as the group stimulus space (except for an overall scale factor that could stretch or contract both dimensions uniformly, leaving their relative saliences unchanged).

The private spaces for subjects 2 and 4 are shown in the lower lefthand and righthand corners of Figure 3, respectively. Subject 2, who weights dimension 1 more highly than dimension 2, has his perceptual space *compressed* along the dimension 2 axis (or what amounts to the same thing, stretched in the dimension 1 direction). The reverse applies to subject 4, who has a higher weight on the second than on the first dimension.

An important property of INDSCAL is one often called 'dimensional uniqueness.' This means that, unlike ordinary (two-way) MDS, the dimensions are uniquely determined, and cannot be rotated or otherwise transformed without changing the solution in an essential way. Psychologically this means that INDSCAL dimensions are assumed to correspond to 'fundamental' physiological, perceptual, or conceptual processes whose strength, or 'salience,' may differ from individual to individual (whether because of genetic or environmental differences, or simply because of differing interpretations of instructions or the like). Mathematically, a rotation or other transformation of a coordinate system will change the family of permissible transformations of the group stimulus space, and thus the family of possible individual metrics. (This can be seen quite clearly in the hypothetical example by noting that, say, a 45° rotation would change the family of private perceptual spaces from one of different rectangles to one of different rhombuses.) Statistically, a rotation or other transformation of axes will generally deteriorate the fit of the data to the INDSCAL model (except in certain special cases, to be detailed later). In the metric case, which we describe later, the goodness of fit is measured by the root mean square correlation (over subjects) between scalar products derived from the input similarities or dissimilarities and scalar products computed from the solution. The solution is, of course, designed to optimize the measure of fit. The property of dimensional uniqueness proves to be one of the most important aspects of INDSCAL, as it obviates the problem of rotation of axes in almost all cases. This problem is one that frequently complicates and often attenuates the utility of MDS analysis (particularly in the case of higher dimensional solutions). The value of this property should become clearer in the discussion of applications.

Distance of a subject from the origin in the subject space is roughly a measure of the proportion of the variance of the data for that subject accounted for by the multidimensional solution (comparable to the concept of communality in factor analysis). Although subjects 3 and 7 have the same pattern of dimension weights, a higher proportion of the variance in subject

3's data could be accounted for by the hypothetical INDSCAL solution. The data for subjects closer to the origin are generally less well accounted for by the INDSCAL analysis, so that the dimensions of the group stimulus space in toto are less salient for them. The lower 'communality' for subjects closer to the origin may be due to idiosyncratic dimensions not uncovered in the r-dimensional solution or to lower reliability (more random error) in their data.

Subject 9, who is precisely at the origin, is completely 'out of' this analysis. Either he was responding completely randomly or he was simply 'marching to a different drummer' (responding reliably to a completely different set of dimensions).

The possibility of zero weights allows as a special case the situation in which two or more groups of subjects have completely different perceptual spaces, with no necessary communality between them. This situation can be accommodated by defining a group stimulus space whose dimensions are the sum total of the dimensions for the two different groups of subjects—each subgroup has nonzero weights on only one subset of dimensions. For example, if one group perceives the stimuli in terms of dimensions A and B, and the second perceives them in terms of C and D, we may accommodate these two completely different 'points of view' (to use Tucker & Messick's, 1963, terminology) by assuming a four-dimensional space comprised of dimensions A, B, C, and D. Thus, the Tucker and Messick 'points-of-view' model can be accommodated as a special case of INDSCAL. INDSCAL is, of course, more interesting and appropriate when there is some communality in perception among the subjects (which is usually the case).

APPLICATIONS OF INDSCAL

A Reanalysis of Helm's Color Distance Data

The difference between the 'points-of-view' and INDSCAL models is illustrated by an example that entails a reanalysis of some data on color perception collected by Helm (1964). Helm and Tucker (1962) had analyzed these data via a 'points-of-view' analysis and found that about ten points of view (that is, ten different perceptual spaces) were required to account for the individual differences among their 14 subjects.

Our analysis in terms of the INDSCAL model, first reported by Carroll and Chang (1970c), has quite nicely accounted for these individual differences in terms of a single two-dimensional solution. Moreover, the unique dimensions from the INDSCAL analysis shown in Figure 4 were interpretable without rotation of axes. Dimension 1 corresponds essentially to a 'blue vs yellow' (or, more accurately perhaps, a 'purple-blue to green-yellow') factor, and

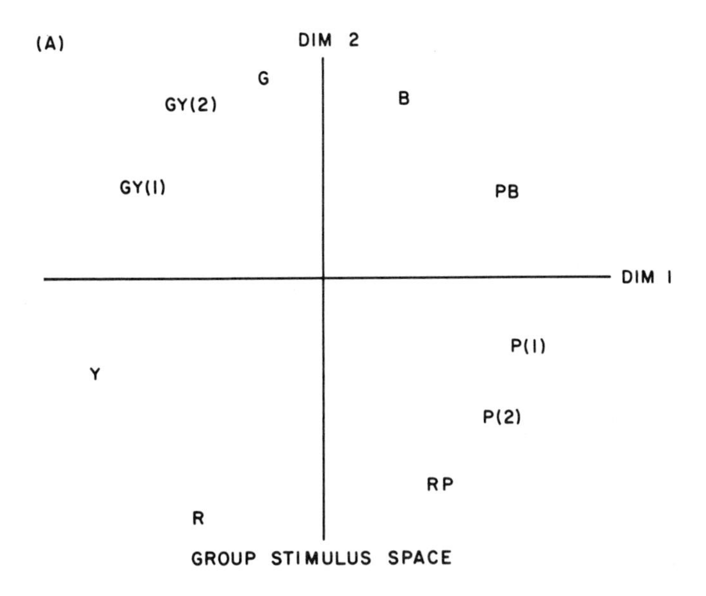

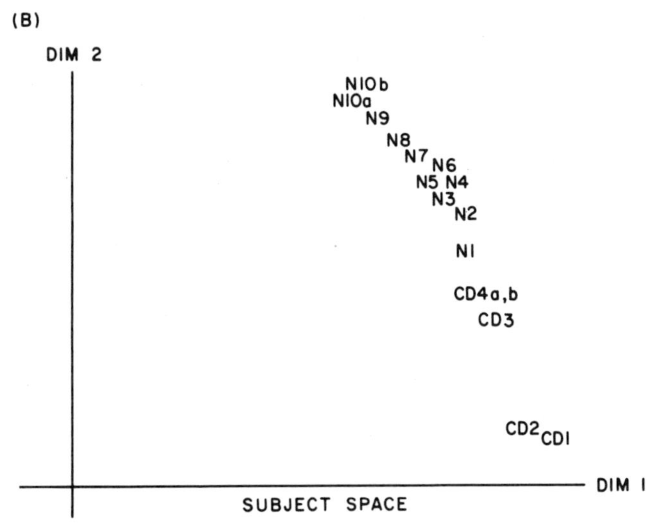

FIGURE 4.
A two-dimensional INDSCAL analysis of Helm's (1964) data on color perception produced the group stimulus space shown in A and the subject space shown in B. Private perceptual spaces for the two color-deficient subjects (CD1 and CD4a) and for two subjects with normal color vision (N10a and N7) were derived from the group stimulus space by applying the subjects' respective weights (shown in the subject space) to the dimensions. The coding of colors (of constant saturation and brightness) is as follows: R = red; Y = yellow; GY(1) = green yellow; GY(2) = green yellow with more green than GY(1); G = green; B = blue; PB = purple blue; P(1) = purple; P(2) = purple with more red than P(1); RP = red purple.

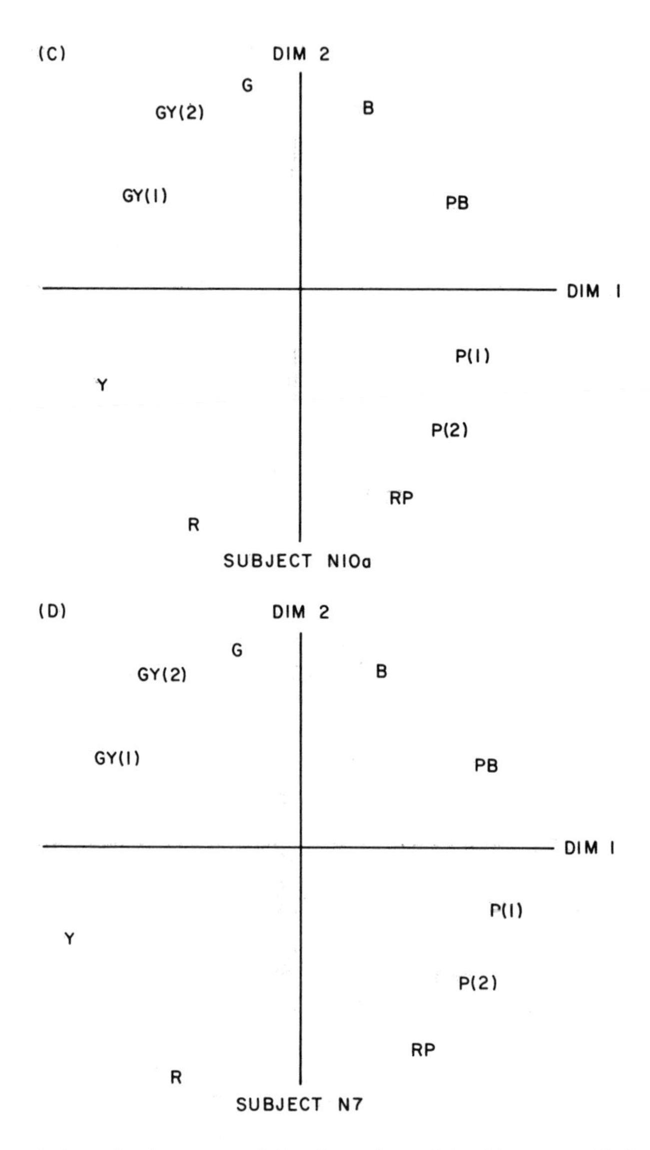

dimension 2 to a 'red vs green' (or 'purple-red to blue-green') factor. This accords very well with physiological and psychophysical evidence strongly suggesting the existence of 'blue-yellow' and 'red-green' receptors.

Included among Helm's subjects were four who were deficient, to various degrees, in red-green color vision. In the INDSCAL analysis this deficiency is reflected in the fact that these subjects all have lower weights for the 'red-green' factor (dimension 2) than do any of the normal subjects. The effect of these differential weights can be seen in Figure 4 by comparing the private

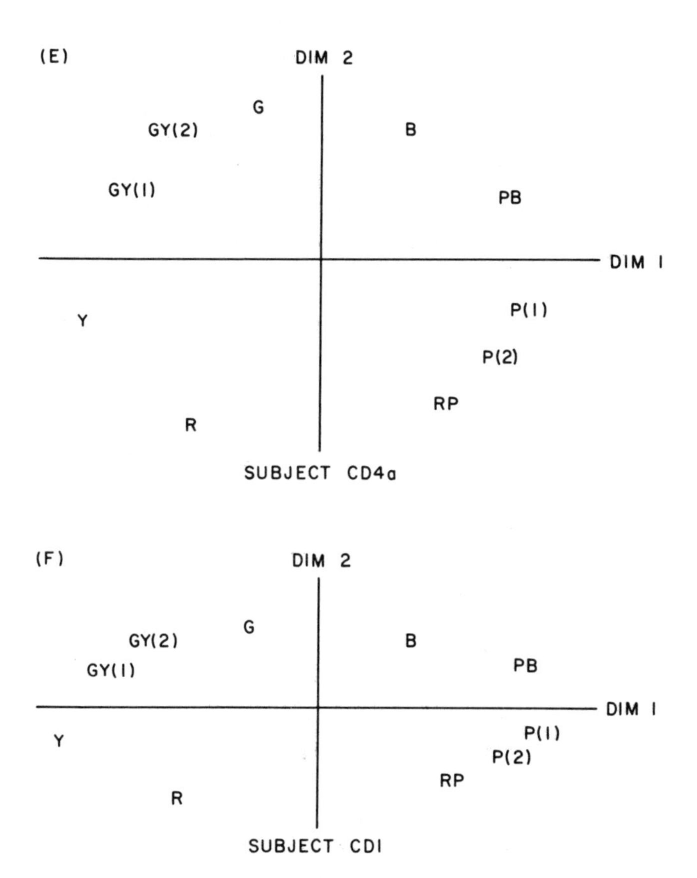

perceptual spaces for the color-blind subjects with those for the normals. The spaces for the color-deficient subjects are compressed in the 'red-green' direction, relative to the spaces for the normals, reflecting the fact that red and green (for example) are much more similar to each other for these subjects than they are for the normals.

Had we not known that these subjects were red-green color deficient, we could have discovered this through inspection of the weights in the subject space. In fact subject N10 seems from this analysis to border on a mild form of 'blue-yellow' deficiency. This is based on the fact that his weight for dimension 1 is about as small as some red-green-deficient subjects' weights for dimension 2. If such a deficiency exists for this subject, this would seem to be a fact determined by the INDSCAL analysis but *not* detected by other measures (in any case, such a deficiency was not reported by Helm).

The inspection of the private spaces provides insight into how INDSCAL uses individual differences in perception to determine a unique orientation of the coordinate axes. If the coordinate axes were oriented in a different way (say, by a 45° rotation of the axes of the group stimulus space), the private spaces for color-deficient subjects could not be compressed along a line from red to green, but would have to be compressed in some other direction. More generally, the *family* of transformations of the stimulus space generated by differential weights varies as a function of the orientation of axes in the group stimulus space. Thus it is that INDSCAL manages to seek out not only the optimal stimulus space, but also an optimal and unique orientation of coordinate axes within that space. It is a matter of empirical observation that this unique orientation is also (in almost every case to date) a highly 'interpretable' one. Thus, INDSCAL avoids the problem of rotation of axes that so often complicates MDS and other multivariate behavioral studies. This is not at all a coincidence, however, since INDSCAL is based on a psychological model that postulates dimensions that are unique in the sense of being modifiable (in salience, or perceptual importance) across, and perhaps even within, individuals. If the INDSCAL model does accurately reflect the underlying psychological processes (at least to a good first approximation), then the coordinate axes from an INDSCAL analysis can be assumed to correspond to 'fundamental' psychological dimensions.

Two other applications of the INDSCAL method, illustrating both the dimensional uniqueness property (the property of unique orientation of coordinate axes) and the utility of the 'subject' space as a way of parameterizing individual subjects or other data sources are now discussed.

Differences in Conceptions of Interpersonal Relations

Although the study of interpersonal relations is at the core of several social science disciplines, relatively little attention has been devoted to determining the dimensions relevant to the conceptions of lay people to such relations. Recently, however, a questionnaire study (Wish, 1973; Wish, Kaplan, & Deutsch, 1973) was undertaken with this aim in mind. Some other goals of the study were (a) to assess the degree and kinds of individual differences in conceptions of these relationships, (b) to relate such differences to other characteristics of the subjects, and (c) to determine how the two directions of various kinds of relations (for example, 'husband to wife' and 'wife to husband') combine psychologically to produce the overall conception of the relationship between the two individuals.

Method. Two sets of stimuli were used in the investigation: one set included 25 two-way relations (for example, 'between supervisor and employee'); the other was comprised of 40 directed relations (for example, 'supervisor to employee' and 'employee to supervisor'). Subjects made several different kinds of judgments with respect to these stimuli—pairwise ratings of similarity, multiple groupings (subjects repeatedly selected subsets of relations with some characteristic in common and identified the shared characteristic), semantic differential ratings, and circling of adjectives in a check list that described each relation. In addition, subjects rated many of their childhood and current relations on numerous bipolar scales. The 136 Columbia University students (mostly from Teachers College) who volunteered for the study were paid $10 each for the three to four hours it generally took to complete the questionnaire.

We describe only the results from an INDSCAL analysis of the data on judged similarities among the two-way, or 'between,' relations. (This task was given to only half of the subjects; the other half made comparable ratings with respect to the directed relations.) However, some semantic differential data are used to support the dimensional interpretations.

The following directions appeared on the instruction sheet, as well as on each page of the questionnaire booklet in which subjects recorded their similarity ratings:

> In making these judgments of similarities, please take into account the ways the two individuals in a relationship typically think and feel about each other, act and react toward each other, talk and listen to each other, and any other characteristics relevant to these relationships.

The response format is illustrated by the following example.

	Very different								Very similar
A. Between guard and prisoner B. Between close friends	1	2	3	4	5	6	7	8	9

Sixty-seven 25 × 25 matrices were derived from these judgments, one for each subject. The entire set of proximity matrices was used as input to a single INDSCAL analysis.

Interpretation of dimensions. An interpretable four-dimensional solution was obtained for these data. Solutions in five and six dimensions provided a negligible increase in the correlation between the original data (converted to scalar products) and the scalar products derived from the INDSCAL dimensions. The interpretations of the dimensions and the stimulus coordinates (normalized so that the sum of squared coordinates on each dimension is 1.00) are shown in Figures 5 and 6.

The choice of labels for dimensions was guided by supplementary results from multiple regression analyses. In Table 1, which provides these results,

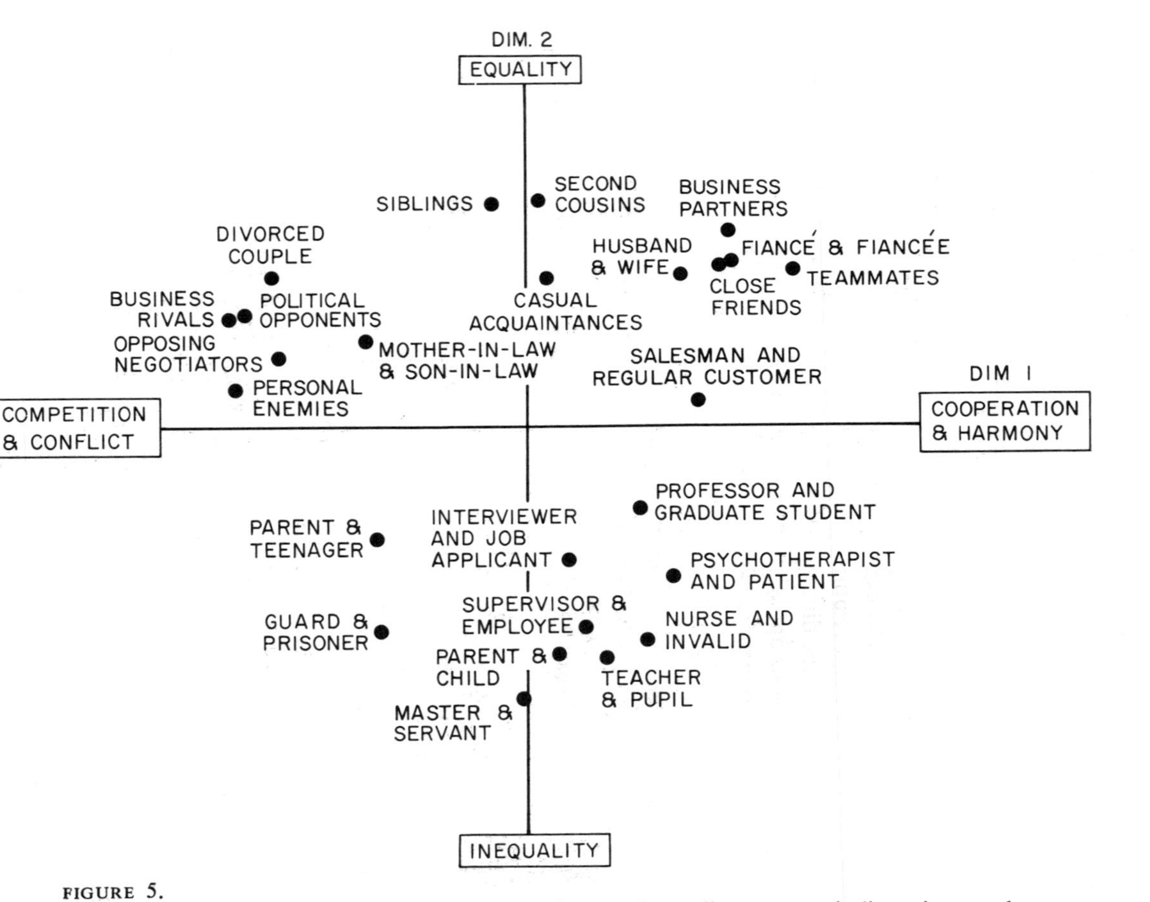

FIGURE 5.
Dimensions 1 and 2 (normalized so that the sum of squared coordinates on each dimension equals 1.00) of the four-dimensional 'group stimulus space' from an INDSCAL analysis of data on perceived similarities among 25 interpersonal relations (Wish, Kaplan, & Deutsch, 1973).

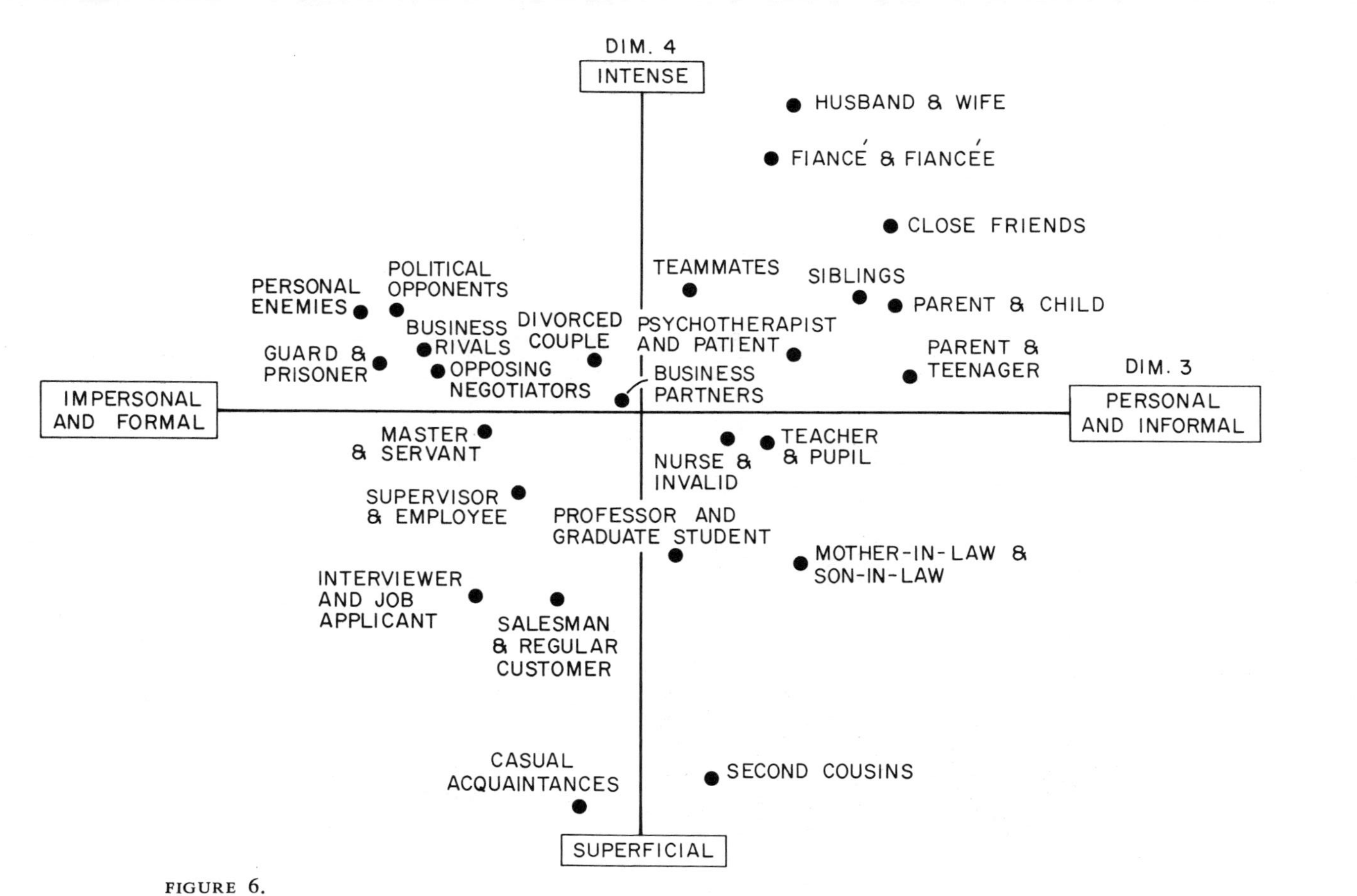

FIGURE 6.
Dimensions 3 and 4 (normalized) of the four-dimensional 'group stimulus space' from an INDSCAL analysis of data on perceived similarities among 25 interpersonal relations (Wish, Kaplan, & Deutsch, 1973).

TABLE 1
Regression weights for predicting mean ratings on semantic differential scales from INDSCAL dimensions

Semantic Differential Scale	Regression Weights				Multiple Correlation
	Dim. 1	Dim. 2	Dim. 3	Dim. 4	
1 Cooperative vs competitive	0.981**	−0.052	0.185	−0.015	0.978
2 Always harmonious vs always clashing	0.976**	0.004	0.141	−0.164	0.967
3 Compatible vs incompatible goals and desires	0.943**	0.161	0.289	0.047	0.982
4 Easy vs difficult to resolve conflicts	0.913**	0.020	0.375	−0.161	0.946
5 Effective vs ineffective communication	0.880**	0.127	0.430	0.159	0.968
6 Productive vs destructive	0.874**	−0.018	0.479	0.081	0.960
7 Fair vs unfair	0.848*	0.423	0.311	0.070	0.950
8 Friendly vs hostile	0.840*	0.145	0.522*	0.029	0.978
9 Altruistic vs selfish	0.721*	0.010	0.682*	0.068	0.957
10 Relaxed vs tense	0.677*	0.309	0.661*	−0.096	0.938
11 Equal vs unequal power in the relationship	0.022	0.986**	−0.087	0.142	0.957
12 Similar vs different roles and behavior	0.111	0.932**	−0.196	0.283	0.946
13 Democratic vs autocratic	0.416	0.887**	0.175	0.099	0.968
14 Power considerations unimportant vs important	0.541*	0.554*	0.508*	−0.377	0.923
15 Personal vs impersonal	0.280	0.125	0.927**	0.216	0.943
16 Pleasure-oriented vs work-oriented	−0.126	0.381	0.906**	0.132	0.841
17 Informal vs formal	0.296	0.331	0.872**	0.206	0.946
18 Emotionally expressive vs emotionally restrained	0.129	0.293	0.831*	0.456	0.914
19 Emotionally close vs distant	0.352	0.222	0.809*	0.415	0.981
20 Sincere vs insincere	0.529*	−0.025	0.760*	0.376	0.966
21 Flexible vs rigid	0.449	0.468	0.759*	−0.063	0.968
22 Difficult vs easy to break off contact	0.151	0.028	0.721*	0.675*	0.964
23 Emotional vs intellectual	−0.276	0.178	0.680*	0.655*	0.811
24 Active vs inactive	0.155	−0.087	0.215	0.960**	0.870
25 Intense vs superficial interaction	−0.097	0.156	0.413	0.892**	0.945
26 Important vs unimportant to society	−0.072	−0.385	0.263	0.882**	0.774
27 Intense vs superficial feelings	−0.314	0.016	0.399	0.861**	0.968
28 Interesting vs dull	−0.096	0.213	0.483	0.844*	0.857
29 Important vs unimportant to the individuals involved	0.501*	−0.111	0.459	0.725*	0.950

** Regression weight greater than 0.850.
* Regression weight between 0.500 and 0.850.

the first four columns indicate the (normalized) regression weights for predicting mean ratings on the semantic differential scales from the four dimensions of the group stimulus space. The multiple regression coefficients in the righthand column show how well mean ratings on a bipolar scale can be predicted (linearly) from the set of optimally weighted dimensions. Geometrically, the regression weights correspond to direction cosines of a vector in the four-dimensional space; projections of stimuli on the vector correlate maximally with mean ratings of stimuli on the associated semantic differential scale. The high multiple correlations for most of the bipolar scales (17 are 0.950 or greater, 7 are between 0.900 and 0.949, and 5 are below 0.900) demonstrate that a wide variety of characteristics of these relations are predictable from the INDSCAL dimensions.

Dimension 1 was interpreted as 'cooperation and harmony vs competition and conflict' since the regression weights associated with it are very high for predicting mean ratings on the 'cooperative vs competitive,' the 'always harmonious vs always clashing,' and the 'compatible vs incompatible goals and desires' scales (0.981, 0.976, and 0.943, respectively). Dimension 2, interpreted as 'equality vs inequality,' is weighted most highly for predicting mean ratings on the 'equal vs unequal power in the relationship' (0.986), the 'similar vs different roles and behavior' (0.932), and the 'democratic vs autocratic' (0.887) scales.

The third dimension, shown in Figure 6, contrasts business-oriented and other impersonal relations with relations involving more personal feelings and interaction. Its interpretation, 'personal and informal vs impersonal and formal,' is supported by the high regression weights that dimension 3 has for predicting mean ratings of the relations on the 'personal vs impersonal' (0.927), the 'pleasure-oriented vs work-oriented' (0.906), and the 'informal vs formal' (0.872) scales. The fourth dimension was labeled 'intense vs superficial' since it distinguishes intense, active relations (such as 'between husband and wife') from those in which the individuals have infrequent and/or superficial contact (such as 'between casual acquaintances' or 'between second cousins'). The scales for which dimension 4 has high regression weights, and for which the multiple correlations are also high, are 'active vs inactive' (0.960), 'intense vs superficial interaction' (0.892), and 'intense vs superficial feelings' (0.861).

Dimension weights for subgroups. Table 2 shows the weight, or relative importance, of each dimension to various subgroups (based on a biographical information sheet). The table gives evidence of systematic differences in conceptions of these dyadic relations among individuals differing in various background characteristics.

The first dimension ('cooperation and harmony vs competition and conflict') is *less* salient (lower mean weight) for younger than for older females

TABLE 2
Dimension weights for subgroups, from an INDSCAL analysis of data on similarities among interpersonal relations

	Dim. 1 Cooperation and Harmony vs Competition and Conflict	Dim. 2 Equality vs Inequality	Dim. 3 Personal and Informal vs Impersonal and Formal	Dim. 4 Intense vs Superficial
Total Group (N = 67)				
Mean	0.374	0.333	0.227	0.253
Root-mean-square	0.394	0.347	0.272	0.260
Standard deviation	0.124	0.096	0.150	0.060
1 Gender and age				
A. Male, 23 or younger (N = 19)	0.345	0.323	0.222	0.255
B. Male, 24 or older (N = 13)	0.337	0.334	0.255	0.238
C. Female, 23 or younger (N = 15)	0.346	0.337	0.267	0.267
D. Female, 24 or older (N = 20)	0.446	0.341	0.182	0.251
2 Marital status				
A. Single (N = 34)	0.360	0.332	0.231	0.250
B. Married (N = 33)	0.388	0.335	0.222	0.256
3 Birth order				
A. Oldest—No siblings (N = 9)	0.402	0.383	0.192	0.268
B. Oldest sibling (N = 31)	0.386	0.323	0.205	0.251
C. Middle sibling (N = 12)	0.407	0.344	0.256	0.251
D. Youngest sibling (N = 15)	0.304	0.316	0.269	0.250
4 Number of siblings				
A. Zero (N = 9)	0.402	0.383	0.192	0.268
B. One (N = 27)	0.367	0.343	0.250	0.243
C. Two (N = 18)	0.376	0.333	0.210	0.285
D. Three or more (N = 13)	0.364	0.280	0.224	0.221
5 Parents' socioeconomic status (when growing up)				
A. Upper middle or above (N = 30)	0.391	0.362	0.226	0.242
B. Lower middle or below (N = 37)	0.360	0.311	0.227	0.262
6 Political orientation[a] (1 = left; 9 = right)				
A. 1 or 2 (N = 18)	0.355	0.373	0.223	0.243
B. 3 or 4 (N = 33)	0.376	0.332	0.226	0.262
C. 5 to 9 (N = 15)	0.391	0.290	0.237	0.241
7 Religion				
A. Jewish (N = 24)	0.353	0.329	0.281	0.259

Note: Based on Biographical Information Sheet.
[a] One subject did not indicate his political orientation.

TABLE 2—continued
Dimension weights for subgroups, from an INDSCAL analysis of data on similarities among interpersonal relations

	Dim. 1 Cooperation and Harmony vs Competition and Conflict	Dim. 2 Equality vs Inequality	Dim. 3 Personal and Informal vs Impersonal and Formal	Dim. 4 Intense vs Superficial
B. Protestant (N = 11)	0.400	0.328	0.219	0.269
C. Catholic (N = 9)	0.393	0.307	0.168	0.256
D. Other (N = 8)	0.352	0.365	0.197	0.223
E. None (N = 15)	0.388	0.344	0.195	0.245
8 Field of study				
A. Psychology (N = 27)	0.366	0.323	0.274	0.240
B. Education (N = 19)	0.368	0.322	0.199	0.267
C. Liberal Arts (N = 12)	0.405	0.333	0.179	0.271
D. Natural Sciences (N = 9)	0.367	0.387	0.206	0.236

and for subjects who are youngest in the family than for those who are not. It is interesting that differences in the perceptual importance of the 'equality vs inequality' dimension (which reflects the distribution of power between the two individuals in a relationship) are associated with political orientation (higher weights, on the average, for subjects with more leftish political views), socioeconomic status (higher weights for subjects who grew up in families of higher 'social class'), and number of siblings (higher weights for subjects with fewer siblings). The patterns of weights on this power-type dimension are similar to results observed in a study of nation perceptions (Wish, Deutsch, & Biener, 1970); that is 'doves' and subjects who were from more 'developed' (higher status) countries had higher weights on an economic development dimension than 'hawks' and subjects from less 'developed' countries.

The relative importance of the third dimension ('personal and informal vs impersonal and formal') is greater for Jewish than for non-Jewish subjects and for students in psychology than for those in other fields. There is also a slightly higher average weight on dimension 3 for subjects who have older siblings than for those who do not. The effects of these three background variables (religion, field of study, and birth order) combine—for example, the average weight on dimension 3 is much greater for Jewish psychologists with older siblings (mean = 0.354) than for non-Jewish nonpsychologists who are the oldest in the family (mean = 0.121). There were no significant differences among the biographical subgroups with respect to the 'intense vs superficial' dimension. (More detailed analyses of subgroups indicate that the differences in Table 2 cannot be attributed to confoundings among the biographical variables.)

The four dimensions revealed by the INDSCAL analysis of the similarities data relate to distinctions that have been repeatedly emphasized in the psychological and sociological literature. Thus, they are central dimensions in the characterization of interpersonal relations by the subjects in this study as well as by researchers interested in personality theory, interpersonal behavior, and group processes.

A Reanalysis of the Miller and Nicely Consonant Confusion Data

In the applications to perceptions of colors and of interpersonal relations, each matrix was associated with the data for one subject. In contrast, the matrices in the application to the Miller and Nicely (1955) data that follow are derived from data in different experimental conditions. The example demonstrates how INDSCAL can be used to compare effects of different experimental conditions or treatments, and how patterns of dimension weights for different conditions can facilitate and perhaps validate the dimensional interpretations.

The subjects in the Miller and Nicely study listened to female speakers read CV syllables, each consisting of 1 of 16 consonants followed by the vowel a, as in ah. Sound spectrograms of typical utterances, by a female speaker, of each stimulus are shown in Figure 7. After each syllable was spoken, subjects wrote down the consonant they heard.

There were 17 experimental sessions, in each of which the speech was degraded in a different way—by varying the signal-to-noise ratio, by low-pass filtering (filtering out acoustical energy above a specified cutoff), and by high-pass filtering (filtering out energy below a cutoff). These degradation conditions are listed in Table 3.

Seventeen matrices of stimulus-response confusions, or errors of identification (the diagonal values indicated frequencies of correct identifications), were derived from the subjects' judgments, one for each condition in Table 3. The entire set of matrices was used as input to a single INDSCAL analysis (Wish, 1970; Wish & Carroll, 1973). (See Shepard, 1972b, for analyses of matrices from the Miller and Nicely study by two-way MDS methods and clustering procedures.)

Interpretations of dimensions. An interpretable six-dimensional solution was obtained for the Miller and Nicely data that accounted quite well for the confusion data from all of the degradation conditions. Stimulus coordinates on these (normalized) dimensions are plotted in Figures 8A (dimensions 1 and 2), 8B (dimensions 3 and 4), and 8C (dimensions 5 and 6). The proportions of variance in the 17 matrices accounted for by dimensions 1 to 6 are, respectively, 0.24, 0.10, 0.13, 0.07, 0.13, and 0.11.

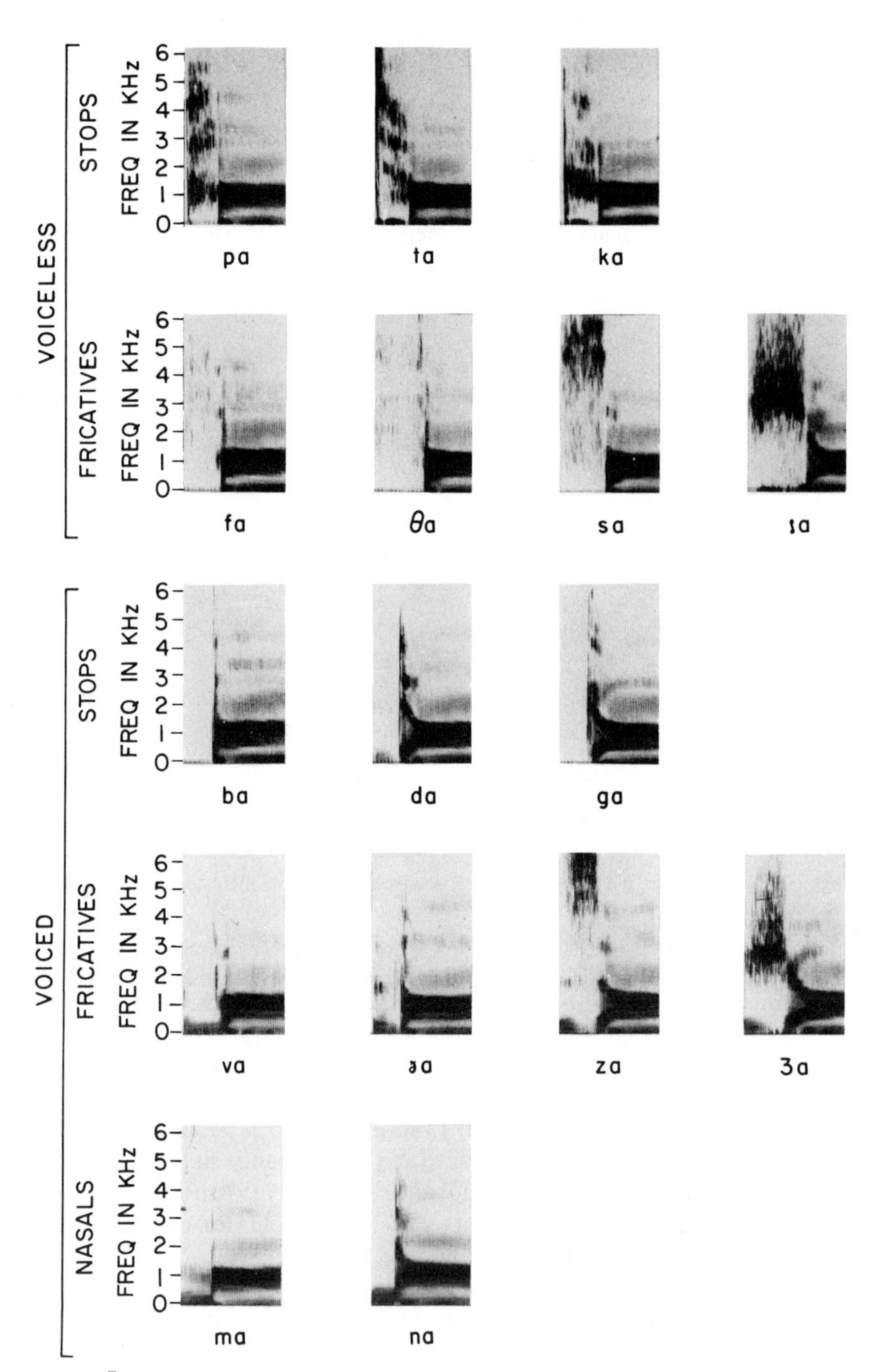

FIGURE 7.
Sound spectrograms of typical utterances, by a female speaker, of the 16 stimuli used in the Miller and Nicely (1955) study of consonant confusability. Time runs from left to right (about a half second per stimulus) in each spectrogram, while frequency in kilohertz (thousands of cycles per second) increases from bottom to top. The consonant portion occupies approximately the leftmost third of each spectrogram. The consonant phonemes θ and ð are the *th* sounds in *th*in and *th*at, respectively, while ʃ and ʒ are the *sh* and *zh* sounds in *sh*ow and *Zh*ivago.

TABLE 3
Degradation conditions in Miller-Nicely study of confusions among consonants

Degradation Condition	Signal-to-Noise Ratio	Frequency Response in hz	Observed Proportions of Confusions
Added noise			
N1 (= L1)	12 db	200–6500	0.092
N2	6 db	200–6500	0.158
N3	0 db	200–6500	0.260
N4	−6 db	200–6500	0.535
N5	−12 db	200–6500	0.730
N6	−18 db	200–6500	0.922
Low-pass filtering			
L1 (= N1)	12 db	200–6500	0.092
L2 (= H1)	12 db	200–5000	0.167
L3	12 db	200–2500	0.272
L4	12 db	200–1200	0.428
L5	12 db	200– 600	0.505
L6	12 db	200– 400	0.592
L7	12 db	200– 300	0.735
High-pass filtering			
H1 (= L2)	12 db	200–5000	0.167
H2	12 db	1000–5000	0.268
H3	12 db	2000–5000	0.494
H4	12 db	2500–5000	0.619
H5	12 db	3000–5000	0.728
H6	12 db	4500–5000	0.787

Note: Based on Miller and Nicely (1955). Higher numbers denote more severe degradations.

Dimensions 1 and 2 were interpreted as 'voicing' and 'nasality,' respectively. The stimuli are quite tightly grouped in this plane into clusters of (a) voiceless consonants (those that do not produce vocal cord vibration when spoken), (b) nasals, and (c) voiced, nonnasal consonants. The interpretation 'sibilance' was assigned to the third dimension since it distinguishes the sibilants from the other consonants, and dimension 4 was interpreted as 'sibilant frequency' since it separates the higher frequency (and narrowband) sibilants, s and z, from the lower frequency (and broadband) sibilants, ʃ (= sh) and ʒ (= zh). Each voiced consonant is close to its voiceless cognate (for example, f and v, or s and z) on both dimensions 3 and 4.

Although another dimension distinguishing the stops from the fricatives might be expected, the fifth dimension makes such a distinction only among voiceless consonants; that is, the voiceless stops and fricatives are at opposite extremes, with the voiced consonants in between. Dimension 6 shows that the discrimination among the voiced consonants is based more on the second formant transition than on the stop-fricative distinction. (This dimension might be affected by the particular vowel that is used.) As shown in Figure 7, the second formant frequency falls (from the consonant to the vowel portion) in two of the voiced stops, da and ga (the stimuli with the highest values on dimension 6), and rises (or at least does not fall) in the other voiced stop, ba, as well as in va and ða (as in that). A possible explanation why the stop versus fricative distinction is more important for voiceless than for voiced consonants is that the voiceless stops and fricatives differ with respect to aspiration as well as affrication. The second formant transition is probably not salient for voiceless consonants since the frequency change, which occurs during aspiration, is likely to be inaudible.

Dimension weights for degradation conditions. Plots of weights for degradation conditions on the corresponding pairs of dimensions are shown in Figures 8D, 8E, and 8F. In these figures, squares, circles, and triangles represent the noise, low-pass filtering, and high-pass filtering conditions, respectively. Larger symbols are used to denote the three most severe degradations of each type. The closer the points for two conditions in Figures 8D, 8E, and 8F (which could be referred to as the 'condition space,' rather than the 'subject space'), the more similar the perceptual effects of the associated speech degradations.

Figure 8D shows that the salience of the 'voicing' dimension increases with greater low-pass filtering or noise (with the exception of N6, where the signal-to-noise ratio is so low that all the stimuli sound about the same), and decreases with greater high-pass filtering. (In general, less variance is accounted for in conditions where the speech degradation is very great or very small than in those in which the degradation is moderate.) The extreme noise and low-pass filtering conditions also have the highest weights on the 'nasality' dimension. In contrast with the 'voicing' dimension, however, there are (a) negligible differences in weights on the 'nasality' dimension among the high-pass filtering conditions, and (b) only small differences in weights between the moderate noise and low-pass filtering conditions on the one hand and the high-pass filtering conditions on the other.

The pattern of dimension weights for the third and fourth dimensions is almost the reverse of that found for the first two; that is, the high-pass filtering conditions have much higher weights than the low-pass filtering and the noise conditions on the 'sibilance' and 'sibilant-frequency' dimensions. This is consistent with the fact that the sibilants differ greatly from the nonsibilants, and the broadband sibilants differ greatly from the narrowband sibilants in the frequency range above 2000 Hz. The very low weights on dimensions 3 and 4 for the extreme low-pass filtering and noise conditions reflect the fact that sibilants are perceptually about as similar to other fricatives as they are to each other in the low-frequency range (the only audible range in these conditions).

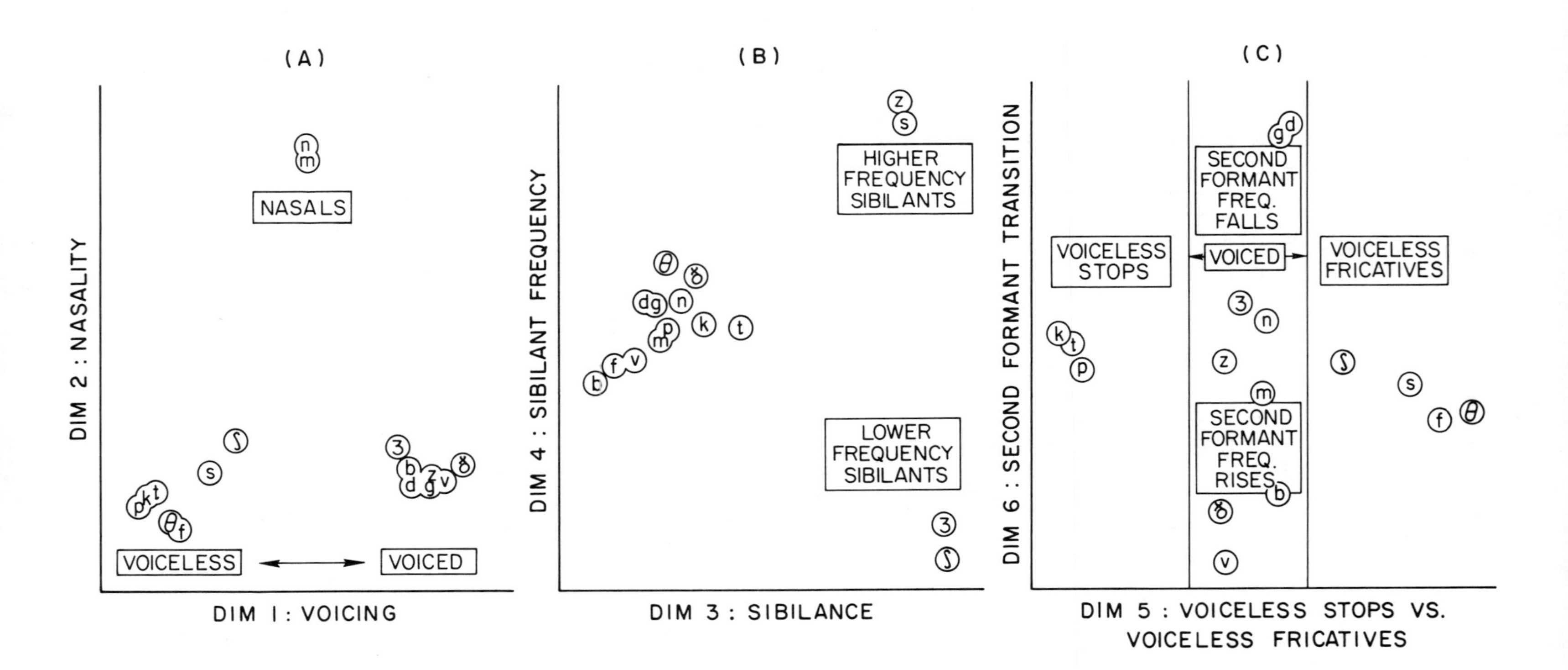
(A)
NASALS
VOICELESS
VOICED
DIM 2 : NASALITY
DIM 1 : VOICING
(B)
HIGHER FREQUENCY SIBILANTS
LOWER FREQUENCY SIBILANTS
DIM 4 : SIBILANT FREQUENCY
DIM 3 : SIBILANCE
(C)
VOICELESS STOPS
SECOND FORMANT FREQ. FALLS
VOICED
SECOND FORMANT FREQ. RISES
VOICELESS FRICATIVES
DIM 6 : SECOND FORMANT TRANSITION
DIM 5 : VOICELESS STOPS VS. VOICELESS FRICATIVES

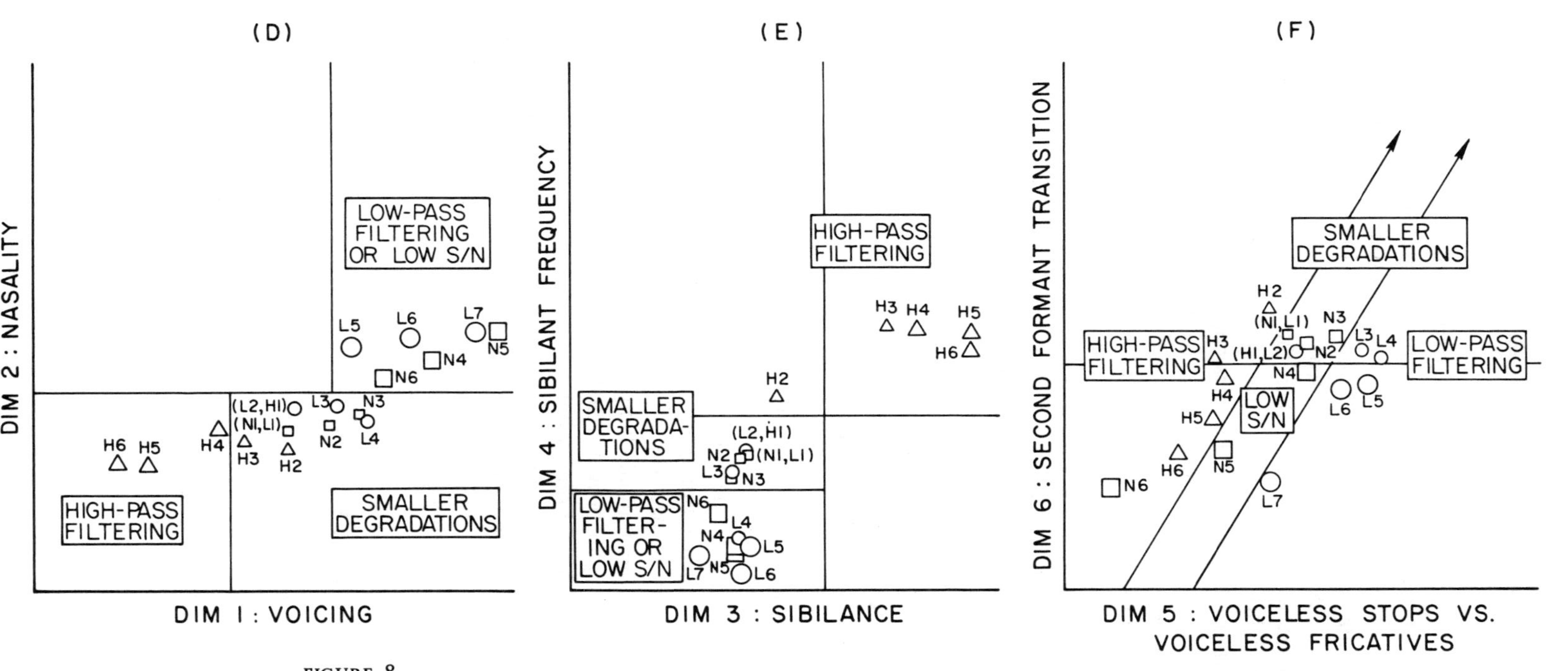

FIGURE 8.
Planes of the ‘group stimulus space’ (A, B, and C) and the ‘condition space’ (which is analogous to a ‘subject space’ in that weights for experimental conditions are plotted) (D, E, and F) from a six-dimensional INDSCAL analysis (Wish, 1970) of the Miller and Nicely (1955) data on confusions among consonants. Noise (or signal-to-noise ratio), low-pass filtering, and high-pass filtering conditions are denoted by squares, circles, and triangles, respectively (see Table 3). Larger symbols are used to represent the three most severe degradations of each type.

Weights on the first four dimensions are quite similar for noise and low-pass filtering conditions having similar confusion rates. Figure 8F shows, however, that the extreme noise conditions are closer to the high-pass than to the low-pass filtering conditions on dimension 5. The low weights on dimension 5 for conditions N5 and N6 show that voiceless stops and fricatives are frequently confused with each other when the signal-to-noise ratio is very low. The high weights on dimension 5 for all low-pass filtering conditions except L7 indicate that the low frequencies contain crucial information for distinguishing voiceless stops from voiceless fricatives.

The sixth dimension is very important for the mildest degradation conditions—the most important dimension for H2 and N1 (= L1) and the second most important dimension for N2 and L2 (= H1)—but it becomes less and less salient as the degree of low-pass filtering, high-pass filtering, or noise increases. This demonstrates the necessity of acoustical information in the intermediate frequencies (where the second formant transition occurs) for distinguishing consonants with a falling versus a rising second formant transition.

These results show that the weights, or perceptual saliences, of the dimensions depend systematically on the degree and kind of degradation the speech has undergone. (They also indicate a limitation in using a single kind of degradation in studies of sensory stimuli, since the stimulus structure may be affected by the particular kind and degree of degradation.) The observed patterns of dimension weights for the degradation conditions make sense acoustically, and tell a great deal about the relative importance of sound energy in different frequency ranges for the perception of consonants. Furthermore, the similarity in patterns of dimension weights for acoustically similar conditions and the gradual change with increasing degrees of degradation show that the weights provide reliable and valid information concerning the perceptual saliences of dimensions.

Some additional applications of INDSCAL can be found in Carroll and Chang (1970a); Wish, Deutsch, and Biener (1970); Green and Rao (1971); Carroll (1972a); Jones and Young (1972); Sherman and Ross (1972); Dobson and Young (1973); and Wish and Carroll (1973).

THE METHOD OF ANALYSIS FOR INDSCAL

So far we have not described how the computer derives the two output matrices defining the 'group stimulus space' and the 'subject space' from the three-way array of data provided as input. Roughly, they are defined on the basis of providing the best r-dimensional approximation to the original three-way array of data. The description of the computer algorithm provided below should make it clear in what sense this statement is 'roughly' true.

Converting the Proximities to Estimated Distances

The first step, as in the 'classical' metric two-way MDS procedure, is to convert the *dissimilarities* to estimated distances. (If the data are similarities, they can be converted into dissimilarities by, for example, simply multiplying all values by -1). This is the process usually referred to as 'estimation of the additive constant.'

Under 'metric' assumptions we may assume, without loss of generality, that

$$d_{jk}^{(i)} \cong \partial_{kj}^{(i)} + c^{(i)}. \tag{7}$$

Dropping the (i) superscript for now, the smallest constant c guaranteeing satisfaction of the *triangle inequality* ($d_{jl} \leq d_{jk} + d_{k\ell}$) for all triples (j, k, ℓ) can be shown as

$$c_{\min} = \max_{(j,k,\ell)} (\partial_{j\ell} - \partial_{jk} - \partial_{k\ell}). \tag{8}$$

(NOTE: $c_{\min}$ *may* be negative.)

This is the 'additive constant' estimation method for converting 'comparative distances' (i.e., interval-scale distance estimates) into absolute distances (i.e., ratio-scale distance estimates) described in Torgerson (1958) as the "one-dimensional subspace" scheme (assuming that at least three points lie exactly on a straight line in the space). It is in many respects the simplest and most straightforward technique for additive constant estimation, and is the one used in the metric INDSCAL procedure.

Conversion of Estimated Distances to Estimated Scalar Products

If $\mathbf{x}_j = (x_{j1}, x_{j2}, \ldots, x_{jr})$ and $\mathbf{x}_k = (x_{k1}, x_{k2}, \ldots, x_{kr})$ are two vectors in r-dimensional space, their *scalar product*, usually written as $\mathbf{x}_j \cdot \mathbf{x}_k$ (which we shall also call b_{jk}) is defined as:

$$b_{jk} \equiv \mathbf{x}_j \cdot \mathbf{x}_k \equiv \sum_{t=1}^{r} x_{jt} x_{kt}. \tag{9}$$

Geometrically, the scalar product can be interpreted as the cosine of the angle between the two vectors multiplied by the product of their lengths. Actually, however, this geometric fact is not really necessary for our purposes. Rather, it is quite sufficient to deal with the algebraic definition in Equation 9.

In most applications of MDS, the *origin* of the space is not of any importance, and thus may be arbitrarily fixed because a shift in origin (defined by adding the same constant vector to all points) leaves (Euclidean) distances unchanged. It has therefore become conventional to place the origin at the *centroid*, or generalized mean, of all the points.

While *distances* remain unchanged with a shift in origin, *scalar products* do not, since the vectors whose lengths and angles are involved are the vectors *from* the origin to the particular point. We can resolve this indeterminacy of scalar products, however, by dealing always with scalar products from an origin at the centroid of all the points. We will henceforth assume that the origin is at the centroid, and will use the symbol b_{jk} to represent scalar products of vectors about such an origin.

As shown by Torgerson (1958), Euclidean distances may be converted into scalar products of vectors about an origin placed (arbitrarily) at the centroid of all the points by Equation 10:

$$b_{jk} = -\tfrac{1}{2}(d_{jk}^2 - d_{\cdot k}^2 - d_{j\cdot}^2 + d_{\cdot\cdot}^2), \tag{10}$$

where

$$d_{\cdot k}^2 = \frac{1}{n} \sum_j d_{jk}^2, \tag{11}$$

$$d_{j\cdot}^2 = \frac{1}{n} \sum_k d_{jk}^2, \tag{12}$$

$$d_{\cdot\cdot}^2 = \frac{1}{n} \sum_j \sum_k d_{jk}^2. \tag{13}$$

This is equivalent to *doubly centering* the matrix of $-\tfrac{1}{2}$ times the *squared* interpoint distances (i.e., subtracting grand mean and row and column main effects, in the analysis of variance sense). A derivation of Equation 10 follows.

Given

$$d_{jk}^2 = \sum_{t=1}^{r} (x_{jt} - x_{kt})^2, \tag{14}$$

assume

$$\sum_{j=1}^{n} x_{jt} = 0 \quad \text{for all } t = 1, 2, \ldots, r. \tag{15}$$

(We may do this without loss of generality, since the origin of the x space is arbitrary, and this just fixes it at the centroid of all n points.) Expanding Equation 14,

$$\begin{aligned} d_{jk}^2 &= \sum_t (x_{jt}^2 - 2x_{jt}x_{kt} + x_{kt}^2) \\ &= \sum_t x_{jt}^2 - 2\sum_t x_{jt}x_{kt} + \sum_t x_{kt}^2 \\ &= \ell_j^2 + \ell_k^2 - 2b_{jk}, \end{aligned} \tag{16}$$

where

$$\ell_j^2 \equiv \sum_t x_{jt}^2, \tag{17}$$

$$b_{jk} \equiv \sum_t x_{jt} x_{kt} \quad \text{(the scalar product).} \tag{18}$$

Because of Equation 15,

$$b_{\cdot k} = b_{j\cdot} = b_{\cdot\cdot} = 0. \tag{19}$$

From Equations 16 and 19, we have

$$d^2_{\cdot k} = \ell^2_{\cdot} + \ell^2_k, \tag{20}$$

$$d^2_{j\cdot} = \ell^2_j + \ell^2_{\cdot}, \tag{21}$$

$$d^2_{\cdot\cdot} = 2\ell^2_{\cdot}, \tag{22}$$

where

$$\ell^2_{\cdot} = \frac{1}{n} \sum \ell^2_j. \tag{23}$$

Then Equations 16, 20, 21, and 22 together imply that

$$d^2_{jk} - d^2_{\cdot k} - d^2_{j\cdot} + d^2_{\cdot\cdot} = -2b_{jk}. \tag{24}$$

Multiplying both sides of Equation 24 by $-\frac{1}{2}$ gives the desired result in Equation 10.

By analogy to this *exact* formula for converting distances to scalar products (about an origin at the centroid), we may apply the same formula to $\hat{d}$s ($\hat{d}_{jk} = \partial_{jk} + c$ min) to get $\hat{b}$s (estimated scalar products). (In matrix terms, $\hat{B}$ is $-\frac{1}{2}\hat{D}^{(2)}$ after double centering, where $\hat{D}^{(2)} = \|\hat{d}^2_{jk}\|$.) We shall generally use a hat ($\wedge$) over a variable to indicate an estimate of the corresponding unhatted variable.

At this point, the 'classical' (two-way) metric MDS procedure a la Torgerson (1958) would simply *factor* this matrix $\hat{B}$ into a product of the form

$$\hat{B} \cong \hat{X}\hat{X}' \tag{25}$$

to get estimates of the coordinates $\hat{x}_{jt}$. ($\hat{X}$ is just the $n \times r$ matrix $\|\hat{X}_{jt}\|$ containing these estimated coordinates; the matrix of Equation 25 is equivalent to the *scalar* equation $\hat{b}_{jk} = \sum \hat{x}_{jt}\hat{x}_{kt}$, the analogue of Equation 9, but with estimates replacing exact values.)

This factorization usually is done by methods closely related to principal components analysis (Hotelling, 1933) or factor analysis (Harman, 1967). Perhaps the best way to describe this is in terms of the procedure of Eckart and Young (1936) for least-squares approximation of an arbitrary matrix by a product of two matrices of smaller rank. When applied to a square symmetric matrix, such as $\hat{B}$ presumably is, the Eckart-Young procedure results in two matrices (such as $\hat{X}$ and $\hat{X}'$) that are simply the transposes of one another. This statement is true as long as appropriate normalizing conventions are observed, and so long as the largest (in absolute value) r *eigenroots* of $\hat{B}$ are nonnegative, which is generally true in practice. Details of this can be found in Torgerson (1958).

The INDSCAL Model in Scalar Product Form

By applying the procedures described above (for converting estimated dissimilarities into estimated scalar products) to each two-way proximities matrix, we arrive at a three-way array of estimated scalar products whose general entry is $\hat{b}_{jk}^{(i)}$ (for subject i and stimuli j and k). We now need to make an analogous conversion of the INDSCAL *model* into scalar product form.

By definition, the (true) scalar products for subject i are defined as:

$$b_{jk}^{(i)} = \sum_t y_{jt}^{(i)} y_{kt}^{(i)}, \tag{26}$$

where the $y^{(i)}$s are defined, in Equation 5, as

$$y_{jt}^{(i)} = w_{it}^{1/2} x_{jt}.$$

Substituting Equation 5 into Equation 26, we have:

$$b_{jk}^{(i)} = \sum_t w_{it} x_{jt} x_{kt}, \tag{27}$$

which is the desired *scalar product form* of the INDSCAL model.

Equation 27 can easily be seen as a *special case* of what we have called the CANDECOMP (for *CAN*onical *DECOMP*osition of N-way tables) model of the form (for $N = 3$):

$$z_{ijk} = \sum_t a_{it} b_{jt} c_{kt}. \tag{28}$$

We get the INDSCAL model (Eq. 27) as a special case of the three-way CANDECOMP model (Eq. 28) by imposing the following constraints:

$$z_{ijk} = b_{jk}^{(i)}, \tag{29}$$

$$a_{it} = w_{it}, \tag{30}$$

$$b_{jt} = c_{jt} = x_{jt}. \tag{31}$$

For the INDSCAL special case of this CANDECOMP (three-way) model, we may, however, ignore the symmetry constraint of Equation 31 and fit the model in its general form. It turns out that the symmetry of the basic data is ordinarily sufficient to guarantee that (after appropriate normalization of the solution) b_{jt} will *in fact* equal c_{jt}.

The Algorithm for Fitting the CANDECOMP Model in the Three-way Case

Given the model

$$\hat{z}_{ijk} \cong \sum_t a_{it} b_{jt} c_{kt} \tag{32}$$

(Eq. 28 with $\hat{z}_{ijk}$ replacing z_{ijk} and $\cong$ replacing $=$) and 'current estimates' of two sets of parameters (say the b_{jt}s and c_{kt}s), we can find an *exact least-squares* estimate of the third set by linear regression methods.

This can be seen by reformulating the problem as

$$\hat{z}^*_{is} \cong \sum_t a_{it} \hat{g}_{st}, \tag{33}$$

where

$$\hat{z}^*_{is} = \hat{z}_{i(jk)} \tag{34}$$

and

$$\hat{g}_{st} = \hat{b}_{jt} \hat{c}_{kt} \tag{35}$$

($\hat{b}_{jt}$ and $\hat{c}_{kt}$ are *current* estimates of b_{jt} and c_{kt}, respectively, and s is a subscript that is a function of j and k, and ranges over all $n_j \cdot n_k$ values of j and k).

By this simple notational device, we have converted this original *trilinear* model into a *bilinear* model, which can be expressed in matrix notation as

$$\hat{Z}^* \cong A\hat{G}'. \tag{36}$$

Note that the matrix $\hat{G}$ incorporates both the $\hat{b}$s and $\hat{c}$s, and the matrix A contains the as. The $\cong$ can be taken to mean, in the present context, that we seek the A providing the best least-squares approximation to Z^*. This proves to be a standard problem, essentially equivalent to least-squares multiple linear regression. In matrix notation, the least-squares estimate of A is

$$\hat{A} = Z^*\hat{G}(\hat{G}'\hat{G})^{-1} \tag{37}$$

(this amounts to postmultiplying both sides of Equation 36 by the right *pseudoinverse* of $\hat{G}'$).

We use a general estimation scheme that Wold (1966) called a NILES (*N*onlinear *I*terative *LE*ast *S*quares) procedure. In the present case this amounts to iterating this least-squares estimation procedure; i.e., estimating the *a*s (with *b*s and *c*s fixed) by least-square methods, then the *b*s (with *a*s and *c*s fixed) and so on round the iterative cycle until convergence occurs. While there is no guarantee that this process will converge to the *overall* least-squares estimates of all three sets of parameters, it does seem to do so in most cases. There is a mild 'local minimum' problem (tendency to converge to estimates such that no *small* change can improve the fit, although a *large* or global change can), but it seems to be very minor.

In practice the method seems 'almost always' to converge to the global optimum solution. In any case the local minimum problem seems to be very slight in comparison with that in nonmetric two-way MDS, where the algorithms are generally based on a gradient, or steepest descent, method. Whether it is the difference in numerical procedures or the difference in the models that is critical is not known at present.

Nonmetric and Quasi-Nonmetric Versions of INDSCAL

While a *fully* nonmetric version of INDSCAL is not currently available,* Carroll and Chang (1970b) have developed a *quasi*-nonmetric version (i.e., an approximation to a fully nonmetric version). This is implemented by alternating the metric version of INDSCAL, as just described, with the use of least-squares monotone regression (using Kruskal's, 1964c, MFIT routine) in an iterative fashion. Given data values $\partial_{jk}^{(i)}$, we first estimate additive constants $c^{(i)}$ to convert to $(0)\hat{d}_{jk}^{(i)}$. Then, the Ith iteration of this 'outer' iterative process (metric INDSCAL provides an 'inner' iterative process) can be described as follows.

Given $\partial_{jk}^{(i)}$ and $_{(I-1)}\hat{d}_{jk}^{(i)}$.

Phase one. (a) Convert $_{(I-1)}\hat{d}_{jk}^{(i)} \longrightarrow {}_{(I-1)}\hat{b}_{jk}^{(i)}$ via Equation 10.

(b) Apply CANDECOMP to the three-way matrix of $\hat{b}$s, and normalize, to get

$$_IX = \|_Ix_{jt}\| \quad \text{and} \quad _IW = \|_Iw_{it}\|. \tag{38}$$

Phase two. (a) Use the weighted Euclidean distance formula of Equation 4 to calculate $_Id_{jk}^{(i)}$, for all i, j, k.

(b) Use the least-squares monotone regression routing (MFIT) to find

$$_I\hat{d}_{jk}^{(i)} = M_i^I(\partial_{jk}^{(i)}) \cong {}_Id_{jk}^{(i)}, \tag{39}$$

(where $\cong$ implies a least-squares fit, and M_i^I is a monotone nondecreasing function).

Increment I by 1 and return to the beginning of phase one. Continue iteratively until no further improvement in fit occurs. Badness of fit, in this case, is measured by a generalization of Kruskal's STRESSFORM2, namely,

---

* *Added in proof.* Fully nonmetric versions of INDSCAL *are* now available. Such algorithms have recently been developed by Carroll and Chang and by deLeeuw. Both optimize STRAIN as defined in Equation 4.

$$\text{STRESS} = \sqrt{\sum_i \left[\frac{\sum_j \sum_k (d_{jk}^{(i)} - \hat{d}_{jk}^{(i)})^2}{\sum_j \sum_k (d_{jk}^{(i)} - \bar{d}^{(i)})^2} \right]}, \tag{40}$$

which is essentially a root mean-square over subjects of STRESSFORM2 computed separately for each subject. STRESS² is the measure actually printed out in what is called the NINDSCAL program, as implemented by Chang (1972).

Typically STRESS will go down for several 'outer' iterations, but will ultimately go up again. This is because the two phases of the algorithm are not optimizing the same criterion. Phase one is least squares in the (derived) *scalar products*, while phase two is least squares in *distances*.

To make the procedure *fully* nonmetric, both phases should be optimizing the *same* criterion. One way is to make phase one least squares in distances. Another, however, which is in some ways more attractive, is to replace the STRESS measure with what Carroll and Chang (1972) have, somewhat whimsically, called STRAIN, defined in this case as:

$$\text{STRAIN} = \sqrt{\sum_i \left[\frac{\sum_j \sum_k (b_{jk}^{(i)} - \hat{b}_{jk}^{(i)})^2}{\sum_j \sum_k (b_{jk}^{(i)})^2} \right]}, \tag{41}$$

where $b_{jk}^{(i)}$ is computed by Equation 27, and the $\hat{b}_{jk}^{(i)}$s are computed from $\hat{d}_{jk}^{(i)}$s by the obvious analogue of Equation 10, with $\hat{d}_{jk}^{(i)} = M_i(\partial_{jk}^{(i)})$. Then phase two would entail finding the M_i optimizing STRAIN (with other parameters held constant); i.e., phase two would be made least squares in scalar products. Carroll and Chang are currently working on such a procedure (which could also be applied to provide a new approach to nonmetric two-way scaling). We understand that Ricardo Dobson and Allen Yates (personal communications) and possibly others have also been working on other approaches to nonmetric versions of INDSCAL.

OTHER THREE-WAY SCALING MODELS

As indicated earlier, other three-way MDS models and methods have been developed, all of which can be considered as generalizations of the INDSCAL model. The psychological validity and/or practical utility of these have not generally been as well established, however, at least in the present authors' opinions. The most general of these models is presented first, then the others (including INDSCAL) in terms of the restrictions each imposes on this most general model. Since these models can be viewed as differing essentially in how they define the distances, $d_{jk}^{(i)}$, for individual (or other data source) i, we focus on the definition of distance for each. The class of models to be considered all involve generalizations of the Euclidean metric.

The IDIOSCAL Model

The most general model in this class has been called (by Carroll & Chang, 1972) the IDIOSCAL (*I*ndividual *D*ifferences *I*n *O*rientation *SCAL*ing) model. It defines distances by the generalized Euclidean distance formula

$$d_{jk}^{(i)} = \sqrt{\sum_{t}^{r} \sum_{t'}^{r} (x_{jt} - x_{kt}) c_{tt'}^{(i)} (x_{jt'} - x_{kt'})}, \tag{42}$$

where $C_i \equiv \|c_{tt'}^{(i)}\|$ is an $r \times r$ symmetric positive definite or semidefinite matrix.

The scalar product form of this model is

$$b_{jk}^{(i)} = \sum_{t} \sum_{t'} x_{jt} c_{tt'}^{(i)} x_{kt'} \tag{43}$$

in matrix notation; this scalar product form can be expressed as

$$B_i = XC_iX'. \tag{44}$$

It will be convenient henceforth to consider the models in their scalar product forms. No loss of generality is entailed in this, since the (generalized) *squared* Euclidean distance between $\mathbf{x}_i$ and $\mathbf{x}_j$ can be defined simply as the (generalized) scalar product between the difference vector ($\mathbf{x}_i$ and $\mathbf{x}_j$) and itself.

Two different ways of decomposing C_i have been proposed. These are, of course, mathematically equivalent, but lead to important differences in interpretation (and, seemingly, in the psychological assumptions underlying the various generalized models).

The Carroll-Chang procedure for decomposing C_i. The decomposition of C_i used by Carroll and Chang is

$$C_i = T_i\beta_iT_i', \tag{45}$$

with T_i orthogonal and β_i a diagonal matrix.

Geometrically, we can think of this as an orthogonal rotation to a new (or IDIOsyncratic) coordinate system for subject i, followed by rescaling or weighting of dimensions of this new coordinate system by the square roots of the diagonal entries of β_i.

Another way to describe this is to define

$$S_i = T_i\beta_i^{1/2}; \tag{46}$$

then

$$C_i = S_i S_i', \tag{47}$$

and we can just say that subject i has subjected the 'group stimulus space' to the linear transformation defined by the matrix S_i. One problem with this interpretation is that this linear transformation is not unique; i.e., given any orthogonal U, we can define

$$S_i^* = S_i U, \tag{48}$$

$$S_i^* S_i^{*\prime} = S_i U U' S_i' = S_i S_i' = C_i, \tag{49}$$

so S_i is defined, in this case, only up to an arbitrary orthogonal transformation.

The Tucker-Harshman procedure for decomposing C_i. The second decomposition has been used most notably by Tucker (1972) and Harshman (1972a). We therefore call it the Tucker-Harshman decomposition of C_i. It is defined as

$$C_i = D_i R_i D_i, \tag{50}$$

where D_i is diagonal and R_i is square symmetric but with *unit diagonal cells.* C_i can be thought of as analogous to a covariance matrix, and R_i to the corresponding correlation matrix. In this analogy, the diagonal values of D_i are the standard deviations that are divided out to convert covariances into correlations. In fact, both Tucker and Harshman interpret R_i as a matrix of correlations, or more properly, of cosines of angles among oblique dimensions. (Oblique dimensions are described by coordinate axes not at right angles to one another.) If the dimensions are in fact orthogonal, then the cosines are all zero. In this case, then, all off-diagonal entries in R_i are zero. Since R_i is an identity matrix, C_i reduces to D_i^2; that is, a diagonal matrix defined as the square of D_i. If this is true for all i, this model will reduce to the INDSCAL model. In this Tucker and Harshman decomposition, the diagonals of D_i can be viewed as weights, or rescaling factors, applied to such oblique dimensions. If C_i is unrestricted, each subject (or other data source) is assumed to have his own particular (or idiosyncratic) matrix of cosines of angles (and thus of angles between coordinate axes) as well as a set of weights. It is important to note that one must project points onto the coordinate axes not by the usual orthogonal projection, but via *oblique* projection. In general, oblique projection is accomplished by projecting points onto each coordinate axis along a line normal (orthogonal) to the hyperplane defined (spanned) by the $r - 1$ remaining coordinate axes.

Equidistance contours for the IDIOSCAL model (plotted, that is, in the group space defined by the matrix X) are ellipses or ellipsoids with arbitrary orientation; in the INDSCAL special case, the axes of the ellipsoid must be parallel to coordinate axes. Because of the arbitrariness of orientation of the equidistance ellipses in IDIOSCAL, the property of unique orientation of the coordinate system that obtains in INDSCAL is lost (at least for this most general model). Carroll and Chang (1970a, 1972) have discussed procedures for fitting this most general model to data.

Some interesting special cases of IDIOSCAL are discussed below. They all can be viewed as putting certain constraints on C_i.

The INDSCAL Model

The INDSCAL model is a special case in which C_i is restricted to be diagonal (or T_i to be an identity transformation, so $C_i = D_i^2$). The properties of INDSCAL have been discussed, but we shall now specify more precisely the exact conditions for 'dimensional uniqueness.'

The uniqueness property for INDSCAL. We shall now state the precise conditions under which dimensional uniqueness holds for INDSCAL. Dimensional uniqueness, it may be recalled, means that the solution is unique up to an extended permutation, that is, a permutation followed by a diagonal, or rescaling, transformation. If we apply the usual normalizing convention of INDSCAL, namely that the centroid of the stimulus space be at the origin and the sum of squared projections on each dimension in that space be one, the solution is unique except for a permutation of dimensions and a possible reflection of each dimension (the possible reflection applies to the stimulus space, but *not* the subject space).

Let us first define a property that we may call a *parallel pattern* of weights for a pair of dimensions. We say that the subject weights for a pair of dimensions, s and t, exhibit the parallel pattern if and only if, for *all* pairs of subjects i and j, $w_{is} \cdot w_{jt} = w_{it} \cdot w_{js}$. What this means geometrically is that in the two-dimensional subspace of the subject space corresponding to dimensions s and t the weights all lie on a straight line passing through the origin. It also means that after an appropriate rescaling of those two dimensions in the *stimulus* space, the same weight can be applied to *both* dimensions for every subject. It is as though the weight is applied to (two-dimensional) squared distances computed in that subspace, rather than to the (one-dimensional) squared distances from each of the two one-dimensional subspaces separately. Since these (Euclidean) squared distances are invariant under orthogonal transformation within the (rescaled) two-dimensional subspace, it is not at all surprising that this parallel pattern of weights should be associated with a failure of dimensional uniqueness. In fact, it turns out that this is generally the only condition under which uniqueness fails.

The INDSCAL solution is unique in the sense specified if the following three conditions hold: (i) the dimensions of the stimulus space are linearly independent; (ii) there are at least two subjects (or matrices derived from other data sources); and (iii) there are no two dimensions for which the parallel pattern of weights property obtains. A formal proof of this has been furnished by Harshman (1972a).

For any pair of dimensions for which the parallel pattern property holds, there is a rotational indeterminacy within the associated two-dimensional subspace. If this property holds for every pair of some subset of r' dimensions, then there is rotational indeterminacy within that r'-dimensional subspace. The r'-dimensional *subspace* itself will still be uniquely determined, however, provided that the parallel pattern does not obtain between those r' dimensions and any other of the $r - r'$ remaining dimensions. (The parallel pattern property is transitive, so if it holds between any one of the r' dimensions and an 'outside' dimension, it would hold for *all* of the r' dimensions and the same 'outside' dimension.)

Conditions (i), (ii) and (iii) together comprise sufficient conditions for uniqueness. Unlike (ii) and (iii), however, (i) is not a necessary condition. It is clear that stimulus dimensions in INDSCAL may be linearly dependent and uniqueness still obtain, but the exact conditions for uniqueness with linearly dependent dimensions have not yet been worked out in detail. One thing that is clear is that the *total* number of stimulus dimensions for n stimuli cannot exceed $n(n - 1)/2$, at least not if stimulus weights are allowed to be negative as well as positive (this is essentially because an $n \times n$ distance matrix has only $n(n - 1)/2$ independent entries, and thus it can be produced as a linear combination of $n(n - 1)/2$ linearly independent distance matrices, each presumably associated with one of the dimensions). This is clearly not a very 'tight' upper bound, however, and it would not hold anyway if a positivity constraint were applied to subject weights. A conjecture (not yet proved) for the number of *subjects* required for uniqueness in the case of linearly dependent stimulus dimensions is that, for r stimulus dimensions of which k are linearly dependent on the other $r - k$, the number of subjects required is at least $2 + k$. (We are assuming, of course, that no two stimulus dimensions are simple multiples of one another, so the number of linearly independent dimensions is at least 2.)

Tucker's Three-mode Scaling Model

Three-mode scaling is Tucker's (1972) procedure for applying three-mode factor analysis to MDS. In this model C_i is assumed to be of the form

$$C_i = \sum_{s=1}^{S} a_{is} G_s, \tag{51}$$

where G_s is the sth 'basis' matrix. Expanding this, we have

$$c_{tt'}^{(i)} = \sum_{s=1}^{S} a_{is} g_{stt'}. \tag{52}$$

$A \equiv \|a_{is}\|$ is the s-dimensional subject space in the three-mode scaling model (S is not necessarily equal to r) while $g_{stt'}$ is an element of the 'core matrix.'

Obviously by making S large enough, any C_i matrix could be produced by Equation 51. The critical number for S turns out, in fact, to be $r(r+1)/2$ (the number of independent entries in C_i). But for $S < r(r+1)/2$, the model generally does put some restriction on the possible family of C_i matrices.

Some characteristics of Tucker's three-mode scaling method are described below.

(i) The orientation of axes is *not* unique. Any two *nonsingular linear transformations* can be applied to the stimulus space and to the subject space, respectively; a new core matrix can then be found giving exactly the same account of the data.

(ii) The subject space and the stimulus space need not have the same dimensionality. The former may have either more or fewer dimensions than the latter. This possibility provides more flexibility (than INDSCAL) in many ways.

(iii) The subject space and the stimulus space alone are not sufficient to describe the structure—the core matrix is also needed. Although the interpretation of the core matrix presents some problems, Tucker (1972) has suggested a way to approach this in terms of an 'ideal individual' conception. The approach is analogous to that used in the Tucker and Messick (1963) 'points-of-view' model.

The three-mode scaling model becomes equivalent to the INDSCAL model if $S \leq r$ and if there exist (separate) linear transformations of the stimulus and subject spaces such that the resulting core matrix (appropriate to these transformed spaces) is comprised of S diagonal $r \times r$ matrices (i.e., $g_{stt'} = 0$ for all S and $t \neq t'$).

Harshman's PARAFAC-2 Model

Harshman's (1972a) PARAFAC (*PAR*allel *FAC*tors) -2 model is another specific case of IDIOSCAL in which C_i is of the form

$$C_i = D_i R D_i \tag{53}$$

(same R for all subjects, but different weights or rescaling factors).

Harshman interprets this as differential weighting of an oblique set of axes.

R is the matrix of cosines of angles among these axes. INDSCAL is the special case of PARAFAC-2 in which the axes are orthogonal (so that R is an identity and simply drops out).

A NILES-type procedure is possible for fitting the PARAFAC-2 model to data. One procedure, based generally on the NILES-type approach, has been worked out by Jennrich, but it takes into account symmetries in data and solution in ways that may not be optimal. (See Harshman, 1972a, for a description of Jennrich's algorithm.) Carroll and Chang have developed a complete NILES procedure for PARAFAC-2 that seems to have some advantage over Jennrich's method.

We now describe the algebraic structure of the PARAFAC-2 model as applied to dissimilarities data. Let X be a matrix, which defines projections of stimuli onto oblique coordinate axes. (X defines the 'group stimulus space' for PARAFAC-2.) Each subject differentially weights these, as represented by postmultiplication by a diagonal matrix D_i. Thus,

$$X_i^* = XD_i. \tag{54}$$

To 'compute' Euclidean distances we need to project the stimulus points for subject i as described by X_i^* onto an orthogonal coordinate system. This projection is given by a linear transformation defined by a matrix T, so that

$$X_i^\circ = X_i^*T = XD_iT, \tag{55}$$

where X_i° is the description of the stimulus points in terms of the orthogonal coordinate system. Computing ordinary Euclidean distances in the space defined by X_i° will lead to the PARAFAC-2 model, with

$$C_i = D_iTT'D_i = D_i^*RD_i^*, \tag{56}$$

where R is a normalized version of TT' $\{R = ETT'E$ with $E = [\text{diag}\,(TT')]^{-1/2}\}$ and $D_i^* = D_iE^{-1}$, with E as defined.

Some characteristics of PARAFAC-2 are described below.

(i) There is ambiguity regarding the extent to which this model exhibits the dimensional uniqueness property. Although Harshman (1972b) has conjectured that the dimensions from PARAFAC-2 are unique, Carroll has proved that the dimensions are *not* unique in the case of only two 'subjects.' Furthermore, Carroll has established that, with r subjects or matrices, an r-dimensional PARAFAC-2 solution is *not* unique if each subject has zero weights for all but one dimension (each subject having a nonzero weight for a different dimension). While the stimulus dimensions and subject weights are uniquely determined in this case, the R matrix is not—its off-diagonal entries could take on any values whatever. However, there is at least one condition in which an r-dimensional solution *is* unique if there are $r + 1$ (or more) subjects. This is true if the pattern of weights for r of the subjects is as described, while the $(r + 1)$st has nonzero weights for all r dimensions. The $(r + 1)$st subject serves to specify R uniquely. This result suggests (but by no means proves) the conjecture that uniqueness (under 'general' conditions) will hold for an r-dimensional solution if and only if there are at least $r + 1$ subjects.

Carroll and Chang have done some Monte Carlo work, however, that brings into doubt the uniqueness property for PARAFAC-2 even when this '$r + 1$' condition is satisfied. In one example with seven subjects and seven stimuli, a solution was obtained (in three dimensions) that differed considerably from the 'true' one, but nonetheless provided essentially a perfect fit ($r > 0.999$). The fit for a case with 14 subjects and 14 stimuli (also three dimensional) was almost as good ($r = 0.994$). There was more agreement between the PARAFAC-2 and the 'true' solution for this example than for the one with seven subjects and seven stimuli. However, there were some noticeable discrepancies, particularly in the subject space and in the R matrix. It seems clear that, at best, there are strong boundary conditions on uniqueness for this model.

(ii) The R matrix provides an estimate of 'subjective intercorrelations' of dimensions, which are assumed in PARAFAC-2 to be the same for all subjects. This would seem to be useful, however, only insofar as the uniqueness property obtains.

(iii) This model would seem to strike a compromise in generality between the INDSCAL model and the IDIOSCAL or three-mode scaling models—it is a generalization of INDSCAL and a special case of IDIOSCAL. It is not, strictly speaking, a special case of three-mode scaling, except when the subject space is of 'maximal' dimensionality ($r(r + 1)/2$, where r is the dimensionality of the stimulus space). In this case the three-mode scaling model becomes equivalent to the fully general IDIOSCAL model. It is clear, however, that PARAFAC-2 is midway in generality between INDSCAL and three-mode scaling.

Relationship of Three-way Models to 'Points-of-view' Analysis

Tucker and Messick's (1963) 'points-of-view' (POV) analysis, referred to earlier in connection with the Helm data, has been a widely used procedure for individual differences analysis of proximities data. While this is perhaps an oversimplification, POV analysis amounts essentially to clustering subjects into relatively homogeneous subgroups (based on the profile similarity of their proximities data), and then developing a separate stimulus space for each subgroup based on the subgroup mean data. In actual practice, an obverse factor analysis of subjects is often used, and distance matrices are developed for 'ideal subjects' in the resulting spaces. These ideal subjects may or may not correspond to clusters of real subjects.

Suppose one of the models discussed here holds, and a POV analysis is attempted. One might raise the question whether there is any necessary relation between (a) the number of 'points of view,' or the total number of dimensions in all 'points of view' and (b) the dimensionality of the stimulus and/or subject space that would result from an analysis by one of these three-way models. The answer is that there is not, so long as the stimulus space is two or more dimensional. If, as suggested, we conceive of different 'points of view' as corresponding to different *clusters* of subjects (for example, clusters based on patterns of subject weights in INDSCAL), then it is clear that there may be an indefinitely large number of clusters even though the dimensionality is quite small. We saw an indication of this in the Helm and Tucker analysis of the Helm color data, where ten 'points of view' were found, although an INDSCAL analysis accounted for the data quite well in two dimensions. (This is not a completely accurate example of what is intended, however, since some of the 'points of view' may have had to do with nonlinearities in the 'distance function.')

Suppose that P 'points of view' are found of dimensionality $r_1, r_2, \ldots, r_p$, respectively. What is the dimensionality of the stimulus space required in the case of (any one of) the models considered here? The best general answer we can give (and this holds for all the models) is that the dimensionality must be bounded by $\max_p r_p$ and $\sum_{p=1}^{P} r_p$. The dimensionality would correspond to the smaller value if each smaller dimensional POV space were 'contained in' the POV space of highest dimensionality—where the meaning of 'contained in' depends on the particular model. In the case of IDIOSCAL, the most general model, this 'contained in' relation would simply mean that the smaller (in dimensionality) is a *subspace* of the larger. In terms of matrices, this means that one coordinate matrix can be derived from the other by an affine transformation—singular if the dimensionalities differ, and nonsingular if they are the same. In the case of the INDSCAL model, 'contained in' would have a more complex meaning; namely, that separate orthogonal transformations of the different POV spaces exist such that the spaces of smaller dimensionality are related (after the transformation) to the one of highest dimensionality by a *diagonal* (or strain) transformation.

While it would be fairly unlikely in practice for the lower-bound dimensionality to hold (for a particular model), we feel it is generally much less likely for the *upper* bound to hold. This would usually be true only if the separate POV spaces had nothing at all in common. In fact, we would argue that the actual dimensionality required for one of these 'generalized Euclidean' models would be considerably closer to the lower than to the upper bound (meaning that some one of these models is likely to be more 'parsimonious' than a POV analysis).

In the case of the IDIOSCAL model (the most general we have considered), we can, in fact, state a quite precise characterization of the required dimensionality. The required dimensionality (of the stimulus space for IDIOSCAL) is simply the dimensionality of the direct sum of the different POV spaces. Unless the individual spaces have nothing at all in common, this dimensionality is likely to be considerably less than the sum of the dimensionalities. (Of course, in an actual analysis involving *real* and *errorful* data, the dimensionality of the sum may indeed equal the sum of the dimensionalities—but presumably there would be many 'small' dimensions, accounting for very little variance, that could be attributed to 'noise' or error.)

HIGHER WAY GENERALIZATIONS

Higher way generalizations have been developed for the INDSCAL model and method (Carroll & Chang, 1970a; Carroll, 1972b) and for three-mode scaling (Tucker, 1972; Carroll, 1972b). These higher way generalizations would be appropriate, for example, if there were two or more sets (matrices) of proximities data for each subject; each matrix for a subject could be based on a different experimental task, condition, occasion, data type, or any characteristic defining an additional 'way' or 'mode' of the multiway data matrix. For example, Wish did a four-way analysis of data from a study on nation perceptions in which subjects made several kinds of proximities judgments with respect to the nations. (It is not necessary that subjects be one of the modes; for example, there could be a single subject who provided proximities data under several conditions on each of a number of occasions.) Stimuli generally constitute two ways of the higher way matrix since there is a stimulus-by-stimulus matrix for each subject (or subject by other way combination). To use a distinction proposed by Tucker, stimuli constitute one *mode*, but two *ways* of the three-way or multiway matrix.

Higher Way Generalization of INDSCAL

The four-way generalization of INDSCAL is expressed in its scalar product form in Equation 57.

$$b_{jk}^{(i\ell)} = w_{it} v_{\ell t} x_{jt} x_{kt}, \tag{57}$$

where $b_{jk}^{(i\ell)}$ is the scalar product between stimuli j and k for the ith 'subject' in the ℓth condition (or other 'way'); and $v_{\ell t}$ is the weight associated with the ℓth condition (or other 'way').

The N-way generalization of INDSCAL is stated in Equation 58.

$$b_{jk}^{(i_1,i_2,\ldots,i_{N-2})} = \sum_t w_{i_1t}^{(1)} w_{i_2t}^{(2)}, \ldots, w_{i_{(N-2)}t}^{(N-2)} x_{jt} x_{kt}, \tag{58}$$

where superscripts on the ws index the different ways.

Carroll and Chang (1970a) have indicated a method for analyzing an N-way table of proximities data in terms of this N-way generalization of INDSCAL. This involves the use of a corresponding N-way generalization of CANDECOMP. The current version of INDSCAL (Chang & Carroll, 1972) is in fact able to deal with such N-way analyses for $N \leq 7$.

Higher Way Generalizations of Three-mode Scaling

The four-way generalization of three-mode scaling is given by Equation 59.

$$b_{jk}^{(i\ell)} = \sum_s \sum_u \sum_t \sum_{t'} a_{is} a_{\ell u}^{*} g_{sutt'} x_{jt} x_{kt'}, \tag{59}$$

with $g_{sutt'} = g_{sut't}$. The four-way analogue of Equation 51 would then be

$$C_{i\ell} = \sum_{s=1}^{S} \sum_{u=1}^{U} a_{is} a_{\ell u}^{*} G_{su}, \tag{60}$$

where $C_{i\ell}$ and G_{su} are symmetric $r \times r$ matrices.

The N-way generalization of three-mode scaling is given in Equation 61.

$$b_{jk}^{(i_1,i_2,\ldots,i_{N-2})} = \sum_{t_1} \sum_{t_2}, \ldots, \sum_{t_{N-1}} \sum_{t_N} a_{i_1t_1}^{(1)}, a_{i_2t_2}^{(2)}, \ldots, a_{i_{N-2}t_{N-2}}^{(N-2)} g_{t_1t_2,\ldots,t_{N-1}t_N} x_{jt_{N-1}} x_{kt_N}, \tag{61}$$

with $g_{t_1t_2,\ldots,t_{N-1}t_N} = g_{t_1t_2,\ldots,t_Nt_{N-1}}$.

Methods for analysis in terms of this model (and the more general N-mode factor-analysis model) have been discussed by Tucker (1964, 1972). No nontrivial higher way generalizations of IDIOSCAL or of PARAFAC-2 have been proposed.

It is not clear when or whether these higher way models are appropriate. In their current forms they impose strong, and usually unrealistic constraints on the data. (Of course, higher way models could be developed that might allow for other more realistic constraints.) In most cases it is more appropriate to simply concatenate the $N - 2$ nonstimulus modes into a single mode, and then to do a three-way analysis. (For example, if there are n_1 subjects and n_2 conditions there would be $n_1 \times n_2$ 'pseudosubjects' defining the third way of the analysis; there would be one two-way matrix for each of those 'pseudosubjects.' Another alternative would be to do separate three-way analyses for each of the n_2 conditions.) Then various other analyses (including some of the three-way or higher way analyses discussed here) could be applied to determine if there were any more basic structure in the weights for the 'pseudosubjects.'

SOME OTHER POSSIBLE THREE-WAY GENERALIZATIONS OF MDS

INDSCAL, IDIOSCAL, and the two intermediate models generalize two-way MDS in terms of what might be called affinely generalized Euclidean metrics (i.e., generalized Euclidean metrics that can be characterized in terms of certain *affine*—or general linear—transformations of a basic coordinate space). We regard the *Euclidean* aspect as largely a pragmatic choice based on considerations of mathematical tractability as well as on the high degree of robustness of the Euclidean metric (not to mention considerable evidence that it is possibly very nearly the 'right' metric, at least for such 'unitary' or unanalyzable stimuli as colors or tones). Obviously all of the models considered here could be generalized by introducing various non-Euclidean (Minkowskian, Riemannian, etc.) metrics, but we shall not pursue this (see Carroll & Wish, 1973, however, for a discussion of this and other generalizations).

Another direction of generalization that seems of more interest would entail *nonlinear* transformations of a common stimulus space. Two general classes of nonlinear transformations that would seem promising are the following.

(i) Separate monotonic transformations of (projections onto) each coordinate axis. This would be of the form

$$y_{jt}^{(i)} = f_{it}(x_{jt}) \tag{62}$$

with f_{it} monotone. INDSCAL is the special case of this model in which $f_{it}(x) = a_{it}x + b_{it}$, with $a_{it} = w_{it}^{1/2}$; that is, in which the fs are all *linear*. In the more general model of Equation 62, the fs could be interpreted as different 'psychophysical transformations' for different individual-dimension combinations.

(ii) Transformations in which the space is distorted as an increasing function of distance from some point in the space (a different point allowed for each subject). One specific model of this kind would involve transformations of the form

$$\mathbf{y}_j^{(i)} = \left(\frac{\mathbf{x}_j - \mathbf{p}_i}{|\mathbf{x}_j - \mathbf{p}_i|}\right) g_i(|\mathbf{x}_j - \mathbf{p}_i|), \tag{63}$$

where $\mathbf{x}_j$ is the vector for the jth stimulus in the common stimulus space, $\mathbf{y}_j^{(i)}$ is the corresponding vector in the 'private space' for individual i, and $\mathbf{p}_i$

is the 'vantage point' of individual i in the common space. (| | means length of the enclosed vector.) (See Coombs, 1964, for a discussion concerning the analogous effect of remoteness of stimuli from an 'ideal point.')

A geographical illustration of this model is provided by the case of the New Yorker who thinks that Los Angeles and San Francisco are very close together (because they are both far away) and that New York and Boston (say) are much farther apart (presumably because both New York and Boston are close to his 'vantage point.') Roger Shepard (personal communication) produced an amusing example of this with real data a few years ago in which he constructed a "Bostonian's map of the United States" based on a kind of proximity measure. W. J. M. Levelt (personal communication) also has some data that seem to be of this general kind based on Dutch students' judgments of distances among cities in The Netherlands. The model defined by Equation 63 is, of course, only one particular realization of the model verbally described above.

Either of these models could be specialized by, for example, placing various restrictions on the class of allowable functions f_{it} or g_i. On the other hand the two could be combined (although this might be pointless for the particular version of model (ii) in Equation 63 since it is already a special case of the general model (i)). Clearly, too, any one of these generalized models could be combined with various non-Euclidean metrics, as well as with different assumptions about the strength of relation of the proximities data to distances.

It does not take a great deal of imagination to generate many other three-way (or higher way) generalizations of two-way MDS. The problem is to devise models that are, first of all, psychologically (or otherwise) meaningful and, second, that are sufficiently mathematically tractable to allow efficient and effective methods of analysis to be developed. We hope some readers of this chapter may be inspired to pursue this quest.

REFERENCES

Abelson, R. P., & Sermat, V. Multidimensional scaling of facial expressions. *Journal of Experimental Psychology*, 1962, **63**, 546–554.

Behrman, B. W., & Brown, D. R. Multidimensional scaling of form: A psychophysical analysis. *Perception and Psychophysics*, 1968, **4**, 19–25.

Bennett, R. S. Intrinsic dimensionality of signal collections. *IEEE Transactions on Information Theory*, 1969, **15**, 517–525.

Carroll, J. D. Individual differences and multidimensional scaling. In R. N. Shepard, A. K. Romney, and S. Nerlove (Eds.), *Multidimensional scaling: Theory and applications in the behavioral sciences*. Vol. 1. *Theory*. New York: Academic Press, 1972. (a)

Carroll, J. D. Some additional procedures for three-way or higher-way scaling. Paper presented at the Workshop on Multidimensional Scaling, Philadelphia, June 1972. (b)

Carroll, J. D., & Chang, J. J. Analysis of individual differences in multidimensional scaling via an *N*-way generalization of Eckart-Young decomposition. *Psychometrika*, 1970, **35,** 283–319. (a)

Carroll, J. D., & Chang, J. J. A "quasi-nonmetric" version of INDSCAL, a procedure for individual differences multidimensional scaling. Paper presented at meetings of the Psychometric Society, Stanford, March 1970. (b)

Carroll, J. D., & Chang, J. J. A reanalysis of some color data of Helm's by the INDSCAL procedure for individual differences multidimensional scaling. *Proceedings, 78th annual convention, American Psychological Association*, 1970, **5,** 137–138. (c)

Carroll, J. D., & Chang, J. J. IDIOSCAL (*I*ndividual *D*ifferences *I*n *O*rientation *SCAL*ing): A generalization of INDSCAL allowing *IDIO*syncratic reference systems as well as an analytic approximation to INDSCAL. Paper presented at meetings of the Psychometric Society, Princeton, N.J., March 1972.

Carroll, J. D., & Wish, M. Multidimensional perceptual models and measurement methods. In E. C. Carterette and M. P. Friedman (Eds.), *Handbook of perception.* Vol. 2. New York: Academic Press, 1974, in press.

Chang, J. J. Notes on NINDSCAL. Unpublished manuscript, Bell Laboratories, 1972.

Chang, J. J., & Carroll, J. D. How to use INDSCAL, a computer program for canonical decomposition of *N*-way tables and individual differences in multidimensional scaling. Unpublished manuscript, Bell Laboratories, 1972.

Chang, J. J., & Shepard, R. N. Exponential fitting in the proximity analysis of confusion matrices. Paper presented at meetings of the Eastern Psychological Association, New York, 1966.

Coombs, C. H. *Theory of data.* New York: Wiley, 1964.

Dobson, R., & Young, F. W. On the perception of a class of bilaterally symmetric forms. *Perception and Psychophysics*, 1973, **13**, 431–438.

Eckart, C., & Young, G. Approximation of one matrix by another of lower rank. *Psychometrika*, 1936, **1,** 211–218.

Ekman, G. Dimensions of color vision. *Journal of Psychology*, 1954, **38,** 467–474.

Ekman, G. Direct method for multidimensional ratio scaling. *Psychometrika*, 1963, **28,** 33–41.

Fillenbaum, S., & Rapoport, A. *Structure in the subjective lexicon.* New York: Academic Press, 1971.

Green, P. E., & Carmone, F. J. *Multidimensional scaling and related techniques in marketing analysis.* Boston: Allyn & Bacon, 1970.

Green, P. E., & Rao, V. R. *Multidimensional scaling—An in-depth comparison of approaches and algorithms.* New York: Holt, Rinehart & Winston, 1971.

Gregson, R. A. M. Theoretical and empirical multidimensional scalings of taste mixture matchings. *British Journal of Mathematical and Statistical Psychology*, 1965, **18,** 59–75.

Guttman, L. General nonmetric technique for finding the smallest coordinate space for a configuration of points. *Psychometrika*, 1968, **33,** 469–506.

Harman, H. H. *Modern factor analysis.* Chicago: University of Chicago Press, 1967.

Harshman, R. A. Determination and proof of minimum uniqueness conditions for PARAFAC 1. U.C.L.A., Working Papers in Phonetics 22, March 1972. (a)

Harshman, R. A. PARAFAC2: Mathematical and technical notes. U.C.L.A., Working Papers in Phonetics 22, March 1972. (b)

Helm, C. E. Multidimensional ratio scaling analysis of perceived color relations. *Journal of the Optical Society of America*, 1964, **54,** 256–262.

Helm, C. E., & Tucker, L. R. Individual differences in the structure of color perception. *American Journal of Psychology*, 1962, **75,** 437–444.

Hotelling, H. Analysis of a complex of statistical variables into principal components. *Journal of Educational Psychology*, 1933, **24,** 499–520.

Indow, T. Applications of multidimensional scaling in perception. In E. C. Carterette and M. P. Friedman (Eds.), *Handbook of perception*. New York: Academic Press, 1974, in press.

Jones, L. E., & Young, F. W. Structure of a social environment: Longitudinal individual differences scaling of an intact group. *Journal of Personality and Social Psychology*, 1972, **24,** 108–121.

Kruskal, J. B. Multidimensional scaling by optimizing goodness of fit to a nonmetric hypothesis. *Psychometrika*, 1964, **29,** 1–27. (a)

Kruskal, J. B. Nonmetric multidimensional scaling: A numerical method. *Psychometrika*, 1964, **29,** 115–129. (b)

Kruskal, J. B. A computer program for monotone regression. Unpublished manuscript, Bell Laboratories, 1964. (c)

Kruskal, J. B., & Carmone, F. How to use M-D-SCAL (version 5M) and other useful information. Unpublished manuscript, Bell Laboratories, 1969.

Kruskal, J. B., & Hart, R. E. Geometric interpretation of diagnostic data from a digital machine: Based on a study of the Morris, Illinois Electronic Central Office. *Bell System Technical Journal*, 1966, **45,** 1299–1338.

Kruskal, J. B., Young, F. W., & Seery, J. B. How to use KYST, a very flexible program to do multidimensional scaling and unfolding. Unpublished manuscript, Bell Laboratories, 1973.

Künnapas, T. Visual memory of capital letters: Multidimensional ratio scaling and similarity. *Perceptual and Motor Skills*, 1967, **25,** 345–350.

Levelt, W. J. M., van de Geer, J. P., & Plomp, R. Triadic comparisons of musical intervals. *British Journal of Mathematical and Statistical Psychology*, 1966, **19,** 163–179.

McGee, V. E. Multidimensional analysis of elastic distances. *British Journal of Mathematical and Statistical Psychology*, 1966, **19,** 181–196.

Messick, S. J. An empirical evaluation of multidimensional successive intervals. *Psychometrika*, 1956, **21,** 367–376.

Messick, S. J. Perceived structure of political relationships. *Sociometry*, 1961, **24,** 270–278.

Miller, G. A., & Nicely, P. E. An analysis of perceptual confusions among some English consonants. *Journal of the Acoustical Society of America*, 1955, **27,** 338–352.

Rosenberg, S., Nelson, C., & Vivekananthan, P. S. Multidimensional approach to structure of personality impressions. *Journal of Personality and Social Psychology*, 1968, **9,** 283–294.

Roskam, E. E. *Metric analysis of ordinal data in psychology*. Voorschoten, Holland: University of Leiden Press, 1968.

Shepard, R. N. Analysis of proximities: Multidimensional scaling with an unknown distance function. I. *Psychometrika*, 1962, **27,** 125–140. (a)

Shepard, R. N. Analysis of proximities: Multidimensional scaling with an unknown distance function. II. *Psychometrika*, 1962, **27,** 219–246. (b)

Shepard, R. N. Analysis of proximities as a technique for the study of information processing in man. *Human Factors*, 1963, **5,** 19–34.

Shepard, R. N. Taxonomy of some principal types of data and of multidimensional methods for their analysis. *Multidimensional scaling: Theory and applications in behavioral sciences*. Vol. 1. *Theory*. New York: Seminar Press, 1972. (a)

Shepard, R. N. Psychological representation of speech sounds. In E. E. David and P. B. Denes (Eds.), *Human communication: A unified view*. New York: McGraw-Hill, 1972. (b)

Shepard, R. N., & Chipman, S. Second-order isomorphism of internal representation: Shapes of states. *Cognitive Psychology*, 1970, **1,** 1–17.

Sherman, R. C., & Ross, L. B. Liberalism–conservatism and dimensional salience in the perception of political figures. *Journal of Personality and Social Psychology*, 1972, **23,** 120–127.

Torgerson, W. S. *Theory and methods of scaling*. New York: Wiley, 1958.

Torgerson, W. S. Multidimensional scaling of similarity. *Psychometrika*, 1965, **30,** 379–393.

Tucker, L. R. The extension of factor analysis to three-dimensional matrices. In N. Fredriksen and H. Gulliksen (Eds.), *Contributions to mathematical psychology*. New York: Holt, Rinehart & Winston, 1964.

Tucker, L. R. Relations between multidimensional scaling and three-mode factor analysis. *Psychometrika*, 1972, **37,** 3–27.

Tucker, L. R., & Messick, S. J. An individual differences model for multidimensional scaling. *Psychometrika*, 1963, **28,** 333–367.

Wish, M. Multidimensional scaling of similarities among nations. Unpublished manuscript, Detroit Area Study Report, Ann Arbor, Michigan, 1964.

Wish, M. A model for the perception of Morse code-like signals. *Human Factors*, 1967, **9,** 529–540.

Wish, M. An INDSCAL analysis of the Miller & Nicely consonant confusion data. Paper presented at meetings of the Acoustical Society of America, Houston, November 1970.

Wish, M. Notes on the variety, appropriateness, and choice of proximity measures. Paper presented at the Workshop on Multidimensional Scaling, Philadelphia, June 1972.

Wish, M. Individual differences in perceptions of dyadic relationships. Paper presented at the American Psychological Association Convention, Montreal, Quebec, Canada, August 1973.

Wish, M., & Carroll, J. D. Applications of INDSCAL to studies of human perception and judgment. In E. C. Carterette and M. P. Friedman (Eds.), *Handbook of perception*. Vol. 2. New York: Academic Press, 1974, in press.

Wish, M., Deutsch, M., & Biener, L. Differences in conceptual structures of nations: An exploratory study. *Journal of Personality and Social Psychology*, 1970, **16,** 361–373.

Wish, M., Kaplan, S. J., & Deutsch, M. Dimensions of interpersonal relations:

Preliminary results. *Proceedings*, 81*st annual convention*, *American Psychological Association*, 1973, **8,** 179–180.

Wold, H. Estimation of principal components and related models by iterative least squares. In P. R. Krishnaiah (Ed.), *Multivariate analysis*. New York: Academic Press, 1966.

Yoshida, M. Dimensions of tactual impressions. Parts A and B. *Japanese Psychological Research*, 1968, **10,** 123–127, 157–173.

Young, F. W., & Torgerson, W. S. TORSCA: A Fortran-4 program for Shepard-Kruskal multidimensional scaling analysis. *Behavioral Science*, 1967, **12,** 498.

Individualized Testing and Item Characteristic Curve Theory

Frederic M. Lord
EDUCATIONAL TESTING SERVICE
PRINCETON, NEW JERSEY

1. INTRODUCTION

In conventional mental testing situations, a group of individuals take the same test. Inevitably, an aptitude or achievement test is too easy for some individuals and too hard for others. Some may obtain perfect or near perfect scores on the test; others may score near zero. If successful random guessing is possible, low scores will be at and below the chance level.

If a test is too easy for some individuals, it will not discriminate effectively among them. A helpful analogy for this situation is a high-jump contest: one would not try to rank the best jumpers by always setting the bar at a level appropriate for mediocre jumpers. Similarly, if a test is too hard for some individuals, it will not discriminate effectively among them either. One would not try to rank poor jumpers by setting the bar at a level where none of them clear it.

If successful random guessing is possible (as it is on almost all objective tests), it is obvious that the test cannot effectively measure an individual who gives random answers to almost all the test questions. The 'noise' on his answer sheet overwhelms the 'signal.' This discussion suggests that for each individual there is an optimal difficulty level at which test questions are most effective for evaluating his performance or 'ability.'

Let us limit further consideration to the common case where all responses to test questions are (treated as) either 'right' or 'wrong,' the individual score being his number of right answers. If there is no guessing, a common rule for effectively measuring performance (this is also the rule to which theory will lead us) calls for a difficulty level such that the individual will answer half the questions correctly and half incorrectly. If questions can be answered correctly by blind guessing, then the optimal difficulty level will be somewhat easier than this.

Clearly it would be desirable to test each individual with questions best suited to his ability level. This is likely to be impractical in ordinary paper-and-pencil testing situations (but, see Lord, 1971a, 1971b, 1971c). Now that many educational institutions have high-speed computers, however, it is becoming practical to have the computer 'tailor' the test for each individual tested, administering only test questions that seem appropriate for his level of ability, as judged from his responses to the questions previously administered.

In order to tailor the test to an individual, the computer must be able (a) to predict from the individual's previous responses how he would respond to various questions not yet administered (these may be more, or less, difficult than any of the questions already administered); (b) to make effective use of this knowledge in picking the question to be administered next; (c) to assign at the end of the testing a numerical score (or interval estimate) somehow representing the 'ability' or overall level of performance of the individual tested.

2. TEST THEORY FOR ITEMIZED TESTS

Classical test theory does not provide an appropriate framework for dealing with any of these three tasks that the computer (or the tailored-test designer) must carry out. Classical test theory is of great practical value in the design, construction, pretesting, scoring, statistical analysis, and interpretation of conventional tests of all kinds. An effective theory for similar purposes is urgently needed for individualized testing. Without careful design and appropriate scoring, individualized testing will often be inferior to conventional testing.

If we are to think meaningfully about 'good' testing procedures and 'inferior' procedures, we first need to be clear about the purpose of testing. The immediate purpose is *not* simply to determine the individual's actual performance on the particular test questions administered. This statement becomes obvious in individualized testing, since here each individual is responding to a different set of test questions, so that no comparisons among individuals are possible in terms of actual performance. Rather, the purpose is to make some inference as to his typical or expected performance on a large class of questions like those administered. In order to have a convenient label, this typical or expected performance will be called the *ability* of the individual in the area represented by the class of test questions.

If the questions in a class are too heterogeneous, 'ability' as just defined has little psychological meaning. Science and understanding will best be served if we choose to work (at least initially) with classes of questions sufficiently homogeneous so that we are happy to describe performance on any one class by a single number, rather than by several. This grouping of questions into homogeneous classes will be assumed in all that follows (however, see Mulaik, 1972, for a model that avoids this assumption). The reader may think of certain spelling tests, vocabulary tests, or tests of spatial abilities, among others, as providing good practical examples of homogeneous grouping of questions.

Note: There is no suggestion that an ability as defined here is in any sense a genetic, anatomical, neurological, or even psychological entity. For example, an 'ability' useful in one set of circumstances as a dimension for describing individuals might in other circumstances be shown to be a composite of several abilities.

Our main problem is to infer the individual's ability (in the area represented by the test) from his performance on certain test questions. In order to do this, it is indispensable to have some idea of how the individual's responses depend upon his ability.

3. THE GUTTMAN SCALE

A simple and appealing model has often been used in the attempt to describe the dependence of examinee response on examinee ability. The test questions are visualized as hurdles, the height of the hurdle being directly related to the difficulty of the question. The ability of the examinee completely determines which hurdles he can clear and which he cannot. In this deterministic model, all questions below a certain difficulty level are answered correctly by a given examinee; all questions above this level are answered incorrectly. A scale of test questions displaying this property for all examinees is called a Guttman scale (see Torgerson, 1958, Ch. 12).

This model is arrived at by asking what we would like a test to do. It would be nice if we could know from the examinee's test score (number of right answers) exactly how he responded to every question in the test. This knowledge can be obtained from a Guttman scale but not from any other kind of test.

Although approximate Guttman scales are of use in sociological work and in attitude measurement, they seem to be of little interest in aptitude and achievement testing. For one thing, in many common situations an ideal aptitude or achievement test should have all items of equal difficulty. According to the deterministic hurdle model, all examinees should obtain either a zero score or a perfect score on such a test. Nothing like this happens in practice, however. The distribution of number-right scores is typically bell shaped, even when we try by every means to obtain a U-shaped distribution.

The Guttman scale assumes that the tetrachoric correlation between scores on any two test questions is 1.00. For two questions of medium difficulty, this would mean a product-moment correlation of approximately 1.00 also. Actually, the tetrachoric correlation between typical aptitude or achievement test questions is not 1.00 but less than 0.35. The product-moment correlation between questions of medium difficulty is less than 0.25.

4. ITEM CHARACTERISTIC CURVE THEORY

If we want a mathematical model capable of fitting typical aptitude or achievement test data, we must use a probabilistic rather than a deterministic model. Denote the probability that individual a will answer a test question correctly by $P_{\boldsymbol{\beta}}(\theta_a) \equiv \text{Prob}\,(U_a = 1 \mid \theta_a, \boldsymbol{\beta})$. Here U_a is a random variable that assumes the value 1 when individual a answers correctly, 0 otherwise. The real number θ_a represents the ability of individual a. The vector $\boldsymbol{\beta}$ contains parameters fully characterizing the test question ('item') administered. The difficulty of the item, for example, is represented by one of the parameters in $\boldsymbol{\beta}$. All this notation serves only to assert that the probability that individual a will answer a question correctly depends only upon the ability of the individual and upon certain characteristics of the test question.

The probability $P_{\boldsymbol{\beta}_i}(\theta_a)$ is interpreted (see next section) as a relative frequency over randomly selected test questions all having the same characteristics $\boldsymbol{\beta} = \boldsymbol{\beta}_i$. There is no consideration of repeated testing—each individual is tested only once.

It is natural to assume that $P_{\boldsymbol{\beta}}(\theta)$ is an increasing function of θ. The higher the ability level, the greater the probability of a correct answer. This is assumed hereafter. Some typical functions $P_{\boldsymbol{\beta}}(\theta)$ (see Lord, 1968) are shown in Figure 1 for illustrative purposes.

We assume that the probability of a correct answer to a question depends *only* on the individual's ability level and on $\boldsymbol{\beta}$, not on any other known characteristic of his, nor on any other characteristics of the question, nor on any other variable available to us. It follows that the probability of an individual's answering correctly is not altered by knowledge of the actual performance of other individuals. Thus, for example, the probability of correct answers to a question by both individuals a and a' is given by the product $P_{\boldsymbol{\beta}}(\theta_a)P_{\boldsymbol{\beta}}(\theta_{a'})$.

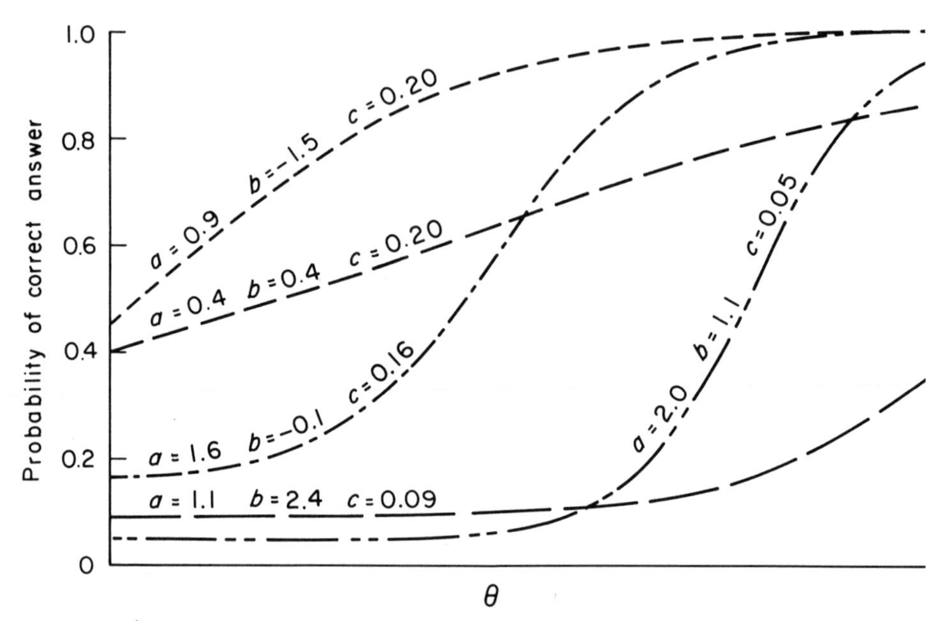

FIGURE 1.
Probability of correct answer as a function of ability, as estimated for five SAT verbal questions.

In addition, it follows that the probability of an individual's answering a question correctly is not altered by knowledge of his actual performance on other questions. Thus, for example, the probability that individual a will answer questions i, i', and i'' all correctly is given by the product $P_\beta(\theta_a)P_{\beta'}(\theta_a)P_{\beta''}(\theta_a)$. This is called the principle of *local independence* (Lazarsfeld, 1959).

It is instructive to see what would happen if local independence did not hold. Suppose that for a certain individual a the probability that he will answer randomly chosen questions i, i', and i'' all correctly is greater than $P_\beta(\theta_a)P_{\beta'}(\theta_a)P_{\beta''}(\theta_a)$. If this is not a unique occurrence, this would mean that there are individuals at ability level $\theta = \theta_a$ who score systematically higher on these test questions than other individuals with the same θ level. Thus these test questions would be measuring some psychological dimension other than θ. This is just the situation that the assumption of local independence is designed to exclude. We want to deal with a test that measures the ability θ; we do not want to deal (at least not at first) with a test score that may represent either of two (or more) psychological dimensions at once.

5. AN ALTERNATIVE MODEL

It seems necessary at this point to mention another model very commonly confused with the $P_\beta(\theta_a)$ model used here. This other model makes assertions about the probability, denoted by $P_i(\theta_a)$, that a specific individual a answers a specific item i correctly. The two models are distinguished, and reasons are given for discarding one of them.

If an individual responds to question i at random, it is clear that his probability of success is the reciprocal of the number of possible responses to question i. There are many questions, however, for which this individual knows the correct answer; for such a question, his probability of answering correctly would seem to be virtually 1. There may be other questions on which this individual is misinformed; for such a question, his probability of answering correctly would seem to be virtually 0.

Consider two individuals, a and b, and two test questions, i and j. Individual a happens to know the answer to question i and to be misinformed on question j. Individual b happens to know the answer to question j and to be misinformed on question i. If we write $P_i(\theta_a)$ for the probability that individual a answers question i correctly, we have $P_i(\theta_a) = 1$, $P_j(\theta_a) = 0$, $P_i(\theta_b) = 0$, $P_j(\theta_b) = 1$, approximately. The first two equations considered together imply that question i is easier than question j, the last two equations imply just the reverse. Thus questions i and j must measure a different ability for individual a than they measure for individual b. This is a possible model (Meredith, 1965) and a possible interpretation, but usually not a fruitful one, since usually we want to compare individuals a and b along the same ability dimension.

In order to avoid the situation just outlined, we use only the model defined in the previous section, which makes no assertions about the probability $P_i(\theta_a)$ that a *specified* individual a answers a *specified* test question i correctly. The model that we use deals instead with $P_\beta(\theta_a)$, which represents the long-run relative frequency of correct answers given by individual a when answering test questions all having the same specified $\boldsymbol{\beta}$. An equivalent statement is that $P_\beta(\theta_a)$ represents the probability that individual a will answer correctly a question chosen at random from all questions having the same $\boldsymbol{\beta}$. When the model holds, the function $P_\beta(\theta)$ of θ is referred to as the *characteristic curve* for each item having parameters $\boldsymbol{\beta}$.

6. SPECIALIZATION, APPLICATION, AND EVALUATION

Empirical checks on the validity and practical utility of the item characteristic curve (icc) model have in large part been delayed for about 25 years because of the difficulty of estimating the characteristic curves of particular items. Recently a number of workers have successfully estimated many iccs and some evidence of the validity and usefulness of the model has been accumulated. The present section is intended to refer the reader to materials relevant for assessing the validity and usefulness of the model; no detailed discussion is possible here.

An approach that estimates iccs without restrictive assumptions about their mathematical form has been described by Lord (1970a). If it can be assumed simply that the iccs differ only by a linear transformation of θ (a common assumption), a computer program implementing Levine (1972) has been found very effective for estimating iccs (Levine, personal communication, 1972).

It has been common to assume that the iccs are normal ogives or logistic curves. If the iccs are logistic and if, for a given test, the curves all have the same slope parameter, the present model can be shown (Birnbaum, 1968, p. 402) to be the same as the well-known Rasch model, which has certain desirable measurement properties (Rasch, 1960, 1961, 1966a, 1966b; Wright, 1968; Wright & Panchapakesan, 1969). Methods for estimating the single item parameter needed in this model and studies evaluating the fit and effectiveness of this model have been reported by Rasch, by Wright and Panchapakesan, as well as by Lawley (1943, 1944); Andersen (1970, 1971a, 1971b, 1972a, 1972b); Anderson, Kearney, and Everett (1968); Choppin (1968); Fischer (1972); Fischer and Scheiblechner (1970); Hambleton (1969); Hambleton and Traub (1971); Panchapakesan (1969); Scheiblechner (1971a, 1971b); Tinsley and Dawis (1972); Urry (1970). Reports on the fit and effectiveness of the one-parameter model range from disapproval to enthusiasm.

If some test questions correlate higher with ability than others, as is commonly the case, a one-parameter model may be inadequate. Whenever correct answers can be obtained by random guessing, even a two-parameter model is likely to be inadequate. Modified normal ogive and logistic models with three parameters are available (Birnbaum, 1968, Ch. 17). The mathematical formulas are

$$P_\beta(\theta) = \gamma + (1 - \gamma) \int_{-\infty}^{\alpha(\theta-\beta)} \frac{1}{\sqrt{2\pi}} \exp\left(-\frac{1}{2}t^2\right) dt \tag{1}$$

for the modified normal ogive, and

$$P_\beta(\theta) = \gamma + \frac{1 - \gamma}{1 + \exp\left[-1.7\alpha(\theta - \beta)\right]} \tag{2}$$

for the modified logistic. The models of Equations 1 and 2 do not differ anywhere by more than 0.01. We need not debate here which model is more nearly correct. Neither, of course, is exactly correct.

The three parameters in $\boldsymbol{\beta} \equiv \{\alpha, \beta, \gamma\}$ may be thought of as

α discriminating power (a measure of the relation between item score and ability),

β difficulty,

γ probability of a correct answer for individuals at lowest ability levels.

A more detailed, practical discussion of these item parameters is given in Lord (1970b).

Lord and Novick (1968, Sec. 16.11) consider for what practical situations the normal ogive model is likely to be appropriate. Studies evaluating the fit of the model to actual test data include Lord (1952, 1970a, 1974) and Indow and Samejima (1962, 1966). Their findings support the model for the data studied. Many more evaluative studies are needed.

Methods for estimating the item parameters have been developed and tried out by Lord (1952, 1968, 1974); Indow and Samejima (1962, 1966); Birnbaum (1968); Bock (1970, 1972); Bock and Lieberman (1970); Kolakowski and Bock (1970); Kolakowski (1969, 1972); and Lees, Wingersky, and Lord (1972). Studies making theoretical or practical use of these models appear in two books by Solomon (1961, 1965). Included among other such studies are those by Brogden (1946), Tucker (1946), Cronbach and Warrington (1952), Lord (1953a, 1953b, 1955, 1970a, 1970b, 1971a, 1971b, 1971c, 1971d, 1971e), Cronbach and Merwin (1955), Anderson (1959), Merwin (1959), Cronbach and Azuma (1962), Paterson (1962), Birnbaum (1968, Chs. 17–20; 1969), Wood and Skurnik (1969), Shiba (1969a, 1969b), Shoemaker and Osburn (1970), Urry (1970), Nishisato and Torii (1971), Bay (1971), Hambleton and Traub (1971), Samejima (1972).

7. PRETESTING

In order to design a test for the specific purpose of measuring the ability of a particular individual, we must have available a large pool of test questions that have been extensively pretested, so that the parameters characterizing each question may be considered known. Note that the item characteristic function $P_{\boldsymbol{\beta}}(\theta)$ does not depend on the distribution of ability in any group of individuals. Consequently, the parameters $\boldsymbol{\beta}$ for a test question can be determined once and for all by pretesting in some convenient group. Of course, reliance on the robustness of the model over wide variations in a group should not be carried to extremes. In practice, the pretest group should resemble the collection of individuals who will later be given the individualized tests.

Corresponding to the invariance of the item parameters $\boldsymbol{\beta}$ over groups of individuals there is an invariance of the ability parameter θ over different tests (cf. Rasch, 1961, pp. 331–333). These invariances are fundamental to the success of item characteristic curve theory in comparison with older item analysis methods. In a leading older method, each item would be characterized by the proportion of correct answers received and by the correlation between item response and total test score. However, these item parameters of the older method would be different for different pretest groups; also, the correlation parameter would change if the test was lengthened or otherwise modified. This lack of invariance limits the usefulness of classical item analysis. The usual kinds of test score for an individual have a similar lack of invariance when the test administered is modified.

8. THE STATISTICAL ESTIMATION OF ABILITY

Once the item parameters $\boldsymbol{\beta}$ have been determined by pretesting, the problem of estimating the ability of an individual from his responses is a straightforward, statistical estimation problem. If his probability of success on question i is $P_{\beta_i}(\theta)$ and his probability of failure is $Q_{\beta_i}(\theta) \equiv 1 - P_{\beta_i}(\theta)$, then the likelihood function for his score ($u_i = 1$ or 0) on question i is simply

$$L_i(\theta) = \begin{cases} P_{\beta_i}(\theta) & \text{if } u_i = 1, \\ Q_{\beta_i}(\theta) & \text{if } u_i = 0. \end{cases}$$

This may be more conveniently written

$$L_i(\theta) = [P_{\beta_i}(\theta)]^{u_i}[Q_{\beta_i}(\theta)]^{1-u_i}.$$

Because of local independence, the likelihood function for the individual's responses to a test of n questions is simply the product of the likelihoods for the separate questions:

$$L(\theta) = \prod_{i=1}^{n} [P_{\beta_i}(\theta)]^{u_i}[Q_{\beta_i}(\theta)]^{1-u_i}.$$

Since the $\boldsymbol{\beta}_i$ are known from pretesting, it is not difficult for a computer, given a mathematical form for $P_{\beta}(\theta)$ such as Equation 1 or 2, to find the maximum-likelihood estimate $\hat{\theta}$ of the individual's ability. ($\hat{\theta}$ is the value of θ that maximizes the likelihood $L(\theta)$ of his observed responses $u_1, u_2, \ldots, u_n$.)

9. A SIMPLER PROCEDURE FOR ESTIMATING ABILITY

There still remains the problem of how to pick the n test questions to be administered to a given individual. One advantage of individualized testing is that testing can be continued until the individual's ability has been estimated with some predetermined degree of statistical accuracy. For the sake of simplicity, however, we consider only the case where n is fixed.

To make matters even more simple, let us select from the pool a large set of pretested questions that differ from each other only in difficulty (β). These questions have identical values of α and γ. If Equation 1 or 2 held with no random guessing, the optimal test for estimating θ with minimum squared error would consist entirely of questions for which $P_\beta(\theta_a) = \frac{1}{2}$, where θ_a is the ability of the individual to be tested. Since we do not know θ_a and cannot estimate it with any accuracy in advance of testing, all this would not give us a method for choosing the n test questions to be administered. Such methods are discussed in Section 10.

Let us assume now (as seems reasonable) that individual ability (θ) and item difficulty (β) are measured along the same dimension, in the sense that for any increment k an increase in ability from θ to $\theta + k$ could hypothetically be exactly offset by an equivalent increase in difficulty from β to $\equiv \beta + k$. In other words, $P_{\{\alpha,\beta,\gamma\}}(\theta)$ and $P_{\{\alpha,\beta+k,\gamma\}}(\theta + k)$ represent exactly the same function of θ. This assumption holds for the models of Equations 1 and 2 and for any other $P_\beta(\theta)$ in which θ and β appear only as their difference $\theta - \beta$. Under this assumption $P_\beta(\theta) \equiv F(\theta - \beta)$ where F is an unspecified monotonic function.

What we have assumed is simply that we have a large set of questions, selected from the pretested pool, whose iccs differ only by a translation along the θ axis. Let us define β^0 as the item difficulty level at which the individual has probability of success $F(0)$. Thus $\theta = \beta^0$. We can determine an individual's ability θ by determining his β^0.

It is possible in practice to find the proportion of correct answers actually given by individual a to test questions at any specified difficulty level. By trial and error, or by better methods discussed in the next section, we can in this way find approximately the difficulty level β_a^0 such that $P_{\{\alpha,\beta_a^0,\gamma\}}(\theta_a) = F(0)$. This difficulty level *is* (approximately) the ability level of individual a, since by definition $\theta_a = \beta_a^0$.

10. STOCHASTIC APPROXIMATION

Clearly what we need now is some method better than trial and error for finding β^0. Stated in this way, the problem is a standard problem in stochastic approximation (Wasan, 1969). Specifically, the stochastic approximation problem is to select a sequence of test questions so that we can conveniently and accurately estimate the individual's ability θ_a from the sequence $u_1, u_2, \ldots, u_n$ of responses. Since for simplicity we have selected from the pretest pool a set of test questions that differ statistically only on their difficulty parameter (for a treatment that avoids this, see Owen, 1970), the problem of selecting a sequence of test questions is simply the problem of selecting a sequence $\beta_1, \beta_2, \ldots, \beta_n$. The resulting sequence of questions constitutes an *individualized test* or *tailored test* designed for effective measurement of the particular individual tested.

The difficulty β_1 of the first question administered can be chosen in the same way that we would choose the average difficulty level of the questions in a conventional test—by subjective judgment or by using a Bayesian prior. If the individual answers the first question incorrectly, we guess that it is too hard for him and choose an easier question to administer next. If he answers the first question correctly, we guess that it is too easy for him and choose a harder question to administer next.

After administering the second question, we could use the statistical method outlined in Section 8 to obtain from his responses to the first two questions an estimate $\hat{\theta}_a^{(2)}$ of the individual's ability. The difficulty of the third question administered could be matched to the individual's estimated ability by choosing $\beta_3 = \hat{\theta}_a^{(2)}$. We could then choose $\beta_4, \beta_5, \ldots$ similarly. However, a procedure that progresses by steps that are individually optimal is not in this case likely to be an optimal procedure overall.

I shall not try here to devise an optimal procedure. Rather, I shall try to find a good, simple procedure that is not only easy to carry out but also easy to evaluate as a procedure for statistical inference.

Under the Robbins-Monro stochastic approximation procedure, the rule for choosing the difficulty of the $(v + 1)$st question is

$$\beta_{v+1} = \beta_v + d_v(u_v - F(0)), \tag{3}$$

where $d_1, d_2, \ldots$ is a suitable decreasing sequence of positive numbers chosen in advance (Robbins & Monro, 1951). If the *step size* d_v is small, the $(v + 1)$st question will be chosen to have nearly the same difficulty as the vth question; if d_v is large, there will be a more substantial change in difficulty. In the Robbins-Monro procedure, the d_v are chosen relatively large initially when little is known about the individual's ability level, allowing substantial readjustments in item difficulty levels. Later when the appropriate difficulty level has been approximated, the chosen d_v are small, eventually approaching zero. Typically $d_v = d_1/v$, $v = 1, 2, 3, \ldots$.

Robbins and Monro's proof shows that when Equation 3 is used with suitable d_v, the item difficulty β_{v+1} is a consistent estimator of the individual's ability θ, in the sense that β_{v+1} converges stochastically to θ as v becomes large.

Formulas leading in some cases to asymptotically optimal choices of the d_v are given by Hodges and Lehmann (1956).

11. THE STAIRCASE METHOD FOR SELECTING THE TEST QUESTIONS

Unfortunately, the Robbins-Monro procedure requires storing 2^n test questions in the computer before testing is begun, where n is the number (here assumed to be fixed in advance) of questions to be administered to the examinee. For most aptitude and achievement tests composed of dichotomously scored questions, $n \geq 25$.

An alternative procedure, keeping the total number of test questions within acceptable limits, is available: the up-and-down method or staircase method, used in testing explosives, in bioassay, in psychophysics, and elsewhere. In the up-and-down method, the rule for selecting questions is still given by Equation 3, but with d_v replaced by a constant *step size d*.

If $F(0) = \frac{1}{2}$, the up-and-down rule becomes

$$b_{v+1} = \begin{cases} b_v + d, & \text{if question } v \text{ is answered correctly;} \\ b_v - d, & \text{if question } v \text{ is answered incorrectly.} \end{cases}$$

This simple form of Equation 3 normally holds only if there is no guessing of correct answers.

For basic discussions of this method, see Dixon and Mood (1943) and Brownlee, Hodges, and Rosenblatt (1953). Some modifications are discussed by Tsutakawa (1963, 1967a, 1967b).

This method requires storing only $n(n+1)/2$ test questions in the computer in advance of testing. This number can be reduced further by taking a few obvious shortcuts.

12. SCORING THE ANSWERS

Consider the following three simple methods for scoring the student's responses to the test questions:[1]

(1) The 'final-difficulty score,' β_{n+1}, the difficulty of the $(n+1)$st question (not actually administered) as defined by Equation 3.

(2) The 'number-right score,' $\sum_{v=1}^{n} u_v$, or the 'proportion-right score,' $\frac{1}{n}\sum_{v=1}^{n} u_v$.

[1] Section 12 and part of Section 11 are a slight revision of material appearing in Lord (1971e).

The former is the score most commonly used in scoring conventional mental tests.

(3) The 'average-difficulty score,' $\bar{\beta} = \frac{1}{n} \sum_{v=2}^{n+1} \beta_v$. This score is simply the average of the difficulty parameters of the questions administered, omitting the first (since the first question is the same for all individuals tested) and including β_{n+1}.

[Before going ahead, the reader may wish to make a guess as to the relative merits of these three scoring methods for the up-and-down (fixed-step-size) procedure.]

When the step size shrinks appropriately as n increases, as in the Robbins-Monro procedure, β_{n+1} is a good estimator of ability. When the step size is fixed, as in the up-and-down method, β_{n+1} is no longer a consistent estimator for θ, nor does its sampling variance approach zero as n becomes large. It turns out that when step size is fixed, number-right score is perfectly correlated with β_{n+1}; so it, too, can be eliminated as an effective method of scoring.

Brownlee, Hodges, and Rosenblatt (1953) have shown that the average-difficulty score is asymptotically equivalent to the maximum-likelihood estimator for θ found by Dixon and Mood (1943) for the up-and-down method. Although no optimum small-sample properties have been proved for the average-difficulty score, it appears at present to be the preferable method of scoring tests administered by the up-and-down method.

It frequently happens that similar groups of students are tested year after year. In this case, an excellent prior distribution for the parameter θ is available, based on records of past performance. In such situations, the careful design of a tailored testing procedure would certainly be based on a Bayesian approach. The Bayesian approach will not be treated here since it is of greater mathematical complexity. The interested reader is referred to Owen (1970), to Freeman (1970), and to Wood (1971).

13. EVALUATION OF TESTING METHODS

The remaining problem for discussion is the evaluation of different stochastic approximation procedures and of different choices of parameters such as d.

Properties of the Robbins-Monro procedure for large n are discussed in the references given. Some properties for small n are treated by Wasan (1969, Ch. 2) and by Cochran and Davis (1965). An improved procedure for small n is suggested by Kesten (1958) and tried out empirically by Odell (1961).

The up-and-down rule for selecting test questions to be administered produces a Markov chain or, more specifically, a random walk for the values of

β_v. The transition probabilities $P_{\{\alpha,\beta_v,\gamma\}}(\theta)$ and $Q_{\{\alpha,\beta_v,\gamma\}}(\theta)$ are stationary. They depend on β_v, but they do not depend on v when β_v and θ are given.

Starting from this, it is not hard to write down a formula for the frequency distribution of $\beta_{(n+1)}$ under the up-and-down method; but $\beta_{(n+1)}$ is not a satisfactory scoring procedure for this method, as already noted. The frequency distribution of the average difficulty score $\bar{\beta}$ is not easily obtained for moderate n, but Brownlee, Hodges, and Rosenblatt have provided recursive formulas from which the mean and sampling variance of $\bar{\beta}$ can be readily calculated numerically by computer for given θ, α, β_1, γ, d, $F(0)$, and for any n likely to be of interest. Given the bias and sampling variance of $\bar{\beta}$ for given θ for each of various testing designs, it is not hard to decide which design is preferable for measuring at a specified ability level.

A variety of testing designs were investigated in this fashion by Lord (1970b, 1971d). Numerical studies of a variety of stochastic approximation methods applicable to individualized testing are reported by Cochran and Davis (1964); Davis (1971); Wetherill (1963); Wetherill and Levitt (1965); and Wetherill, Chen, and Vasudeva (1966). Other empirical studies of individualized testing include Bayroff and Seeley (1967); Ferguson (1971); Hansen and Schwarz (1968); Linn, Rock, and Cleary (1969, 1972); Paterson (1962); Seeley, Morton, and Anderson (1962); Urry (1970); Waters (1964); Waters and Bayroff (1971); and Wood (1969).

14. RELATION TO PSYCHOPHYSICAL METHODS

The up-and-down method is often used in bioassay. According to Guilford (1954), it originally was developed for the study of explosives. When used in psychophysical studies, it is known as the staircase method (Cornsweet, 1962).

The psychophysicist does not need to know the precise mathematical form of the psychometric function. To elucidate comparison with icc theory, let us assume that the psychometric function actually is given by Equation 1 or 2 with $\gamma = 0$.

Whereas the mental tester controls α and β (by using pretested items) while trying to estimate the value of θ, the psychophysicist (or bioassayist) controls θ while trying to estimate the value of β and, sometimes, the value of α. Note that θ and β play reversed roles for the mental tester and for the psychophysicist. For the latter, θ might represent the physical intensity of the various stimuli presented under experimental control. Then, β would be the 'threshold' at which the subject says "Yes, I detect the stimulus" $F(0)$ of the time; α would be the precision of the psychometric function. The psychophysicist chooses the stimulus level θ_1, administers this stimulus, and records the response $u_{i1} = 0$ or 1. He then chooses another stimulus level θ_2, administers this stimulus, records $u_{i2} = 0$ or 1, and continues in this way.

In mental testing, we are interested only in the relative values of θ for different examinees; θ is, at best, measured on an interval scale. The unit and zero point of this scale have little ready meaning for most other scientists concerned with mental measurement. The psychophysicist, on the contrary, usually estimates the absolute value of β on some standard scale having a unit and origin well known to physicists and other scientists.

Avoiding bias in his estimates is therefore of crucial importance for the psychophysicist. In mental testing, any linear transformation of θ is as valid as any other. Bias is usually of no importance to the mental tester so long as it affects all scores equally.

The fact that the psychophysicist has two unknown parameters, α and β, creates a further problem. It is not possible for him to choose the step size d optimally without knowing α. A poor choice of d leads either to excessive standard error or bias in the estimated threshold, or else to experiments that are unnecessarily lengthy.

Often the psychophysicist can obtain observations cheaply. It may be easy for him to obtain a thousand or ten thousand responses from a single subject. The mental tester cannot do this. The objective situation forces the mental tester to use reasonably efficient testing and estimation methods. For the psychophysicist, statistically efficient procedures may be unnecessary and distinctly uneconomical.

In addition to the staircase method, the psychophysicist sometimes uses block up-and-down methods (Stuckey, Hutton, & Campbell, 1966; Tsutakawa, 1963, 1967a, 1967b; Cochran & Davis, 1964) and unequal-step-size 'sequential' methods (Taylor & Creelman, 1967; Pollack, 1968). The time-honored constant-stimulus method corresponds in part to conventional (not individualized) mental testing; the scoring methods are different in the two applications, however.

The indicated correspondence between individualized testing and certain psychophysical experiments is clear and instructive whenever the mental tester can work with items all having the same α and γ accurately determined by pretesting. Such situations do not really exist at present, however. Not enough items are usually available to do practical work with a pool of items all having the same α and γ (proponents of Rasch's method may disagree).

Present work in icc theory and practice is concerned with estimating item and examinee parameters simultaneously. This is very different from the typical psychophysical problem. An outstanding current problem is how to carry out individualized testing using test items characterized by a variety of inaccurately estimated item parameters. A recent article by Dupač and Král

(1972) is relevant for individualized testing with fallibly estimated values of β_i.

ACKNOWLEDGMENT

Preparation of this chapter was supported in part by Grant GB-32781X from the National Science Foundation.

REFERENCES

Andersen, E. B. *Conditional inference for multiple choice questionnaires.* Report No. 8. Copenhagen: Copenhagen School of Economics and Business Administration, 1970.

Andersen, E. B. *A goodness of fit test for the Rasch model.* Report No. 9. Copenhagen: Copenhagen School of Economics and Business Administration, 1971. (a)

Andersen, E. B. *Conditional inference and models for measuring.* Copenhagen: Copenhagen School of Economics and Business Administration, 1971. (b)

Andersen, E. B. The numerical solution of a set of conditional estimation equations. *Journal of the Royal Statistical Society*, Series B, 1972, **34,** 42–54. (a)

Andersen, E. B. *A computer program for solving a set of conditional maximum likelihood equations arising in the Rasch model for questionnaires.* Research Memorandum 72-06. Princeton, N.J.: Educational Testing Service, 1972. (b)

Anderson, J., Kearney, G. E., & Everett, A. V. An evaluation of Rasch's structural model for test items. *British Journal of Mathematical and Statistical Psychology*, 1968, **21,** 231–238.

Anderson, T. W. Some scaling models and estimation procedures in the latent class model. In U. Grenander (Ed.), *Probability and statistics.* New York: Wiley, 1959.

Bay, K. S. An empirical investigation of the sampling distribution of the reliability coefficient estimates based on alpha and KR20 via computer simulation under various models and assumptions. Unpublished doctoral dissertation, University of Alberta, 1971.

Bayroff, A. G., & Seeley, L. C. *An exploratory study of branching tests.* Technical Research Note 188. Washington, D.C.: U.S. Army Behavioral Science Laboratory, 1967.

Birnbaum, A. Some latent trait models and their use in inferring an examinee's ability. In F. M. Lord and M. R. Novick (Eds.), *Statistical theories of mental test scores.* Reading, Mass.: Addison-Wesley, 1968.

Birnbaum, A. Statistical theory for logistic mental test models with a prior distribution of ability. *Journal of Mathematical Psychology*, 1969, **6,** 258–276.

Bock, R. D. Estimating multinomial response relations. In R. C. Bose (Ed.), *Contributions to statistics and probability essays in memory of Samarendra Nath Roy.* Chapel Hill, N.C.: University of North Carolina Press, 1970.

Bock, R. D. Estimating item parameters and latent ability when responses are scored in two or more nominal categories. *Psychometrika*, 1972, **37,** 29–51.

Bock, R. D., & Lieberman, M. Fitting a response model for *n* dichotomously scored items. *Psychometrika*, 1970, **35,** 179–197.

Brogden, H. E. Variation in test validity with variation in the distribution of item difficulties, number of items, and degree of their intercorrelation. *Psychometrika*, 1946, **11,** 197–214.

Brownlee, K. A., Hodges, J. L., Jr., & Rosenblatt, M. The up-and-down method with small samples. *Journal of the American Statistical Association*, 1953, **48,** 262–277.

Choppin, B. H. An item bank using sample-free calibration. *Nature*, 1968, **119,** 870–872. Reprinted in R. Wood and L. S. Skurnik (Eds.), *Item banking*. Slough: National Foundation for Educational Research in England and Wales, 1969.

Cochran, W. G., & Davis, M. Stochastic approximation to the median effective dose in bioassay. In J. Gurland (Ed.), *Stochastic models in medicine and biology*. Madison: University of Wisconsin Press, 1964.

Cochran, W. G., & Davis, M. The Robbins-Monro method for estimating the median lethal dose. *Journal of the Royal Statistical Society*, Series B, 1965, **27,** 28–44.

Cornsweet, T. N. The staircase method in psychophysics. *American Journal of Psychology*, 1962, **75,** 485–491.

Cronbach, L. J., & Azuma, H. Internal-consistency reliability formulas applied to randomly sampled single-factor tests: An empirical comparison. *Educational and Psychological Measurement*, 1962, **22,** 645–665.

Cronbach, L. J., & Merwin, J. C. A model for studying the validity of multiple-choice items. *Educational and Psychological Measurement*, 1955, **15,** 337–352.

Cronbach, L. J., & Warrington, W. G. Efficiency of multiple-choice tests as a function of spread of item difficulties. *Psychometrika*, 1952, **17,** 127–148.

Davis, M. Comparison of sequential bioassays in small samples. *Journal of the Royal Statistical Society*, Series B, 1971, **33,** 78–87.

Dixon, W. J., & Mood, A. M. A method for obtaining and analyzing sensitivity data. *Journal of the American Statistical Association*, 1943, **43,** 109–126.

Dupač, V., & Král, F. Robbins-Monro procedure with both variables subject to experimental error. *The Annals of Mathematical Statistics*, 1972, **43,** 1089–1095.

Ferguson, R. L. *Computer assistance for individualizing measurement*. Technical Report. Pittsburgh, Pa.: University of Pittsburgh, 1971.

Fischer, G. H. *Conditional maximum-likelihood estimation of item parameters for a linear logistic test-model.* Research Bulletin No. 9. Vienna: Psychologisches Institut der Universität Wien, 1972.

Fischer, G. H., & Scheiblechner, H. H. *Two simple methods for asymptotically unbiased estimation in Rasch's measurement model with two categories of answers.* Research Bulletin No. 1. Vienna: Psychologisches Institut der Universität Wien, 1970.

Freeman, P. R. Optimal Bayesian sequential estimation of the median effective dose. *Biometrika*, 1970, **57,** 79–89.

Guilford, J. P. *Psychometric methods.* (2nd ed.) New York: McGraw-Hill, 1954.

Hambleton, R. K. An empirical investigation of the Rasch test theory model. Unpublished doctoral dissertation, University of Toronto, 1969.

Hambleton, R. K., & Traub, R. E. Information curves and efficiency of three logistic test models. *British Journal of Mathematical and Statistical Psychology*, 1971, **24,** 273–281.

Hansen, D. N., & Schwarz, G. *An investigation of computer-based science testing.* Tallahassee: Institute of Human Learning, Florida State University, 1968.

Hodges, J. L., Jr., & Lehmann, E. L. Two approximations to the Robbins-Monro process. In J. Neyman (Ed.), *Proceedings of the third Berkeley symposium on mathematical statistics and probability.* Vol. 1. Berkeley: University of California Press, 1956.

Indow, T., & Samejima, F. LIS *measurement scale for non-verbal reasoning ability.* Tokyo: Nihon-Bunka Kagakusha, 1962. (In Japanese.)

Indow, T., & Samejima, F. *On the results obtained by the absolute scaling model and the Lord model in the field of intelligence.* Yokohama: Psychological Laboratory, Hiyoshi Campus, Keio University, 1966. (In English.)

Kesten, H. Accelerated stochastic approximation. *Annals of Mathematical Statistics*, 1958, **29,** 41–59.

Kolakowski, D. Maximum likelihood estimation of item parameters and latent ability by generalized probit analysis. Paper presented at the spring meeting of the Psychometric Society, Princeton, N.J., 1969.

Kolakowski, D. An investigation of bias in the estimation of latent ability and item parameters under the normal ogive model. Paper presented at the meeting of the Psychometric Society, Princeton, N.J., March 1972.

Kolakowski, D., & Bock, R. D. *A Fortran* IV *program for maximum likelihood item analysis and test scoring: Normal ogive model.* Educational Statistics Laboratory Research Memo No. 12. Chicago: University of Chicago, 1970.

Lawley, D. N. On problems connected with item selection and test construction. *Proceedings of the Royal Society of Edinburgh*, 1943, **61,** 273–287.

Lawley, D. N. The factorial analysis of multiple item tests. *Proceedings of the Royal Society of Edinburgh*, 1944, **62-A,** 74–82.

Lazarsfeld, P. F. Latent structure analysis. In S. Koch (Ed.), *Psychology: A study of a science.* Vol. 3. New York: McGraw-Hill, 1959.

Lees, D. M., Wingersky, M. S., & Lord, F. M. *A computer program for estimating item characteristic curve parameters using Birnbaum's three-parameter logistic model.* Office of Naval Research Technical Report, Contract No. N00014-69-C-0017. Princeton, N.J.: Educational Testing Service, 1972.

Levine, M. V. Transforming curves into curves with the same shape. *Journal of Mathematical Psychology*, 1972, **9,** 1–16.

Linn, R. L., Rock, D. A., & Cleary, T. A. The development and evaluation of several programmed testing methods. *Educational and Psychological Measurement*, 1969, **29,** 129–146.

Linn, R. L., Rock, D. A., & Cleary, T. A. Sequential testing for dichotomous decisions. *Educational and Psychological Measurement*, 1972, **32,** 85–95.

Lord, F. M. A theory of test scores. *Psychometric Monograph*, 1952, No. 7.

Lord, F. M. An application of confidence intervals and of maximum likelihood to the estimation of an examinee's ability. *Psychometrika*, 1953, **18,** 57–75. (a)

Lord, F. M. The relation of test score to the trait underlying the test. *Educational and Psychological Measurement*, 1953, **13,** 517–548. (b)

Lord, F. M. Some perspectives on "The Attenuation Paradox in Test Theory." *Psychological Bulletin*, 1955, **52,** 505–510.

Lord, F. M. An analysis of the Verbal Scholastic Aptitude Test using Birnbaum's three-parameter logistic model. *Educational and Psychological Measurement*, 1968, **28,** 989–1020.

Lord, F. M. Item characteristic curves estimated without knowledge of their mathematical form—a confrontation of Birnbaum's logistic model. *Psychometrika*, 1970, **35,** 43–50. (a)

Lord, F. M. Some test theory for tailored testing. In W. H. Holtzman (Ed.), *Computer-assisted instruction, testing, and guidance*. New York: Harper & Row, 1970. (b)

Lord, F. M. The self-scoring flexilevel test. *Journal of Educational Measurement*, 1971, **8,** 147–151. (a)

Lord, F. M. A theoretical study of the measurement effectiveness of flexilevel tests. *Educational and Psychological Measurement*, 1971, **31,** 805–813. (b)

Lord, F. M. A theoretical study of two-stage testing. *Psychometrika*, 1971, **36,** 227–242. (c)

Lord, F. M. Robbins-Monro procedures for tailored testing. *Educational and Psychological Measurement*, 1971, **31,** 3–31. (d)

Lord, F. M. Tailored testing, an application of stochastic approximation. *Journal of the American Statistical Association*, 1971, **66,** 707–711. (e)

Lord, F. M. *Estimation of latent ability and item parameters when there are omitted responses. Psychometrika*, 1974, **39,** in press.

Lord, F. M., & Novick, M. R. *Statistical theories of mental test scores*. Reading, Mass.: Addison-Wesley, 1968.

Meredith, W. Some results based on a general stochastic model for mental tests. *Psychometrika*, 1965, **30,** 419–440.

Merwin, J. C. Rational and mathematical relationships of six scoring procedures applicable to three-choice items. *Journal of Educational Psychology*, 1959, **50,** 153–161.

Mulaik, S. A. A mathematical investigation of some multidimensional Rasch models for psychological tests. Paper presented at the meeting of the Psychometric Society, Princeton, N.J., March 1972.

Nishisato, S., & Torii, Y. Assessment of information loss in scoring monotone items. *Multivariate Behavioral Research*, 1971, **6,** 91–103.

Odell, P. L. *An empirical study of three stochastic approximation techniques applicable to sensitivity testing*. NAVWEPS Report 7837. Albuquerque, N.M.: U.S. Naval Weapons Evaluation Facility, 1961.

Owen, R. J. *Bayesian sequential design and analysis of dichotomous experiments with special reference to mental testing*. Ann Arbor, Mich.: Author, 1970.

Panchapakesan, N. *The simple logistic model and mental measurement*. (Doctoral dissertation, University of Chicago) Chicago: University of Chicago Library, Dept. of Reproduction, 1969.

Paterson, J. J. An evaluation of the sequential method of psychological testing. Unpublished doctoral dissertation, Michigan State University, 1962.

Pollack, I. Methodological determination of the PEST (Parameter Estimation by Sequential Testing) procedure. *Perception and Psychophysics*, 1968, **3,** 285–289.

Rasch, G. *Probabilistic models for some intelligence and attainment tests.* Copenhagen: Danmarks Paedogogishe Institut, 1960.

Rasch, G. On general laws and the meaning of measurement in psychology. In J. Neyman (Ed.), *Proceedings of the fourth Berkeley symposium on mathematical statistics and probability.* Vol. 4. Berkeley: University of California Press, 1961.

Rasch, G. An individualistic approach to item analysis. In P. F. Lazarsfeld and N. W. Henry (Eds.), *Readings in mathematical social science.* Chicago: Science Research Associates, 1966. (a)

Rasch, G. An item analysis which takes individual differences into account. *British Journal of Mathematical and Statistical Psychology*, 1966, **19,** 49–57. (b)

Robbins, H., & Monro, S. A stochastic approximation method. *The Annals of Mathematical Statistics*, 1951, **22,** 400–407.

Samejima, F. A general model for free-response data. *Psychometric Monograph Supplement*, 1972, No. 18.

Scheiblechner, H. H. CML-*parameter-estimation in a generalized multifactorial version of Rasch's probabilistic measurement-model with two categories of answers.* Research Bulletin Nr. 4/71. Vienna: Psychologisches Institut der Universität Wien, 1971. (a)

Scheiblechner, H. H. *A simple algorithm for* CML-*parameter-estimation in Rasch's probabilistic measurement model with two or more categories of answers.* Research Bulletin Nr. 5/71. Vienna: Psychologisches Institut der Universität Wien, 1971. (b)

Seeley, L. C., Morton, M. A., & Anderson, A. A. *Exploratory study of a sequential item test.* Technical Research Note 129. Washington, D.C.: U. S. Army Personnel Research Office, 1962.

Shiba, S. Information transmission rate of psychological tests I. *Japanese Journal of Psychology*, 1969, **40,** 68–75. (a)

Shiba, S. Information transmission rate of psychological tests II. *Japanese Journal of Psychology*, 1969, **40,** 121–129. (b)

Shoemaker, D. M., & Osburn, H. G. A simulation model for achievement testing. *Educational and Psychological Measurement*, 1970, **30,** 267–272.

Solomon, H. (Ed.) *Studies in item analysis and prediction.* Stanford, Calif.: Stanford University Press, 1961.

Solomon, H. (Ed.) *Item analysis, test design, and classification.* Project No. 1327. Stanford, Calif.: Stanford University, U.S. Office of Education, Cooperative Research Program, 1965.

Stuckey, C. W., Hutton, C. L., & Campbell, R. A. Decision rules in threshold determination. *Journal of the Acoustical Society of America*, 1966, **40,** 1174–1179.

Taylor, M. M., & Creelman, C. D. PEST: Efficient estimates on probability functions. *Journal of the Acoustical Society of America*, 1967, **41,** 782–787.

Tinsley, H. E. A., & Dawis, R. V. *A comparison of the Rasch item probability with three common item characteristics as criteria for item selection.* Technical Report No. 3003. Minneapolis, Minn.: The Center for the Study of Organizational Performance and Human Effectiveness, University of Minnesota, 1972.

Torgerson, W. S. *Theory and methods of scaling*. New York: Wiley, 1958.

Tsutakawa, R. K. Block up-and-down method in bio-assay. Unpublished doctoral dissertation, University of Chicago, 1963.

Tsutakawa, R. K. Random walk design in bio-assay. *Journal of the American Statistical Association*, 1967, **62,** 842–856. (a)

Tsutakawa, R. K. Asymptotic properties of the block up-and-down method in bio-assay. *The Annals of Mathematical Statistics*, 1967, **38,** 1822–1828. (b)

Tucker, L. R. Maximum validity of a test with equivalent items. *Psychometrika*, 1946, **11,** 1–13.

Urry, V. W. A Monte Carlo investigation of logistic mental test models. Unpublished doctoral dissertation, Purdue University, 1970.

Wasan, M. T. *Stochastic approximation*. Cambridge: Cambridge University Press, 1969.

Waters, C. J. Preliminary evaluation of simulated branching tests. Technical Research Note 140. Washington, D.C.: U.S. Army Personnel Research Office, 1964.

Waters, C. W., & Bayroff, A. G. A comparison of computer-simulated conventional and branching tests. *Educational and Psychological Measurement*, 1971, **31,** 125–136.

Wetherill, G. B. Sequential estimation of quantal response curves. *Journal of the Royal Statistical Society*, Series B, 1963, **25,** 1–38.

Wetherill, G. B., Chen, H., & Vasudeva, R. B. Sequential estimation of quantal response curves: A new method of estimation. *Biometrika*, 1966, **53,** 439–454.

Wetherill, G. B., & Levitt, H. Sequential estimation of points on a psychometric function. *British Journal of Mathematical and Statistical Psychology*, 1965, **18.** 1–10.

Wood, R. The efficacy of tailored testing. *Educational Research*, 1969, **11,** 219–222.

Wood, R. Computerized adaptive sequential testing. Unpublished doctoral dissertation, University of Chicago, 1971.

Wood, R., & Skurnik, L. S. *Item banking*. London: National Foundation for Educational Research in England and Wales, 1969.

Wright, B. D. Sample-free test calibration and person measurement. *Proceedings of the 1967 invitational conference on testing problems*. Princeton, N.J.: Educational Testing Service, 1968.

Wright, B., & Panchapakesan, N. A procedure for sample-free item analysis. *Educational and Psychological Measurement*, 1969, **29,** 23–48.

Foundations of Fechnerian Psychophysics

J. C. Falmagne
NEW YORK UNIVERSITY

1. INTRODUCTION

The history of classical (or Fechnerian) psychophysics is an interesting case for the student of philosophy of sciences. In 1860, Gustav Theodor Fechner raised the question, *Is it possible to explain discrimination data by the notion of 'sensation differences'?* He had in mind sensory continua, e.g., loudness, brightness, heaviness. His treatment of the question, his methods, and the final answer he gave were almost universally accepted for three quarters of a century.[1] Then criticisms and dissent came that were as extreme as the praise had been (for a sample, see Johnson, 1929, 1930; Cobb, 1932; Stevens, 1957; Luce & Galanter, 1963a). Today, many probably feel that Fechner's approach to psychophysics is, at best, of historical interest. Still, it should be realized that no adequate theory of Fechner's fundamental idea has yet been proposed, and that (consequently) the basic data are still lacking. This chapter is an attempt to remedy, in part, this very unsatisfactory situation. Let us be clear about what is intended here. No review of Fechner's whole scheme of psychophysics, nor of the objections, are given. No new data are reported. Rather, this chapter is devoted to a systematic discussion of his original

[1] James (1890) is a noteworthy exception.

question. As pointed out by Luce (1959, p. 39), this question involves, in fact, two essentially distinct issues:

(1) Can we assign to each stimulus a number, in such a way that, whenever two pairs of stimuli are equidiscriminable, the corresponding pairs of numbers have equal differences? This question involves the representation of pairs of stimuli by differences of numbers and, accordingly, is referred to as the *representation problem.*

(2) Assuming this is possible, is the number assigned to a stimulus a measure of the magnitude of the sensation evoked by the stimulus? This is referred to as the *interpretation problem.*

Each problem is discussed. I shall begin with an informal introduction to each of them.

1.1. The Representation Problem

Let X be a sensory continuum, say a set of pure tones of a given frequency, differing only in their intensities; let M be a discrimination index, say for x, y in X, $M(x, y)$ is the probability that the subject answers 'yes' to the question "is x louder than y?" Formally, the representation problem amounts to finding, if it exists, a scale u such that

$$M(x, y) = M(x', y') \text{ iff } u(x) - u(y) = u(x') - u(y'). \tag{1}$$

('iff' is used as an abbreviation for 'if and only if.')

Under this more specific form, the representation problem is sometimes referred to as Fechner's Problem (Luce & Galanter, 1963a; Falmagne, 1971). Its rationale is that, if solvable for some function u, this function can be considered as a possible candidate for measuring sensation magnitudes. This is the interpretation problem, and it is discussed later. It is well known that Fechner's Problem has a solution if Weber's Law holds: u takes then the familiar form of the so-called Fechner's Law. Weber's Law, however, is by no means a necessary condition. A large number of hypothetical empirical laws can be conceived, which would ensure the existence of a scale u satisfying Equation 1. One example is given by Krantz (1971). Another is discussed in some detail later in this chapter (Sec. 3). This remark is important in view of the fact that Weber's Law is often not fulfilled by the data. It would be poor strategy, however, to guess at some specific empirical law, which would provide a better fit than Weber's to existing data, in order to derive the desired function u; the appropriate techniques exist (Luce & Edwards, 1958; Krantz, 1971; Falmagne, 1971). This would be a mere replication of Fechner's own discussion of the problem with a different law, and would not be appropriate for two reasons. First, considering the ever-growing evidence of the complexity of sensory mechanisms, it seems unlikely to me that we shall end up with psychophysical laws that, by some stroke of luck, would have a simple analytical form. Second, Fechner's original question is uncommitted to any such law. Assuming one would involve an essential change of the problem at hand. What is required is a sufficiently general theory of discrimination that would involve the solvability of Fechner's Problem without necessarily implying Weber's Law, or for that matter, any other specific law.[2] The main purpose of this chapter is to present such a theory. This theory will not be restricted to discrimination indices derived from threshold methods, such as the constant stimuli. It will be applicable to any reasonable discrimination index. (An example provided by a reaction-time method is described in Sec. 2.) This is desirable since Fechner's question makes no reference to experimental methods. Moreover, various objections have been raised against Fechner's own methods. Whether or not these objections are justified, it would not seem a good idea to restrict our discussion to them. In itself, this theory will say nothing about sensation. A first attempt in that direction was made in Falmagne (1971). The analysis of Fechner's Problem there was not quite satisfactory however, since it involved some conditions that are not satisfied by the most commonly used discrimination indices. (For related material, see also Levine, 1971.)

1.2. The Interpretation Problem

In addition to Fechner's, at least three distinct positions have been expressed with regard to the problem of measuring sensation by psychophysical methods. The first position, held by Stevens (1957) and his followers, argues that the sensation evoked by a stimulus is essentially a phenomenological event involving the conscious experience of the subject. Accordingly, it could, and should, be measured directly, that is, by methods involving the subject's judgment. (In the sequel, we shall refer to these as *judgmental methods.*) This position is rather controversial. Various objections have been raised, which will not be reviewed here (see Luce & Galanter, 1963b; Zinnes, 1969; Krantz, 1972). What I see as the main difficulty is that it leads us in fact to reject discrimination data as being irrelevant to the question of sensation. This is not a very unifying viewpoint. It will perhaps become obvious in the future that the sensory mechanisms underlying discrimination are essentially differ-

[2] In this discussion I use the term 'law' in an informal but frequent sense, meaning: some constraint on the data symbolized by an *analytical expression.* Strictly speaking, of course, any empirical theory involves empirical laws. For instance, Axioms F4 through F6 of the theory of discrimination described in Section 2 are genuine empirical laws but do not have an analytical expression.

ent from those involved in the judgmental methods. At present, I know of no compelling argument or data that would indicate that this is the case. Moreover, even if this happens, it would still not be clear that the reasonable decision is to reserve the label sensation for the sensory mechanisms involved in the subject's judgments about the stimuli.[3] More will be said later on this question.

The second position (Luce & Galanter, 1963a; Falmagne, 1971) is somewhat less faithful to philosophical tradition.[4] Its main concern is economy of thought, or more precisely, economy of explanatory concepts. The decision to assign the label 'sensation' to a particular sensory scale should not be taken lightly. As expressed by Luce and Galanter (1963a, p. 207),

> If a scale serves no purpose other than as a compact summary of the data from which it was calculated, if it fails to predict different data or to relate apparently unrelated results, that is, if it is not an efficient theoretical device, then, it is worth but little attention and surely we should not let it appropriate such a prized word as 'sensation.' If, however, a scale is ever shown to have a rich theoretical and predictive role, then the scientific community can afford to risk the loss of a good word.

A third position, recently expressed by R. Shepard in an unpublished manuscript and by Krantz (1972; see also Stevens, 1957; Kristof, 1965, 1967), is that sensation is basically relational. There is no such thing as "the sensation evoked by a stimulus." Rather, a sensation 'ratio' results from the comparison (conscious or not) of two stimuli. In Krantz (1972), this takes the form of a completely formalized theory, which is intended to apply, primarily, in cross-modality matching and magnitude estimation experiments. It is assumed that to each pair (x, y) of stimuli, corresponds a sensation 'ratio' $\psi(x, y)$. The axioms of the theory imply the existence of a mapping f of the sensory continuum into the reals such that

$$\psi(x, y) = \psi(x', y') \text{ iff } \frac{f(x)}{f(y)} = \frac{f(x')}{f(y')}. \tag{2}$$

As noted by Krantz, this representation is essentially equivalent to the representation embodied in Equation 1. Whether we decided to represent pairs of stimuli by ratios, or by differences, is arbitrary. The difference between the Shepard-Krantz position and, say, Fechner's position, does not lie in the representation, but rather, in the interpretation. As far as I can see, only the representation is empirically testable. (At this point, the reader might argue that, since Fechner's Problem and Krantz' theory involve equivalent representations, the latter could conceivably be considered as a discrimination

[3] A similar viewpoint is expressed in Zinnes (1969) and Krantz (1972).
[4] However, see Pears (1967, Ch. 11).

theory. This is not so, for somewhat technical reasons that are discussed in Sec. 2.)

Both the positions of Stevens and Shepard-Krantz involve a commitment with regard to what the sensation continuum is. Here, I shall take and discuss in some detail the more cautious position 2, that a sensory scale should be called a scale of sensation only if it explains a large body of sensory data, collected with a variety of methods. Anyone acquainted with the psychophysical literature would probably agree that, at present, no scale meets this requirement. The reason is not so much that existing data lead to rejection of all the known sensory scale, but rather that the most appropriate questions have not been raised. The question whether the same scale rises from different methods has been given, in recent years, the following interpretations: do we obtain a logarithmic or a power function, using one or the other of the judgmental methods? Do we obtain the same power function, in particular, the same value of the exponent, using different judgmental methods? These questions aroused an astonishingly large number of experimental studies. The results, however, were generally disappointing. If a power function (of some sort) is often found to give a reasonable fit to the data, the value of the exponent clearly varies with the experimental conditions (see Zinnes, 1969, for a pessimistic account and extensive references; see also Ross & di Lollo, 1971; Teghtsoonian, 1971). Some of the reasons that could account for the failure are discussed in Stevens (1971). I shall not review them here, but shall make two remarks:

(1) The analysis of the failure is difficult, if not practically impossible, without a firm theoretical foundation for the psychophysical methods involved.[5]

(2) This approach, in part, begs the question in assuming that sensation magnitude is either a logarithmic function or a power function on the sensory continuum.

How the more general question of the existence of a common sensory scale, for different psychophysical methods, could be approached is suggested by the two following examples.[6]

Example 1. Consider, for the same sensory continuum, two discrimination indices M,M^* (say M is derived from a threshold method and M^* from a reaction-time method). Under what conditions on M,M^* does there exist a common scale u, such that Fechner's Problem can be solved simultaneously for M,M^*? Assuming that the appropriate conditions are found, are they empirically satisfied? Showing that a variety of discrimination methods yield the same sensory scale, in the sense of Fechner's Problem, would be a very

[5] At the time I am writing, Shepard and Krantz' papers are still unpublished, and had no influence on these works.

[6] Some other examples are discussed in Krantz (this vol.).

important step. Some results seem to indicate that such a general discrimination scale might perhaps exist. The available evidence, however, is scarce (Falmagne, 1971). In any event, even if this discrimination scale was shown to exist, this should not be considered as conclusive. For this scale to deserve the label 'sensation scale,' it is desirable that it also explain data collected by judgmental methods. Our second example suggests how the question could be formulated in the particular case of the so-called bisection method.

Example 2. I shall assume that this method, which was invented by Plateau a century ago, is familiar to the reader (see Stevens, 1971, for a brief description and a recent evaluation of the results). Let us denote by $x \circ y$, the stimulus appearing equally distant from x and y. An interesting representation for this method follows. Assume that there exists a scale u, and a number p, $0 < p < 1$ such that

$$u(x \circ y) = p\, u(x) + (1 - p)\, u(y). \tag{3}$$

This representation relies on the notion that the subject is comparing the differences between the scale values of the stimuli. The number p allows for some order or position bias. (It is generally observed that $x \circ y$ is not the same stimulus as $y \circ x$.) A system of axioms, due to Pfanzagl (1959, 1968), implies the existence of a scale satisfying this representation. An empirical test of the most critical axiom was performed by Cross (1965; see also Coombs, Dawes, & Tversky, 1970), with reasonably positive results. In line with our first example, we raise the following question: let M be some discrimination index, and $\circ$ be the above bisection index. Under what conditions on $M, \circ$ does there exist a scale u such that Equations 1 and 3 are simultaneously satisfied?

Theoretical results on the questions raised by these two examples are given in Section 3 with some related material. In particular, in view of the undisputable fact that judgmental data can, in most cases, be summarized fairly well by a power function, it is natural to ask whether a power function can also be fitted to discrimination data. For example, let M be some discrimination index, and assume that

$$M(x, y) \leq M(x', y') \text{ iff } \frac{x^\beta + k}{y^\beta + k} \leq \frac{x'^\beta + k}{y'^\beta + k},$$

for some constant $\beta > 0$ and k. This representation is consistent with Equation 1, (taking $u(x) = \log (x^\beta + k)$) and also in line with the usual interpretation of the results obtained with judgmental methods. It will be shown that this leads to a generalized form of Weber's Law. Moreover, the exponent β can be estimated from the data with a reasonable degree of accuracy. Experimental results from Knudsen (1923) and Miller (1947) have been reanalyzed, and the obtained values of β were close to the values obtained with judgmental methods, as reported in the literature.

The organization of this chapter is as follows. Section 2 is devoted to a detailed presentation of a theory of discrimination. In Section 3, the question of the comparison of methods is discussed along the lines just given. A critical viewpoint is taken in Section 4. Possible weak points of the theory are examined and some generalizations are briefly discussed.

2. A THEORY OF DISCRIMINATION

First, the basic notions of the theory are presented and leisurely discussed. Next, a compact summary of the axioms and of the main results are given. The end of the section is devoted to a number of remarks regarding the practical construction of the scale, the special case of the constant stimuli method, and the feasibility of empirical tests of the theory.

2.1. Basic Notions

Our discussion relies, in part, on the two following examples of methods.

(1) *The constant stimuli method.* In this well-known case, a discrimination index is defined from a choice probability P. For instance, $P(x, y)$ is the probability of choosing x over y.

(2) *Reaction time to change.* Each trial begins with the presentation of a preparatory stimulus y (y is some energy value). After a constant delay, a sudden increase of energy of Δ units is applied, yielding a stimulus $x = y + \Delta$. The subject is required to react to the presentation of x by pressing a key. It is assumed that Δ is large enough so that the subject can reliably detect the presentation of x. The latency of the reaction is recorded; let $L(x, y)$ be the (average) latency. Some catch trials are introduced to prevent anticipations. This method has been used, for example, by Piéron (1936). A more detailed discussion of these methods can be found in Falmagne (1971).

In the constant stimuli method, the experiment can be performed, in principle, with any pair of stimuli. However, only some of these pairs are of interest. Indeed, P is sensitive to variations of x and y only if x and y are close to each other. Otherwise, $P(x, y) = 1$ or $P(x, y) = 0$. Similar remarks apply to the second method. (But here, the stimuli cannot be too close.) In general, in terms of Fechner's Problem, this means that the 'if' part of Equation 1 can be assumed to hold at most on a specific subset, say D, of the set of all pairs of stimuli, depending on the method. (This is why Krantz' theory cannot be applied without important modifications to discrimination methods.)

The theory thus concerns a sensory continuum X, a subset D of $X \times X$, and a real-valued function M, a discrimination index, defined on D. When (x, y) is in D, I shall write xDy and sometimes say that x *is comparable to* y. Some idealization of the results typically obtained in applying the methods described are made. (For example, any experiment is obviously finite, yet our theory supposes that the domain D of M is infinite.)

To begin with, I assume that X is a real interval. This is reasonable since X is the domain of some measure of stimulus energy.

There are some obvious differences between methods 1 and 2. For example, P is increasing in the first variable and decreasing in the second, but for L the reverse situation clearly holds. This is a minor point, however. A remedy is to take $1/L$, rather than L, to define a discrimination index. I shall assume that

M is continuous on D, strictly increasing in the first variable and strictly decreasing in the second variable.

There can certainly be no quarrel about the assumption of continuity. As for the strict monotonicity, it is implied by the existence of a scale u solving Fechner's Problem, if we also assume, as is customary, that u is strictly increasing on X. Another difference between both methods is that $P(x, y)$ may be defined with $x < y$, $x = y$, or $x > y$, while $1/L(x, y)$ is defined only if $x > y$ and far enough from it. Notice, however, that P is generally considered redundant: $P(x, y) = 1 - P(y, x)$. There would thus be no less information in taking P as defined only for pairs (x, y) such that $x \geq y$. Still $P(x, x)$ is, in principle, defined for all x in X, while $L(x, x)$ is not. In other words, the constant stimuli method explores the subject's ability to discriminate between stimuli in the neighborhood of each other, when the reaction-time method applies to stimuli that are far apart. This is the essential difference between both methods. Our next assumption accommodates both cases:

if x is comparable to y, then $x \geq y$.

This assumption tells us something about the domain of the discrimination index M. As indicated by the above discussion, the converse proposition—$x \geq y$ implies x comparable to y—cannot be assumed to hold. Our theory, however, cannot go very far without specifying D more precisely. An interesting assumption about D results from the following idea. A stimulus x may fail to be comparable to a stimulus y only in two cases. This might occur if the stimuli are too close; an example of how this might happen is provided by the reaction-time method. Or x may be too far from y, in which case the method fails to be sensitive. Except for these two cases, $x \geq y$ implies x comparable to y. Notice, however, that the notion of x being close to y, or far from it, is ambiguous. Do we mean close on the physical scale, or close on the scale u of Fechner's Problem? Clearly, the scale u is what matters here.

This line of argument leads us to the following 'convexity' assumption:

if x is comparable to y, x'' is comparable to y'',
and $u(x) - u(y) \geq u(x') - u(y') \geq u(x'') - u(y'')$,
then x' is comparable to y'.

This assumption is in agreement with the most appealing interpretation of the scale u: $u(x') - u(y')$ represents the distance between the sensations evoked by x' and y'. Under this form, this assumption is not very convenient, since it involves the scale u, whose existence is what should be derived from the axioms of the theory. Moreover, it is rather strong. I shall only suppose that this convexity assumption holds 'locally.' Suppose, for example, that $x \geq x' \geq x''$, and $y'' \geq y' \geq y$. If a scale u exists, strictly increasing on X, we get $u(x) \geq u(x') \geq u(x'')$ and $u(y'') \geq u(y') \geq u(y)$, yielding $u(x) - u(y) \geq u(x') - u(y') \geq u(x'') - u(y'')$. If x is comparable to y, and x'' to y'', it is reasonable to require that x' be comparable to y'. Figure 1 summarizes this discussion. We obtain

> *if $x \geq x' \geq x''$, $y'' \geq y' \geq y$, x is comparable to y, x'' is comparable to y'', then x' is comparable to y'.*

By and large the above assumptions are 'technical.' That is, empirical tests of the theory will not, in general, be carried on to validate them. Let us turn to the core assumptions of the theory. Suppose that $M(x, y) = M(x', y')$ and $M(y, z) = M(y', z')$. Again, if u satisfies Equation 1, we have

$$u(x) - u(y) = u(x') - u(y'),$$
$$u(y) - u(z) = u(y') - u(z').$$

Adding these two equalities yields

$$u(x) - u(z) = u(x') - u(z'),$$

and, assuming that x is comparable to z, and x' to z', we conclude that $M(x, z) = M(x', z')$. We can also start with $M(x, y) = M(x', y')$, $M(x, z) = M(x', z')$ and derive, by a similar argument, $M(y, z) = M(y', z')$. Summarizing, we get

> *if $M(x, y) = M(x', y')$, x and y are comparable to z, and x', y' are comparable to z', then $M(x, z) = M(x', z')$ iff $M(y, z) = M(y', z')$.*

There are two more assumptions of this type. Since the necessity of each of them is easily verified, using simple arithmetic along the lines just exemplified, I simply state them without comment.

> *If $M(x, y) = M(y', z')$, x and y are comparable to z, and x' is comparable to y' and z', then $M(x, z) = M(x', z')$ iff $M(y, z) = M(x', y')$.*
>
> *If $M(x, y) = M(z, w)$, $M(x, t) = M(s, w)$, s is comparable to y, and z is comparable to t, then $M(s, y) = M(z, t)$.*

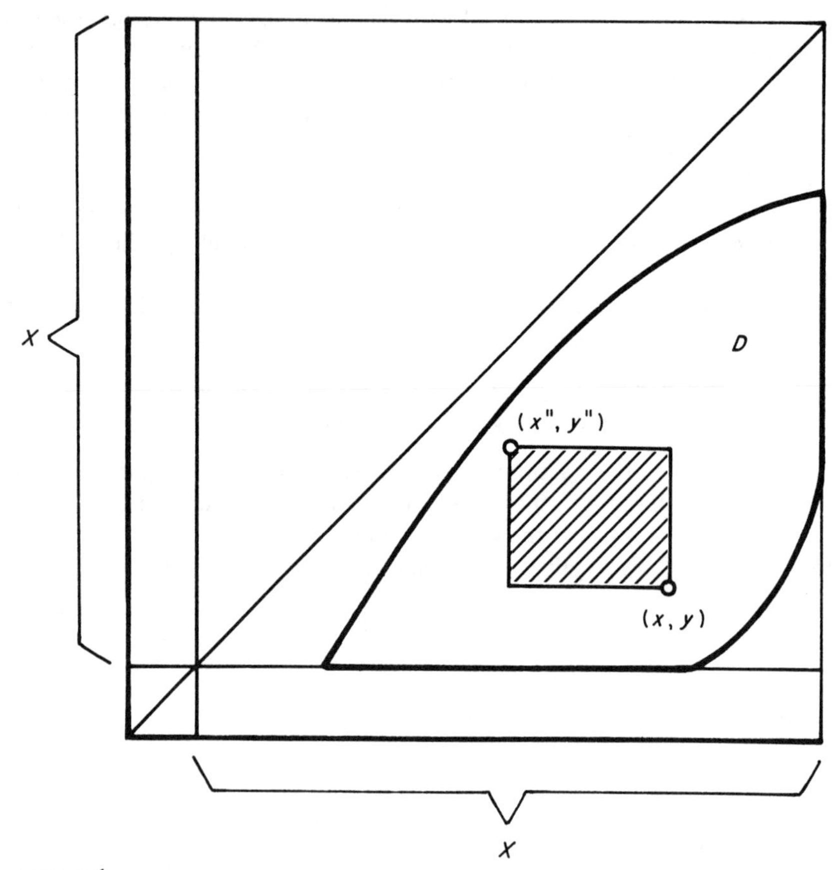

FIGURE 1.
Convexity assumption. If (x, y) and (x'', y'') are in D, all the points of the shaded area are in D.

Finally, we add two more 'technical' assumptions (see Axioms F7 and F8). The first assumption states, essentially, that the set D of pairs of comparable stimuli cannot be too small. As it turns out, the second asserts that the strong convexity assumption must hold on some part of D. To avoid repetition, only a formal statement of these conditions is given. (More detailed comments regarding these axioms can be found in Doignon & Falmagne, 1972.) We have the following result: if a discrimination index satisfies all the conditions mentioned previously, then Fechner's Problem can be solved for this index. That is, there exists a scale u satisfying Equation 1. Moreover, u is an interval scale. Some geometrical insight on what is achieved by the transformation u of the physical scale is provided by Figure 2. For any pair of points (x, y), the set of all points (x', y') in D, equidiscriminable to (x, y) is, under the

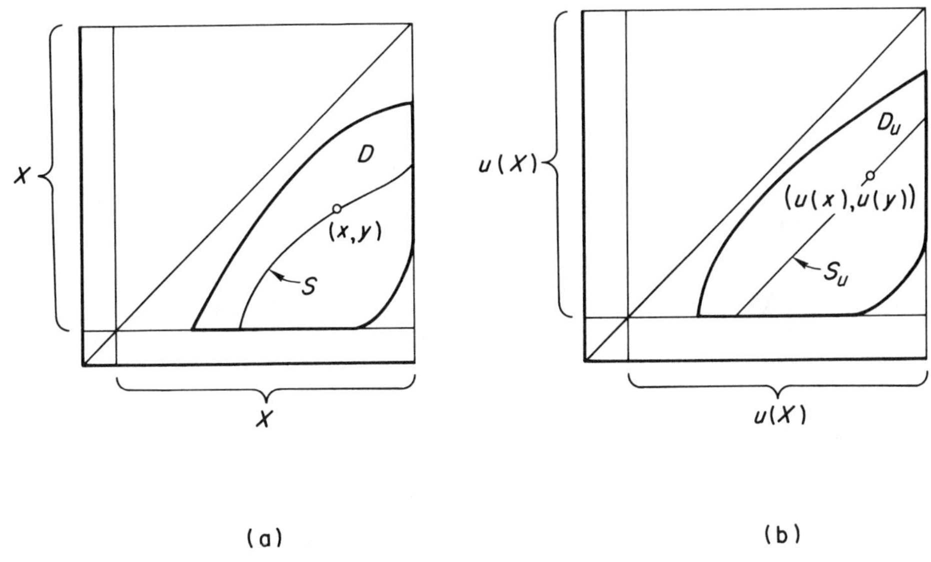

FIGURE 2.
Result of the transformation u. The set D is mapped onto $D_u = \{(u(x), u(y)) \mid xDy\}$, and the curve $S = \{(x', y') \mid M(x', y') = M(x, y)\}$ is mapped onto the segment $S_u = \{(u(x'), u(y')) \mid M(x', y') = M(x, y)\}$.

axioms of the theory, a curve in D. The transformation u maps this set onto (a segment of) a straight line parallel to the first bisector. In the sequel the theory just presented will be referred to as discrimination theory. A compact statement of the basic notions of the theory follows.

2.2. Summary

Primitives

X, a nondegenerate, real interval;
D, a subset $X \times X$; (we write xDy iff $(x, y) \in D$);
M, a real-valued function on D.

Axioms

(F1) M is continuous on D, strictly increasing in the first variable, strictly decreasing in the second;

(F2) if xDy, then $x \geq y$;

(F3) if $x \geq x' \geq x''$, $y'' \geq y' \geq y$, xDy, and $x''Dy''$, then $x'Dy'$;

(F4) if $M(x, y) = M(x', y')$, xDz, $x'Dz'$, yDz, and $y'Dz'$, then $M(x, z) = M(x', z')$ iff $M(y, z) = M(y', z')$;

(F5) if $M(x, y) = M(y', z')$, xDz, yDz, $x'Dy'$, and $x'Dz'$, then $M(x, z) = M(x', z')$ iff $M(y, z) = M(x', y')$;

(F6) if $M(x, y) = M(z, w)$, $M(x, t) = M(s, w)$, sDy, and zDt, then $M(s, y) = M(z, t)$;

(F7) there exist five distinct points $x_1, x_2, x_3, x_4, x_5 \in X$ such that $x_{i+1}Dx_i$ for $1 \leq i \leq 4$, x_5Dx_1, and $M(x_{i+1}, x_i) = M(x_2, x_1)$ for $2 \leq i \leq 4$. Moreover, if $M(x, y) = M(x', y')$, $M(y, z) = M(y', z')$ and $M(x, z) \leq M(x_5, x_1)$, then $x'Dz'$;

(F8) if $M(x_2, x_1) \leq M(x, y) \leq M(x_5, x_1)$, and $x \leq z$ (respectively, $z \leq y$), then there exists w such that $M(z, w) = M(x, y)$ (respectively, $M(w, z) = M(x, y)$).

THEOREM 1. *Suppose that X, D, M satisfy Axioms F1 through F8; then there exists a real-valued increasing, continuous function u on X such that for all x, y, x', y' ∈ X*

$$M(x, y) \leq M(x', y') \text{ iff } xDy,\ x'Dy', \text{ and } 0 \leq u(x) - u(y) \leq (x') - u(y'). \tag{4}$$

Moreover, if u' is another function satisfying these conditions, then $u' = Au + b$ *for some constant* $A > 0$ *and B; that is, u is an interval scale.*

In the sequel, the writing will be simplified. When $\mathfrak{D} = (X, D, M)$ is such that there exists a real-valued, strictly increasing function u on X, satisfying Equation 4 for all $x, y, x', y' \in X$, I shall say that $\mathfrak{D}$, or sometimes M, is u-Fechnerian. Notice that the above result is somewhat stronger than what was announced: $M(x, y)$ is strictly increasing with $u(x) - u(y)$, and u is continuous. This result is an immediate consequence of Theorems 2 and 3 in Doignon and Falmagne (1972), where additional technical comments and specific references can be found. The argument, although straightforward, is too long to include here. Rather, I shall outline the various practical steps of a construction of the scale in Section 2.4.

2.3. Special Cases

In some instances the theory can be given a somewhat simpler form. Take, for example, the constant stimuli method. Suppose that for two stimuli x, y we have $\frac{1}{2} \leq P(x, y) < 1$. Then, if $y \leq y' \leq x' \leq x$, it is reasonable to expect that $\frac{1}{2} \leq P(x', y') < 1$. This leads to the following strengthening of the convexity Axiom (F3):

(F3*) if xDy and $y \leq y' \leq x' \leq x$, then $x'Dy'$.

This condition leads to some simplifications of the theory. Namely, Axioms F5 through F8 can be dropped and replaced by the following new axioms:

(F8*) if $M(w, z) \leq M(x, y)$, there exists x', y', such that $M(w, z) = M(x', y) = M(x, y')$;

(F9) if $y \leq x$, there exists a finite set $y = y_1 \leq y_2 \leq \cdots \leq y_i \leq \cdots \leq y_N = x$, such that $y_{i+1} D y_i$ for all i, $1 \leq i \leq N - 1$.

Clearly, each of these axioms is satisfied by the constant stimuli method. The first one is a strengthening of Axiom F8. We then have the following result:

THEOREM 2. *Suppose that* $\mathfrak{D} = (X, D, M)$ *satisfies Axioms F1, F2, F3*, F4, F8*, and F9; then* $\mathfrak{D}$ *is u-Fechnerian, with u continuous, and an interval scale. Moreover, whenever* $0 \leq u(x') - u(y') \leq u(x) - u(y)$ *and* xDy, *then also* $x'Dy'$.

The net result is that Axioms F5 and F6 have been dropped, which is important since these are two of the three structural axioms. (Incidentally, no systematic study of the independence of the axioms has yet been made. The possibility that one of Axioms F4, F5, or F6 can also be dropped in Theorem 1 cannot be ruled out.) Theorem 2 is an easy consequence of Theorem 5 in Doignon and Falmagne (1972). No proof is given here. The last part of the theorem is rather obvious. It reminds us, however, of the strong convexity property discussed in Section 2.1, namely:

if $u(x) - u(y) \geq u(x') - u(y') \geq u(x'') - u(y'')$ *and* xDy, $x''Dy''$, *then* $x'Dy'$.

If we interpret $u(x') - u(y')$ as the difference between the sensation magnitudes of x' and y', this property appears as a natural requirement. It means indeed that two stimuli are comparable as long as the differences between their sensation magnitudes remain between bounds. Neither Axiom F3 nor F3* implies, in the presence of the other relevant axioms, this strong convexity property. An appropriate axiom is as follows:

(F3**) if xDy and $x \leq x'$ (respectively, $y' \leq y$), then there exists y' (respectively, x') such that $M(x, y) = M(x', y')$.

Thus, if $\mathfrak{D} = (X, D, M)$ satisfy Axioms F1, F2, F3**, and F4 through F8, then $\mathfrak{D}$ is u-Fechnerian, with u a continuous interval scale satisfying the strong convexity property. In other words, the set D_u in Figure 2b is a regular trapezoid, with two sides parallel to the first bisector. In itself, this result adds very little to what we already know. It is mentioned here only for the reason that Axiom F3** is of relevance in later developments.

2.4. Construction of the Scale

Various methods for construction of the Fechnerian scale have been proposed (Luce & Edwards, 1958; Krantz, 1971; Falmagne, 1971). In general, these methods are appropriate when the discrimination index is derived from a threshold method, or from any formally equivalent method. The critical point is that, in such cases, it can be assumed that there exist stimuli x, y such that the difference $u(x) - u(y) > 0$ is as small as desired. These methods would then not be applicable to, say, the reaction-time method of Example 2. The method outlined that follows will do in all cases satisfying Axioms F1 through F8 of discrimination theory. The reader will notice that this method is not based on a principle of addition of Jnds.

Step 1. Construct a set $\cdots < x_{-n} < \cdots < x_1 < \cdots < x_5 < \cdots < x_n < \cdots$ such that $x_1, x_2, \cdots, x_5$ are as in Axiom F7, $M(x_{n+1}, x_n) = M(x_2, x_1)$ for all x_n, x_{n+1} in the set, and for all $y \in X$, if either $M(y, x_n) = M(x_2, x_1)$ or $M(x_n, y) = M(x_2, x_1)$ for some x_n in the set, then y is also a member of the set. Define $u(x_n) = n$.

Step 2. Take any $w \in X$ and suppose that $x_n \leq w \leq x_{n+1}$ and that x_{n-1} exists. Construct two sequences (y_m), (z_m) satisfying the following conditions (a glance at Figure 3 might be helpful): $x_{n-1} = y_1 < y_2 < \cdots < y_m < \cdots < x_n = z_1 < z_2 < \cdots z_m < \cdots < x_{n+1}$; y_2 is sufficiently close to y_1; $M(z_m, y_m) = M(x_n, x_{n-1})$ and $M(z_{m+1}, y_m) = M(z_2, y_1)$ for all indices m. Let p be the largest integer such that $z_{p+1} \leq w$; let q be the largest integer such that $z_{q+1} \leq x_{n+1}$. We have then

$$u(w) - u(x_n) \doteq p[u(z_2) - u(z_1)],$$
$$u(x_{n+1}) - u(x_n) \doteq q[u(z_2) - u(z_1)]$$

(where $\doteq$ means 'approximately equal to'), which yields, after dividing and replacing $u(x_{n-1})$, $u(x_n)$, $u(x_{n+1})$ by their values,

$$u(w) = n + \frac{p}{q}.$$

If x_{n-1} does not exist, then x_{n+2} exists and $u(w)$ can be defined by a similar procedure. (I leave this to the reader.)

Step 3. It remains to define u for the stimuli w, if they exist, such that for all x_n, either $w > x_n$ or $w < x_n$. Consider the first case, and let $x_M \geq x_n$ for all the elements x_n. Then x_{M-1} and x_{M-2} exist, with $x_M < w$, wDx_{M-1} and $M(w, x_{M-1}) < M(x_M, x_{M-2})$. Also, there exists w' such that $M(w', x_{M-2}) = M(w, x_{M-1})$. The appropriate definition is then clearly $u(w) = u(w') + 1$. The second case is dealt with similarly.

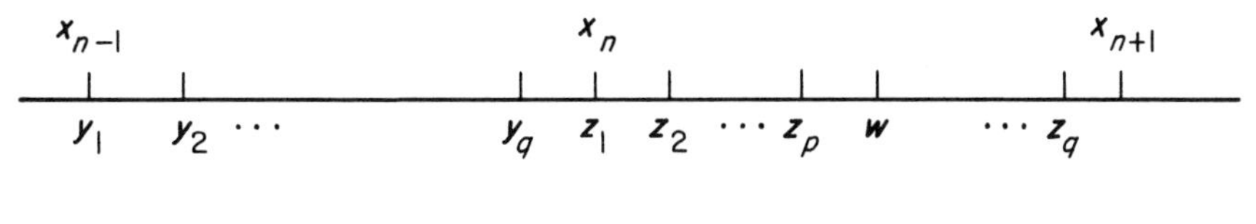

$$M(z_m, y_m) = M(x_n, x_{n-1}), M(z_{m+1}, y_m) = M(z_2, x_{n-1}).$$

(a)

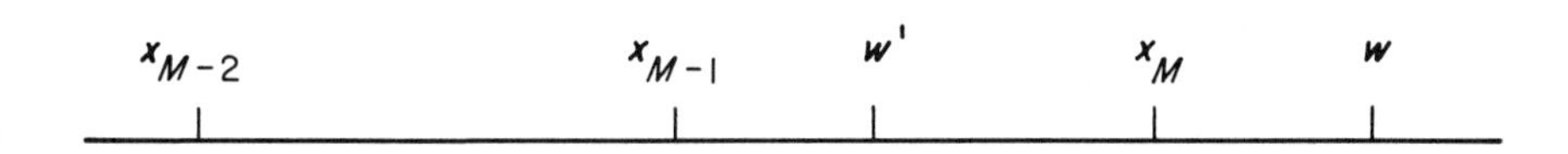

$$M(w', x_{M-2}) = M(w, x_{M-1}) < M(x_M, x_{M-2}).$$

(b)

FIGURE 3.
Construction of u.

2.5. Remarks on Empirical Tests of the Theory

As mentioned before, if Weber's Law holds for some discrimination index M, then Fechner's Problem can be solved. Consequently, the main axioms of the theory, such as F4 through F6, are satisfied. More precisely, define

$$\Delta(y, r) = x - y \text{ iff } M(x, y) = r. \tag{5}$$

The function Δ is usually referred to as Weber's Function. It is well defined by Equation 5 if, for example, Axiom F1 holds. Weber's Law is then symbolized by the expression

$$\Delta(y, r) = (y + k)a(r), \tag{6}$$

in which a is a strictly increasing function, and k is a constant. Equation 6 implies that M is u-Fechnerian, with $u(x) = \log(x + k)$. Indeed, it is easy to show that Equation 6 is equivalent to

$$M(x, y) \leq M(x', y') \text{ iff } \frac{x - y}{y + k} \leq \frac{x' - y'}{y' + k},$$

which, in turn, is equivalent to

$$M(x, y) \leq M(x', y') \text{ iff } \log(x + k) - \log(y + k) \leq \log(x' + k) - \log(y' + k),$$

yielding $u(x) = \log(x + k)$, the so-called Fechner's Law, as a solution to Fechner's Problem. Weber's Law thus provides an empirical test of discrimination theory. The trouble with this test, is that, as remarked earlier, Weber's Law is not a consequence of discrimination theory. A more direct test follows.

If some of the axioms of the theory are to be tested, they should be F4 and F5. Axioms F1, F2, F3, F3*, F8*, and F9 will appear either as reasonable generalizations or as obviously wrong after a mere inspection of the data. Axioms F3**, F7, and F8 may perhaps lead to more elaborate investigations. Extensive tests of these axioms will prove to be difficult and will probably not be carried out. Axiom F6 is essential and should lead to systematic tests. This axiom, however, is a 'local' property, to the extent that all the points involved are, in some sense, close to each other. (For example, z in F6 is comparable to y and t and cannot be far from s, w, and z, since s is comparable to y, w is comparable to s, and z is comparable to t.) Consequently, this axiom might perhaps be insensitive to violations of the theory. For these reasons, F4 and F5 stand out as the key axioms of the theory. Unsurprisingly, F4—more precisely, a slightly weaker form of it—was also an essential part of the Shepard-Krantz theory for magnitude estimation and cross-modality matching. A practical test of Axiom F4 is outlined below.

Step 1. Choose stimuli $z_1 < z_2 < \cdots < z_N$ in such a way that the whole range of X is more or less covered. For each i, $1 \leq i \leq N$, construct or estimate the 'psychometric function' $M(., z_i)$.

Step 2. Choose two values r_1, r_2 in the range of M, in such a way that there are $x_1, x_2, \ldots, x_N$ and $y_1, y_2, \ldots, y_N$ with $M(y_i, z_i) = r_1$, $M(x_i, z_i) = r_2$, and x_i comparable to y_i, for all i, $1 \leq i \leq N$.

Step 3. Check whether $M(x_i, y_i)$ is approximately constant for all i.

Some of these steps, of course, raise statistical questions. I do not believe, however, that the problem involved is more formidable than in other cases of algebraic theories. Tests of Axioms F5 and F6 could easily be constructed along the same lines.

3. COMPARISON OF METHODS AND THE INTERPRETATION PROBLEM

Assuming that the discrimination theory is supported by the data, say, for some discrimination index M with domain $D \subset X \times X$, is the obtained scale a measure of sensation magnitude? As stated before, I would favor a positive answer only if the same scale were derived from a number of essentially different methods, providing thus an explanation for a large body of sensory data. Evidence in this respect would come from two sources: other discrimination methods; judgmental methods. Each case is discussed in some detail. The special case in which the scale is a power function of the physical intensity is also considered.

3.1. Comparison of Discrimination Methods

Let M^* be some other discrimination index, with domain $D^* \subset X \times X$. Suppose that both (X, D, M) and (X, D^*, M^*) satisfy Axioms F1 through F8. An obvious possibility then is to construct the two corresponding scales u, u^*, and to check whether they are (approximately) linearly related. The trouble is that the actual construction of a scale is a difficult task. The procedure in Section 2.4 is based on the assumption that the data are error free. In practice, delicate problems of estimation will have to be solved. It would thus be desirable to have some more direct way of comparing the two methods. A number of direct tests are investigated.

Suppose first that $D^* \subset D$. We want to check whether the same scale, say u, can explain both sets of data. Thus whenever xD^*y, $x'D^*y'$, we should have

$$M(x, y) \leq M(x', y') \quad \text{iff } u(x) - u(y) \leq u(x') - u(y')$$

and

$$M^*(x, y) \leq M^*(x', y') \quad \text{iff } u(x) - u(y) \leq u(x') - u(y').$$

The obvious conditions are then

$$M(x, y) \leq M(x', y') \quad \text{iff } M^*(x, y) \leq M^*(x', y'),$$

whenever xD^*y, $x'D^*y'$. In other words, M and M^* must induce the same ordering on D^*. This condition, however, is only of limited interest since $D^* \subset D$ is a strong requirement, which is not fulfilled, for example, by the two methods described in Section 2.1. (In this case, we might even have $D \cap D^* = \varnothing$.)

A more generally applicable condition would be as follows. Suppose that we have two sequences $(x_i)_{1 \leq i \leq N}$, $(x'_i)_{1 \leq i \leq N}$ such that for all i with $1 \leq i \leq N - 1$, either $M(x_{i+1}, x_i) = M(x'_{i+1}, x'_i)$ or $M(x_i, x_{i+1}) = M(x'_i, x'_{i+1})$. If M is u-Fechnerian, we have thus $N - 1$ equations

$$u(x_{i+1}) - u(x_i) = u(x'_{i+1}) - u(x'_i).$$

Adding the corresponding members of these equations yields

$$u(x_N) - u(x_1) = u(x'_N) - u(x'_1),$$

and, for any $x \leq x_N$, we have

$$u(x) - u(x_1) \leq u(x'_N) - u(x'_1).$$

If M^* is also u-Fechnerian, and xD^*x_1, $x'_N D^* x'_1$ we have necessarily

$$M^*(x, x_1) \leq M^*(x'_N, x_1).$$

This suggests the following axiom.

(CF$_1$) Suppose that there exist a point x and two sequences $(x_i)_{1 \leq i \leq N}$, $(x'_i)_{1 \leq i \leq N}$ such that (a) $M(x_i, x_{i+1}) = M(x'_i, x'_{i+1})$ or $M(x_{i+1}, x_i) = M(x'_{i+1}, x'_i)$ for all i, $1 \leq i \leq N - 1$; (b) xD^*x_1, $x'_N D^* x'_1$. Then, $M^*(x, x_1) \leq M^*(x'_N, x'_1)$ iff $x \leq x_N$.

This argument shows that Axiom CF$_1$ is a necessary condition for M^* to be u-Fechnerian, when M is u-Fechnerian. In the presence of a small number of technical axioms, CF$_1$ is also a sufficient condition. We have, for example, the following result.

THEOREM 3. *Suppose that* (a) $\mathfrak{D} = (X, D, M)$ *is u-Fechnerian and satisfies Axioms F3** and F9;* (b) $\mathfrak{D}^* = (X, D^*, M^*)$ *satisfies Axiom F2; and* (c) $\mathfrak{D}$ *and* $\mathfrak{D}^*$ *jointly satisfy Axiom CF*$_1$. *Then,* $\mathfrak{D}^*$ *is u-Fechnerian.*

The particular choice of axioms, for this theorem, is best understood if we think of M as being derived, for example, from the constant-stimulus method, and M^* as being derived from the reaction-time method. Clearly, various other combinations of axioms are possible.

Proof. Suppose that $M^*(x, y) \leq M^*(x', y')$ and $u(x') - u(y') < u(x) - u(y)$. We proceed by constructing the two sequences of Axiom CF$_1$. Let $y'_1 = y' \leq y'_2 \leq \cdots \leq y'_N = x'$ be the set of Axiom F9 (Axiom F2 is applied here to M^*, to ensure $y' \leq x'$). *Case 1.* Assume $y \leq y'$. Set $y_1 = y$ and consider $M(y'_2, y')$. Since $y \leq y'$, there exists z such that $M(z, y) = M(y'_2, y')$, using Axiom F3**. Set $y_2 = z$ and notice that $y_2 \leq y'_2$, $y_2 < x$, and $u(x') - u(y'_2) < u(x) - u(y_2)$. If $y'_2 \neq x'$, we consider $M(y'_3, y_2)$ and define y_3 by the same method. Clearly, we can construct a set $y_1 = y < y_2 < \cdots < y_N < x$ such that $M(y_{i+1}, y_i) = M(y'_{i+1}, y'_i)$ for all i, $1 \leq i \leq N - 1$. Applying Axiom CF$_1$ we obtain $M^*(x', y') < M^*(x, y)$, a contradiction. *Case 2.* If we assume $y' < y$, we must have $x' < x$. In this case, using the method of Case 1, we first define a set $x_1 = x \geq x_2 \geq \cdots \geq x_N > y$, with $M(x_i, x_{i+1}) = M(y'_{N-i+1}, y'_{N-i})$. We set then $y_1 = y$ and apply Axiom F3** to the points x_N, x_{N-1}, y to define y_2 such that $M(y_2, y_1) = M(x_{N-1}, x_N) = M(y'_2, y'_1)$. If $y'_2 \neq x'$, we define $y_3, \ldots$ and so on, and a contradiction follows here also (we leave the details to the reader). The proof is essentially the same if we assume $M^*(x', y') < M^*(x, y)$ and $u(x) - u(y) \leq u(x') - u(y')$. Q.E.D.

It was argued in Section 1 that some degree of arbitrariness was involved in the representation of Equation 1: we could choose to represent pairs of stimuli by ratios, rather than by differences. This arbitrariness, however, is partly removed when the information provided by two indices is available, if we assume that the same scale is obtained in both cases: if one of the representations is fixed, the other one is no longer arbitrary. An interesting question, in this connection, is whether a sensory scale could yield a difference representation with one method and a ratio representation with another method. Suppose that M, M^* are two discrimination indices for the same sensory continuum. Under what conditions on M, M^* does there exist a sensory scale u, such that M is u-Fechnerian and M^* is $\log u$-Fechnerian? Essentially the same question was raised by Krantz, Luce, Suppes and Tversky (1971, pp. 152–154; see also Pfanzagl, 1968) in the framework of judgmental methods. It is easy to show that if M is u-Fechnerian and M^* is u^*-Fechnerian a necessary condition for $u^* = A \log u + B$ for some constants $A > 0$ and B is that

(CF_2) for all $x, y, z, x', y', z' \in X$, if $M^*(x, x') = M^*(y, y') = M^*(z, z')$, and xDy, yDz, $x'Dy'$, $y'Dz'$, then $M(x, y) \geq M(y, z)$ iff $M(x', y') \geq M(y', z')$.

I did not investigate the sufficiency. So far as I know, neither Axiom CF_1 nor CF_2 has ever been tested empirically. Positive results regarding the existence of a common sensory scale for two different discrimination indices have been obtained by Piéron (1936) and Roufs (1963). The evidence, however, is far from conclusive. (See Falmagne, 1971, for a detailed discussion of the results.)

3.2. Comparison Between a Discrimination Method and a Judgmental Method

We now discuss the case of the bisection method. We ask under what conditions will this method yield the same scale as a given discrimination method—a question that obviously presupposes a psychophysical theory for the bisection method. Let $(x, y) \longrightarrow x \circ y$ be a binary operation on an open interval X, and suppose that the data are so constrained that there exist a number p, $0 < p < 1$, and a real-valued, strictly increasing continuous function $f > 0$ on X, such that

$$f(x \circ y) = pf(x) + (1 - p)f(y),$$

is satisfied for all $x, y \in X$. (For example, we assume that Pfanzagl, 1968, axioms are satisfied.) When $(X, \circ)$ satisfies these conditions, we shall say it is a *bisection system with a representation* (p, f). We also consider a discrimination index M, for the same sensory continuum, assumed to be u-Fechnerian. We ask, when do we have $u = Af + B$? Our summarized results follow.

THEOREM 4. *Suppose that (X, D, M) is u-Fechnerian and that $(X, \circ)$ is a bisection system with a representation (p, f); then $u = Af + B$, for some constant $A > 0$ and B, iff for all $x, y, z, w \in X$*

$$M(x \circ z, y \circ z) = M(x \circ w, y \circ w) \tag{7}$$

whenever $x \circ z D y \circ z$, $x \circ w D y \circ w$.

A similar theorem was proved by Krantz et al. (1971, p. 492), with an extensive measurement system in lieu of the bisection system.

Proof. (Necessity.) Since M is u-Fechnerian, there exists a (continuous, strictly increasing) function F, defined on the set of all real numbers of the form $u(x) - u(y)$, with xDy, such that $M(x, y) = F[u(x) - u(y)]$. Assume $x \circ z D y \circ z$, $x \circ w D y \circ w$, and $u = Af + B$. Then, successively

$$\begin{aligned} M(x \circ z, y \circ z) &= F[u(x \circ z) - u(y \circ z)] \\ &= F[Af(x \circ z) + B - Af(y \circ z) - B] \\ &= F\{A[pf(x) + (1 - p)f(z)] - A[pf(y) + (1 - p)f(z)]\} \\ &= F\{A[pf(x) + (1 - p)f(w)] - A[pf(y) + (1 - p)f(w)]\} \\ &= F[Af(x \circ w) + B - Af(y \circ w) - B] \\ &= F[u(x \circ w) - u(y \circ w)] \\ &= M(x \circ w, y \circ w). \end{aligned}$$

(Sufficiency.) We can assume, without loss of generality, that 0 is a limit point of the range of f. Let $\phi = u \circ f^{-1}$, and assume

$$M(x \circ z, y \circ z) = M(x \circ w, y \circ w);$$

then successively,

$$\begin{aligned} u(x \circ z) - u(y \circ z) &= u(x \circ w) - u(y \circ w), \\ \phi[f(x \circ z)] - \phi[f(y \circ z)] &= \phi[f(x \circ w)] - \phi[f(y \circ w)], \\ \phi[pf(x) + (1 - p)f(z)] & \\ - \phi[pf(y) + (1 - p)f(z)] &= \phi[pf(x) + (1 - p)f(w)] \\ & \quad - \phi[pf(y) + (1 - p)f(w)], \end{aligned}$$

and, with $pf(x) = a$, $qf(z) = b$, $pf(y) = c$, $qf(w) = d$, we obtain

$$\phi(a + b) - \phi(c + b) = \phi(a + d) - \phi(c + d) \tag{8}$$

with $a, b, c, d > 0$. Take a sufficiently small e in the domain of ϕ (that is, in the range of f) and define

$$g(a) = \phi(a + e) - \phi(e),$$

for all a such that $a + e$ is the domain of ϕ. Notice that Equation 8 implies

$$\phi(a + b + e) - \phi(b + e) = \phi(a + e) - \phi(e), \left(\text{use } e = \frac{e}{2} + \frac{e}{2}\right).$$

Hence,

$$\begin{aligned} g(a + b) &= \phi(a + b + e) - \phi(e) \\ &= \phi(b + e) - \phi(e) + \phi(a + e) - \phi(e) \\ &= g(a) + g(b). \end{aligned}$$

Standard techniques apply (see, for example, Krantz et al., 1971, p. 45, Theorem 4), and yield $g(a) = Aa$, for some constant $A > 0$, and all a such that $\phi(a + e)$ is defined. Hence, for all $b > e$, if $b = a + e$, we have

$$\begin{aligned} \phi(b) &= \phi(a + e) \\ &= g(a) + \phi(e) \qquad \text{(by def. of } g) \\ &= g(a + e) - g(e) + \phi(e) \\ &= Ab + B, \end{aligned}$$

with $B = \phi(e) - g(e)$, and it is clear that A, B cannot depend upon the choice of e. (Take $e' < e$, yielding two constants $A' > 0$, B' with $\phi(b) = A'b + B'$ for all $b > e'$. Then if $b, b' > e$, we have $\phi(b) = Ab + B = A'b + B'$, and $\phi(b') = Ab' + B = A'b' + B'$, yielding $A'(b - b') = A(b - b')$.) Consequently, $\phi(b) = Ab + B$ for all b in the range of f. We have thus

$$u(x) = \phi[f(x)] = Af(x) + B \qquad \text{for all } x \in X. \quad \text{Q.E.D.}$$

3.3. Discrimination Theory and the Power Law

It has sometimes been suggested that the only essential theoretical difference between Fechner's position and Stevens' position involves the choice of representing pairs of stimuli by differences, rather than by ratios. Surprisingly, this interesting possibility cannot be eliminated on the basis of existing data. Since this might not be obvious to all readers, I shall discuss the matter in some detail. We shall start from the fact that a power function often gives a good fit to the data, say, in magnitude estimation. In the framework of discrimination theory, this leads to investigation of the following assumption: if xDy, $x'Dy'$, then,

$$M(x, y) \leq M(x', y') \text{ iff } \frac{x^\beta + k}{y^\beta + k} \leq \frac{x'^\beta + k}{y'^\beta + k}.$$

The constant k is of critical importance. If we assume $k = 0$, β is not specified. And our ambition, of course, is to compare the values of β obtained from judgmental methods and from discrimination methods. Let us investigate the consequences of this assumption. It implies that there exists a strictly increasing function h, such that, with $M(x, y) = r$, successively

$$\begin{aligned} h[M(x, y)] &= h(r) \\ &= \frac{[y + \Delta(y, r)]^\beta + k}{y^\beta + k}, \end{aligned}$$

where Δ is the Weber's function defined by Equation 5, yielding

$$\Delta(y, r) = [h(r)(y^\beta + k) - k]^{1/\beta} - y. \tag{9}$$

This equation suggests three remarks. First, Equation 9 provides an example of a hypothetical empirical law different from Weber's, which nevertheless entails the existence of a solution to Fechner's Problem, namely $u(x) = \log(x^\beta + k)$. Second, if $\beta = 1$, Equation 9 reduces to

$$\Delta(y, r) = (y + k)[h(r) - 1], \tag{10}$$

that is, to Weber's Law (in the form of Eq. 6). Thus, Equation 9 generalizes Weber's Law. Third, dividing by y in both sides of Equations 9 and 10 gives, respectively, after rearrangement,

$$\frac{\Delta(y, r)}{y} = \left\{\frac{k}{y^\beta}[h(r) - 1] + h(r)\right\}^{1/\beta} - 1, \tag{11}$$

and

$$\frac{\Delta(y, r)}{y} = \left(1 + \frac{k}{y}\right)[h(r) - 1]. \tag{12}$$

The left member of Equations 11 and 12 is often referred to as Weber's fraction, and is interesting to consider since data are often reported under that form. A glance at Equation 11 indicates that, when y gets large, $\frac{k}{y^\beta}[h(r) - 1]$ is negligible. That is, for a fixed value of $r = r_0$, and large y, $\frac{\Delta(y, r_0)}{y}$ is approximately constant, equal to $h(r_0)^{1/\beta} - 1$. This shows that, if a reasonable number of very small stimuli are not used in the experiment, it will be difficult to distinguish Equation 11 from 12 (if β is not very different from 1). Unfortunately, this is the general situation. In most cases, only a few experimental points (three or four) are taken in the nonflat region of the Weber fraction.

In any event, it was worth trying to fit Equation 9 to existing discrimination data. Our first choice was Miller's (1947) data (white noise). Since $r = r_0$, a constant, we write Equation 11 as

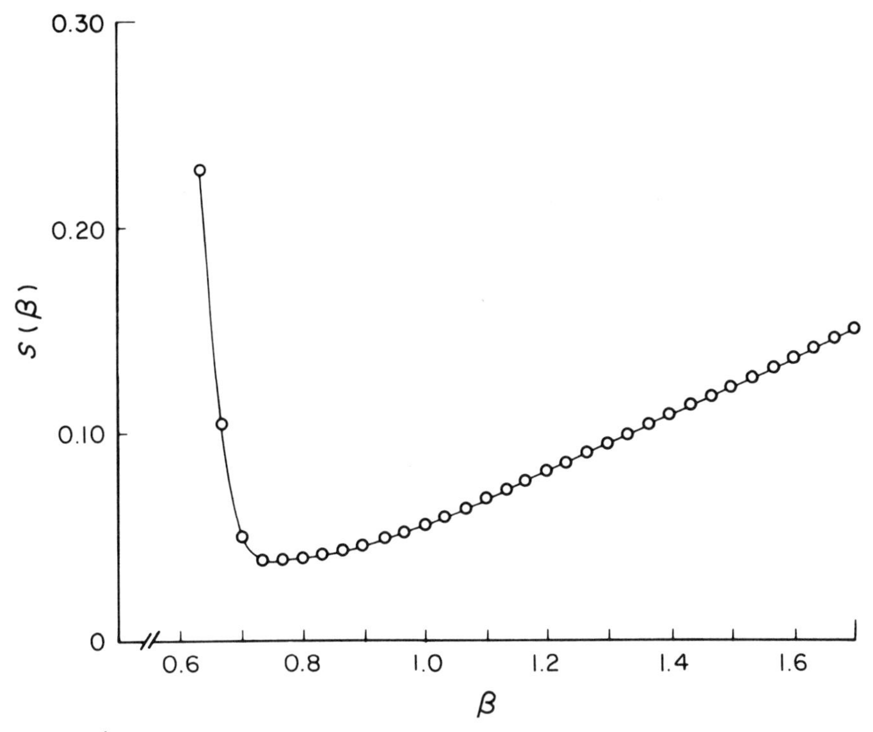

FIGURE 4.
Minimum of the squares of the deviations of the theoretical values of Equation 13 from the experimental values in Miller's (1947) data, for various values of β.

$$\frac{\Delta(y)}{y} = \left(\frac{A}{y^{\beta}} + B\right)^{1/\beta} - 1, \tag{13}$$

with $A = k[h(r_0) - 1]$, $B = h(r_0)$. Equation 13 was fitted to Miller's data, (S G.M.), using the Stepit program.[7] We were especially interested in assessing how accurately β could be estimated. To this effect, we fixed β successively at various, regularly spaced values between 0.1 and 1.7, and we minimized the sum of the squares of the deviations of the theoretical points from the experimental points to obtain estimates of the two remaining parameters. The results are shown in Figure 4. The value of β is in abscissa. The ordinate is the final value S of the sum of the squares of the deviations, after minimization. The minimum of S for the computed points is at $\beta = 0.73$, which gives $S(0.73) = 0.038$, $A = 1.02$, $B = 1.71$. $\left(\text{Hence, } k = \frac{A}{B - 1} = 1.43.\right)$ The in-

[7] Chandler (1969). I am greatly indebted to A. Dwivedi and J. Moss for adapting the Stepit program and performing the two analyses reported here.

teresting question is, of course, whether this analysis leads us to believe that $\beta \neq 1$. One might argue that, for $0.6 \leq \beta \leq 1$, $S(\beta)$ is fairly stable. In fact, percentagewise the improvement from $\beta = 1$ to $\beta = 0.73$ is not negligible: $\frac{S(0.73)}{S(1.0)} = \frac{0.038}{0.056} = 0.60$. (When we fix $\beta = 1$, we obtain $A = 2.2$ and $B = 2.1$.) If on the basis of these results, we would have to make a guess for the best estimate of β, it would be some number between 0.7 and 0.8. Interestingly enough, this is also the range of the values of the exponent of the power law for noise obtained by judgmental method, as reported in the literature (Stevens, 1955, 1971). As might be expected, the approximation to Miller's data provided by Equation 13, using $\beta = 0.73$, $A = 1.02$, and $B = 1.71$, is quite good. By eye, however, it is difficult to distinguish these predictions from those obtained with $\beta = 1$, $A = 2.20$, and $B = 2.10$. The details of the comparison are contained in Table 1.

TABLE 1
Comparison of best-fitting predictions for Miller's data

y (Energy Above Threshold)	$\frac{\Delta(y)}{y}$ (Weber Fraction)	Predictions from Equation 13	
		$\beta = 1.0$	$\beta = 0.73$
0.199×10	2.089	2.200	2.166
0.316×10	1.995	1.800	1.843
0.100×10^{2}	1.309	1.321	1.401
0.316×10^{2}	1.216	1.173	1.217
0.100×10^{3}	1.119	1.125	1.139
0.316×10^{3}	1.112	1.104	1.106
0.316×10^{4}	1.096	1.103	1.086
0.316×10^{5}	1.102	1.103	1.082
0.316×10^{6}	1.094	1.103	1.081
0.100×10^{8}	1.094	1.103	1.081
0.316×10^{9}	1.079	1.103	1.081
0.100×10^{11}	1.067	1.103	1.081

Note: Comparison of best-fitting predictions for $\beta = 1$ [$A = 2.20$, $B = 2.10$, $S(1.0) = 0.056$] and $\beta = 0.73$ [$A = 1.02$, $B = 1.71$, $S(0.73) = 0.038$] using Equation 13 for Miller's data (1947, Subject G.M.).

Essentially, the same analysis was performed on Knudsen's data (1923, pure tones, 1000 cycles). The obtained values were $\beta = 0.60$, $A = 0.807$ and $B = 1.11$, with $S(0.60) = 0.257 \times 10^{-3}$. (See Fig. 5 and Table 2.) The improvement of the fit from $\beta = 1$ to $\beta = 0.60$ is more spectacular here: $\frac{S(0.60)}{S(1.0)} = 0.13$! (It should be mentioned, however, that Knudsen's data are

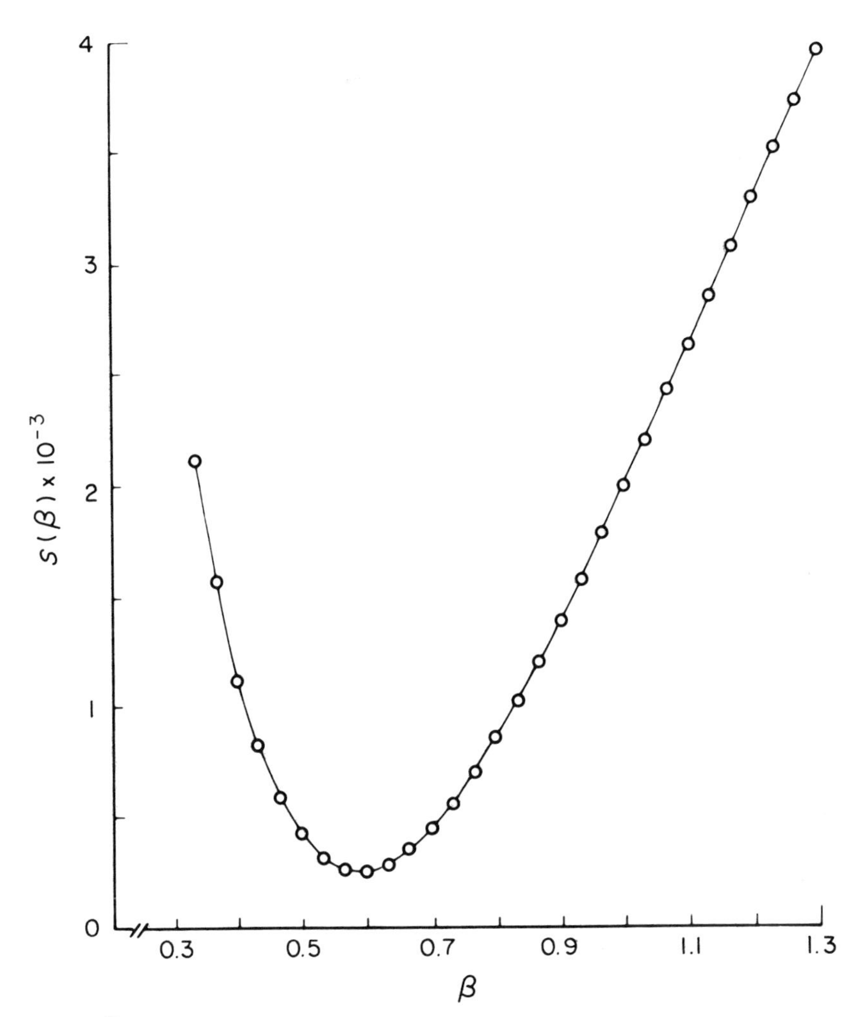

FIGURE 5.
Minimum of the squares of the deviations of the theoretical values of Equation 13 from the experimental values of Knudsen's (1923) data, for various values of β.

only reported on the form of a graph. We had to estimate the numerical values of the experimental points.)

The value reported for the exponent of the power law, for this particular continuum, is of the order of 0.30 (Stevens, 1961), thus much smaller than the value of β obtained. In fact, $\beta = 0.60$ is close to the value obtained above for white noise. A possible explanation is that, as already remarked by Miller (1947), the subjects in Knudsen's experiment reacted to the 'noise' produced by the short onset of his tones.

TABLE 2
Comparison of best-fitting predictions for Knudsen's data

y (Energy Above Threshold)	$\frac{\Delta(y)}{y}$ (Weber Fraction)	Predictions from Equation 13	
		$\beta = 1$	$\beta = 0.60$
0.400×10	0.300	0.3123	0.2972
0.920×10	0.215	0.1982	0.2192
0.211×10^2	0.167	0.1487	0.1730
0.640×10^2	0.142	0.1231	0.1386
0.256×10^3	0.127	0.1136	0.1183
0.102×10^4	0.115	0.1112	0.1095
0.410×10^4	0.110	0.1106	0.1057
0.164×10^5	0.105	0.1105	0.1040
0.655×10^5	0.100	0.1104	0.1033
0.262×10^6	0.100	0.1104	0.1030
0.105×10^7	0.099	0.1104	0.1029
0.419×10^7	0.099	0.1104	0.1028
0.168×10^8	0.099	0.1104	0.1028

Note: Comparison of best-fitting predictions for $\beta = 1$ [$A = 0.807$, $B = 1.11$, $S(1.0) = 0.197 \times 10^{-2}$] and for $\beta = 0.60$ [$A = 0.249$, $B = 1.06$, $S(0.60) = 0.257 \times 10^{-3}$] using Equation 13 for Knudsen's data (1923, 1000 cycles).

These results suggest that discrimination data and judgmental data might possibly be reconciled within a common psychophysical theory. The irony of the situation is that if Weber had formulated his law on the form of Equation 9, rather than on the form of Equation 6, then Fechner's integration procedure would have given $u(x) = \log(x^\beta + k)$, rather than $u(x) = \log(x + k)$, and no controversy would probably have arisen (see Krantz, 1971 for a discussion of this point).[8]

4. DISCUSSION

Here, we discuss discrimination theory from a critical viewpoint. We analyze some of its weak points, examine some of the foreseeable reasons for failure,

[8] Incidentally, if we assume $M(x, y) \leq M(x', y')$ iff $x^\beta - y^\beta \leq x'^\beta - y'^\beta$, we obtain

$$\frac{\Delta(y)}{y} = \left(\frac{A}{y^\beta} + 1\right)^{1/\beta} - 1,$$

that is, a special case of Equation 13.

in particular, in connection with the interpretation problem, and consider generalizations that bypass the difficulties.

One difficulty of discrimination theory is that it is an algebraic, rather than a probabilistic theory. In practice, the values $M(x, y)$ to which the theory will be applied result from the computation of some measure of central tendency of empirical distributions. For instance, in one of our examples involving the constant stimuli method, $M(x, y)$ would be estimated from the frequency of 'yes' responses to the question "Is x louder than y?" The objection to this is not that the variability of the data makes an algebraic theory unappropriate. Indeed, the $M(x, y)$ might conceivably be fairly stable, if computed from a large number of observations. Rather, it is that the axioms of the theory should reasonably involve random variables, since the subject's behavior is not constant on a given stimulus situation. In other words, what is called for is, for example, a generalization of Thurstone's Law of comparative judgment. How such a probabilistic version of discrimination theory could be constructed, in the general case, is outlined next.

4.1. A Probabilistic Version of Discrimination Theory

With X and D as before, let $(\mathbf{M}_{xy})_{(x,y)\in D}$ be a family of real random variables. Thus, when a pair of stimuli (x, y) is presented to the subject, his response is assumed to be the random variable $\mathbf{M}_{xy}$. This variability, I assume, is due to (a) a stimulus factor, and (b) a response factor. To this effect, let $(\mathbf{U}_x)_{x\in X}$ be a family of real random variables, not necessarily pairwise independent; let $\mathbf{R}$ be a real random variable, independent of $\mathbf{U}_x$ for all $x \in X$. Thus, when stimulus x is presented, its effect on the subject's sensory system is the random variable $\mathbf{U}_x$; the variability of the subject's (motor) response is represented by $\mathbf{R}$. The central assumption of the theory is that there is some function $\mathbf{F}$ of random variables such that, for all $(x, y) \in D$

$$\mathbf{M}_{xy} = \mathbf{F}(\mathbf{U}_x - \mathbf{U}_y; \mathbf{R}), \tag{14}$$

where $\mathbf{F}$ is (strictly) increasing in the first argument. This assumption is clearly a generalization of Thurstone's Law of comparative judgment and also the natural probabilistic version of Equation 1. If no further assumption is made, it is not clear how Equation 14 can be tested. Pursuing the parallel with Thurstone, a natural question arises. What constraints (on $\mathbf{M}_{xy}$, $\mathbf{U}_x$, etc.) involve the existence of a scale u, and of some measure $M(x, y)$ of central tendency of $\mathbf{M}_{xy}$ such that, considered as a discrimination index, M is u-Fechnerian? If acceptable constraints could be found, we would be justified in applying the discrimination theory described in Section 2 to the measure of central tendency, M. We did not investigate this problem in detail, and restrict ourselves to some remarks. From Equation 14, we derive

$$\mathbb{P}(\mathbf{M}_{xy} > m) = \mathbb{P}[\mathbf{F}(\mathbf{U}_x - \mathbf{U}_y; \mathbf{R}) > m], \tag{15}$$

where $\mathbb{P}$ denotes the probabilities. Suppose that $(\mathbf{U}_x)_{x \in X}$ is a family of random variables related by a shift. In other words, there exist some random variable $\mathbf{U}$ and a real-valued function u on X such that $\mathbf{U}_x = u(x) + \mathbf{U}$ for all $x \in X$. Let us also assume that u is strictly increasing. We obtain

$$\mathbb{P}(\mathbf{M}_{xy} > m) = \mathbb{P}[\mathbf{F}(u(x) - u(y) + \mathbf{U} - \mathbf{U}'; \mathbf{R}) > m], \tag{16}$$

in which $\mathbf{U}$ and $\mathbf{U}'$ are identically distributed. For a fixed m, the right side of Equation 16 is (under mild conditions) strictly increasing in $u(x) - u(y)$. That is, $\mathbb{P}(\mathbf{M}_{xy} > m)$ is a u-Fechnerian discrimination index. But, of course, this implies that the expectation of $\mathbf{M}_{xy}$ is also u-Fechnerian. Thus, if we assume that Equation 14 holds and that the $\mathbf{U}_x$ have parallel distributions, the discrimination theory presented here can reasonably be applied to the means of the empirical distributions.

This assumption that the $\mathbf{U}_x$ have parallel distributions is strong, and may seem undesirable. This condition is not necessary however, as suggested by the following example involving Thurstone's Case III. Consider the constant stimuli method again, and let $\mathbf{M}_{xy}$ be a dichotomous random variable with values 1 and 0. As a particular case of Equation 14 (with $\mathbf{F}$ increasing but not strictly increasing), assume

$$\mathbf{M}_{xy} = \begin{cases} 1, & \text{if } \mathbf{U}_x - \mathbf{U}_y \geq 0; \\ 0, & \text{if } \mathbf{U}_x - \mathbf{U}_y < 0. \end{cases}$$

Assume that each $\mathbf{U}_x$ has a normal distribution with mean $u(x)$ and variance $\sigma^2(x)$, and that Thurstone's Case III holds (that is, the $\mathbf{U}_x$ are pairwise independent random variables).

We obtain

$$\mathbb{P}(\mathbf{M}_{xy} = 1) = N\left\{\frac{u(x) - u(y)}{[\sigma(x)^2 + \sigma(y)^2]^{1/2}}\right\},$$

in which N is the unit normal cumulative, that is, the normal distribution function with mean 0 and variance 1. If we further assume that $\sigma(x) = Au(x)$, for some $A > 0$ and all $x \in X$, we get

$$\mathbb{P}(\mathbf{M}_{xy} = 1) = N\left\{\frac{\dfrac{u(x)}{u(y)} - 1}{A\left[\dfrac{u(x)^2}{u(y)^2} + 1\right]^{1/2}}\right\},$$

that is,

$$\mathbb{P}(\mathbf{M}_{xy} = 1) = F[\log u(x) - \log u(y)],$$

for some strictly increasing function F (since $\frac{t-1}{(t^2+1)^{1/2}}$ is strictly increasing for $t \geq 0$). Hence, $\mathbb{P}(\mathbf{M}_{xy} = 1)$ is log u-Fechnerian and the $\mathbf{U}_x$ are linearly related. I did not investigate this question any further. The relation between Thurstone's theory and Fechner's Problem is discussed in more detail in Krantz and Falmagne (in preparation).

4.2. An Algebraic Generalization

A likely reason for the failure of discrimination theory is that order effects are often present in the data. It is a well-known fact that, in the constant-stimuli method, a 'constant error' is observed in some conditions. In other words, in the notation of the preceding paragraph, we cannot assume $\mathbb{P}(\mathbf{M}_{xx} = \text{yes}) = \frac{1}{2}$: the presence of one stimulus affects in some sense the perception of the second. How such effects could be explained is discussed by Luce and Galanter (1963a) (among others) where some modification of Thurstone's theory, and of the choice theory, is considered. These possibilities will not be reviewed here.

A natural extension of the theory discussed in Section 2 that might be appropriate in such cases is as follows. Assume that there exists a pair (u, g) of scales such that the following representation holds

$$M(x, y) = M(x', y') \text{ iff } u(x) - g(y) = u(x') - g(y'). \tag{17}$$

This representation could be analyzed without major difficulty along the lines of Section 2.1, leading to a theorem analogous to Theorem 1. An example of a typical constraint on the data involved by Equation 17 is the following:

if $M(x, y) = M(x', y')$, $M(x', z) = M(x, z')$, $M(w, y') = M(w', y)$, and wDz, $w'Dz'$, then $M(w, z) = M(w', z')$.

We leave it to the reader to check the arithmetic.

The representation of Equation 17 would raise some problem for the interpretation question: we would have to choose between u and g as possible measures of sensation magnitude, that is, unless the relation between u and g is particularly simple, of the type $g = Au$ for some $A > 0$. In that case, it is irrelevant which of u or g is taken, as a consequence of the uniqueness of both u and g: at best, the representation of Equation 17 leads to interval scales for both u and g. Incidentally, the representation

$$M(x, y) = M(x', y') \text{ iff } u(x) - Au(y) = u(x') - Au(y') \tag{18}$$

is also empirically testable. Suppose that $M(x, y) = M(x', y')$, $M(x, z) = M(x', z')$, and $M(y, w) = M(y', w')$. Then, if u and A exist satisfying Equation 18, we have

$$u(x) - Au(y) = u(x') - Au(y'), \tag{19}$$

$$u(x') - Au(z') = u(x) - Au(z), \tag{20}$$

$$u(y) - Au(w) = u(y') - Au(w'). \tag{21}$$

Adding Equations 19 and 20 gives, after dividing by A and changing signs,

$$u(z) + u(y') = u(z') + u(y), \tag{22}$$

and Equations 21 and 22 finally yield

$$u(z) - Au(w) = u(z') - Au(w').$$

Hence, if z is comparable to w, and z' to w', we obtain $M(z, w) = M(z', w')$. In other words, Equation 18 implies the following condition:

> if $M(x, y) = M(x', y')$, $M(x, z) = M(x', z')$, $M(y, w) = M(y', w')$, and zDw, $z'Dw'$, then $M(z, w) = M(z', w')$.

4.3. A Fundamental Assumption of Discrimination Theory

Essentially, this assumption is that the effect of a simple stimulus—such as a tone or the flash of a light—on the organism can be summarized by one number. This is critical not solely for the discrimination theory discussed here, but in fact for the very notion of psychophysical scaling.[9] There are various reasons why this may lead into trouble. Some were considered in the two preceding paragraphs and require no essential changes in the theory. What I have in mind involves a more serious difficulty. Suppose that stimulus x is presented at time t_0. Let $G_x(t)$ stand for the effect of x on the organism at time $t \geq t_0$. For illustrative purposes, suppose that $G_x(t)$ is the electrical response to x, recorded at some part of the brain. In some cases, $G_x(t)$ might appear as in Figure 6, which simply involves the notions that $G_x(t)$ is continuous in t, is increasing up to some maximum value $\overline{G}_x$, and then gradually decreasing. This example suggests that the effect of x on the organism is a function of time, and it is not clear that it makes sense to summarize this function by one number, the sensory scale value of x. Let us, however, pursue the discussion of this example a little further. The core of the objection, it seems, is not addressed to the representation aspect of discrimination theory, but rather to the interpretation aspect: different aspects of $G_x(t)$ may influence the subject's behavior in different tasks. Take, for example, a simple reaction-time task. A reasonable possibility is that the subject reacts as soon as $G_x(t)$

[9] This objection to discrimination theory was first pointed out to me by G. Sperling.

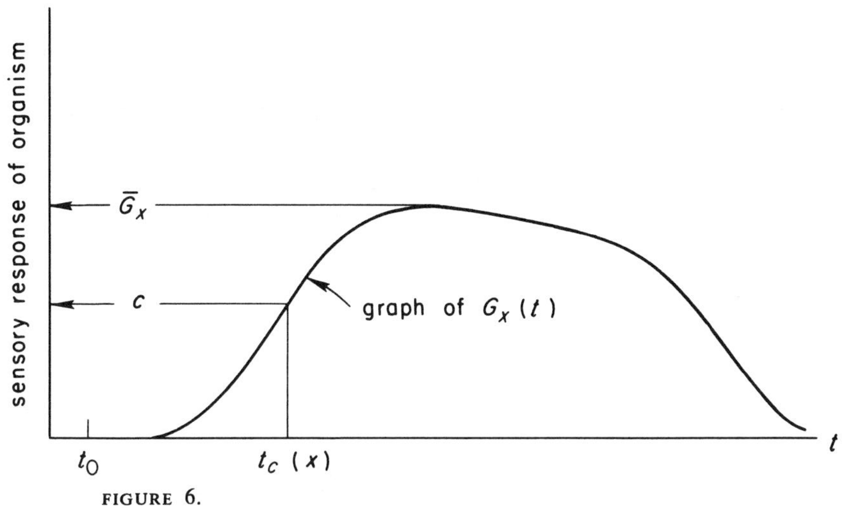

FIGURE 6.
Effect of x on the sensory system of the subject (hypothetical data).

reaches some criterion. The observed latency will then be equal to $t_c(x) + k$, in which k represents the latency of the nonsensory part of the response, and the meaning of $t_c(x)$ is made clear by Figure 6. With $x \geq y$, let $M(x, y)$ stand for the difference between the reaction time to y and the reaction time to x. We are led to the following representation:

$$M(x, y) \leq M(x', y') \text{ iff } t_c(y) - t_c(x) \leq t_c(y') - t_c(x').$$

(Statistical questions are left aside in this discussion. If one wishes, $M(x, y)$, $t_c(x)$, etc., may be considered as averaged values.) Thus, M is $-t_c$-Fechnerian. In some other methods, such as the constant stimuli, the subject may be asked to compare the subjective intensity of the stimuli. It is possible that, in this case, the maximum $\bar{G}_x$ of $G_x(t)$ will serve as a basis for the subject's decision. Conceivably, this could lead to the following representation:

$$M^*(x, y) \leq M^*(x', y') \text{ iff } \bar{G}_x - \bar{G}_y \leq \bar{G}_{x'} - \bar{G}_{y'}.$$

We obtain thus two sensory scales, $-t_c$, $\bar{G}$, and only if the $G_x(t)$ satisfy some special conditions will $-t_c$ and $\bar{G}$ be essentially the same scale. (What these conditions are is easy to guess: t_c and $\bar{G}$ should be linearly related.) The moral of the story is this—two different discrimination methods will lead to the same scale only if either (a) the operation on the $G_x(t)$ required from the subject is the same in both cases (for example, in both cases the discrimination is based on the maxima $\bar{G}_x$) or (b) it is not the same, but then the $G_x(t)$ satisfy specific conditions with respect to the operations involved.

ACKNOWLEDGMENTS

Research was supported by National Science Foundation Grant GB-33681 to New York University. I am grateful to Eric Holman and Michael Levine for their comments on a previous draft of this chapter and to Stephen Cohen and Micha Razel for their discovery of a couple of critical misprints in the final draft.

REFERENCES

Chandler, J. P. STEPIT: Finds local minima of smooth function of several parameters. (CPA 312) *Behavioral Science*, 1969, **14,** 81–82.

Cobb, P. W. Weber's Law and the Fechnerian muddle. *Psychological Review*, 1932, **39,** 533–551.

Coombs, C. H., Dawes, R. M., & Tversky, A. *Mathematical psychology: An elementary introduction.* Englewood Cliffs, New Jersey: Prentice-Hall, 1970.

Cross, D. V. *An application of mean value theory to psychological measurement.* Progress Report No. 6, Report Number 05613-3-P. Ann Arbor: The Behavioral Analysis Laboratory, University of Michigan, 1965.

Doignon, J. P., & Falmagne, J. C. A theorem for difference measurement and other results. Mimeographed report, 1972.

Falmagne, J. C. The generalized Fechner Problem and discrimination. *Journal of Mathematical Psychology*, 1971, **8,** 22–43.

James, W. The measure of discriminative sensibility. In *Principles of psychology*. Vol. 1. New York: Holt, 1890. (Republished by Dover, New York, 1950.)

Johnson, H. M. Did Fechner measure 'introspectional' sensations? *Psychological Review*, 1929, **36,** 257–284.

Johnson, H. M. Some properties of Fechner's intensity of sensation. *Psychological Review*, 1930, **37,** 113–123.

Knudsen, V. O. The sensibility of the ear to small differences in intensity and frequency. *Psychological Review*, 1923, **21,** 84–103.

Krantz, D. H. Integration of just-noticeable differences. *Journal of Mathematical Psychology*, 1971, **8,** 591–599.

Krantz, D. H. A theory of magnitude estimation and cross-modality matching. *Journal of Mathematical Psychology*, 1972, **9,** 168–199.

Krantz, D. H., Luce, R. D., Suppes, P., & Tversky, A. *Foundations of measurement.* Vol. 1. New York: Academic Press, 1971.

Kristof, W. *On the theory of unidimensional scaling.* Research Bulletin RB-65-14. Princeton, N.J.: Educational Testing Service, 1965.

Kristof, W. *A foundation of interval scale measurement.* Research Bulletin RB-67-22. Princeton, N.J.: Educational Testing Service, 1967.

Levine, M. V. Transformations that render curves parallel. *Journal of Mathematical Psychology*, 1971, **7,** 410–444.

Luce, R. D. *Individual choice behavior.* New York: Wiley, 1959.

Luce, R. D., & Edwards, W. The derivation of subjective scales from just noticeable differences. *Psychological Review*, 1958, **65,** 227–237.

Luce, R. D., & Galanter, E. Psychophysical scaling. In R. D. Luce, R. R. Bush, differences. *Psychological Review*, 1958, **65,** 227–237.

Luce, R. D., & Galanter, E. Discrimination. In R. D. Luce, R. R. Bush, and E. Galanter (Eds.), *Handbook of mathematical psychology*. Vol. 1. New York: Wiley, 1963. (a)

Luce, R. D., & Galanter, E. Psychophysical scaling. In R. D. Luce, R. R. Bush, and E. Galanter (Eds.), *Handbook of mathematical psychology*. Vol. 1. New York: Wiley, 1963. (b)

Miller, G. A. Sensitivity to changes in the intensity of white noise and its relation to masking and loudness. *Journal of the Acoustical Society of America*, 1947, **19,** 609–619.

Pears, D. F. *Bertrand Russell and the British tradition in philosophy*. New York: Random House, 1967.

Pfanzagl, J. *Die axiomatischen Grundlagen einer allgemeinen Theorie des Messens*. Würzburg, 1959.

Pfanzagl, J. *Theory of measurement*. New York: Wiley, 1968.

Piéron, H. Recherche sur la latence de perception des accroissements de luminosité. *Année Psychologique*, 1936.

Ross, J., & di Lollo, V. Judgment and response in magnitude estimation. *Psychological Review*, 1971, **6,** 515–527.

Roufs, J. Perception lag as a function of stimulus luminance. *Vision Research*, 1963, **3,** 81–91.

Stevens, S. S. The measurement of loudness. *Journal of the Acoustical Society of America*, 1955, **27,** 815–820.

Stevens, S. S. On the psychophysical law. *Psychological Review*, 1957, **64,** 153–181.

Stevens, S. S. The psychophysics of sensory function. In W. H. Rosenblith (Ed.), *Sensory communication*. New York: Wiley, 1961.

Stevens, S. S. Issues in psychophysical law. *Psychological Review*, 1971, **5,** 426–450.

Teghtsoonian, R. On the exponents in Stevens' Law and the constant in Ekmans' Law. *Psychological Review*, 1971, **1,** 71–80.

Zinnes, J. L. Scaling. *Annual Review of Psychology*, 1969, **20,** 447–478.

Measurement Theory and Qualitative Laws in Psychophysics

David H. Krantz
UNIVERSITY OF MICHIGAN

The measurement of distances, by applied trigonometry, depends on the accuracy of a theory (plane Euclidean geometry). This theory adequately describes the empirical observations of such relations as equidistance, collinearity, and perpendicularity. For large distances on the earth's surface the theory in fact fails and must be replaced by spherical geometry. Likewise, in physics, measurement is a by-product of theory, as well as a tool that facilitates the development of new theory.

Not all theory gives rise to measurement. Physiological theories, for example, depend heavily on physical measurement, but do not give rise to any new or special physiological quantities. Blood pressure, for example, is simply *physical* pressure, measured in a physiological context. Since not all theories yield new measurement scales as by-products, there is no a priori reason to expect that development of psychological theory should give rise to measurement of special psychological quantities.

Historically, however, the idea of measuring special mental quantities has had great appeal, and nowhere more than in psychophysics, where our impressions of stimuli can often be naturally described in terms of variations of certain perceptual attributes, or psychological dimensions, in an ordered and perhaps a quantitative way. Most people will agree, for example, that sunlight, reflected off snow, is more than twice as bright as the light from a

3-watt nightlamp coming through a crack under a door. This statement (more than twice as bright) refers to our mental impressions, not to any observations of physical quantities such as watts.

Such intuitive appeal of a quantitative perceptual attribute does not lead automatically to measurement. The principal goal, in my view, should not be the construction of a measure, but the development of theories. Measurement scales of brightness, for example, will occur as by-products of theories of brightness.

Much of this chapter is devoted to theories of brightness and to their measurement by-products. I use the plural, 'theories,' because there are several different data structures, with separate laws, and therefore, at least initially, separate theories. Of course, one also hopes to understand the relationships between these different judgments, thus arriving at a unified theory.

SIX MEASUREMENT STRUCTURES INVOLVING BRIGHTNESS

Which six? Rather than have the reader guess or look ahead, I provide a quick overview. This list is not exhaustive of tasks involving brightness; it merely selects six families of tasks that illustrate a fairly wide range of different theories and different measurement scales.

(1) *Matching:* gives rise to an equivalence relation $\sim_B$, where $a \sim_B b$ if lights a and b appear equally bright.

(2) *'Ratio' judgment:* gives rise to an ordering relation $\gtrsim_B$ of pairs of lights, where $(a,b) \gtrsim_B (c,d)$ if the brightness 'ratio' of a to b is at least as great as that of c to d.

(3) *Discrimination:* gives rise to a probability function $P_B(a,b)$, the probability that light a is judged brighter than b; or to a receiver operating characteristic (ROC; Green & Swets, 1966) for the pair (a,b).

(4) *Dissimilarity:* gives rise to an ordering relation $\gtrsim$ of pairs of lights, where $(a,b) \gtrsim (c,d)$ if the dissimilarity of a to b is at least as great as that of c to d, where (a,b) and (c,d) may be dissimilar in brightness and/or in other perceptual attributes.

(5) *Contrast:* gives rise to a cross-context equivalence relation $\sim_{ab}$, where $c(\sim_{ab})d$ if c, in context a, matches d, in context b. Here, one studies the way in which the relation $\sim_{ab}$ varies as a function of the contextual brightnesses, determined by a and b. (The nature of the match between c and d is left open—it could bc a brightness match, a color match, etc.)

(6) *Veiling:* a matching relation for some other attribute of lights (e.g., saturation, or redness) is studied as a function of the brightness of the lights.

It will be noted that the first three structures seem to involve brightness more or less directly in the judgment task (brightness equality, brightness 'ratio,' brighter than) whereas the latter three study the effect of brightness on other kinds of judgments such as dissimilarity or saturation of colors. In the case of contrast, one is concerned with the brightness of contextual stimuli, rather than focal stimuli (Jameson & Hurvich, 1959), and in the other two cases, one is concerned with the brightness of the focal stimuli though some judgment other than brightness is made.

Some superficial interrelations among these structures are obvious. Matching, for example, could be regarded as a special case of both 'ratio' judgments and discrimination. Presumably, $a \sim_B b$ if and only if for all c, $(a, c) \approx_B (b, c)$ (here, $\approx_B$ means that $\gtrsim_B$ holds in both directions), and again, $a \sim_B b$ if and only if $P_B(a, b) = \frac{1}{2}$. These interrelations may provide a start toward a unified theory.

Recent developments in foundations have led to the study of a great variety of abstract measurement structures. I shall now sketch the descriptions of six abstract measurement structures, relevant in turn to the six kinds of empirical relations listed above. These are as follows: Grassmann structures, algebraic-difference structures, representable ROCs, additive conjoint structures, cross-context Grassmann structures, and distributive conjoint structures.

1. *Grassmann structure.* This consists of a set A with an addition operation $\oplus$, an operation $*$ of multiplication by positive real numbers, and an equivalence relation $\sim$. The operations $\oplus$ and $*$ satisfy the formal properties of vector operations (restricted, say, to the positive orthant of an n-dimensional coordinate space, since there is no subtraction or multiplication by zero or negative scalars), and in addition, $\sim$ satisfies Grassmann's (1853) additivity laws, which can be written as follows:

$$a \sim b \quad \text{iff} \quad a \oplus c \sim b \oplus c. \tag{1}$$

$$\text{If } a \sim b, \quad \text{then } t * a \sim t * b. \tag{2}$$

The measurement theorem based on these laws (Krantz, 1974) states that any Grassmann structure can be represented by a convex cone in a vector space over the reals, where $\sim$ is represented by $=$ (vector equality). That is, each a in A is represented by a vector $\phi(a)$, with

(i) $\phi(a \oplus b) = \phi(a) + \phi(b)$ (vector addition)

(ii) $\phi(t * a) = t \cdot \phi(a)$ (scalar multiplication)

(iii) $a \sim b$ iff $\phi(a) = \phi(b)$.

The representation is unique up to vector-space isomorphisms (nonsingular linear transformations).

In the application to color, A is the set of colored lights, $\oplus$ is additive color mixture, and $*$ is an overall change in energy (multiply the radiant energy at each wavelength by the same factor t). Thus, A, $\oplus$, $*$, and their properties involve only physics. The psychological input is the relation $\sim$. This can be either complete color matching (metameric colors a, b look identical in all perceptual attributes of color) or partial matching, with respect to one or more attributes, e.g., brightness matching, $\sim_B$.

In the case of complete color matching, Grassmann's additivity laws hold extremely well (see Brindley, 1970, pp. 214–216) and have profound import for physiological mechanisms of color coding (Brindley, 1970, pp. 210–214). In this case, the representing cone of vectors is three dimensional (law of trichromacy, which has even more basic physiological import). This three-dimensional vector representation is the basis of the international standards of color measurement (CIE system).

For brightness matching, the additivity laws are known as Abney's law, or Grassmann's fourth law, and their validity is a subject of dispute. Certainly, additivity is grossly violated when $\sim_B$ is established by direct comparison of the brightness of lights differing in wavelength composition. For example, in (1), let a, b, c be lights of wavelength 450 nm, 650 nm, and 520 nm, respectively, and let their intensities be adjusted so that all three are equally bright, i.e., $a \sim_B b \sim_B c$. Then (1) predicts that

$$a \oplus c \sim_B b \oplus c \sim_B 2 * c = c \oplus c,$$

but in fact (Guth, Donley, & Marrocco, 1969) the mixture of 450 and 520 nm is brighter than twice the 520 nm stimulus (enhancement), while the sum of 650 and 520 nm (red and green) is not bright enough. Adding a small amount of 520 nm light to 650 nm can even decrease the brightness!

However, other methods of establishing a 'brightness' equivalence relation do yield good approximations to additivity. These involve minimizing flicker when the two lights being compared are rapidly alternated (LeGrand, 1968, pp. 67–72, 128–129) or minimizing the apparent contrast at a border (Boynton & Kaiser, 1968). Obviously, different methods are getting at somewhat different perceptual attributes of color; the term 'brightness' must be replaced by more precise names. If additivity does hold, then the resulting cone of representing vectors is one dimensional, i.e., $\phi(a)$ is a positive number. (In the case of the flicker method, the additive scale $\phi(a)$ corresponds closely to the *luminance* scale.)

The preceding discussion illustrates extremely well the way in which measurement scales (CIE tristimulus system, or luminance) depend on the lawfulness of empirically determined structures.

Below, I shall discuss in detail two other one-dimensional Grassmann structures for perceptual attributes of color, which lie at the foundation of opponent-colors theory.

2. *Algebraic-difference structure.* This consists of a set A and an ordering of pairs in A, $\succsim$. The ordering must satisfy the following two laws (reversibility and weak monotonicity):

$$\text{If } (a, b) \succsim (c, d), \text{ then } (d, c) \succsim (b, a). \tag{3}$$

$$\text{If } (a, b) \succsim (a', b') \text{ and } (b, c) \succsim (b', c'), \text{ then } (a, c) \succsim (a', c'). \tag{4}$$

Besides these, two technical axioms are needed: solvability, which assures that the set of stimuli is sufficiently rich, and the Archimedean axiom, which assures that no interval (a, b) is infinitely large relative to the 'mesh' defined by some other interval.

The measurement theorem based on these laws (Krantz, Luce, Suppes, & Tversky, 1971, p. 151) asserts that an algebraic-difference structure can be represented by an ordering of numerical differences, i.e., to each a in A one assigns a number $\phi(a)$ such that

$$(a, b) \succsim (c, d) \quad \text{iff} \quad \phi(a) - \phi(b) \geq \phi(c) - \phi(d).$$

Similar theorems have been established previously by many people: see Pfanzagl (1968, p. 147), Krantz et al. (1971, p. 150), and especially Falmagne (this vol.).

If we define $\psi = \exp \phi$, the difference representation is converted into a ratio one:

$$(a, b) \succsim (c, d) \quad \text{iff} \quad \psi(a)/\psi(b) \geq \psi(c)/\psi(d).$$

The function ψ is unique up to transformations of form $\psi' = \alpha\psi^{\beta}$, where α, β are positive constants; that is, ψ is a logarithmic interval scale.

The application to brightness of these ideas is very direct: all that is needed is a method of obtaining ordinal judgments of brightness pairs (a, b), (c, d), etc. If the ordering $\succsim_{\text{B}}$ satisfies laws (3) and (4), then, since the technical axioms pose no problem for brightness, the resulting brightness function ψ_{B} can be constructed. For example, one can construct a *standard sequence* of stimuli, $a_0, a_1, a_2, \ldots, a_m, \ldots$, increasing in brightness, with equal brightness 'ratios' in the sense that $(a_2, a_1) \approx_{\text{B}} (a_1, a_0)$, $(a_3, a_2) \approx_{\text{B}} (a_2, a_1)$. Then for any positive constants α, β, the function

$$\psi_{\text{B}}(a_i) = \alpha e^{\beta i}$$

is a suitable ratio representation for $a_0, a_1, \ldots$. The weak monotonicity property, law (4), plays a key role: it guarantees that $(a_2, a_0) \approx_{\text{B}} (a_3, a_1)$, etc., and more generally, that $(a_i, a_j) \approx_{\text{B}} (a_k, a_\ell)$ for $i - j = k - \ell > 0$. Thus, ψ_{B} ratios describe all brightness 'ratio' properties for this sequence. (Law (3), reversibility, is needed for the reversed pairs.)

With minor changes, the same abstract structure can be construed as describing the results of the classic direct-estimation experiments of Stevens (1957, 1966) using the techniques of magnitude estimation and cross-modality matching. This development, described in more detail below, is due to R. Shepard (see Krantz, 1972b).

In comparing the kinds of results obtained from brightness matching and ordering of brightness 'ratios,' we see why it is superficial to dismiss matching as just the special case of a unit ratio. The measurement structure based on matching centers around the additive structure, $\oplus$ and $*$, and the additivity laws, (1) and (2), are crucial. Corresponding to the additivity laws are physiological questions, i.e., how do different wavelengths of light contribute additively to brightness? The 'ratio' structure ignores the additive structure of lights; in fact, brightness 'ratio' tasks typically employ pairs of lights that have minimal or no differences in spectral composition. The key law is (4), which asserts that the stimulus 'ratio' in fact does have the qualitative property of a numerical difference or ratio. If a physiological matter can be raised, it is the question of what sort of neural network is used to compute an effect of a stimulus pair that serves as a basis for such judgments. Law (4) constitutes a constraint on that network.

3. *Representable ROCs.* We have a set A; and for each pair (a, b) in A we have a continuous and strictly increasing function f_{ba}, from the unit interval $[0, 1]$ onto itself. This function is the ROC curve determined by discrimination of b from a: if p is the probability that the subject (incorrectly) identifies a as 'b,' then $q = f_{ba}(p)$ is the probability of a (correct) response 'b' given that b was really presented. Holding a, b constant, p and q vary with the subject's criterion for saying 'b.' The desired representation postulates random variables X_a, X_b corresponding to stimuli a and b; in discriminating b from a (where, e.g., b is on the average slightly brighter than a) the subject chooses some criterion x and responds 'b' if his observation exceeds x. Thus,

$$p = P[X_a > x],$$
$$q = P[X_b > x].$$

From results of Marley (1971) (applied to the case where the functions f_{ba} are continuous and strictly increasing from $[0, 1]$ onto $[0, 1]$) the necessary and sufficient condition for this representation by random variables is that the following law hold:

$$f_{cb}(f_{ba}(p)) = f_{ca}(p). \tag{5}$$

That is, if (p, q) is on the ROC for (a, b), and (q, r) is on the ROC for (b, c), then (p, r) must be on the ROC for (a, c).

The uniqueness result is relatively weak. The distribution functions of the representing random variables are determined only up to a common monotonic transformation.

The additional laws that must hold, of ROCs are to be represented by random variables whose distributions are identical except for shifts, or except for shifts and changes of scale, were discovered by Levine (1970, 1972). One circumstance in which the representation is by distributions that differ only by shifts is when the relation $\gtrsim$, defined by

$$(a, b) \gtrsim (c, d) \quad \text{iff} \quad f_{ba}(p) \geq f_{dc}(p),$$

yields an algebraic-difference structure on the set A. The main force of this is that the ROC curves must be uncrossed (Levine, 1970), i.e., if $f_{ba} \geq f_{dc}$ for one p, the same holds for all p. For then, law (4) for algebraic-difference structures follows from (5).

There seem to be no good data that test whether ROCs are uncrossed or whether law (5) holds.

4. *Additive conjoint structures.* We have m sets, $A_1, \ldots, A_m$ and form the set of m-tuples, $A = A_1 \times \cdots \times A_m$. We also have an ordering relation, $\gtrsim$, on A. This ordering satisfies laws sufficient to guarantee the constructibility of m functions, $\phi_1, \ldots, \phi_m$, on $A_1, \ldots, A_m$, such that

$$(a_1, \ldots, a_m) \gtrsim (b_1, \ldots, b_m) \quad \text{iff} \quad \sum_{i=1}^{m} \phi_i(a_i) \geq \sum_{i=1}^{m} \phi_i(b_i).$$

In the application to dissimilarity, A_i is the set of differences on the ith perceptual attribute. Thus, if $\delta(a, a')$ denotes the dissimilarity between a and a', and a is identified with its values $a_1, \ldots, a_m$ on m dimensions, we have:

$$\delta(a, a') \geq \delta(b, b') \quad \text{iff} \quad \sum_{i=1}^{m} \phi_i(a_i, a_i') \geq \sum_{i=1}^{m} \phi_i(b_i, b_i');$$

or, in other words,

$$\delta(a, a') = F\left[\sum_{i=1}^{m} \phi_i(a_i, a_i')\right], \tag{6}$$

where F is a strictly increasing function. A special case is the Euclidean model of ordinal multidimensional scaling (Shepard, 1966), where

$$\delta(a, a') = F\left[\sum_{i=1}^{m} |\phi_i(a_i) - \phi_i(a_i')|^2\right]. \tag{7}$$

Various sets of sufficient laws for the additive conjoint representation have been given (see Krantz et al., 1971, Ch. 6), and the applications of additive conjoint structures to multidimensional scaling have been discussed by Beals, Krantz, and Tversky (1968) and Tversky and Krantz (1970). A full treatment of the Euclidean model also requires analysis analogous to that of algebraic-difference structures, since difference-representation terms, $|\phi_i(a_i) - \phi_i(a_i')|$, occur.

The most important qualitative laws in additive conjoint structures are *independence laws* (Krantz & Tversky, 1971). For clarity, let me explain these by a three-factor illustration from an experimental test (Tversky & Krantz, 1969). As shown in Figure 1, the stimuli were schematic faces varying in shape of face (long vs wide), eyes (empty vs filled), and mouth (straight vs curved). *Independence* states that if the same shape difference occurs in two pairs of faces, then the dissimilarity ordering of those two pairs is determined by the differences in the other attributes and does not depend on the particular constant level of shape difference that occurs. An example of such a prediction is shown in Figure 2. If a combination of mouth difference and zero eye difference (left-hand pairs) produces more dissimilarity than a combination of eye difference and zero mouth difference (right-hand pairs), then this must hold regardless of whether the constant shape difference is set at zero difference (top) or nonzero difference (bottom).

FIGURE 1.
Eight faces varying on three binary attributes, used to test whether the differences on the three attributes contribute additively to overall dissimilarity.

FIGURE 2.
A test of joint-factor independence for the schematic faces of Figure 1. The dissimilarity ordering of mouth difference combined with zero eye difference (left) versus eye difference combined with zero mouth difference (right) should be independent of the constant level of shape difference (zero in the top row, nonzero in the bottom one).

If we designate the two kinds of shape by a, a', the two eye types by b, b', and the two mouths by c, c', then Figure 2 can be transcribed as:

$$[(a, a), (b', b'), (c, c')] > [(a', a'), (b', b), (c, c)]$$
$$\text{iff}$$
$$[(a, a'), (b', b'), (c, c')] > [(a, a'), (b', b), (c', c')].$$

This formulation makes clear that the prediction involves not only the independence law (the joint effects of the latter two factors is independent of the constant level of the first factor), but also another axiom, namely, that (a, a) and (a', a') both represent the zero level of the shape-difference factor, while (c, c) and (c', c') represent the zero level of the mouth-difference factor.

If three or more factors contribute nontrivially to the ordering,[1] and if all possible independence laws (of the kind just illustrated) are satisfied, then with the addition of technical axioms like those for algebraic-difference

[1] With only two factors, an additional qualitative law must be assumed to guarantee additivity. See Krantz et al. (1971) for various *cancellation laws* that can be used for the purpose.

structures, the additive conjoint scales $\phi_1, \ldots, \phi_m$ can be constructed. (For proof, see Debreu, 1960; Krantz et al., 1971.)

The application to brightness measurement involves dissimilarity judgments of colors varying in hue, saturation, and brightness. For example, if a model of form (7) could be made to fit such judgments, where a_1, a_1' denote different (ordinal) levels of brightness, then $\phi_1(a_1), \phi_1(a_1')$ could be regarded as interval-scale values for brightness. These scale values would be by-products of a satisfactory theory about contribution of brightness differences to dissimilarity.

5. *Cross-context Grassmann structures.* Here, we regard a, b as context stimuli and have equivalence relations $\sim_{ab}$ on a set A. For each a, $\sim_{aa}$ (identical context) may satisfy the additivity laws for a Grassmann structure. (In fact, for complete color matching, $\sim_{aa}$ is approximately independent of a: metameric color matches are context-invariant.) Can these additivity laws be extended to $\sim_{ab}$, for $a \neq b$? An extension suggested[2] by Krantz (1968a) is the following:

$$\begin{aligned} \text{If} \quad & c_1 \sim_{ab} d_1, \quad (c_1 \oplus c) \sim_{ab} e_1, \\ & c_2 \sim_{ab} d_2, \quad (c_2 \oplus c) \sim_{ab} e_2, \\ \text{then} \quad & d_1 \oplus e_2 \sim d_2 \oplus e_1. \end{aligned} \tag{8}$$

The explanation is that the addition of c to c_1 produces a change in appearance in context a (c_1 vs $c_1 \oplus c$) that is matched by the change from d_1 to e_1 in context b. Similarly, adding c to c_2 produces a context-a change that is matched, in context b, by the change from d_2 to e_2. The conclusion of the axiom is that the 'vector differences' $e_1 - d_1$ and $e_2 - d_2$ are equal, or rigorously, $d_1 \oplus e_2$ matches $e_1 \oplus d_2$ (with identical context, e.g., $\sim_{bb}$; the context is suppressed when equal).

The above law, plus technical assumptions, yields the result that context change can be represented by an affine transformation in the vector space derived from the usual Grassmann structure. For example, in the three-dimensional color space,

$$c \sim_{ab} d \quad \text{iff} \quad \phi(d) = f_{ba}[\phi(c)],$$

where f_{ba} is an affine transformation, representable by a 3×4 matrix:

$$\phi_i(d) = \alpha_{i0}(b, a) + \sum_{j=1}^{3} \alpha_{ij}(b, a)\phi_j(c), \quad i = 1, 2, 3.$$

[2] The formulation given here combines the general assumption of context-invariance (Krantz, 1968a, pp. 10–12) with the special use of addition transformations (p. 29).

There is in fact moderately strong evidence that color-appearance changes due to context can be fit by affine transformations (Burnham, Evans, & Newhall, 1957), although a nonlinear model (Jameson & Hurvich, 1964) is a strong competitor.

If the above structure is valid, then the next step is to relate the 12 functions $\alpha_{ij}(b, a)$ to characteristics of b and a, such as brightness. However, the matrix α_{ij} depends on the coordinate system adopted for the vectors $\phi(c)$, etc. Also, the context stimuli, a and b, have representations $\phi(a)$, $\phi(b)$ in the same space. Introducing the Grassmann structure directly into the set of context stimuli suggests the following additional law:

$$d \sim_{ab} e \quad \text{iff} \quad d[\sim_{(a \oplus c, b \oplus c)}]e.$$

This law leads to a situation where α_{ij} is a function of the vector difference, $\phi(b) - \phi(a)$. One might hope that a suitable coordinate system, incorporating brightness or luminance as one coordinate, would lead to an especially simplified matrix $\alpha[\phi(b) - \phi(a)]$; for example, one in which brightness scales for b, a bear simple relations to the transformation f_{ba} that represents the effects of context.

6. *Distributive conjoint structures.* Here we have three sets, A_1, A_2, A_3, and an ordering $\gtrsim$ of triples (a_1, a_2, a_3) in $A_1 \times A_2 \times A_3$, as in a three-factor additive conjoint structure. The difference is that the representation involves measurement scales ϕ_1, ϕ_2, ϕ_3 on A_1, A_2, and A_3, such that:

$$(a_1, a_2, a_3) \gtrsim (b_1, b_2, b_3) \quad \text{iff}$$
$$[\phi_1(a_1) + \phi_2(a_2)]\phi_3(a_3) \geq [\phi_1(b_1) + \phi_2(b_2)]\phi_3(b_3).$$

That is, $\phi = (\phi_1 + \phi_2)\phi_3$ is monotone with the ordering of $A = A_1 \times A_2 \times A_3$.

Discussion of distributive structures can be found in Coombs and Huang (1970), Krantz (1968b), Krantz et al. (1971, Ch. 7), and Krantz and Tversky (1971).

The qualitative laws that characterize distributive conjoint structures are a bit more complicated than those for the three-factor additive case. One needs to have two sets of laws similar to the ones for two-factor additivity: one centering around addition, one around multiplication. But something more (the so-called distributive cancellation law) is needed to make sure that addition and multiplication mesh properly. (See Krantz & Tversky, 1971, for flowcharts of qualitative laws for three-variable polynomial models.)

Note that the independence law of additive structures holds in the distributive case for $A_1 \times A_2$ effects, with the A_3-effect constant (except if ϕ_3 changes sign) but the other two independence laws, for $A_1 \times A_3$ effects or $A_2 \times A_3$ effects, need not hold, and indeed, will usually be violated where the distributive model is correct.

The distributive model may be useful for representing the influence of redness, yellowness, and brightness on the saturation of orange lights (or greenness, yellowness, and brightness on the saturation of yellow-green lights, etc.). The chromatic components may add, while the inverse of brightness acts as a multiplicative factor. This hypothesis is consistent with the Hurvich and Jameson (1955, 1957) opponent-colors treatment of saturation. Thus, study of the 'veiling' of chromatic components by brightness, studied via saturation judgments, might yield yet another scale for brightness, ϕ_3^{-1}.

ADVANTAGES OF THE MEASUREMENT-THEORY FORMULATIONS

One advantage, emphasized also by Anderson (1970) and Krantz (1972a), is that measurement is no longer regarded as something that must be struggled with prior to theory construction; rather, it is a consequence of the discovery of quantitative and/or qualitative laws. In numerous cases, illustrated especially by the work of Anderson (this vol.) and his collaborators, simple additive laws describe the effects of different attributes on perception or psychophysical judgment.

Second, the formulation of a theory in terms of a few qualitative laws, the intuitive meanings of which are fairly transparent, aids the testing and the development of theory. It concentrates observations on critical points, where small but systematic deviations may be theoretically meaningful—where systematic violations may suggest alternative qualitative laws.

Third, by making the tests of the measurement model more sensitive to small systematic deviations, one promotes the viewpoint that psychometric techniques are not theoretically neutral; rather, they impose a theoretical structure, whose psychological content deserves analysis. Nowhere is this more necessary than in multidimensional scaling, where there are numerous and widely applied 'data-reduction techniques.' These work well enough on a gross level, because of the excellent analogy between psychological distance and geometric distance. If a is close to b, and b is close to c, then a and c cannot be too distant, etc. But the question raised by measurement theory concerns the finer grain of the data reduction: are the details of the Euclidean or other geometric model a good psychological theory?

Fourth, in many cases, the formulation of qualitative laws suggests experiments that are intrinsically simple and interesting, quite apart from the measurement motivation, but which might be very hard to arrive at otherwise. This is illustrated nicely in Falmagne (this vol.). In my own work, each of the six structures described above has led at least to the contemplation of an experimental question that had not been tackled before, despite its apparent simplicity; and in three of the six cases, I have performed the experiment. A list of six experimental questions follows:

1. (*Grassmann structure for opponent-colors theory*). Given a yellow and a blue light, each of which appears neither reddish nor greenish, is the additive color mixture neither reddish nor greenish?

2. (*Algebraic-difference structure for sensation 'ratios'*). Is the ordering of pairs of lights the same when the subject is told to judge brightness 'ratios' vs brightness 'differences,' and how are these orderings related to single-stimulus judgments with 'ratio' or 'difference' instructions?

3. (*Representable ROCs*). Does the Weber fraction determine the entire ROC curve, i.e., are ROCs uncrossed?

4. (*Additive conjoint structure for color dissimilarity*). Given two lights of different brightness, but identical hue and saturation, does the overall dissimilarity rating vary with the particular hue and saturation selected?

5. (*Cross-context Grassmann structure*). Is the vector difference in tristimulus coordinates of the lights that match c_1 and $c_1 \oplus c$ independent of c_1, when c_1 and $c_1 \oplus c$ are viewed with a colored surround and the matching lights are viewed in a dark surround?

6. (*Distributive conjoint structure for saturation judgments*). Are saturation judgments, at fixed brightness, related additively to the judgments of hue components of the stimuli in question?

A fifth advantage is that measurement theory provides an alternative, or at least a useful precursor, to reductionist or process-oriented theories. Qualitative laws of psychophysical judgment may lead more or less directly to more elaborate and powerful theories; or at least, they provide strong constraints on other types of theory. They tell the physiologist what to look for, rather than vice versa.

Last, the abstract character of the measurement structures facilitates transfer of experimental and theoretical ideas across disparate subject matters. This is illustrated by Anderson's numerous applications of additivity; by geometric models for psychological dissimilarity; by the varied examples of distributive conjoint structures (Krantz & Tversky, 1971); by the fact that the expected-utility model can also be applied to perceived risk, time preferences, social-welfare judgments, and linearity analysis of biological systems. But the best example of all is one that I shall briefly describe below, the analogy between force measurement and color measurement, where the interplay of similarities and differences between the two fields has led to new insights about both.

This final advantage, of easy transfer, is not an unmixed blessing. In some cases, the pursuit of one of the known abstract measurement structures may be a sterile research strategy. Perhaps the best guarantee that this is not happening, in a particular instance, is when the experiments dictated by such pursuit are appealing ones that should be done anyhow.

PERCEPTUAL ATTRIBUTES OF HUMAN COLOR VISION

The Force-Table Analogy

There is a superficial analogy between force and color: both are represented by three-dimensional vectors. The empirical basis for the color representation is well understood: as explained above, metameric color matching yields a trichromatic Grassmann structure, $\langle A, \oplus, *, \sim \rangle$. The validity of the vectorial representation of force is even better established, but paradoxically its foundations are less well understood, because of confusing debates over the logical status of the concept of 'force.' But in fact, the empirical structure for vectorial measurement of force is isomorphic to that for color—so the analogy is not so superficial, after all. This basis is illustrated (for the two-dimensional case) by the force table pictured in Figure 3. Various weights, attached to the central ring by strings running over pulleys, form the *force-magnitude configuration* acting on the small central ring. The force-magnitude configuration, specified by its weight at each angle around the table, is analogous to the colored light, specified by its radiant power at each wavelength of the visible spectrum. Combining two configurations is performed by combining their weights, angle by angle, just as additive color mixture combines radiant power, wavelength by wavelength. Two physically distinct configurations can have the same 'resultant' force, i.e., they can be brought into equilibrium (no motion of the central ring) when the same third configuration (opposite to the 'resultant') is added to them. Analogously, two physically distinct lights can have the same color, constituting a metameric match.

Furthermore, the qualitative laws are identical: in particular, Grassmann's laws of additivity, (3) and (4) above, hold also for equivalences of force-magnitude configurations. Three dimensionality (or, for the force table, as for dichromatic vision, two dimensionality) is similarly isomorphic in the two situations. Just as any four colors yield a metameric match (three with one, or two with two) after adjustment of the overall energy of three of them, so any four force configurations (in the three-dimensional analog of the force table) can be linearly combined into two equivalent configurations, after appropriate rescaling of three of them.

Thus far, the analogy is interesting, but to me at least, it does more to clarify physics (by separating the observable force-magnitude configuration from its measurement representation, the resultant-force vector) than psychophysics. But there is a further aspect, which lies still deeper (Krantz, 1974c).

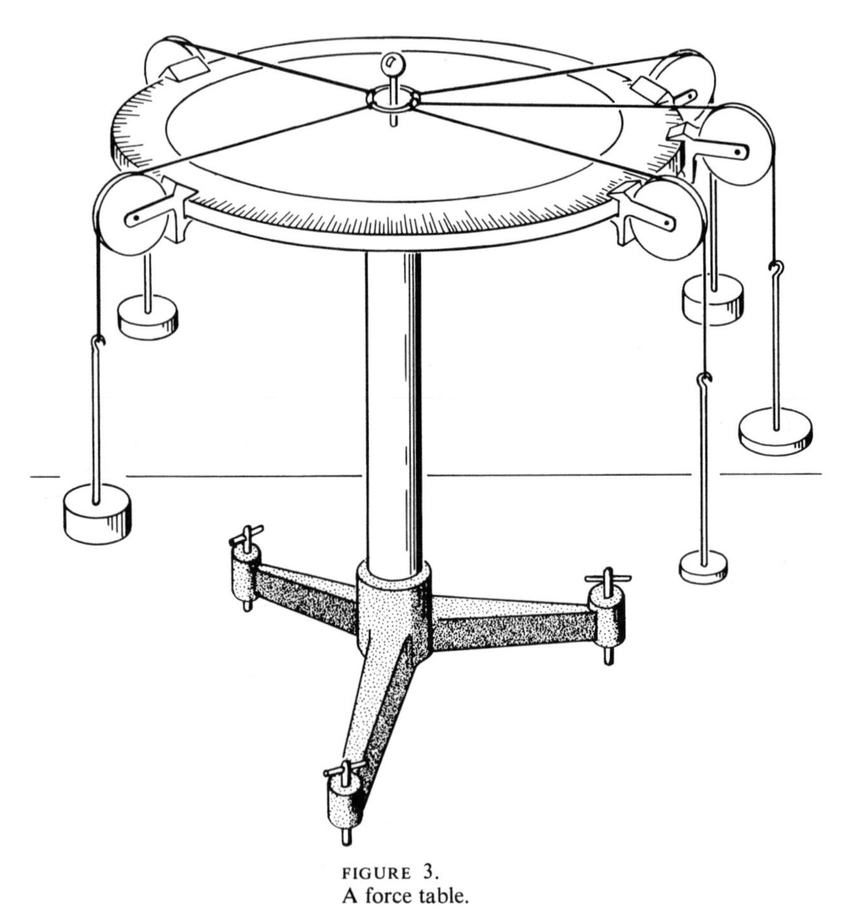

FIGURE 3.
A force table.

Trichromacy of color is usually considered to be a startling law, with profound physiological implications (three receptor systems, etc.). But tridimensionality of force is taken for granted, because motion occurs in three-dimensional physical space. In fact, it is a mistake to take three dimensionality of force for granted. It is logically possible for equilibrium of forces to be defined by the absence of acceleration in two- or three-dimensional physical space, and for force-magnitude configurations to form an m-dimensional Grassmann structure relative to that definition of equilibrium, where m is any integer whatsoever. (A physical system with that property can be built.)

The reason for the tridimensionality of force lies in the fact that we observe not only acceleration but *components of acceleration in particular directions*. For example, let γ denote some direction in (three-dimensional) physical space. We can put a constraining rail into the system that permits the central object, whose motion is observed, to move only in direction γ, and then, for any force-magnitude configuration, we can evaluate its component in direction γ by seeing what it takes to equilibrate it with respect to motion in that direction. Two configurations that are equilibrated for γ by the same thing are equivalent with respect to their components in direction γ; this relation can be denoted $\sim_\gamma$. The structure $\langle A, \oplus, *, \sim_\gamma \rangle$ is a one-dimensional Grassmann structure; the grand equivalence of configurations, $\sim$, amounts to $\sim_\gamma$ holding with respect to each and every direction γ; and *because acceleration is a three-dimensional vector, if* $\sim_\gamma$ *holds between two configurations for each of three linearly independent directions* γ, *then it holds for every direction, i.e., the two configurations satisfy the relation* $\sim$. It can be proved that the three dimensionality of the vector representation of force configurations follows from these statements. (Even more follows: the second part of Newton's second law, that the directions of acceleration and resultant force coincide.)

This clarification of why the dimensionality of the force representation coincides with the dimensionality of physical space also has its counterpart in color perception. Just as we can observe components of motion by introducing a constraining rail, so also we can observe components, or unidimensional codes for color, by constraining the observer's judgment or by obtaining a unidimensional physiological measure. The three classical attributes of color are hue, brightness, and saturation; each involves judgment of a single perceptual attribute. There are other attributes that can be judged, e.g., redness or yellowness (if the light is in the orange range), and there are also various relevant physiological measures, e.g., the total quantum catch from the light by one of the cone pigments. Given a system of three-primary metameric matching, the amounts of any one of the primaries can also be considered as a unidimensional code.

The point is that this system, consisting of all the various attributes and codes of color, is intrinsically three dimensional. For example, if two lights match in hue, in saturation, and in brightness, then they are metameric; they match on all other attributes, and the quantum-catch is equal for them, by each of the cone pigments. The same holds for any three independent attributes or codes for color.

Color Theory in a Measurement Framework

The force-table analogy focuses attention on the observations of perceptual attributes of color, which are analogous to directional components of motion. But this analogy can be carried a step further. The vector representation of a Grassmann structure (for force or color) is not uniquely determined; it can be transformed by any nonsingular linear transformation of the representing space, without losing any part of its meaning. But if we demand that the force vector also represent all the directional Grassmann structures, based on relations $\sim_\gamma$, we no longer have this arbitrariness. The requirement that two configurations that have matching components in direction γ also have equal force-vector components in that direction ties the coordinates for force to the coordinates for physical space, and makes the force representation unique except for choice of unit. The goal of color theory is similar: to find a special coordinate system in which the attributes and codes of colors have simple or natural representations.

In the case of force, the structure of physical directions is linear and isotropic (no preferred directions), and so all the relations $\sim_\gamma$ have a simple representation. The vector $\phi(a) = [\phi_1(a), \phi_2(a), \phi_3(a)]$ representing a configuration a satisfies

$$a \sim b \quad \text{iff} \quad \phi(a) = \phi(b),$$

$$a \sim_\gamma b \quad \text{iff} \quad \sum_{i=1}^{3} \gamma_i \phi_i(a) = \sum_{i=1}^{3} \gamma_i \phi_i(b),$$

where $(\gamma_1, \gamma_2, \gamma_3)$ are direction cosines of γ.

In the case of color, however, the structure of perceptual attributes is nonlinear and nonisotropic. The empirical relations that need to be represented by numerical relations include those derived from many of the classical observations of color perception. They certainly include some or all of the ones listed above in connection with brightness: $\sim_B$ from brightness matching (recall that there are at least two different perceptual attributes related to 'brightness,' one satisfying additivity and one not), $\gtrsim_B$ from brightness 'ratios,' the ROC curve f_{ba} from brightness discrimination, $\gtrsim$ from overall color similarity, $\sim_{ab}$ from cross-context matching, and matching relations based on various other color attributes, including saturation, redness, etc. Besides these, one might consider equivalence relations based on matches of color-blind individuals, and possibly other empirical relations.

Below, I shall describe a special vector representation for colors, in which the coordinates are tied to special 'pure' attributes of opponent-colors theory, each coordinate being unique except for choice of unit. Then I will discuss the problems of representing other empirical color relations in terms of these opponent-color coordinates.

The idea that a color theory consists of a special coordinate system, in which various psychophysical relations have a simple description, goes back at least to the work of Helmholtz (1896) and König and Dieterici (1892). Helmholtz tried to derive the special coordinate system from the representation of wavelength-discrimination data. He assumed that a version of Weber's law describes discrimination along each coordinate and that the contribution of the three coordinates to overall discrimination is described by a Euclidean metric. König and Dieterici, on the other hand, tried to derive the 'fundamental sensations,' as the special coordinates were called, from color-blind matching relations, assuming that each of the major types of dichromatism involved one missing coordinate of the three. And since, there have been numerous other attempts, placing emphasis on various psychophysical phenomena. Some of these were summarized by Judd (1951).

The special contribution of modern measurement to such a program of color theory stems from the ability to diagnose quantitative representations from the qualitative generalizations satisfied by particular psychophysical phenomena. To take just one example, suppose that one asks whether it is possible to find a system of linear coordinates, representing the Grassmann structure of colors, such that the Euclidean distances in these orthogonal coordinates yield an ordinal representation of color dissimilarity or discriminability. That is, if $\succsim$ is the dissimilarity ordering,

$$(a, b) \succsim (c, d) \quad \text{iff} \quad \sum_{i=1}^{3} [\phi_i(a) - \phi_i(b)]^2 \geq \sum_{i=1}^{3} [\phi_i(c) - \phi_i(d)]^2.$$

A measurement analysis of this idea uncovers the following qualitative law, which is obviously necessary (though certainly not sufficient) for such a representation:

$$(a, b) \approx (a \oplus c, b \oplus c).$$

This is necessary because, by linearity, $\phi_i(a \oplus c) - \phi_i(b \oplus c) = \phi_i(a) - \phi_i(b)$. This qualitative law can practically be rejected by a thought experiment: let a be orange, b be red-orange, and let c be a very intense red. Adding c to both a and b will very likely wash out most of the dissimilarity. Such a violation immediately suggests an alternative general form for the numerical representation of dissimilarity: the difference $|\phi_i(a) - \phi_i(b)|$ will have to be weighted inversely by a factor that increases with the absolute magnitudes of $\phi_i(a)$ and $\phi_i(b)$, e.g., $|\phi_i(a)| + |\phi_i(b)|$.

One further recognizes, in the above case, that there are all too many possibilities for the precise form of the numerical representation of dissimilarity. Unless qualitative laws can be found, which constrain the choice, dissimilarity is not a promising jumping-off place for constructing special coordinates. The construction will involve arbitrary assumptions about the right form of the numerical representation. Rather, it is preferable to choose a point of departure where the known qualitative laws already provide strong constraints. After deriving a special coordinate system, one can then examine directly the way in which ordering of discriminabilities or of perceived dissimilarities varies with the values of these coordinates.

We see, then, that measurement theory provides a unifying standpoint, from which all the special empirical relations surveyed earlier for brightness can be integrated. The goal of a unified theory of color psychophysics is to discover a relatively unique coordinate system, and simple representations for all the various phenomena in terms of those coordinates. One presumes that such coordinates will be closely tied to physiological mechanisms, and that the nature of the numerical representations for various empirical relations will reflect the nature of color coding. These representations will be established only when the qualitative laws satisfied by the empirical relations are established. Unlike the situation for force, where all the added observable relations $\sim_\gamma$ have a very simple structure, the more complex structure of perceptual attributes of color forms the main subject of investigation in color theory.

In the remainder of this chapter, I shall briefly survey three topics that point toward realization of the above program for an integrated color theory. These are: linear codes of opponent-colors theory based on one-dimensional Grassmann structures, transformation to nonlinear codes via representation of sensation 'ratios,' and color dissimilarity. In these three topics, the measurement analysis has led me to new experiments (experiments 1, 2, and 4, alluded to in the preceding section).

LINEAR CODES OF OPPONENT-COLORS THEORY

Opponent-colors theory (Hering, 1878, 1920) assumes that three perceptual attributes constitute a distinguished, or unique set of directions in the space of perceptual attributes. The three are bipolar or opponent pairs: redness/greenness, yellowness/blueness, and whiteness/blackness. The assumption is based on (introspective) analysis of sensations. Orange consists of redness, yellowness, and whiteness; light violet consists of redness, blueness, and whiteness; 'olive' consists of greenness, yellowness, and perhaps neutrality between whiteness and blackness; brown often consists of redness, yellowness, and blackness; etc. Redness and greenness are bipolar opponents since no homogeneous color sensation is analyzed into both redness and greenness; and the same holds for the other two opponent pairs.

Jameson and Hurvich (1955) were the first to evaluate systematically the amount of redness in reddish colors, by determining how much of a standard green had to be added to bring the color to equilibrium with respect to the redness-greenness attribute. Similarly, they evaluated greenness, yellowness, and blueness by equilibration with an opponent light. Note that this technique is precisely analogous to the evaluation of a directional component of a force, by constraining the motion to one direction and equilibrating. In measurement of redness by opponent-hue cancellation, the judgment is constrained to redness vs greenness of the mixed light. To measure the redness of a light a, one determines the scale factor t for a standard green b_1 that must be added to reach equilibrium. If t is too small, the mixture $a \oplus (t * b_1)$ appears reddish; if t is too large, then $a \oplus (t * b_1)$ appears greenish; and the value of the measurement is defined to be that value of t where the light appears neither reddish nor greenish, i.e., $a \oplus (t * b_1)$ is unique (whitish) yellow, or unique (whitish) blue, or white.

Two lights equilibrated by the same thing with respect to redness-greenness stand in relation $\sim_1$, and a representation function, $\phi_1(a)$, can be taken to be proportional to the value of t needed for equilibration. Thus, if $\phi_1(a) = \phi_1(a')$, then $a \sim_1 a'$. A similar relation $\sim_2$ can be introduced for yellowness/blueness equilibration, with corresponding representation ϕ_2.

One can immediately ask whether the relations $\sim_1$ and $\sim_2$ determined by opponent-cancellation measures of redness/greenness and yellowness/blueness yield one-dimensional Grassmann structures, like the directional relations $\sim_\gamma$ in force measurement. In fact, the methods used by Jameson and Hurvich implicitly assumed that such is the case.[3] They went on to test their method by testing whether the hue components ϕ_1 and ϕ_2 measured in this way could in fact be given a linear representation in the vector space derived from the three-dimensional Grassmann structure of metameric color matching.

Figure 4 shows values of ϕ_1 as a function of wavelength, as measured by Jameson and Hurvich (1955) for their own eyes, compared with a linear function of the CIE standard-observer space proposed by Judd (1951) to represent the Hering theory. Figure 5 shows similar comparisons for the other opponent attribute ϕ_2. It is obvious that the gross agreement is excellent, considering the individual differences and the fact that the Judd transformations were not chosen to fit these particular data.

In view of differences among real observers, and between them and the CIE standard observer, it is clear that this question deserves a much closer examination. The great benefit of the axiomatic measurement analysis is that a conclusive test can be formulated that does not require complete determination of individual color-matching representations. It suffices simply to determine whether the relations $\sim_i$, for individual observers, do yield Grassmann structures.

Let A_1 denote the set of unique yellow-blue or achromatic lights, i.e., the set of red/green equilibria (or endpoints of red/green cancellation), and let A_2 denote the set of unique red/green or achromatic lights, i.e., the set of yellow/blue equilibria (or endpoints of yellow-blue cancellation). It turns out that $\sim_i$ yields a Grassmann structure ($i = 1$ or 2) if and only if every light can be equilibrated by some other light with respect to attribute i (this

---

[3] Only if this holds are the measurements ϕ_i guaranteed to be independent of the choice of equilibrating light, and only in that case can these measurements be rescaled to give curves of red-green and yellow-blue response for an equal-energy spectrum.

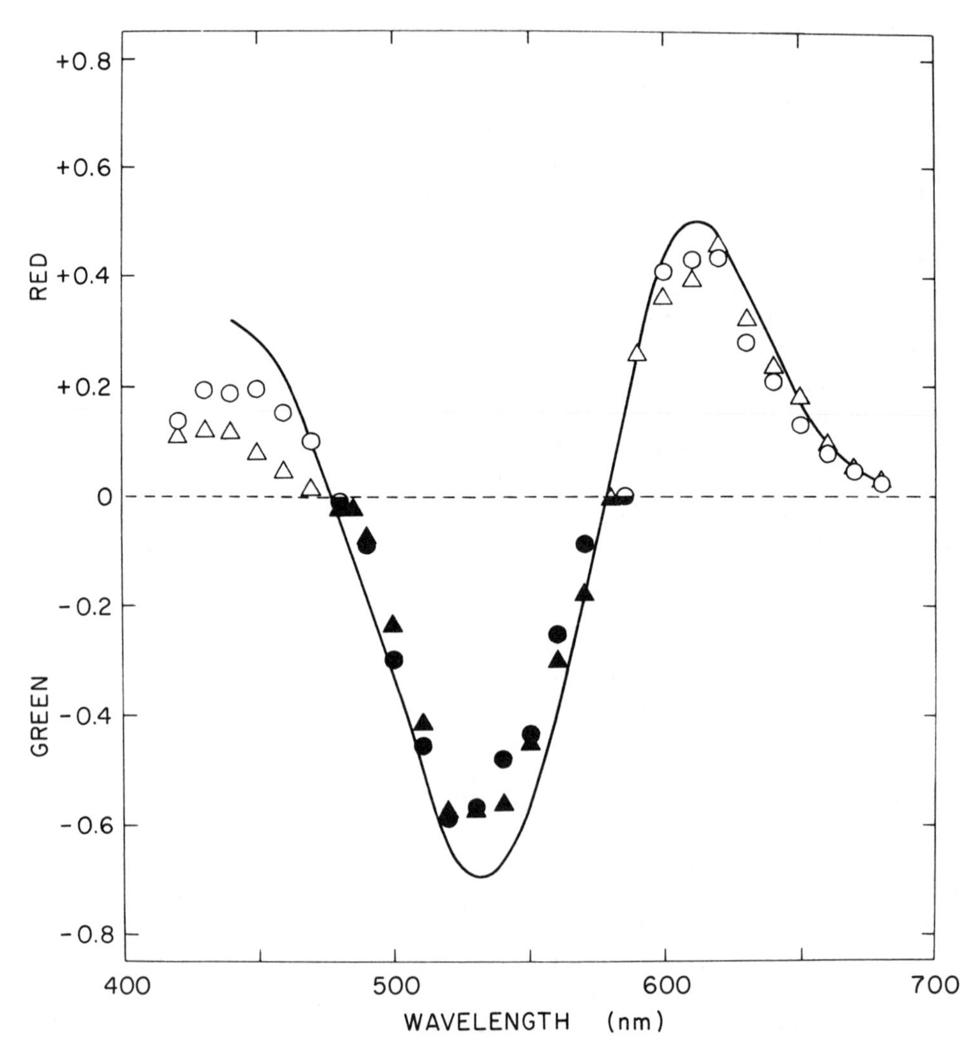

FIGURE 4.
Opponent-cancellation measures of redness (open symbols) and greenness (filled symbols) for observers H (circles) and J (triangles), for an equal-energy spectrum, compared with a linear function of the CIE standard observer (solid line). Data are from Jameson and Hurvich (1955); CIE function is from Judd (1951).

certainly holds, in fact) and the following linearity properties are satisfied:

$$\text{if } a \in A_i, \text{ then } t * a \in A_i; \tag{9}$$

$$\text{if } a \in A_i, \text{ then } b \in A_i \quad \text{iff} \quad a \oplus b \in A_i. \tag{10}$$

Moreover, since $\sim_i$ includes $\sim$ [proof: if $a \sim a'$ (metameric match) and $a \oplus (t * b_1) \in A_1$ (equilibration) then $a' \oplus (t * b_1) \in A_1$, since $a \oplus (t * b_1)$ and

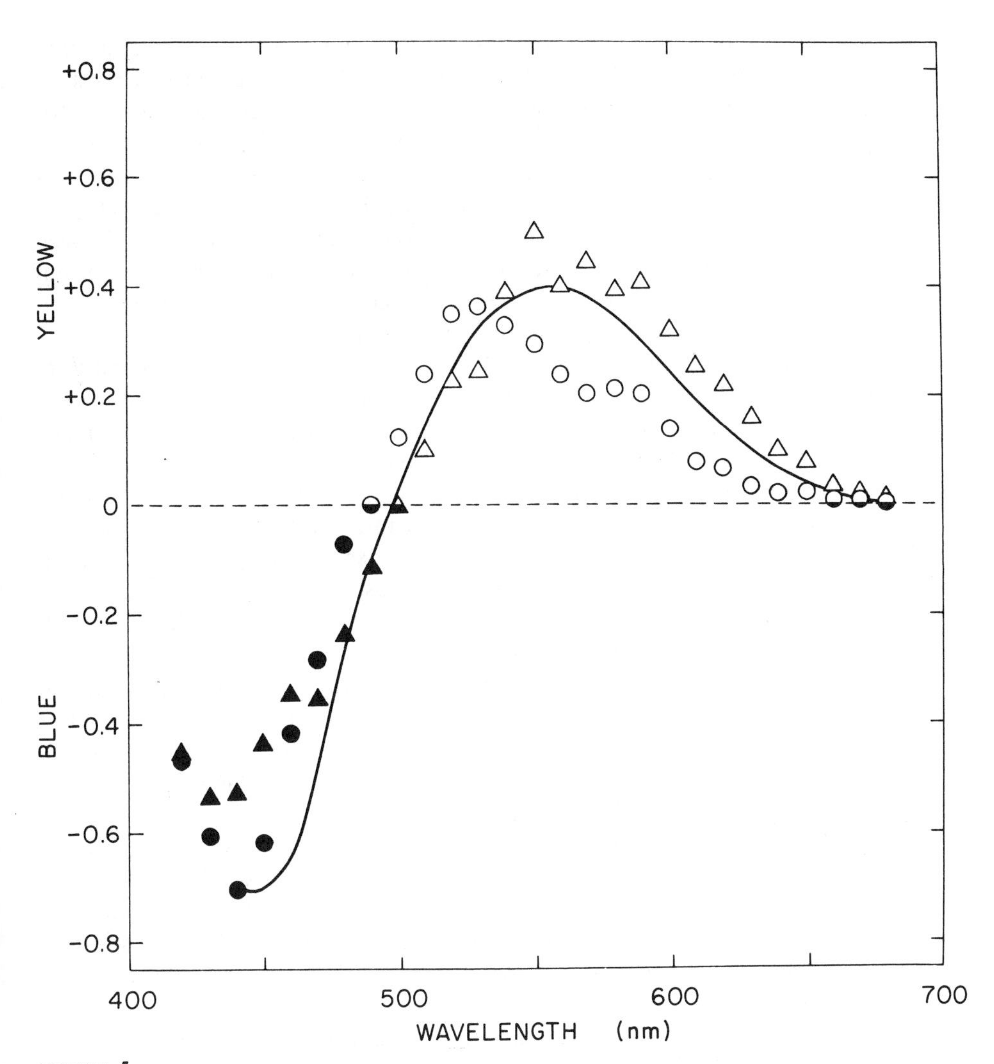

FIGURE 5.
Opponent-cancellation measures of yellowness (open symbols) and blueness (filled symbols), compared with a linear function of the CIE standard observer. Other details as in Figure 4.

$a' \oplus (t * b_1)$ look alike by Grassmann's law of additivity (1)] the one-dimensional representation for $\sim_i$ can be proved to be linearly related to the three-dimensional representation for $\sim$, i.e., the linearity hypothesis tested by Jameson and Hurvich (Figs. 4 and 5) must be valid. The above relationships, proved by Krantz (1974b), show the importance of testing the qualitative laws (9) and (10).

As Figure 4 shows, there are two wavelengths in A_1 for normal color observers: equilibrium yellow, around 580 nm, and equilibrium blue, around 475 nm. If (9) and (10) hold, then these equilibrium spectral loci must be invariant with luminance (9), and their additive mixtures must be in A_1 for all luminance combinations (10). Moreover, since in that case any light in A_1 will be metameric to a linear combination of these two wavelengths, it is reasonably straightforward to obtain a quite exhaustive test of (9) and (10) for A_1. Testing these laws for A_2 is a bit more complicated because, as can be seen from Figure 5, there is only one equilibrium spectral wavelength (equilibrium green, around 495 nm) in A_2. To obtain an equilibrium red, one must use a mixture of long-wavelength and short-wavelength light.

Larimer, Krantz, and Cicerone (1974a) have conducted experimental tests of (9) and (10) for A_1. These experiments involved determinations of unique yellow and blue wavelengths, by a staircase method, at different luminance levels, and determinations of mixtures of yellow and blue in A_1.

The subjects were dark-adapted and there was no fixation point.[4] Stimuli were exposed for 1 sec, with an intertrial interval of 20 sec (in darkness). When the stimulus was exposed, the subject fixated it and judged whether it was reddish or greenish. Two kinds of staircases were used: variable yellow, with stimuli varying mostly in the 565–585 nm region, and variable blue, with stimuli varying mostly between 465 and 485 nm. In a variable-yellow staircase, the variable wavelength was reduced (toward 565 nm—yellow-green) after a 'too red' response, and increased (toward the 585 nm orange) after a 'too green' response. In a variable-blue staircase, the variable wavelength was moved toward blue-green (485 nm) or toward violet (465 nm) after 'too red' or 'too green' responses, respectively.

The stimulus was either monochromatic (the variable wavelength alone) or a mixture of the variable wavelength with some fixed addend from the other spectral region under study. For example, in variable-yellow staircases with a high-luminance blue addend and low-luminance yellows, the stimulus looked bluish white; with reversed luminance relations, the stimulus was yellowish white.

Some of the main results are shown in Figures 6 and 7. Figure 6 shows the means of equilibrium yellow and equilibrium blue wavelengths (monochromatic stimuli) at different luminances, for five observers, together with 80 percent confidence intervals based on between-session variability. The open symbols in Figure 7 show wavelength pairs whose additive mixture was in A_1 (neither reddish nor greenish), while the filled symbol in each section of Figure 7 has for coordinates the mean equilibrium yellow and the mean equilibrium blue wavelength for the observer in question. Different open symbols in Figure 7 denote different luminance combinations for the yellow

---

[4] These conditions were employed, because of the possibility that adaptation or colored fixation points could induce nonlinearities.

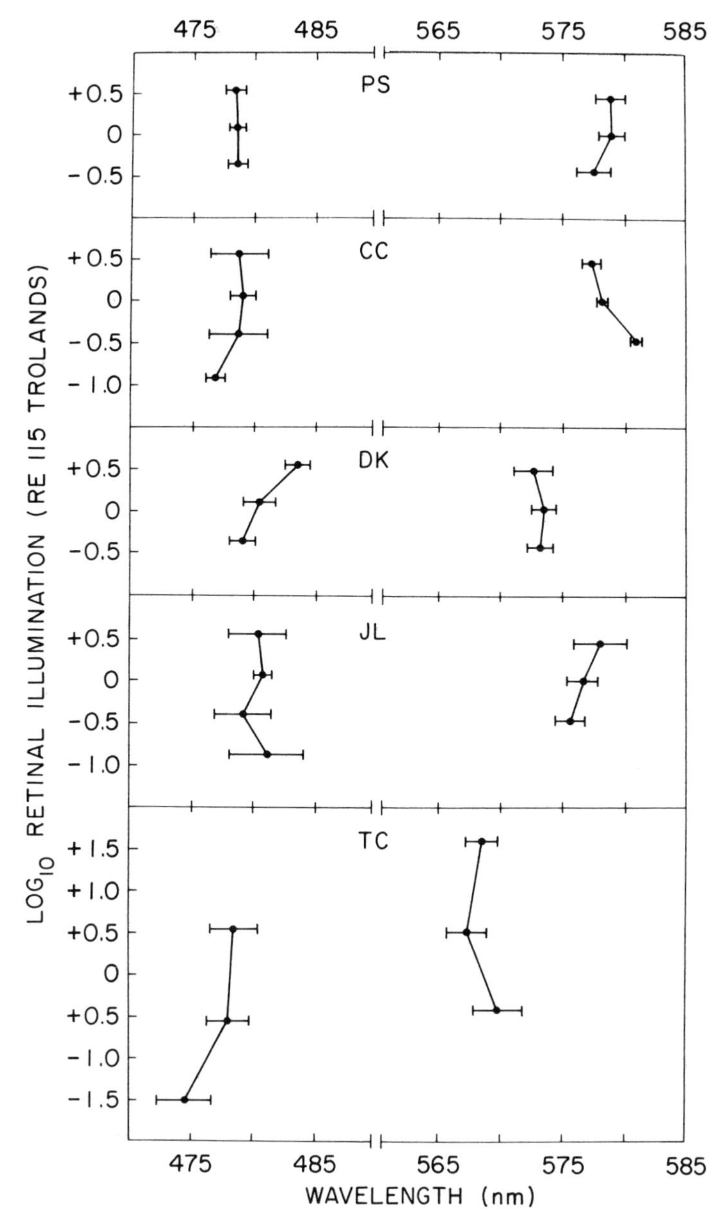

FIGURE 6.
Means of determinations of equilibrium blue and equilibrium yellow wavelengths, at various luminance levels, for five observers, with 80 percent confidence intervals based on between-session variability.

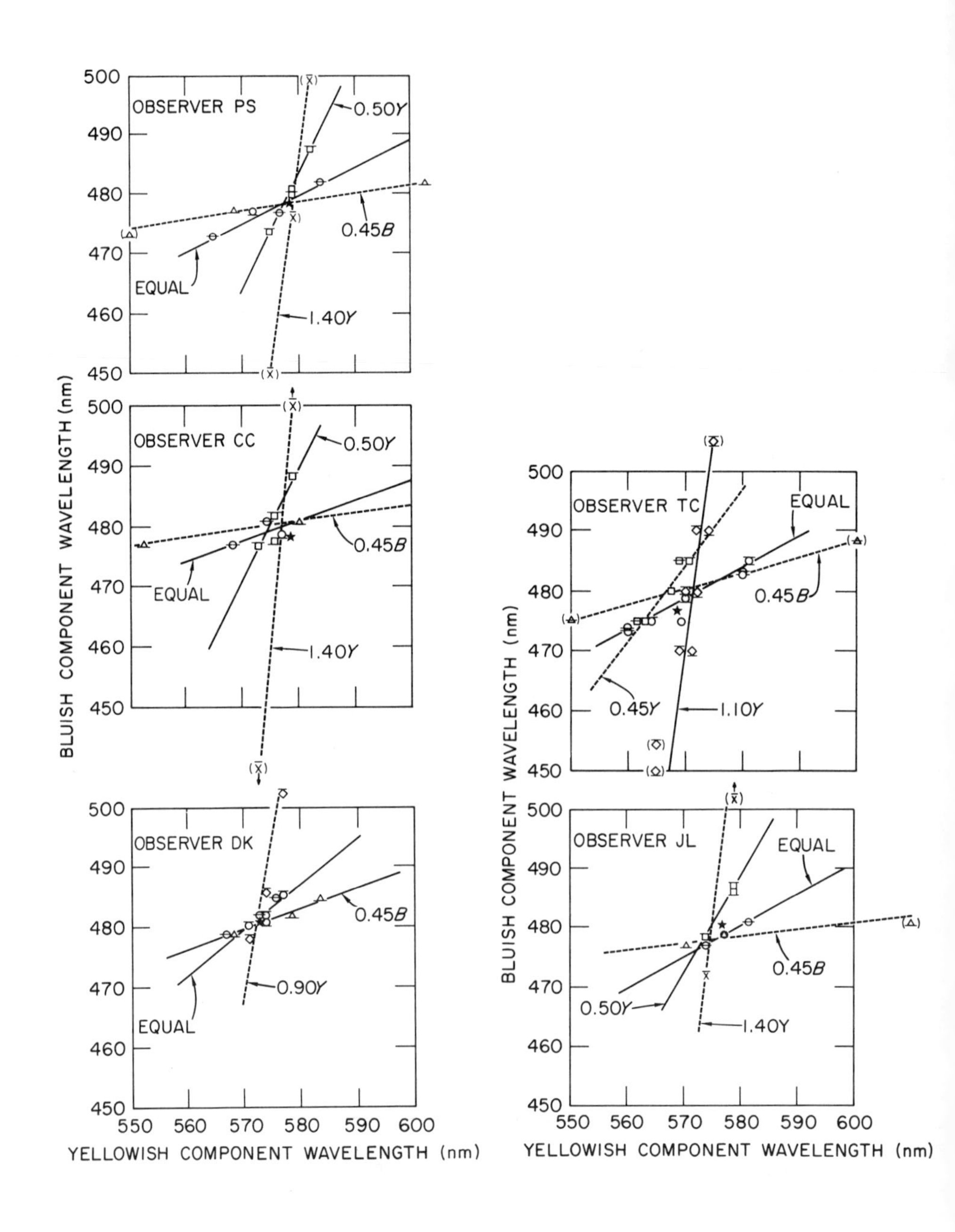
OBSERVER PS
0.50Y
0.45B
EQUAL
1.40Y
OBSERVER CC
0.50Y
0.45B
EQUAL
1.40Y
OBSERVER DK
0.45B
0.90Y
EQUAL
OBSERVER TC
EQUAL
0.45B
0.45Y
1.10Y
OBSERVER JL
EQUAL
0.45B
0.50Y
1.40Y
BLUISH COMPONENT WAVELENGTH (nm)
YELLOWISH COMPONENT WAVELENGTH (nm)
500
490
480
470
460
450
550
560
570
580
590
600

and blue wavelengths; the straight lines are fit to open symbols denoting wavelength pairs with approximately constant luminance ratio.

The facts that the equilibrium yellow and blue wavelengths vary only slightly with luminance (Fig. 6), and that different luminance combinations having the same luminance ratio yield wavelength pairs that fit the same curve (the same straight lines in Fig. 7) support law (9). The fact that the equilibrium wavelength (filled) point in Figure 7 lies at or near the intersection of the straight lines implies that additivity (10) is supported. Even for observer CC, who deviates most, the actual wavelength differences are small and could be due either to shifts in the criterion for 'neither reddish nor greenish' across lights differing widely in the brightness and yellowness/blueness dimensions or to very small nonlinearities.

In short, the hypothesis that (9) and (10) are satisfied, so that $\sim_1$ yields a Grassmann structure, with representation ϕ_1 (the cancellation red/green measure) linearly related to the representation ϕ for color matching, is quite well supported. Tests of the analogous hypothesis for A_2 and $\sim_2$ show a small but reliable nonlinearity (Larimer, Krantz, & Cicerone, 1974b).

If one assumes that $\sim_1$, $\sim_2$ (cancellation equivalences), and $\sim_3$ (luminance matching by flicker photometry, say) yield Grassmann structures, then one can prove (see Krantz, 1974b, for details) that there is an additive vector representation $\phi = (\phi_1, \phi_2, \phi_3)$ such that $\phi_1(a)$ represents cancellation redness/greenness, $\phi_2(a)$ represents cancellation yellowness/blueness, and $\phi_3(a)$ represents luminance. These would be the main linear perceptual attributes.

FIGURE 7.
Loci of yellow-plus-blue mixtures that are neither reddish nor greenish. Shape of symbol denotes approximate ratio of yellow to blue.

Symbol	$\log_{10}$ [luminance Y/luminance B]
×	1.40
◇	0.90–1.10
□	0.45–0.50
○	0.00
△	−0.45

Position of bar on symbol denotes approximate retinal illumination from yellow component. Upper bar: high intensity; middle bar: medium intensity; lower bar: low intensity. (Respectively, 324, 115, or 41 trolands.) Straight lines were fit by eye to each set of points having a fixed luminance ratio. Poorly determined points are indicated by parentheses. Heavy solid lines are more reliably determined than light dashed ones. The unique monochromatic wavelengths (average from Fig. 6, for each observer) are the ordinate and abscissa of the point indicated by the large star. The linearity hypothesis predicts that the locus for a given luminance ratio is independent of the luminance of the yellow component; that the different loci intersect at the star; and that the ratios of their slopes at the star are equal to the ratios of the corresponding luminance ratios.

NONLINEAR CODES; SENSATION 'RATIOS'

Brightness

The linear luminance function (ϕ_3, above) is based on a 'brightness' matching procedure that satisfies the laws of a Grassmann structure (Abney's law of additivity, chiefly). Suppose, however, that one uses a nonlinear transformation, say,

$$\psi_3 = \phi_3^{\beta},$$

where β is the exponent of Stevens' Law. The nonlinear function ψ_3 preserves $\sim_3$, the additive matching relation: for it is still true that both

$$a \sim_3 b \quad \text{iff} \quad \psi_3(a) = \psi_3(b), \quad \text{and}$$
$$a \sim_3 b \quad \text{iff} \quad a \oplus c \sim_3 b \oplus c.$$

What is lost is the representation of the empirical operation $\oplus$ by addition $(+)$ of ψ_3 values: we have

$$\begin{aligned} \psi_3(a \oplus b) &= [\phi_3(a \oplus b)]^{\beta} \\ &= [\phi_3(a) + \phi_3(b)]^{\beta} \\ &= [\psi_3(a)^{1/\beta} + \psi_3(b)^{1/\beta}]^{\beta}. \end{aligned}$$

The loss of a simple representation of $\oplus$ may be compensated for by the fact that ψ_3 gives a simple representation of some other kind of relation, e.g., brightness 'ratio' or 'difference' judgments.

There is nothing special about power functions in the above discussion; any strictly increasing function, $\psi_3 = h(\phi_3)$, could be employed, if it were useful in representing quantitative brightness judgments.[5]

The same remarks naturally hold for the other linear perceptual codes, cancellation redness/greenness and cancellation yellowness/blueness. Judged redness, for example, may be a nonlinear function of cancellation redness, with no contradiction of any qualitative law; one merely replaces ϕ_i by $\psi_i = h_i(\phi_i)$, giving up simple vector additivity of the representation ϕ in order to represent quantitative judgments of colors.

Closer examination reveals, however, that even arbitrary monotone functions h_i do not suffice to describe the transformation between linear perceptual codes and nonlinear judgments. This is most obvious and best documented in the case of brightness: the brightness-matching relation $\sim_B$ is simply not

---

[5] This is analogous to using nonadditive velocity in special relativity theory, in order to have a velocity scale that is simple related to distance and time. See Krantz et al. (1971, pp. 91–102).

the same as $\sim_3$, quite apart from quantitative judgments: for $\sim_B$ violates Abney's law. One is forced to postulate a more general nonlinear relation,

$$\psi_3 = h_3(\phi_1, \phi_2, \phi_3). \tag{11}$$

Since now ψ_3-equality is not the same as ϕ_3-equality, and h_3 is nonlinear, ψ_3-matches can be nonadditive.

A question that may occur to the reader is whether this degree of complication is justified. Why bother with nonadditive brightness matches? Indeed, it is sometimes suggested that the nonadditivity of brightness matches is an artifact of a confused criterion—saturated lights are judged brighter than less saturated mixtures of the same luminance. Moreover, heterochromatic brightness matches are quite variable, which suggests the possibility of criterion variability.

To ignore brightness, however, is to turn one's back on some striking regularities. Guth, Donley, and Marrocco (1969) found quite similar nonadditivity in threshold determinations and in brightness matches; and in fact when threshold and brightness are determined under identical viewing conditions, one obtains identical luminous efficiency curves (Jameson, 1970; Wilson, 1964).

Figure 8 shows relative energy needed for threshold and for equal brightness, as a function of wavelength, for one of Wilson's observers. The inset shows the stimulus configuration: one-half the bipartite field was illuminated with 550 nm light, and the large surround was illuminated with broad-band light, both at a fixed luminance (10 mL). The other half of the bipartite field was filled with light of another wavelength, whose visibility threshold was measured, or which was matched in brightness to the 550 nm light.

Figure 9 shows the logarithm of brightness magnitude estimates, for 17 different wavelengths, ranging from 420 to 700 nm. The abscissa is the logarithm of energy, expressed for each wavelength in units of its threshold energy (or equal-brightness energy, since they are the same). These data, also from Wilson (1964), demonstrate two points: first, that the form of the brightness magnitude-estimation function is independent of wavelength (also shown by Ekman, Eisler, & Künnapas, 1960) and second, that brightness magnitude estimates are consistent with heterochromatic brightness matching. Two lights that have the same energy, relative to their thresholds, and that are viewed under identical conditions, will be judged equally bright, and they will yield the same brightness magnitude estimates, on the average.

A further validation of brightness judgments comes from cross-context matching. Figure 10 contains the results of an experiment by Jameson (1970), in which brightness magnitude estimates were obtained with both bright and dark surround, and in addition, cross-context brightness matches were made (haploscopically, with one eye viewing a fixed luminance in a bright surround,

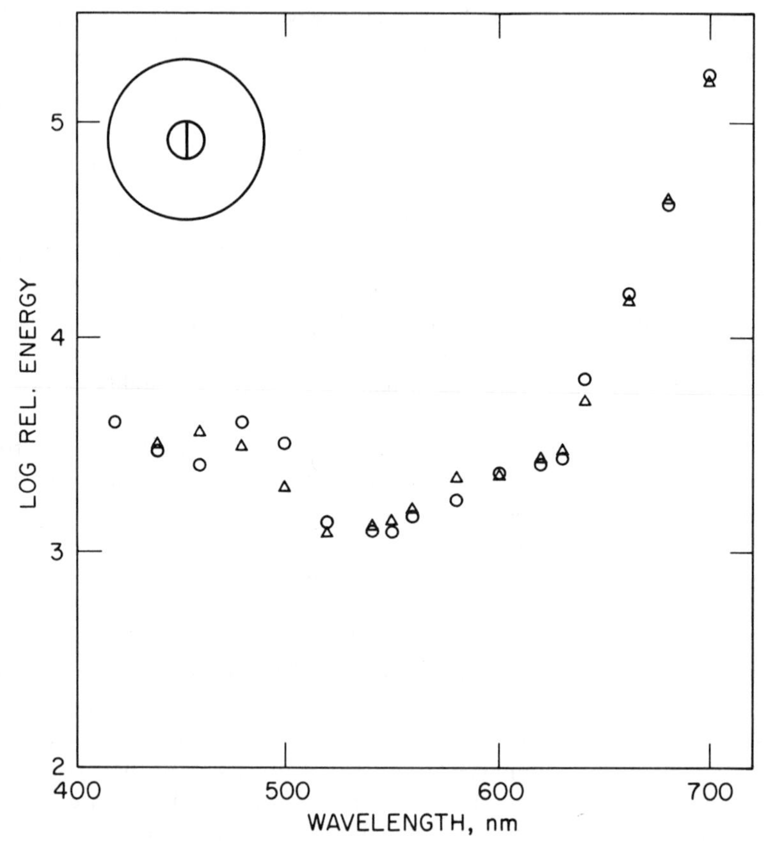

FIGURE 8.
Relative energy for threshold (circles) and equal brightness (triangles), for one observer in a study by Wilson (1964). The inset shows the stimulus configuration used to achieve identical viewing conditions. The surround was illuminated with 5500° K light, and one-half the bipartite field was illuminated with 550 nm light, both at 10 mL luminance, during both threshold and brightness-match measurements. (Reprinted from Jameson, 1970, by permission of Musterschmidt, Göttingen.)

the other eye a variable luminance in a dark surround). The brightness estimates for the bright surround (circles) and the haploscopic matches predict the brightness estimates for the dark surround. The predictions are shown by crosses; the obtained magnitude-estimation function in the dark surround was well fit by the dashed line.

These results establish beyond much doubt the validity and importance of the perceptual attribute brightness. The differences between this attribute and luminance are shown by the nonadditivity of brightness, as well as by the great difference between Figure 8 and the usual CIE photopic luminance

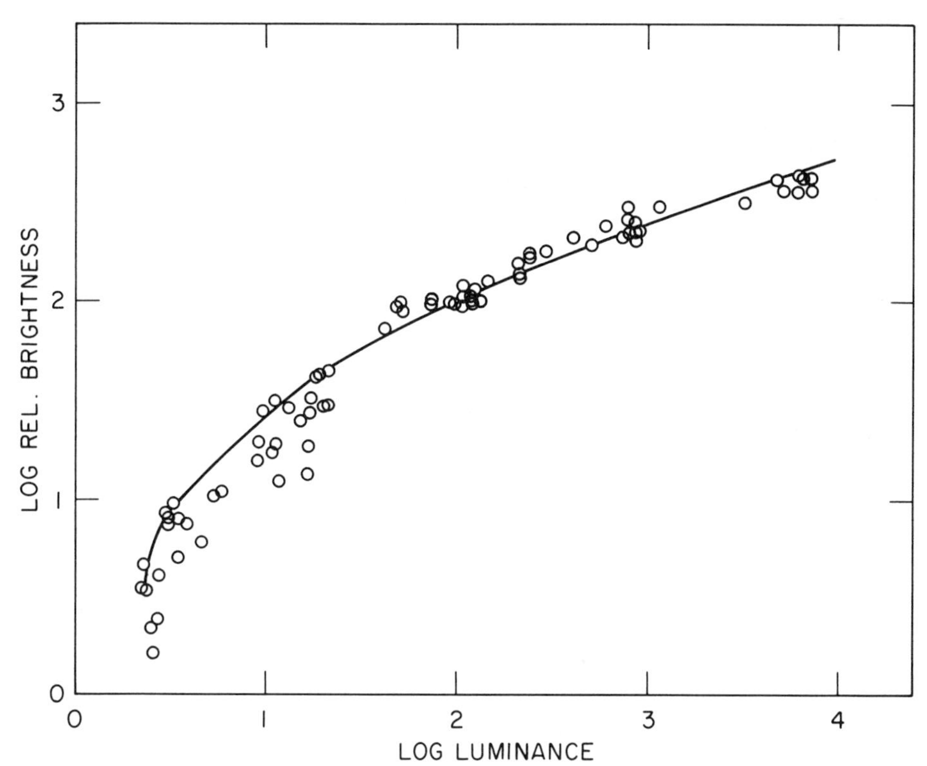

FIGURE 9.
Brightness magnitude estimates as a function of stimulus energy (in threshold-energy units) for 17 wavelengths, from Wilson (1964). (Reprinted from Jameson, 1970, by permission of Musterschmidt, Göttingen.)

function. (The curve in Fig. 8 is quite typical of the results of heterochromatic brightness matching from many studies.)

We can now return to the general relationship (11) with some confidence, and ask what can be done to specify more precisely the relationship between ψ_3 and the linear codes ϕ_1, ϕ_2, and ϕ_3. There are two possible approaches to this problem. One is to ask whether the relationship satisfies qualitative laws that characterize a known conjoint structure. For example, it is possible that we have an additive conjoint structure, in which case we can find a function ψ_3 that represents brightness ordering judgments, i.e.,

$$a \gtrsim_3 b \quad \text{iff} \quad \psi_3(a) \geq \psi_3(b),$$

and, simultaneously, satisfies

$$\psi_3 = h_{31}(\phi_1) + h_{32}(\phi_2) + h_{33}(\phi_3). \tag{12}$$

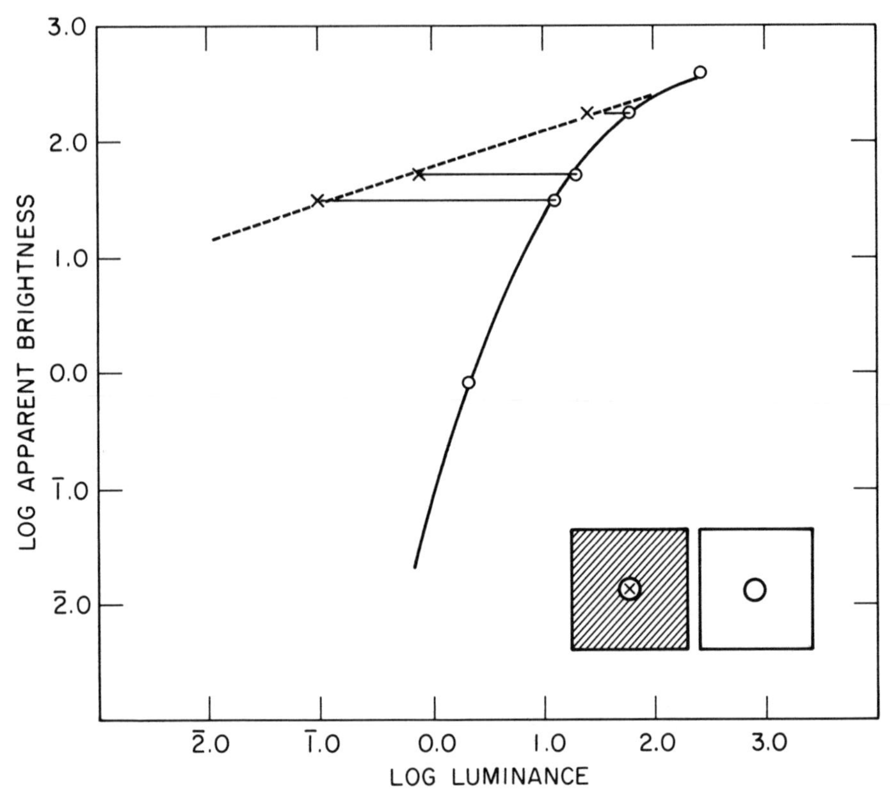

FIGURE 10.
Brightness magnitude estimates (open circles) for lights viewed in a bright surround and predicted magnitude estimates (crosses) for lights viewed in a dark surround. The predictions are based on cross-context matching: the luminances shown on the abscissa for points joined by horizontal lines were judged equally bright haploscopically (left eye, dark surround; right eye, bright surround). The dashed line is the best-fitting line to actual brightness magnitude estimates with dark surround. (From Jameson, 1970; reprinted by permission of Musterschmidt, Göttingen.)

Such a structure is characterized by independence laws, as described above. For example, if luminance (ϕ_3) is constant, the contributions of cancellation codes ϕ_1 and ϕ_2 to the brightness ordering should be the same, independent of the fixed level of ϕ_3; or if cancellation redness (ϕ_1) is held constant, the joint contributions of ϕ_2 and ϕ_3 to the brightness ordering should not depend on the fixed level of ϕ_1.

If one or more independence laws are violated, then the additive form (12) must be rejected, and some more complicated form sought. Qualitative laws characteristic of other conjoint structures may or may not be helpful.

The other tack is to use numerical judgments, or brightness 'ratio' judgments, to establish a function ψ_3 directly, satisfying

$$(a, b) \gtrsim_B (c, d) \quad \text{iff} \quad \psi_3(a)/\psi_3(b) \geq \psi_3(c)/\psi_3(d).$$

The relation $\gtrsim_B$ would have to satisfy the properties of an algebraic-difference structure, as described earlier. The function ψ_3 could then be studied numerically in relation to ϕ_1, ϕ_2, and ϕ_3.

These two approaches might both work, and might or might not lead to the same function ψ_3. If two different functions resulted, they would of course have to be monotonically related. The form of such a monotonic relation might be of interest in modeling the mechanism relating sensory 'ratios' to other codes.

Hue and Saturation

The regularities of (nonlinear) perceived brightness have also been found in judgments of relative or absolute amounts of various hues. The most extensive study was done by Indow and Stevens (1966). They found that the saturation of red, yellow, green, and blue lights is well described as power functions of colorimetric purity, with exponents respectively equal to 1.7, 2.9, 1.7, and 1.4. As in the case of brightness, these judgments had to be validated by heterochromatic matching of saturation. The prediction is that the purity of the matching light will be a power function of the purity of the test light, with the exponent equal to the ratio of the two magnitude-estimation exponents. The agreement with this prediction is shown in Figure 11, reprinted from Indow and Stevens' paper.

Since saturation is the judgment of relative subjective amounts of chromaticness and whiteness, one would expect that the judged saturation of a relatively pure red, for example, would depend nonlinearly on both ϕ_1 (cancellation redness) and ϕ_3 (luminance). But even if only redness were judged, rather than the relation of redness to whiteness, one still cannot expect a simple formula of the form $\psi_1 = h_1(\phi_1)$. Rather, the perceived redness will still depend on whiteness or brightness, and in the case of reddish-blue colors, it will also depend on blueness. That is, we must have in this case also an equation of the general form

$$\psi_1 = h_1(\phi_1, \phi_2, \phi_3). \tag{13}$$

The reason for this is that perceived redness is 'washed out'—I prefer the term 'veiled'—by high intensities of whiteness, blackness, blueness, or yellowness. This is shown by the experiment of Hurvich and Jameson (1951) on whiteness, in which a light that was chromatic (e.g., reddish-yellow tungsten light) at low luminance appeared achromatic (white) at higher luminance, despite the fact that the cancellation redness and cancellation yellowness must have actually increased with luminance. There are also several other facts

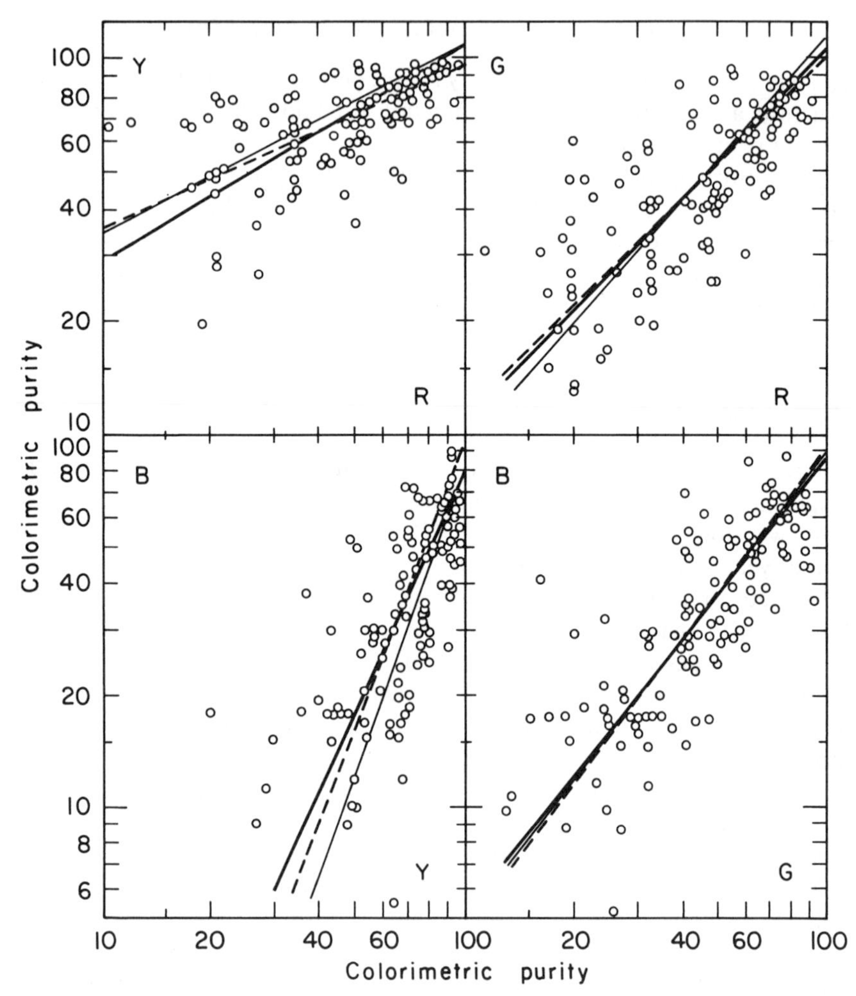

FIGURE 11.
Four samples of data from the heterochromatic matching of four hues at a time. Each point represents a single match. The scatter is due mainly to individual differences among the ten observers. The dashed lines are the perpendicular regression lines based on the least-square distances of the points from the lines. The thin continuous lines are the predicted matching functions based on the results of magnitude estimation. The heavy lines are the double-regression functions based on the geometric mean of the slopes of the two regression lines. (Reprinted from Indow & Stevens, 1966, by permission of Psychonomic Press.)

that point to this conclusion, including the difficulty subjects have in deciding whether a light wave 5–10 nm from unique blue is reddish or greenish. Such a decision becomes trivial when the blue is cancelled by yellow: the amount of cancellation redness or greenness, that far from unique blue, is very considerable.

Similar considerations, leading to an equation analogous to (13), probably hold for perceived yellowness and blueness. Another qualitative phenomenon, which must be kept in mind in thinking about these nonlinear attributes, is the Bezold-Brücke phenomenon: the shift in perceived hue as a function of luminance. This phenomenon operates as though yellowness and blueness come in at higher luminances and then grow faster than redness and greenness.

As in the case of brightness, closer specification of the form of (13) can be tackled in two ways—via conjoint-measurement analysis, or by obtaining ψ_1 to represent redness 'ratio' and greenness 'ratio' judgments, and examining its relation to the linear codes, ϕ_i.

Theory of 'Ratio' Scaling

The potential value of sensation 'ratio' judgments for specification of nonlinear color attributes makes all the more urgent an analysis of the 'ratio' judgments themselves. For example, if the additivity equation (12) holds, relating brightness, ψ_3, to the linear codes ϕ_i, then the left side of that equation, ψ_3, is uniquely determined except for origin and unit. But why, on what *theoretical* ground, should the subject be able to report directly the particular function ψ_3 for which additivity holds?

Elsewhere (Krantz, 1972b) I have reviewed two theories of 'ratio' scaling: mapping theory, based on the assumption of a common sensory-magnitude dimension, and relation theory, which assumes that the fundamental judgment is an ordering of stimulus pairs, or relations. (Levine, this vol., presents a new theory that combines some of the better aspects of both mapping and relation theories.) Neither theory gives an immediate answer to the above question (though one possible relation between ratio and additive representations has been analyzed by Krantz et al., 1971, pp. 491–493). At any rate, the relation-theory formulation led me to explore (in collaboration with Edward Pugh) some simple empirical questions. Is the ordering of pairs different when subjects judge brightness 'ratios' vs brightness 'differences'? (And if so, are the two orderings interlocked in the qualitative way that corresponds to numerical ratios and differences—see Krantz et al., 1971, pp. 152–154?) How does either ordering of pairs relate to numerical estimates of single stimuli, with 'ratio' or 'difference' instructions?

Our experiment did not produce definitive answers to these questions, partly because we did not pursue it long enough to be sure that our tentative answers would hold up. Just for illustration, however, I show in Figure 12 the data from one observer. The bottom half of the figure is based on 'ratio' judgments: the dashed curve is the brightness magnitude-estimation function, using instructions very much like Stevens', while the solid curves are brightness estimates inferred from brightness 'ratio' judgments of stimulus pairs

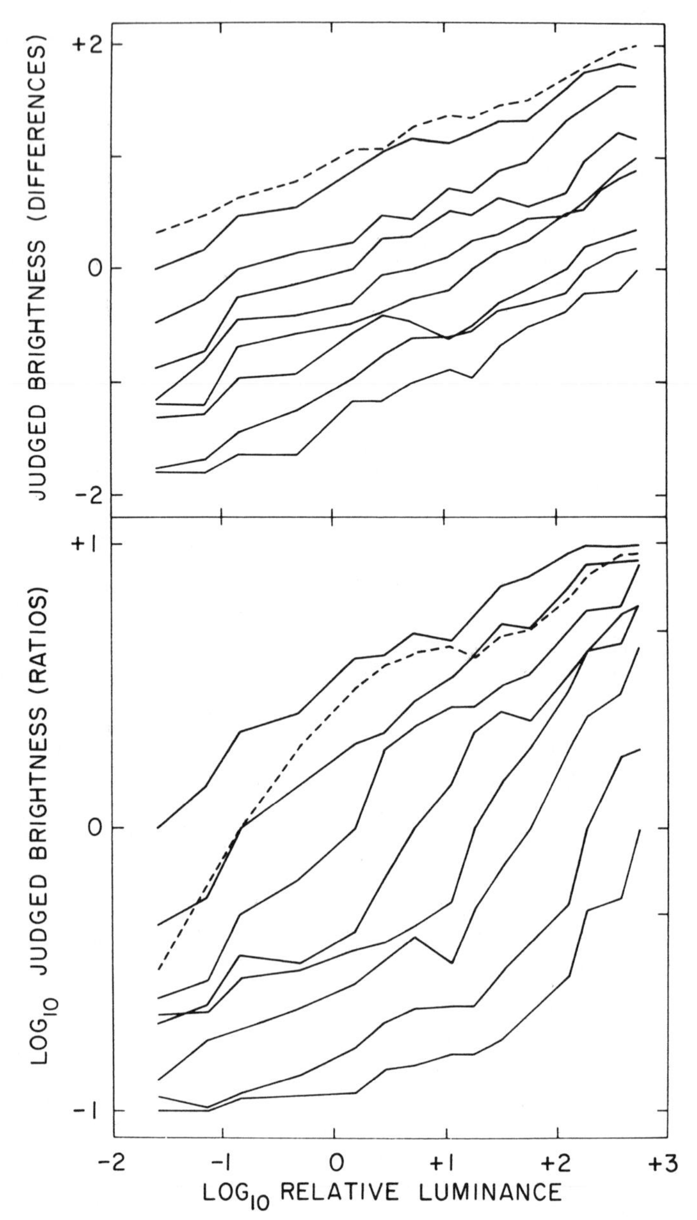

FIGURE 12.
Logarithms of brightness 'ratio' judgments (bottom) and ordinary brightness 'difference' judgments (top), plotted against logarithm of stimulus luminance, for one observer. The dashed lines are single-stimulus judgments. The solid lines are pair judgments, with luminance of one member of the pair constant and the other member varying.

(simultaneous presentation). To plot a single one of these latter curves, we held one stimulus luminance fixed and plotted the (log) 'ratio' judgment against the (log) luminance of the other stimulus in the pair.[6] The curves in the top half of the figure are based on the same stimuli, same observer, and same data analysis; what varied were the instructions, which now required judgments of brightness 'differences' rather than 'ratios.' In this case the judgments themselves (rather than their logarithms) were plotted against (log) stimulus luminance.

The fact that the solid curves in the bottom (ratio) half change systematically from negatively to positively accelerated (moving from top to bottom) and are not parallel to the dashed curve shows that there is no function $\phi(a)$ such that the pair judgment for (a, b) is $\phi(a)/\phi(b)$. The data suggest an 'anchoring' effect in which ratios near unity are expanded and extreme ratios contracted. Such an effect might be eliminable by a monotone transformation of the pair judgments, so that the ordering of ϕ-ratios represents the ordering of pair judgments. Tests of the key monotonicity axiom (4) for algebraic-difference structures showed 19 failures of the predicted conclusion out of 98 cases where a prediction was made. (In 105 cases, there was no prediction, because one or the other antecedent condition of (4) was not satisfied.) If this level of violation is due to sampling error, then an ordinal ratio representation might be quite sensible. In fact, the magnitude-estimation function itself does moderately well: the rank correlation between pair judgments and corresponding ratios of magnitude estimates is 0.73 (Kendall τ). (The higher range of the magnitude-estimation function is a power function with an exponent about 0.2).

In the top half of Figure 12, all the 'difference' judgments are quite parallel; denoting the single-stimulus judgment as $d(a)$ and the pair judgment as $d(a, b)$, we have approximately

$$d(a, b) = d(a) - d(b)$$

and, incidentally, $d(a)$ is a logarithmic function of the luminance.

To calibrate the sampling error in the 'ratio' case, we note that in the case of 'difference' judgments, 20 monotonicity predictions failed, 125 passed; there were 58 no-test cases. The discrepancy between 20 out of 145 and 19 out of 98 is not statistically significant. Also, the rank correlation between $d(a, b)$ and $d(a) - d(b)$ is 0.70.

One thing is obvious: the 'ratio' and 'difference' judgments are not ratios and differences of a common psychophysical scale. In fact, the exponentials of the 'difference' judgments are closely related to the 'ratio' judgments: the rank correlation between $\exp[d(a) - d(b)]$ and the pair 'ratio' ordering is 0.79.

---

[6] For clarity, every second solid curve was omitted from both halves of Figure 12.

These data illustrate both the orderliness of quantitative judgments of brightness and the need for a theoretical framework in which factors affecting the judgment can be understood.

Obviously, the study of nonlinear perceptual attributes has not progressed so far as the study of Grassmann structures in color vision. What the measurement-theory formulation can offer here is mainly a guiding point of view: that the quantitative interrelations can best be determined by establishing qualitative generalizations involving the various perceptual phenomena.

PERCEPTUAL ATTRIBUTES AND MULTIDIMENSIONAL SCALING

One of the most attractive ideas in psychophysics is that perceptual attributes of stimuli can be inferred from the dimensional representation of the stimuli in a geometric model. From a measurement standpoint, the geometric model must be regarded as a *theory* of dissimilarity judgments. In the context of such a theory, a perceptual dimension can be *defined* as a function bearing certain simple relations to the dissimilarity judgment.

For example, the Euclidean (ordinal model, Eq. 7 above),

$$\delta(a, a') = F\left[\sum_{i=1}^{m} |\phi_i(a_i) - \phi_i(a_i')|^2\right],$$

requires that a dimension satisfy a subtractive law (similar to the laws of an algebraic-difference structure), that different dimensions combine additively (independence laws) and that *any* two stimuli a, a' define some (subtractive) dimension that can be embedded in a set of mutually additive dimensions (rotational isotropy of Euclidean space). Other geometric models make weaker assumptions about perceptual dimensions (see Tversky & Krantz, 1970).

One way to test this idea is to see whether such properties of perceptual dimensions are valid for perceptual attributes defined by other criteria. For example, does brightness obey such laws with respect to judgments of dissimilarity of colors?

In an (as yet unpublished) experiment, I found that color dissimilarity cannot be described by any equation of form

$$\delta(a, a') = F[\phi_{12}(a_1, a_2, a_1', a_2'), \phi_3(a_3, a_3')],$$

where (a_1, a_2) is the chromaticness vector (hue and saturation) of stimulus a, a_3 is its brightness, and F is a function that is strictly increasing in each of its two arguments, ϕ_{12} and ϕ_3. The reason is that an independence law is violated: if a_3, a_3' are fixed and the dissimilarity ordering is examined as a function of chromaticness vectors, then different orderings are obtained for different, fixed brightness pairs (a_3, a_3'). In fact, it seems that blueness differences contribute much less at high brightness [pair (a_3, a_3), with a_3 large] than at low brightness [pair (a_3', a_3') with a_3' small]. Also, the same brightness difference (a_3, a_3') contributes more to dissimilarity for yellow colors than for blue ones.

While far from conclusive, this experiment does little to encourage the idea that perceptual attributes enter in simple ways into geometric models of perception. The contribution of measurement theory in this case is negative: by specifying what qualitative laws need to hold, for a given geometric model, we are enabled to perform very simple experimental tests—in this case, dissimilarity judgments of selected critical stimulus pairs—that reject that model. The systematic character of the deviation (effects of blueness and yellowness, cited above) may suggest, or at least constrain, a more complicated model relating brightness to color dissimilarity.

ACKNOWLEDGMENTS

Preparation of this chapter was supported by NSF Grant GB 8181 to the University of Michigan. The very helpful comments of Jean-Claude Falmagne, on the first draft, are gratefully acknowledged.

REFERENCES

Anderson, N. H. Functional measurement and psychophysical judgment. *Psychol. Rev.* 1970, **77,** 153–170.

Beals, R., Krantz, D. H., & Tversky, A. Foundations of multidimensional scaling. *Psychol. Rev.* 1968, **75,** 127–142.

Boynton, R. M., & Kaiser, P. K. Vision: the additivity law made to work for heterochromatic photometry with bipartite fields. *Science*, 1968, **161,** 366–368.

Brindley, G. S. *Physiology of the retina and visual pathway.* (2nd ed.) London: Edward Arnold, 1970.

Burnham, R. W., Evans, R. M., & Newhall, S. M. Predictions of color appearance with different adaptation illuminations. *J. Opt. Soc. Amer.* 1957, **47,** 35–42.

Coombs, C. H., & Huang, L. C. Polynomial psychophysics of risk. *J. Math. Psychol.* 1970, **7,** 317–338.

Debreu, G. Topological methods in cardinal utility theory. In K. J. Arrow, S. Karlin, and P. Suppes (Eds.), *Mathematical methods in the social sciences,* 1959. Stanford: Stanford Univ. Press, 1960.

Ekman, G., Eisler, H., & Künnapas, T. Brightness scales for monochromatic light. *Scand. J. Psychol.* 1960, **1,** 41–48.

Grassmann, H. Zur Theorie der Farbenmischung. *Poggendorffs Ann. Phys.* 1853, **89,** 69–84. [Engl. transl. in *Phil. Mag.* (London) 1854, Ser. 4, **7,** 254–264.]

Green, D. M., & Swets, J. A. *Signal detection theory and psychophysics.* New York: Wiley, 1966.

Guth, S. L., Donley, N. J., & Marrocco, R. T. On luminance additivity and related topics. *Vision Res.* 1969, **9,** 537–575.

Helmholtz, H. v. *Handbuch der physiologischen Optik.* (2nd ed.) Leipzig: Voss, 1896.

Hering, E. *Zur Lehre vom Lichtsinne.* Vienna: C. Gerolds Sohn, 1878.

Hering, E. *Grundzüge der Lehre vom Lichtsinn.* Berlin: Springer, 1920. (Engl. transl. by L. M. Hurvich & D. Jameson, *Outlines of a theory of the light sense*, Harvard Univ. Press, 1964.)

Hurvich, L. M., & Jameson, D. A psychophysical study of white. I. Neutral adaptation. *J. Opt. Soc. Amer.* 1951, **41,** 521–527.

Hurvich, L. M., & Jameson, D. Some quantitative aspects of opponent-colors theory. II. Brightness, saturation, and hue in normal and dichromatic vision. *J. Opt. Soc. Amer.* 1955, **45,** 602–616.

Hurvich, L. M., & Jameson, D. An opponent-process theory of color vision. *Psychol. Rev.* 1957, **64,** 384–404.

Indow, T., & Stevens, S. S. Scaling of saturation and hue. *Percept. and Psychophys.* 1966, **1,** 253–271.

Jameson, D. Brightness scales and their interpretation. In *International Colour Meeting Stockholm* 1969. Göttingen: Musterschmidt, 1970.

Jameson, D., & Hurvich, L. M. Some quantitative aspects of an opponent-colors theory. I. Chromatic responses and spectral saturation. *J. Opt. Soc. Amer.* 1955, **45,** 546–552.

Jameson, D., & Hurvich, L. M. Perceived color and its dependence on focal, surrounding, and preceding stimulus variables. *J. Opt. Soc. Amer.* 1959, **49,** 890–898.

Jameson, D., & Hurvich, L. M. Theory of brightness and color contrast in human vision. *Vision Res.* 1964, **4,** 135–154.

Judd, D. B. Basic correlates of the visual stimulus. In S. S. Stevens (Ed.), *Handbook of experimental psychology*. New York: Wiley, 1951.

König. A., & Dieterici, C. Die Grundempfindungen in normalen und anormalen Farbensystemen und ihre Intensitätvertheilung im Spectrum. *Zeitsch. Psychol. Physiol. Sinnesorgane* 1892, **4,** 241–347.

Krantz, D. H. A theory of context effects based on cross-context matching. *J. Math. Psychol.* 1968, **5,** 1–48. (a)

Krantz, D. H. A survey of measurement theory. In G. B. Dantzig & A. F. Veinott, Jr. (Eds.), *Mathematics of the decision sciences, part* 2. Lectures in applied mathematics. Vol. 12. Providence: Amer. Math Soc., 1968. (b)

Krantz, D. H. Measurement structures and psychological laws. *Science* 1972, **175,** 1427–1435. (a)

Krantz, D. H. A theory of magnitude estimation and cross-modality matching. *J. Math. Psychol.* 1972, **9,** 168–199. (b)

Krantz, D. H. Color measurement and color theory. I. Representation theorem for Grassmann structures. *J. Math. Psychol.*, 1974, in press. (a)

Krantz, D. H. Color measurement and color theory. II. Opponent-colors theory. *J. Math. Psychol.*, 1974, in press. (b)

Krantz, D. H. Fundamental measurement of force and Newton's first and second laws of motion. *Philos. of Sci.*, 1974, in press. (c)

Krantz, D. H., Luce, R. D., Suppes, P., & Tversky, A. *Foundations of measurement.* Vol. 1. *Additive and polynomial representations.* New York: Academic Press, 1971.

Krantz, D. H., & Tversky, A. Conjoint-measurement analysis of composition rules in psychology. *Psychol. Rev.* 1971, **78,** 151–169.

Larimer, J., Krantz, D. H., & Cicerone, C. M. Opponent-process additivity. I. Red/green equilibria. *Vision Res.*, 1974, in press. (a)

Larimer, J., Krantz, D. H., & Cicerone, C. M. Opponent-process additivity. II. Yellow/blue equilibria and nonlinear models. Submitted to *Vision Res.*, 1974. (b)

LeGrand, Y. *Light, colour, and vision.* (2nd ed.) London: Chapman & Hall, 1968.

Levine, M. V. Transformations that render curves parallel. *J. Math. Psychol.* 1970, **7,** 410–444.

Levine, M. V. Transforming curves into curves with the same shape. *J. Math. Psychol.* 1972, **9,** 1–16.

Marley, A. A. J. Conditions for the representation of absolute judgment and pair comparison isosensitivity curves by cumulative distributions. *J. Math. Psychol.* 1971, **8,** 554–590.

Pfanzagl, J. *Theory of measurement.* New York: Wiley, 1968.

Shepard, R. N. Metric structures in ordinal data. *J. Math. Psychol.* 1966, **3,** 287–315.

Stevens, S. S. On the psychophysical law. *Psychol. Rev.* 1957, **64,** 153–181.

Stevens, S. S. Matching functions between loudness and ten other continua. *Percept. and Psychoph.* 1966, **1,** 5–8.

Tversky, A., & Krantz, D. H. Similarity of schematic faces: A test of interdimensional additivity. *Percept. and Psychophys.* 1969, **5,** 124–128.

Tversky, A., & Krantz, D. H. The dimensional representation and the metric structure of similarity data. *J. Math. Psychol.* 1970, **7,** 572–596.

Wilson, B. C. *An experimental examination of the spectral luminosity concept.* (Doctoral dissertation, New York University) Ann Arbor, Mich.: Univ. Microfilms, 1964.

Geometric Interpretations of Some Psychophysical Results

Michael V. Levine
UNIVERSITY OF PENNSYLVANIA
AND
EDUCATIONAL TESTING SERVICE

1. INTRODUCTION

Many people, especially certain psychologists, speak and act as if their impressions, perceptions, and opinions were like numbers. Otherwise reasonable people say such things as "The football team seemed twice as strong when Jackson was coach." They behave as if addition, subtraction, multiplication, and division were defined for mental objects. The articles in our technical journals have ample evidence: Psychophysicists commonly set willing subjects to such tasks as adjusting a light until it looks half as bright or deciding whether one tone is louder than the difference in loudness of two other tones.

These curious phenomena are taken as the starting point for a highly speculative theory of psychophysics. The theory extends the commonsense, classical mapping theories which recently have been sharply attacked (Krantz, 1972).

The new theory enlarges the classical theory in three general ways: (a) The multidimensional nature of perception is made explicit; (b) the transformations of the theory are interpreted geometrically; (c) the attributes are distinguished from sensations and only partially ordered.

I shall try to show that with the enlarged theory and some elementary geometry one may use a few ideas to explain many qualitative results, to give a credible and precise account of quantitative properties of data, and to design experiments that seem worth doing whether the theory is valid or not. In the process of developing the new theory, the major criticisms of the classical theory are answered.

1.1. Orientation and Organization of the Chapter

Throughout most of experimental psychology, theories are needed to predict new experimental results. However, the area of magnitude estimation and direct measurement is unusual in that experimental sophistication very greatly exceeds theoretical knowledge. Magnitude estimation, bisection, and related procedures have attracted a lot of interest, quite apart from their value as measurement procedures, as remarkable forms of information processing. There are now many replicable results and adequate descriptive models for most of them. There have also been extensive experimental applications of direct measurement as measurement procedures. Direct measurement has proven very easy to use, sufficiently sensitive to permit the study of phenomena usually studied by more elaborate techniques and free from many of the interactions that complicate other measurement techniques (Marks, 1973).

In spite of the success of the experimental work there has been widespread dissatisfaction with the theoretical work. Ad hoc psychological hypotheses and exotic assumptions from sensory physiology and other disciplines are regularly introduced to explain particular experimental findings. This is especially disturbing for some psychologists since the key theoretical problem, the relation between complex perceptions and unidimensional attributes, is a very old, purely psychological problem. Even so, it has recently (Krantz, 1972) been shown that classical psychology has completely failed to provide an acceptable interpretation of some of the most basic findings.

These considerations have determined several aspects of this work. It seems more important to make sense of what is already known than to predict new experimental results. The main theoretical problem is to explain why responses have some of the algebraic properties of numbers. However to show that the theory is not ad hoc and that it is capable of integrating diverse experimental results, it has been necessary to simultaneously consider several other findings. Finally, with few exceptions, only constructs from classical psychology have been used.

The following sections of this introductory part of the chapter give further details on the above points. Section 1.2 delimits the scope of the paper and is intended to make obvious that the main problem is purely psychological. Section 1.3 gives further details on the discrepancy between experiment and theory.

Section 2 is also introductory. Except for the last paragraph, it can be skipped on first reading by readers familiar with correspondences between arithmetic statements and geometric constructions. The basic theory is presented in Section 3. Details and applications are given in Section 4.

1.2. Scope of the Chapter

Three aspects of the typical psychophysical measurement experiment must be distinguished to clearly define the scope of this chapter. For concreteness, consider an observer's viewing a small illuminated panel and judging brightness. The three aspects are (a) the relation between physical stimuli and perceptions, (b) the relation between complex perceptions (for example, the subjective representation of an illuminated panel having a definite texture, distance from the observer, area) and simpler attributes (for example, brightness or perceived size), and (c) the relation between attributes and the responses the observer uses to report the level of an attribute. This paper is limited to (b), the relation between perceptions and attributes. The other aspects are deliberately treated in a schematic, noncommittal way.

Although the theory has been extended to category and rating scales (Levine, 1974c), only magnitude estimation and related measurement procedures are discussed here. The scope of this chapter is further limited to phenomena that have analogues in several sense modalities.

1.3. Theoretical Background: The Failure of Earlier Theories

The essence of our classical, commonsense theories of direct measurement seems to be this: there is a functional relationship between stimuli and the observer's internal states. Numbers, or more properly number names, correspond to internal states just as do physical stimuli. When performing in a measurement experiment, the observer responds by selecting stimuli mapping onto approximately the same internal state. Following Krantz (1972) these theories are called mapping theories in this chapter.[1]

Unfortunately when the details of mapping theories are made explicit, they often seem highly implausible. For example, consider the following analysis of an observer's selecting a light appearing half as bright as a given light. The observer first selects a number mapping onto approximately the same internal state or magnitude as the light. He divides the number by two. Finally he selects a light with the same magnitude as the halved number. Even naive

[1] See Krantz (1972) for a finer classification of the classical theories.

experimental subjects consider this analysis implausible, because they find the bisection response easier and faster than mental arithmetic and because they are not cognizant of performing complex mental operations.

In spite of this implausibility, mapping theories have been remarkably fruitful. An impressive number of replicable findings have been discovered by experimenters evidently using such theories.

A detailed and penetrating analysis of this situation is given in Krantz's (1972) recent theoretical study of magnitude estimation. Krantz lists a number of empirical generalizations that are based on the direct measurement experiments. He then surveys the available classical interpretations of the generalizations. Each interpretation involves unlikely mental arithmetic, complicated sequences of matches, or is otherwise unacceptable.

Krantz's solution has been to discard the classical framework and to develop a special new theory for direct measurement experiments called relation theory. For a clear and persuasive exposition of relation theory with references to Kristof's and Shepard's earlier formulations, see Krantz (1972).

The solution offered in this chapter retains the classical framework. By appropriately specifying the relationship between perceptions and attributes, one obtains a number of simple, intuitive interpretations of the generalizations. See Section 4.7 for further details.

These interpretations from the new theory are free of all the defects of earlier classical theories. The new theory seems able to integrate a much wider range of data than the alternative relation theories. In fact, as noted in Section 2.1, this new version of the classical theory may even be able to accommodate some important parts of relation theory.

2. SOME INFORMAL CONSIDERATIONS INTRODUCING GEOMETRIC INTERPRETATIONS OF ARITHMETIC AND THE MAIN HYPOTHESIS OF THE THEORY

2.1. Comments on Analogue Mental Arithmetic

Ordinarily when we think of arithmetic, we think of the operations on whole numbers we were taught in school. However, various experimental results have led psychologists to conjecture that people are able to solve arithmetic problems as analogue computers solve them: by representing numbers as continuous quantities such as lengths or electrical potentials and operating on them.

In this section, one experimental finding is used to introduce geometry as a language for reasoning about the sort of analogue devices and phenomena that interest psychologists. The goal is to introduce the geometric ideas needed for psychophysics rather than to explain the experimental result. Consequently, details about experiments are omitted.

One of the simplest results to consider geometrically is an unpublished study of Restle (1969a). A related published study is Restle (1969b). Restle gave his subjects some carefully chosen mental arithmetic problems, placed them under time pressure and measured their accuracy and rates of responding. He found that the speed and accuracy of deciding whether a number b is closer to a (equal to zero in this study) or c depends largely upon the proportion $(b - a)/(c - b)$ rather than upon the digits of b. This suggests (at least to Restle and me) that the numbers were encoded like lengths and that the encoded numbers rather than the digits were manipulated and compared.

All of the elementary statements of arithmetic can be translated into geometric constructions. These constructions[2] are well understood. A few examples are given in this section and the next.

From the psychologist's point of view, it is suggestive that a multidimensional space is required for the constructions and that there is little correlation between the complexity of the arithmetic statement and the complexity of a geometric interpretation. For example, in most constructions the distinction between rational and irrational numbers is irrelevant.

Here is a construction relevant to the Restle study. Consider three parallel lines called vertex, quantity, and intensity arranged as in Figure 1. Although

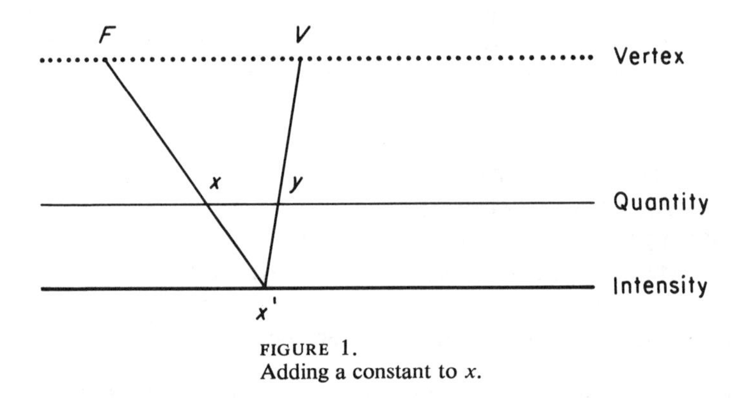

FIGURE 1.
Adding a constant to x.

the middle line will be used to represent numbers, all of the metric properties on all of the lines are deliberately ignored. The points on each of the lines (vertices, quantities, intensities) will be regarded as ordered from left to right. After this discussion it should be clear that only the ordering on one of the

[2] For a lucid discussion of these constructions, see Coxeter (1955). That introductory text contains much more geometry than is needed in this chapter.

lines, say the intensity line, need be ordered. The transformations we consider will induce the orderings on the other lines.

Let two points F and V (for fixed and variable) be selected on the vertex line. One can define a transformation of the quantity line by passing lines through V and F so that the intersection x' of these lines lies on the intensity line. It can be shown that the length of the segment yx is independent of the point x. (Use similar triangles FVx' and xyx' to show that the ratio of the lengths of segments xy to FV is independent of x.) Consequently, we may regard the mapping $x \longrightarrow y$ as the addition of a constant mapping, $x \longrightarrow x + k$. By elaborating the diagram we can interpret various statements about addition and subtraction.

One way to use this construction to understand the Restle result is as follows: Each of the stimuli a, b, c is encoded as points on the quantity line also denoted a, b, c. The fixed vertex F places each of the quantities in correspondence with intensities a', b', c' as in Figure 2. The observer under time

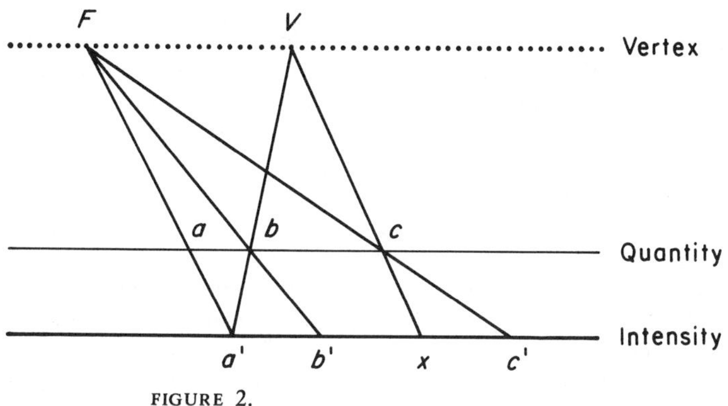

FIGURE 2.
Geometric interpretation of an experiment.

pressure quickly selects the vertex V on the same line as a' and b. By projecting through c he creates an intensity x. Simply by comparing b' and x he is able to decide whether $b - a$ is greater than $c - b$. For $c - b$ is greater if and only if x is to the right of b'. In this way the observer could behave as if he were doing arithmetic without actually counting or manipulating digits.

Before relating these observations to psychophysics there are several qualifying remarks that must be made.

1. Many other constructions might have been used to formulate process models for mental arithmetic. The one given above was chosen for its simplicity and usefulness in introducing considerations needed in the discussions of psychophysics. For more plausible models of mental arithmetic, see Restle (1969b) and his references.

2. Some of the alternative constructions are compatible with the relation theories now being proposed as alternatives for mapping theory. For details see the concluding paragraph of this section.

3. Whether or not these geometric constructions can be realized by the components of hypothetical neural nets now accepted by physiologists and anatomists seems to be as irrelevant a consideration for psychologists as, for example, the physiological realizability of Hering's[3] now acceptable, previously rejected, psychological ideas on color. The relationship between attributes and perceptions seems to be a purely psychological problem. However, to make the calculations required by the constructions, little more is needed than neural nets with output $y(t)$ accepting positive bounded signals $x(t)$ such that if $x(t)$ rapidly converges to x, then $y(t)$ rapidly converges to $\ell(x)$, where ℓ is a linear fractional transformation.

Some details follow. They are indented and printed in smaller type. This device is used throughout the chapter to indicate material that can be skipped on first reading.

On alternative constructions: As indicated in the discussion of Figure 1, each vertex V to the right of F is associated with an increment of quantity of length k. (k depends on V and the ordering of these ks is the same as the ordering from left to right of the Vs.) In this sense it may be legitimate to interpret V as a sense distance of size k. Then each pair of quantities is associated with a unique sense distance $V(a,b)$. To solve Restle's problem the subject need only check to see whether sense distance $V(a,b)$ is to the left of $V(b,c)$. To modify this construction in order to obtain the sense ratios that play an important role in the Shepard-Krantz relation theory, simply rotate the vertex line through F until it passes through the point on the quantity line used to represent the number zero. Then the transformations analogous to those in the preceding discussion, except with vertices on the rotated line, will define multiplicative transformations $x \longrightarrow kx$. Just as vertices on the parallel line were interpreted as sense distances, these new vertices behave like sense ratios. More is said about this construction and geometric interpretations of multiplication in the next section.

2.2. The Central Hypothesis of the Theory

Consider an experiment in which an observer adjusts a light until it appears half as bright. There are many published studies of this kind. One of the

---

[3] See Hurvich (1969) for a lively account of this story. Hering's psychological ideas were rejected when introduced decades ago because they were judged to be in conflict with contemporary physiological knowledge. Now physiology has changed and Hering's color theory is widely accepted. Although Professor Hurvich does not agree in conversation or in print, the physiological realizability of Hering's psychological ideas is irrelevant in the twentieth century as it was in the nineteenth.

earliest careful fractionation studies is Hanes (1949). For an exceptional bibliography and a description of current experimental procedures, results, and theories of direct measurement, see L. E. Marks' *The New Psychophysics of Sensory Processes* (1973).

In the opinion of many experimental psychologists, subjects can behave consistently and reliably *as if* they were converting stimuli to numbers, multiplying these numbers by 0.5, and finding a stimulus that maps onto the product. Experimental evidence for numberlike encodings with interpretable ratios has been cited for many different kinds of physical stimuli and also for stimuli (such as crimes of varying seriousness) for which there is no established physical measurement procedure (S. S. Stevens, 1966).

Here is one way to consider the ability to perform in bisection experiments. Suppose people represent numbers on a quantity line parallel to an intensity line in a plane having a fixed vertex F placing quantities in correspondence with intensities, as in the discussion of Figure 1. However, now consider the vertex line passing through the encoding z of zero and its corresponding intensity i_0 as in Figure 3. For a vertex M on the vertex line, a correspondence $x \longrightarrow x' \longrightarrow y$ can be defined as illustrated in Figure 3. This new correspondence $x \longrightarrow y$ can be considered multiplication by a constant in the sense that the segment between y and the representation of zero is easily shown to be a constant fraction of the segment xz. (Use similar triangles Mzy, Mi_0x' and Fi_0x', Fzx with common segment i_0x'.) Changing the vertex M only changes the multiplicative constant. There is exactly one M such that the constant is one half.

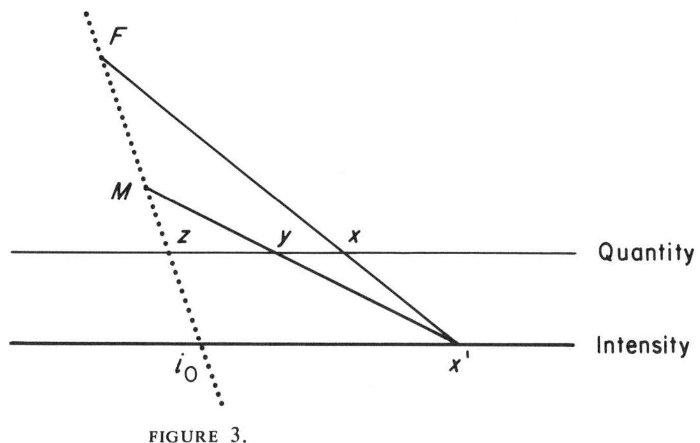

FIGURE 3.
A geometric interpretation of fractionation.

To relate this construction to bisection experiments, suppose that the experimental series of lights is represented by the observer as a series of points in the same plane. Then the vertices F and M induce a transformation of lights analogous to the halving transformation just discussed.

Notice that once the vertices F and M are specified, the observer is independent of numbers. There is absolutely no reason for him to map lights to numbers, halve the numbers, and return to the corresponding lights. The transformation of lights is produced by an observer's operating on lights and their intensities only.

I propose that we acquire[4] the ability to halve sensations by first learning to do this geometric arithmetic. Then we use the same apparatus for operating on sensory continua. There is nothing special about the number continuum. The psychological continuum of length or time, for example, could play the same role. The important point is that the vertices, once defined for any dimension in which arithmetic is natural, become available for an arithmetic of visual brightness or of seriousness of criminal offenses.

The main hypothesis of this chapter is that we have essentially multidimensional perceptions, that we use processes like the projections considered above to abstract unidimensional attributes and that the regularities of data suggesting mental calculation actually appear as a consequence of these processes.

3. THE BASIC PSYCHOPHYSICAL THEORY

3.1. Outline and Introductory Remarks

The theory is called projective theory. Its components are

Physical Space
- physical events
- physical continua
- parameterizations

Psychological Space
- appearances
- straight psychological continua
- intensities
- projections

---

[4] The word 'acquire' is used advisedly. The reader may choose to think of the ability as acquired in the course of the experiment, in developmental time or evolutionary time. The equations of the theory are independent of the choice.

Magnitudes
  points of view
  intensity curves and measures

In this section they are briefly discussed. The purpose is to provide a vocabulary for mapping theory interpretations of psychophysical phenomena. Enough structure is specified to provide geometric interpretations and information processing schemes for arithmetic tasks with sensations and with numbers.

The most closely related alternative theory seems to be that of Ross and Di Lollo (1968).[5] In their paper they use multidimensional representations and parallel projections for some of the same purposes as multidimensional representations and not necessarily parallel projections are used here.

The main problem in classical psychophysics has been to describe the relationship between physical events and their mental representations. The view implicit in mapping theory, namely that the psychological representation is a function of the physical event, is used here. However, an attempt will be made to regard all psychophysical regularities as imposed by the observer rather than transmitted from the physical world. For this reason only the crudest physical relations will be acknowledged. This is made explicit in the discussion of physical space.

3.2. Physical Space

Physical space is simply the set of all physical environments that the experimenter may wish to present to his subject. The points in the set are called *physical events.*

In a typical experiment the psychophysicist adjusts a bit of hardware to obtain a one-parameter subset of physical space. For example, when studying brightness, he may turn a dial through λ degrees to control the amount of electricity passing through the filament of a light bulb. These one-dimensional subsets will be called *physical continua.* More precisely, a physical continuum is a parameterized set of physical events $\{X_\lambda\}$ where the index λ ranges over a set (generally interval) of numbers. The mapping $\lambda \longrightarrow X_\lambda$ is called a *parameterization.*

From the point of view adhered to in this chapter, physical space is just a set. All of the quantitative regularities of data are studied from a psychological point of view.

[5] See Ross (1972) for more recent references.

3.3. Psychological Space

Psychological space, the space in which physical events are represented by the observer, is taken as a multidimensional space (Euclidean space) of low dimensionality. Each physical event is represented by the observer as a point in psychological space called an *appearance*. At any given time, for any given observer, there is exactly one appearance for one physical event. Some transformations and extra structure are defined in later sections. The usual topology in Euclidean space is used.

It will not be necessary to refer to coordinates in psychological space. The psychological considerations that are ordinarily dealt with by coordinate systems are handled by projections (Sec. 3.5) and points of view (Sec. 3.7).

The set of appearances corresponding to a physical continuum is called a *psychological continuum*. The function between a physical continuum and its corresponding psychological continuum is called a *psychophysical function.* Since every physical continuum has a real parameterization, every psychological continuum also has a real parameterization. Although this may occasionally be confusing, the real parameterization of a psychological continuum will also be called a psychophysical function. To restrict attention to the sort of parameterizations psychophysicists study, only continuous psychophysical functions are considered.

If the appearances in a psychological continuum fall on a line segment in psychological space, then the continuum is called *straight*. The main geometric fact dictating the use of this word is the fact that two lines can meet in at most one point. However, a curve can cross a line twice.

Not every psychological continuum will be straight. The example that follows will be clearer after the sections on projections and points of view have been read. The reader may wish to return to this example after reading those sections.

Since Newton's time, spectral colors have often been represented by points on a curve. Consider the physical continuum that one gets by selecting a narrow segment of sunlight analyzed by a prism. If one were to use this theory to understand the scaling of redness one would be forced to conclude that the psychological continuum corresponding to these stimuli is not straight. There would be equally red-appearing stimuli at both ends of the spectrum. This can only happen if some line intersects the psychological continuum twice.

Some continua—those that appear to vary in exactly one salient attribute—may be considered straight. Finding suitable continua seems to be a matter of experimental skill and psychological sophistication. After the continuum has been chosen there will be experimental implications of straightness that can be tested.

It is curious that a great many psychophysical experiments can be analyzed without considering spaces of high dimensionality. This may reveal a significant limitation on the observer's ability to process information, or it may simply be a consequence of the experimenter's successful attempts to design experiments in which all but a few dimensions of experience are irrelevant. In the absence of direct experimental evidence the more conservative latter alternative is used, and the low dimensional spaces are considered as quotient spaces of higher dimensional spaces.

3.4. Intensities and the Interpretation of Matching

In cross-modality matching experiments, subjects attempt to equate levels of different attributes. For example, they may attempt to adjust a light until it appears as bright as a sound is loud. Responses are variable, both within and between observers. But averaged results have a consistency that has strengthened belief in the following interpretation:

> *Traditional mapping-theory interpretation of matching:* An appearance has several qualities such as brightness or yellowness and has each quality to a definite extent. The extent to which an appearance has brightness can be represented by a number or something numberlike such as neural excitation or length. The same numberlike continuum is used for brightness and for loudness. The observer matches brightness to loudness by equating neural excitation, length, magnitude, or whatever the continuum that carries ordinal information for attributes is called.

I do not accept this interpretation, except as an approximation. Criticism and an alternative interpretation are offered in Sections 3.6, 3.7, and 4.8. However, the traditional interpretation will be used frequently for two reasons. It gives a convenient approximation of the central tendencies predicted by the alternative interpretation. And, being familiar to most readers, it simplifies the exposition of the parts of the theory for which the alternative and traditional views agree. For these reasons a special line is now considered in the following assertion about psychological space.

There is a distinguished line in psychological space, the points of which are called *intensities*. No appearance[6] is an intensity.

---

[6] Instead of introducing an abstract continuum in psychological space some theorists have assumed that there are mappings from one psychological continuum to another. For example, lights of varying brightness might be compared with sounds of varying loudness by referring each to a continuum of lines of varying length. Physical laws relating distance to a source with physical intensity are sometimes used to defend the correlation and induce quantitative relationships. This view is not used here. In addition to the objections of Krantz (1972) there is a vulnerability to rejection in experiments using the following considerations:

3.5. Projections

Appearances must somehow be placed in correspondence with intensities if this theory is to be related to the traditional mapping-theory interpretation of matching. There are certain qualitative properties demanded by data. For example, the whiteness ordering on a psychological continuum must be the opposite of the blackness ordering. The simplest mappings from a vector space to one of its lines with all of the appropriate qualitative properties and with sufficient generality to include the analogue arithmetic schemes of Section 1 are the projections.

Projections give this theory its geometric and intuitive character. Although they are a part of classical algebraic geometry, a few illustrations and definitions are included to show some of the qualitative phenomena they describe and how they function like coordinate systems.

Projections in a two-dimensional space were illustrated in Sections 2.1 and 2.2. To define a projection from an appearance in a two-dimensional space onto the intensity line, one begins by selecting a point in the space not on the intensity line called a *vertex*. If an appearance is not on the line through the vertex parallel to the intensity line, then the projection of the appearance is the unique intensity on the line through the vertex and the appearance. Just as the ratio of some numbers is undefined, the projection of an appearance on the parallel line is not defined.

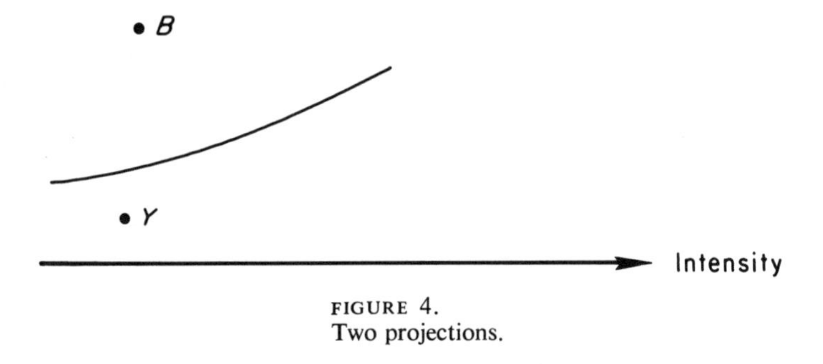

FIGURE 4.
Two projections.

For illustrative purposes, suppose a continuum of lights is presented, and that the observer's processing can, for simplicity, be adequately represented in a two-dimensional space. (See Fig. 4.) A vertex B (for brightness) induces one ordering on the appearances when appearances with greater projected

---

with procedures such as selective adaptation, the correspondence between length and perceived length, as well as the relationship of appearances on a length continuum, may be systematically changed. These changes should affect the cross-modal matches in predictable ways. For this reason, a new continuum is introduced.

intensities are regarded as greater. Another vertex Y (for yellowness) induces the opposite ordering for these appearances. As Figure 4 is drawn, there is very little change in the projection through the yellowness vertex for one extreme of the psychological continuum.

The two projections function as coordinates in the sense that each appearance x can be specified exactly by a pair of intensities x_Y, x_B, where x_Y is the projection through the yellowness vertex and x_B is the projection through the brightness vertex. This is true not only for appearances on the continuum but also for all appearances in the same plane, provided the projections are defined.

In three dimensions, instead of taking a point as a vertex to define a projection, one takes a line (not in a plane with the intensity line) as vertex. Then the projection of an appearance not in the plane containing the vertex parallel to the intensity line is the unique intensity on the same plane as the vertex line and the appearance. To develop intuitions quickly, it may be helpful to visualize a large book with its spine on the vertex. The projection of an appearance is the unique intensity on the same page as the appearance. The same remarks about coordinates remain valid, however three rather than two projections are needed to locate an appearance.

There is no need to consider more than three dimensions in this chapter, however, for completeness a vertex in psychological n-space is an n-2 affine subspace not in the same hyperplane as the intensity line. A projection onto the intensity continuum is the mapping from appearances to intensities such that corresponding points lie on the same hyperplane containing the vertex.

3.6. Motivation for Points of View

This section contains some informal remarks needed to motivate an alternative to the traditional mapping-theory interpretation of matching.

Both intensities and appearances are points in psychological space. Yet there seems to be a fundamental difference between the appearance of an object and the degree to which it possesses an attribute. One intuitively obvious difference is the definiteness of perception and the uncertainty of psychophysical judgment. It is common to see an object clearly and still be very uncertain about one of its attributes.

For a simple demonstration, I scattered seven pennies on a sheet of paper and eight on another. Two quarters, three nickels, and a dime were added to both displays to break up easily counted clusters. People passing my office were invited to guess which paper had more pennies. They were asked not to count or place the coins in one-to-one correspondence. Although it is easy to see a few coins on a white sheet of paper in a well-lighted room, four out of ten first guesses were wrong. (With practice people became very good at the task.)

In the currently fashionable jargon, 'perception is categorical.' If the perceptual system is functioning normally and processing familiar objects, then, when we see, we see something in particular. But when we are forced to abstract and judge the intensity of an attribute such as numerosity, a vagueness becomes evident.

Consider matching brightnesses and loudnesses once again. A moderately loud tone is held constant. A very bright light is clearly more intense and a very weak light is clearly less intense. Brightness appears to change continuously with luminance, so we expect to find one light that exactly matches the sound. Instead, there is a broad range of equally acceptable lights. Within the range, one light matches the tone as well as another.

It is not sufficient to dispatch this uncertainty with the usual observation that there are always errors of measurement. These errors have an orderliness that indicates that something basic has been omitted from the traditional interpretation. For example, with myself as subject in informal experiments I have observed that the acceptable range of lights matched to lights is considerably narrower than the range of lights matched to tones. A probabilistic device unreliably ordering points on an intensity line would give equal ranges. A semiorder structure (Suppes & Zinnes, 1963, Sec. 3.2) on the intensity line again would give equal ranges.

Undoubtedly there are many ways to account for these aspects of the intensity of experience. I wish to do so without losing two attractive features of the traditional mapping interpretation: (a) the traditional interpretation is deterministic rather than probabilistic, and (b) magnitudes of different qualities are comparable in the traditional interpretation. My solution, given in the next section, is to make appearances and the degree to which an appearance has an attribute fundamentally different entities. The appearance of a light remains a point in psychological space. But the brightness will be a curve in a plane. In addition to yielding an intuitive theory capable of dealing with the considerations above, the analysis gives a new explanation (see Sec. 4.8) of a systematic departure from transitivity observed in matching.

3.7. Points of View and Magnitudes

This section may be skimmed on first reading. The ideas presented are not needed for predicting the main effects in most experiments.

The traditional mapping-theory interpretation of matching has each appearance in correspondence with a definite point on the intensity line prior to matching. The alternative to the traditional view offered in this section has each appearance associated with a definite curve. The goal is to let the shape of these curves (henceforth called intensity curves) carry the information intensity points failed to carry.

Each intensity curve will be the graph of a number-valued function of intensity. The values of the function may be thought of as the extent to which an intensity is characteristic of an appearance. Once these curves are defined, it will be easy to reason with precision about the problems raised in the preceding section.

In the following paragraphs intensity curves are discussed informally. Then a generalization of the vertices used to define correspondences between appearances and intensities is introduced. The generalization, called points of view, is used to place appearances in correspondence with intensity curves.

Intensity curves will be specified in such a way as to be generally unimodal and zero except on an interval. Physically intense stimuli will have modes over a high intensity. Assumptions will be made to force the shape of the curve and the length of the supporting interval to change smoothly as an appearance is varied along a psychological continuum.

Instead of comparing points as in the traditional interpretation the observer will be considered to be comparing curves. As discussed in the final section of Falmagne (this vol.), there are many alternative ways to order curves. The details of the actual processing carried out by the observer will depend on minor experimental details such as instructions. For example, an observer instructed to bracket (see Sec. 4.8) is likely to do so. Fortunately, it is possible to derive many predictions by simultaneously considering a large class of plausible processing schemes. The constraints[7] defining these schemes are given in the next paragraph.

It is assumed that the observer is equipped with a device for comparing intensity curves. When given a pair of curves u, v the device responds that u is greater than v, that v is greater than u, or that it is unable to choose. For the purposes of this section it need only be assumed that the device has the following properties (see Note 7 for a discussion of these properties): If the supporting intervals of u and v are disjoint then it always responds that the

---

[7] An analogy may make the criteria for an acceptable device clearer. Consider an individual with imprecise knowledge of the distribution of heights in various populations who is asked to make decisions of the following kind. An individual is selected at random from two populations. The decision-maker is required to predict which population contributes the taller individual. Three critical comparisons are (a) mice with men, (b) men who have just taken their shoes off with men who have just put their shoes on, and (c) horses with men. The decision-maker is predictable in (a) since the distributions do not overlap. The decision-maker attempting to maximize the proportion of correct predictions is predictable in case (b) since the unknown distributions will have very nearly the same shape. Although the crucial statistic may be much larger in the last case than in the second case, the decision-maker is unpredictable because he lacks the information he needs to make a rational decision; the distributions have different shapes and they overlap.

curve with the interval on the right is greater. If the curves have exactly the same shape in the sense that it is possible to obtain the curve u by shifting the curve v to the right, then it always chooses u. If the curves have approximately the same shape, then it applies some complicated rule which generally selects the curve to the right as greater. When the curves have very different shapes and overlapping supports, the device responds that it is unable to choose.

To see some of the implications of assuming that the observer uses a device with these properties, consider matching experiments with lights and tones. A very bright light with supporting interval to the right of a moderately loud tone will be judged more intense. Lights with approximately the same appearance will have approximately the same shape intensity curves and so slightly different lights can be reliably ordered. Since the loudness curves need not have the same shape as brightness curves, the ordering device would be expected to fail to choose over a larger range for matching lights to tones than matching lights to lights. Some other qualitative phenomena are considered in Section 4.8.

The alternative to the traditional view will be entirely deterministic. There is no need for probability at this level of the theory. However, the quickest way to communicate the alternative theory is to appeal temporarily to the reader's intuitions about probability.

Suppose that instead of specifying the vertex of a projection precisely, a probability distribution over vertices is given with most of its mass concentrated on a particular vertex. Then each appearance defines a distribution of intensities in the following way. The probability of a set of intensities is the probability of the set of vertices projecting the appearance into the set. If suitable restrictions are made on the probability distribution over the vertices, then the intensity of the appearance can be thought of as a random variable with a continuous density.

A point of view is simply a measure defined on vertices. Only measures sufficiently well behaved to permit the definition of the analogue of the density in the preceding paragraph are considered. But measures other than probability measures are permitted. The extra generality (signed measures, measures with total mass not equal to one) is not especially important. What is important is to retain a deterministic theory with a partially ordered space of magnitudes sufficiently rich to account for effects such as those considered in the last section.

Magnitudes in the traditional interpretation were zero-dimensional intensity points. In this alternative they are taken to be the one-dimensional intensity curves or, equivalently, the measure on the intensity line used to define the intensity curve. The intensity curves will be used when it is desirable to reason geometrically about psychological phenomena, and the measure will be used when it is desirable to consider properties that are invariant under smooth transformations of the intensity line.

The brightness of an appearance is simply the intensity curve (or measure) induced by a point of view characteristically associated with brightness, and the loudness of an appearance is simply that induced by a loudness point of view. When magnitudes are defined in this way, there is no difference in kind between the brightness and the loudness of appearances. They differ only in the manner in which they are generated. To the extent that curves with different shapes can be ordered, these brightnesses and loudnesses can be ordered. In particular, when overlapping curves of different shapes are compared, the observer's behavior is expected to be variable.

To return to the distinction between a probabilistic and deterministic theory, a magnitude is an intensity curve or measure. It is not a random variable having a particular measure or density. An analogous physical situation is described in the next paragraph to show that this is more than a verbal distinction.

A stubborn man might refuse to acknowledge the fact that his right index finger is extended in space. In order to avoid being incorrect when asked questions about its location, he might explain that the location is not *really* a point in space, but a *random* point in space. We know enough about electricity to quickly enlighten such a man. Not even a random point can be in two places at the same time, and it is safe for an ungrounded man to touch one of the terminals of a light socket.

3.8. Judgmental and Purely Sensory Effects

To recapitulate, physical events are represented by psychophysical functions as appearances in psychological space. When the observer wishes to judge the degree to which an appearance has an attribute he selects an appropriate point of view and generates (by a process like projection) an intensity measure or curve.

This leaves only two ways for accounting for the effects of experimental manipulations that change the relation between stimuli and magnitudes. There are purely sensory changes in which only the psychophysical functions change, and there are purely judgmental changes in which points of view change.

In simple studies the two kinds of effects are likely to be confounded. For example, if an observer is asked to judge brightness and then is asked to judge the yellowness of a series of lights, the most obvious theoretical interpretation is a change from brightness point of view to a yellowness point of view. However, the data may indicate that asking the subject to attend to a different aspect of the lights changes their appearance. This is an empirical question and can be translated into a standard statistical question of the form: Does a model give a significantly better fit of data than a submodel?

For an opposite example, suppose the subject is exposed to an intense blue light prior to viewing the stimuli for yellowness judgments. The appearances of the lights are certainly affected by the blue light. But, as discussed in Section 4.7, there may be judgmental effects in addition to the purely sensory effects. Even though the yellowness point of view remains constant, other points of view are involved in producing the response. Some of these may change when the distribution of appearances in psychological space changes. These considerations are examined in greater detail in Section 4.7.

In the remainder of the chapter, only effects of changes of point of view are considered in detail. The calculation of psychophysical functions and the quantification of changes in psychophysical functions are outside the scope of this chapter. Some references to my work in this area and some comments on results being prepared for publication are included in Footnotes 10 and 11.

4. ELABORATIONS OF THE BASIC THEORY AND SOME ILLUSTRATIVE APPLICATIONS

In this section the basic theory is elaborated and applied to published data. The elaborations needed are the following:

1. Degenerate points of view for the approximation of central tendencies in published data.
2. Embedding functions to generate graphical displays of experimental results.
3. Number appearances to place magnitude estimation, cross-modality matching, and rating experiments in the same framework.
4. Inhomogeneities in the intensity continuum to study the consequences of assuming psychological space is bounded.
5. General factors influencing the subject's selection of a point of view.
6. Symmetry in points of view to simplify the calculation of regression effect and Ekman's law.

Most of the elaborations are incorporated in the discussion of experimental results.

In the following paragraphs it can be seen that the theory is consistent with several essentially different explanations of published experimental findings. The data for choosing between alternative explanations are not yet available. But each of the alternative explanations has testable implications. In Section 3.6 I try to show that in at least one case the experiments suggested by the theory are intrinsically interesting.

The theory is formulated as a description of individual information processing. Most of the data considered are averaged over groups of observers. Fortunately, there are recent theoretical and experimental developments that should help overcome some of the problems[8] involved in obtaining useful data from individual subjects. At this time appropriate individual data are not available.

4.1. Degenerate Points of View

Points of view in their full generality are needed only in the final applications of this section. In the other applications a trivialization of the point of view called a degenerate point of view is used to approximate central tendencies in published data.

If a point of view has all but a very small amount of its mass concentrated on a single vertex, then the intensity curve for an appearance is a sharp spike over a narrow range of intensities. The limiting case, a point of view with all of its mass concentrated on a single vertex, will be called a *degenerate point of view*. The vertex induces a projective correspondence between intensities and appearances. When the point of view is degenerate the distinction between magnitudes and intensities is ignored.

4.2. Embedding Functions

The appearance of a physical event is a point in a vector space rather than a number or a number-measured sensation. And so it seems that the convenience of real functions in alternative mapping theories is lost. In this section it is shown that this is not generally so.

Consider a physical continuum $\{x_t\}$ carried by its psychophysical function to a straight continuum $\{X_t\}$. For example, consider dial settings t parameterizing the luminance of a light source in a brightness study. To define a real function, simply select two different dial settings t and s having appearances A equal to X_t and B equal to X_s. Since every point on the line through A and B can be written in the form $[1 - \alpha]A + \alpha B$ for exactly one number α, we can represent the psychophysical function $t \longrightarrow X_t$ by a real function $t \longrightarrow u(t)$ with $X_t = [1 - u(t)]A + u(t)B$. Any such function is called an *embedding function*.

[8] An example of a major problem is the effect of repeated exposures to the stimuli on the observer. After many exposures the observer knows the range of stimuli that will be used. This can have complex effects on magnitude estimates. For an example of some important recent results, see Teghtsoonian (1973).

The definition of an embedding function for a psychophysical function $t \longrightarrow X_t$ depended upon the choice of arbitrary parameter values t and s. Embedding functions are interval scales in the following sense. A different embedding function v is defined if different t and s are selected. But it is easy to show that there will always be a pair of numbers a and b such that for all t, $v(t) = au(t) + b$. In particular, when s and t are interchanged, v is $1 - u$. For this reason every strictly monotonic embedding function can and will be assumed to be increasing.

As illustrated in the sequel, many questions can be considered in detail in terms of the number-valued embedding functions rather than the vector-valued psychophysical functions. It is in this sense that the convenience of real functions is retained in this version of mapping theory.

4.3. Some Qualitative Facts About Projective Transformations

A few easily proven, elementary geometric facts are needed for relating qualitative descriptions of psychological processes to quantitative aspects of data. Proofs[9] and detailed discussions can be found in introductory textbooks.

A projective or linear fractional transformation is a function f defined for all numbers and the symbol ∞. It is defined by four numbers a, b, c, d such that ad and bc are different. For the numbers x such that $cx + d$ is not zero, $f(x)$ is defined by

$$f(x) = (ax + b)/(cx + d).$$

If $cx + d$ is zero, then $f(x)$ is equal to ∞. If c is zero, then $f(\infty)$ is ∞. If c is not zero, then $f(\infty)$ is equal to a/c. The word 'infinity' is sometimes used instead of ∞ below.

The transformation f defined with numbers a, b, c, d has an inverse denoted by f^{-1} and defined with numbers $d, -b, -c, a$ in the place of a, b, c, d. The projective transformations form a group with products fg defined by function composition

$$(fg)(x) = f[g(x)].$$

Projective transformations enter the theory in the following way. Suppose a psychological continuum and vertex are given. When another continuum and vertex are chosen, a correspondence between appearances on the two continua is obtained by pairing appearances projecting onto the same intensity. If the first continuum has embedding function u and the second has

---

[9] In fact, almost every assertion can be proven directly from the definition by routine algebraic manipulations.

embedding function v then there will always be a unique projective transformation f relating u and v in the following sense: if the appearance of x on the first continuum is paired with the appearance of y on the second, then $v(y)$ equals $f[u(x)]$.

The study of variations of points of view is expedited by writing f as a product $f = pq$ of two projective transformations. This can always be done so that a change of the first point of view only changes q and a change of the second only changes p.

The decomposition of f into p and q can be done in many ways. A device to force uniqueness upon the decomposition is to label the intensities with numbers in the manner that embedding functions were used to label appearances with numbers. For a fixed pair of intensities A, B any intensity can be written uniquely in the form $iA + (1 - i)B$ for exactly one number i.

The factors p and q are specified exactly by the following condition: if the appearances of both x and y project to the intensity $iA + (1 - i)B$ then $qu(x)$ equals i.

> Some statements using these number assignments on the following pages may strike the reader as a retrogression to the earlier theories that failed to distinguish numbers from nonnumerical psychological processes. Expressions such as 'the intensity i_0' are written in place of more precise or clearly nonnumerical expressions such as 'the intensity $i_0A + (1 - i_0)B$' or 'the projected intensity of the appearance of the weakest stimulus.' However, the numbers are never used in an essential way. Every argument with numbers can be replaced by a nonnumerical statement. The translation is routine.

A basic property of projective transformations is that if $f(x_n)$ equals $g(x_n)$ for three different values of x_n then f equals g. One implication is that unless f is the identity transformation, $x \longrightarrow x$, there can be no more than two solutions to the equation, $f(x) = x$.

A number or infinity is a *fixed point* of f if it satisfies the equation $f(x) = x$. If infinity is a fixed point of f, then f is called an *affine transformation* and can be written in the form $f(x) = ax + b$ for some real constants a and b.

The affine transformations f are the only projective transformations satisfying the monotonicity condition

x less than y implies $f(x)$ less than $f(y)$
for all real numbers x and y.

If f and g have exactly the same fixed points then f and g commute, i.e., fg equals gf. As a special case, two affine transformations with a common real-fixed point commute.

As another special case, if f and g both have zero and infinity as fixed points, then they are both in the commutative group of similarity transformations $\{x \longrightarrow ax\}$. In particular, if f and g both carry x_0 to zero and x_+ to infinity, then for some constant a, $g(x) = (gf^{-1})[f(x)]$ equals $af(x)$.

4.4. Magnitude Estimation as a Matching Experiment

In this section a simple interpretation of magnitude estimation is offered. The goal is to introduce some ideas needed in the sequel. In the process, an interpretation of Stevens's power law is given. Later (Sec. 4.7), the evidence for the power law is reconsidered and a different interpretation is given. This section concludes with comments on some extra assumptions used to relate projective theory to the power law.

As is now common among psychologists, perceived numbers are given approximately the same theoretical status as perceptions of traditional stimuli. The physical events representing particular numbers are assumed to give rise to appearances falling on a straight line in psychological space and an increasing embedding function u is assumed to relate numbers to number appearances.

Some complications tangential to the interpretation of the magnitude estimation experiment as a matching experiment are avoided by treating the set of number appearances as if it were topologically equivalent to a line segment. No matter what topological structure the set of number *appearances* is assumed to have, a subject with a fixed set of rules for naming appearances and a definite rate of speaking must select his *responses* in an experiment from a finite set of number names. In Section 4.7 the discreteness of the set of responses is acknowledged and used.

Numbers are assumed to have a special feature as a consequence of being extremely familiar and abstract: The observer can generate and operate upon a number appearance in the absence of an obviously appropriate stimulating physical event.

In the magnitude estimation experiment, the observer attempts to choose numbers with the same magnitude as auditory, visual, or other appearances. For concreteness, consider a continuum of lines of varying length. Magnitudes are obtained from number appearances with a number point of view and from line appearances with a length point of view. If both points of view are degenerate, then it will be possible to match magnitudes exactly.

To make contact with the power law, suppose the length continuum is straight with embedding function v. Then, as noted in the preceding section, there will be constants a, b, c, d such that number y matches line x if and only if

$$u(y) = [av(x) + b]/[cv(x) + d]. \qquad (1)$$

S. S. Stevens's power law asserts that for matching pairs (x, y), y is a power function of x; i.e., y equals ax^n for real constants a and n. Stevens and many others get good fits of data with linear functions when mean $\log y$ is plotted against $\log x$. This generalization holds for length and a very large number of other physical continua. The fit is best when x is large.

The most straightforward way to relate the power law to projective theory is to adjoin assumptions similar to those made by Stevens and his associates:

1. The number embedding function u is linear.
2. The physical continuum embedding function v is a power function.
3. (Monotonicity) If x matches y, x' matches y' and x' is greater than x, then y' is greater than y.

The power law data can be easily deduced from these assumptions. Since affine functions are the only monotonic projective functions, c must be zero in Equation 1. Consequently, for large x, $\log y$ is approximately equal to a linear function of $\log x$.

None of these assumptions are essential parts of the theory. They are intended as convenient by expendable specializations of the theory. The monotonicity assumption (3) is relaxed in Section 4.7. Some experimental evidence is considered there.

Assumptions (1) and (2) restricting the form of the embedding functions have been systematically studied by D. Curtis and S. Rule in the context of another mapping theory. Some of their findings are summarized in a recent symposium paper (Curtis & Rule, 1972). Using a general curve-fitting algorithm of Kruskal they have computed the best-fitting monotonic embedding functions for direct measurement data. In partial support of the assumptions used here, they find the physical-continuum embedding function to be a power function. The number-embedding function is also a power function, but with exponent close to one.

Curtis and Rule's work shows that it is not necessary to assume particular functional forms for the embedding functions. The functions can be calculated from experimental data. Some alternative methods for obtaining embedding functions from psychological data will be published separately.[10]

Projective theory is neutral with respect to the controversy over the validity of the power law and the competing refinements of the law. The functional form of embedding functions is also a peripheral concern for the theory. The

---

[10] Section 4 of the earlier version of this chapter is to be submitted for publication separately to the *Journal of Mathematical Psychology*. It contains an outline of some unpublished results on functional equations. Some related published papers are Levine (1970) and Levine (1972).

theory is concerned with the quantification of the relationships between functions and the psychological processes generating functions rather than with the shape of any particular curve.[11]

4.5. On the Size-Weight Illusion and Concurrent Graphs

J. C. Stevens and Rubin (1970) have recently published a remarkably simple and precise study of the size-weight illusion. Magnitude estimates of weights of various size were made under carefully controlled conditions. To a very high degree of accuracy, the logarithm of the (geometric mean of the) estimates of weights of fixed size was a linear function of the logarithm of weight measured in grams. When the slope and intercept of the lines were estimated by the standard least squares procedure applied to each of the sizes separately, a rather odd relation was discovered: The various slopes were a definite function of the intercepts. Geometrically, the extrapolations of the separately fitted lines all intersected at one point. This point, coincidentally, corresponded to the heaviest weight the subject could lift in the experimental position.

J. C. Stevens (1972) recently showed that this is not an isolated occurrence. Concurrent magnitude estimation functions have been observed many times, in many laboratories, with many dimensions of stimulation. In nearly all of the published cases cited by Stevens the point of intersection was interpretable.

In order to further elaborate the theory and illustrate its application, I shall attempt an interpretation of the size-weight illusion. No attempt will be made to deduce the fact that the curves are linear on logarithmic paper. Instead I shall attempt to deduce the existence of some transformations having the property of logarithm transformation. In the concluding paragraph of the section it is shown that the Stevens-Rubin finding can be expressed algebraically and tested without fitting special functions or extrapolating.

An essential part of the theory needed in the interpretation of the Stevens-Rubin finding is the boundedness of psychological space. Nearly all psychological continua definitely seem bounded. Few of us have any conception of the heaviness of a 500-pound weight. The range of intensities is also bounded. It is probably unnecessary to consider negative intensities. The

---

[11] Some tools for this quantification are presented in Levine (1974a, 1974b). The first paper suggests using psychophysical data to define local semigroups with essentially unique matrix representations. The entries in the matrices quantify the relations between curves. The second paper introduces a method for using Fourier analysis to compute the quantifying matrices with great precision from matching functions defined on small ranges of stimuli. In collaboration with D. Saxe, a computer program to do this calculation is being prepared for public use.

bounds of the intensity continuum are probably not sharp. More likely our ability to use intensities varies over the range of intensities so that only in the middle range do we process and compare quickly and accurately. The adjustments of point of view considered below and in greater detail in Section 4.7 are a sort of calibration enabling us to operate over ranges where we are most efficient.

The subject's behavior in the experiment is considered as a process with five stages: estimation of heaviness prior to contact, the initial attempt to lift, revised calculation of heaviness, compensatory muscular adjustment and judgment.

One idealization suggested by the Stevens-Rubin finding is this: there is a usable range of intensities extending from a lower intensity i_0 to an upper intensity i_+. There is a straight effort continuum in psychological space corresponding to the appearance of muscular effort in the lifting position ranging from e_0 for no effort to e_+ for the greatest effort the subject can or will produce. The number continuum is straight and its point of view constant throughout the experiment. The effort point of view is also constant. All points of view are assumed degenerate. Effort e_0 projects to i_0; e_+ projects to i_+.

The subject views the stimulus. From its size and his guess about its density he estimates its appearance on a straight continuum. He selects his heaviness point of view such that he will be able to deal with whatever stimulus the experimenter offers. In particular, he chooses so that zero weight projects to i_0 and a barely liftable object projects to i_+.[12] Then his best estimates of the appearance project into the usable range of intensities.

It will now be assumed that all the heaviness vertices are independent of irrelevant changes in appearance in this sense: Weights differing in color, size, or texture say, but equally resisting the subject's efforts to lift, upon being lifted, have appearances projecting by such vertices to equal intensities.

This extreme position implies that all of the effects of size upon heaviness are indirect, that size affects the heaviness only by affecting the point of view. Experimental implications of this are developed in Section 4.6.

It follows from straightness and the degeneracy of the points of view that the intensity of a weight of x pounds and size λ will be of form

$$p_\lambda[v(x)]A + [1 - p_\lambda v(x)]B,$$

where A and B are points on the intensity continuum, p_λ is a projective transformation, and $v(x)$ is a number-valued function of weight only.

To make an initial pull, the subject chooses an effort with intensity curve closest to the heaviness curve of the estimated weight. After he makes con-

---

[12] Note that in order for there to be more than one such vertex, either the two extreme appearances, i_0 and i_+, span a three-dimensional space or two of the four points coincide.

tact and has more information, he adjusts the strength of contraction and completes the lifting. Finally he judges by choosing a number projecting onto the same place on the intensity continuum as the final heaviness.

This gives

$$qu(y) = p_\lambda v(x) \tag{2}$$

for matching number y and weight x pairs with size parameter λ. Here q is a projective transformation from the conversion of number appearances into intensities; u is a number-embedding function. As noted in Section 4.3, only p_λ changes when the heaviness point of view is changed.

Equation 2 may be rewritten as

$$squ(y) = sp_\lambda r^{-1}[rv(x)],$$

where r and s are any increasing projective transformations such that $rv(0) = s(i_0) = \infty$. For instance there are

$$r(x) = 1/[v(0) - x] \quad \text{and}$$
$$s(x) = 1/[i_0 - x].$$

Each of the p_λ must satisfy

$$p_\lambda[v(0)] = i_0 \quad \text{and}$$
$$p_\lambda[v(x_+)] = i_+.$$

Therefore, if we plot $squ(y)$ against $rv(x)$, we obtain a portion of the graph of the projective transformation $sp_\lambda r^{-1}$. But for each λ this transformation has ∞ as a fixed point and consequently is affine. Thus for some numbers a_λ and b_λ,

$$sp_\lambda r^{-1}(x) = a_\lambda x + b_\lambda.$$

Thus each graph is a straight line. Furthermore, each projective transformation $sp_\lambda r^{-1}$ must take $rv(x_+)$ to the number $s(i_+)$. Since this condition is independent of λ, all of the straight lines must converge on the point of the plane corresponding to the heaviest liftable weight and its corresponding number.

In this way the Stevens-Rubin result can be interpreted geometrically.

Before concluding this section it seems advisable to call attention to the fact that the Stevens-Rubin result can be experimentally tested without curve fitting or extrapolating. There is an unusual property of the graphs of concurrent linear functions, which is not affected by nonlinear transformations and which can be tested without considering the values of the function near the point of concurrence. To define this property, consider the functions $f = f_\lambda$ such that $f(x)$ is the number matched to weight x of size λ. For any two such functions f and g that are strictly increasing and have overlapping domains and ranges it is possible to define a transformation of a portion of the domain of g denoted $\bar{f}g$ and defined by

$$\bar{f}g(x) = y \quad \text{if} \quad f(y) = g(x).$$

If $f, g, h, \ldots$ can simultaneously be transformed to concurrent linear functions then the various transformations $\bar{f}g, \bar{f}h, \bar{g}h$, etc., will commute. Commutativity is a property that can be experimentally tested even when there are no data points near the point of concurrence. It is in this sense that the Stevens-Rubin result can be reformulated without curve fitting or extrapolation. For further details on functions that can be transformed to concurrent linear functions see Levine (1972, Sec. V).

4.6. Experiments and the Size-Weight Analysis

There will generally be several alternative interpretations of an experimental result like the Stevens-Rubin data in the theory. But each has testable experimental implications. In this section some implications of the preceding analyses are discussed.

All of the size-weight illusion was attributed to the point of view. Consequently if the point of view could be controlled, the illusion could be controlled.

The crucial part of the analysis is the selection of the heaviness point of view prior to contact with the weight. There are two ways to exploit this. A complicated way would be to incorporate estimates of heaviness made prior to contact or measures of the force exerted by the subject during his initial pull on the weight in a model and to mathematically infer the point of view. An easier and more amusing way would be to use an experimental technique developed by Nielsen (1963) to separate the feel and the sight of the lifted object.

Nielsen showed that it is possible to convince an observer that he is viewing his own body when in fact he is viewing another person's. In a study of free will he instructed each subject to move his hand along a painted line. A white glove obscured identifying marks on the hand. The line was placed in a box containing a mirror system such that the subject was actually viewing a confederate's hand. Despite the fact that the confederate repeatedly moved his hand off the line and did not respond to the compensatory motions of the subject, the illusion persisted. Eighteen of Nielsen's 20 subjects regarded the viewed hand as their own pulled by 'magnets' or otherwise not completely under control.

It seems possible to adapt Nielsen's technique to obtain control of the point of view. Consider a subject's lifting an object of unknown size. The subject places his hand in a poorly illuminated box. He lifts one object and views a confederate lifting another. When the object is lifted and held steady in a standard position for a full second, the illumination in the box is increased so that the subject can easily see the object in the confederate's hand.

One further manipulation is needed. In order to be certain that the subject does not change his point of view prior to making his judgment, he is instructed to hold his hand steady. If he fails to obey these instructions or wishes to terminate the trial or otherwise moves his hand, the light in the box is automatically extinguished. There are two anticipated effects of this manipulation. First, if the subject changes his heaviness point of view, then, as a consequence of the hypothesized relation between strength of pull and heaviness, there should be a compensatory movement of the subject's hand. The prompt extinguishing of the light gives the subject the feedback needed to learn not to change his initial point of view. Second, the manipulation may protect the illusion that the subject is viewing his own hand. When he moves his hand he promptly sees an effect; when he does not move his hand, the viewed confederate's hand remains steady.

Under these conditions the theory predicts a greatly diminished size-weight illusion. The correlation between viewed size and judged weight should be insignificant or positive rather than negative.

4.7. Changes of Modulus and Changes of Point of View

In his critique of mapping theory, Krantz (1972) lists a number of generalizations about direct measurement. He argues that the alternatives to relation theory are implausible because they require mental arithmetic or complicated mental processes. This is no longer a valid criticism. Within projective theory the invariances inherent in the generalizations follow from a well-known fact of geometry: projective transformations with the same fixed points commute. Projective theory imposes projective transformations upon data; plausible (see below) psychological assumptions restrict the fixed points. Full details will be given in a separate paper (Levine, 1974c). The basic idea is sketched in an interpretation of the effects of change of modulus in this section.

In some magnitude estimation studies the experimenter instructs the observer to assign a particular number to a particular stimulus. In cross-modal matching of lights to tones, the experimenter will sometimes instruct the subject to pair a particular light with a particular tone. Changing the initial pairing of number to stimulus or light to tone is called changing the modulus.

Krantz's major objection to mapping theory is that it fails to give a plausible interpretation of some invariances associated with the change of modulus. One of the central generalizations of direct measurement is this: If $f(x)$ is the average number matched to physical stimulus x in a magnitude estimation study with one modulus and $g(x)$ with another modulus, then there is a constant such that, for all of the stimuli, $g(x)$ is the constant times $f(x)$. Consequently, ratios of number responses are said to be invariant. There is an analogous invariance assumed for cross-modality matching. (The exponents computed from cross-modality matching data are independent of modulus.)

The main purpose of this section is to develop the notion of having several points of view for a single dimension of experience. In the process it is shown that this mapping theory quite easily accounts for changes of modulus. Even the simplest versions of the theory yield a plausible account of the invariances. No mental arithmetic or complicated ideation of any kind is required.

After the invariances are dealt with, some steps are made towards a less idealized description of the experiments. Hopefully these more realistic descriptions will lead to a specification of the conditions under which the invariances are observed and to an account of some of the systematic departures from the generalizations about the experiments. The section is concluded with a list of some of the factors that may influence the selection of a point of view.

In the following discussion, only degenerate points of view are considered. The notations i_0 and i_+ of Section 4.5 are used for extreme intensities.

Suppose that instructing the subject to change his modulus in a magnitude estimation study only causes him to change his number point of view. The psychophysical function and the magnitudes of the appearances of the physical stimuli are unchanged. Further, suppose that each number point of view pairs i_0 with the appearance of zero and i_+ with the appearance of the subject's largest number appearance. Then there is a simple geometric fact that can be related to the invariances. If p and q are projective transformations such that for two different numbers x_1 and x_2

$$p(x_1) = q(x_1) = \text{zero} \quad \text{and}$$
$$p(x_2) = q(x_2) = \text{infinity},$$

then there is a constant such that for all x, $q(x)$ equals the constant times $p(x)$. Consequently for any numbers y and z such that $x_1 < y < z < x_2$ the ratios $p(y)/p(z)$ and $q(y)/q(z)$ are equal.

To avoid reference to infinity one may assume $p(x_2) = q(x_2) = M$ for some large positive M instead of $p(x_2) = q(x_2) =$ infinity. (M may be thought of as the largest number that is psychologically significant for the observer.) Then the cross ratios

$$\frac{p(y)}{p(z)} \cdot \frac{M - p(y)}{M - p(z)} \quad \text{and} \quad \frac{q(y)}{q(z)} \cdot \frac{M - q(y)}{M - q(z)}$$

are equal. If M is very large relative to $p(z)$ and $q(z)$, then the ratios $p(y)/p(z)$ and $q(y)/q(z)$ are very nearly equal. This follows from

$$\frac{M - p(y)}{M - p(z)} = 1 + \frac{p(z) - p(y)}{M - p(z)} \quad \text{and}$$

$$0 < \frac{p(z) - p(y)}{M - p(z)} < \frac{1}{M/p(z) - 1}.$$

To apply these observations to magnitude estimation, recall that the interpretation of magnitude estimation as matching experiment (Sec. 4.4) implies that the number matched to a physical stimulus is a projective transformation of the physical continuum-embedding function followed by a second projective transformation relating intensities to number appearances. When the number-embedding function u is linear, the invariance follows from an elementary calculation with projective transformations. (Take $u(0)$ equal to zero for interval scale u.)

The invariance of cross-modality matching with changes of modulus can be interpreted similarly. The key to the argument is that the changes following the change of modulus leave the correspondence of a pair of intensities with a pair of appearances invariant. This can be interpreted as i_0 and i_+ corresponding to particular appearances as above or it can be interpreted as the prolongation of a straight psychological continuum either intersecting or being parallel to the intensity line. No matter which of these alternatives is chosen, the invariances can now be interpreted in several ways, none of which require mental arithmetic, complicated ideation, or other implausible processes.

The device of varying the point of view removes the major criticism of mapping theory by providing a number of simple processing schemes which imply the invariances. This device also seems useful in accounting for other aspects of direct matching experiments, especially experiments using a large range of physical stimuli. It seems profitable to consider the observer functioning as an automatic gain-control device using inputs and a portion of its own response to adjust its characteristics in order to operate over an optimal range. This could be achieved by adjustments of the number point of view or by adjustment of the points of view relevant to the experimenter-controlled events. For the present time, only number points of view are considered.

As a special case, suppose that the observer has exactly two number points of view, labelled Lo and Hi. Both can be thought of as projecting a grid of numbers upon the intensity line. Assume that Lo projects many more of the number appearances for which the observer has number names into a neighborhood of i_0 than Hi so that Lo is better suited for assigning different names to different magnitudes of weak stimuli. Lo consequently will give steeper matching functions for these weak stimuli when it is used. On any given trial the subject uses either Lo or Hi. The averaged response to stimulus x will be

$$f(x) = P_{\text{Lo}}(x)L(x) + P_{\text{Hi}}(x)H(x),$$

where P_{Lo} and P_{Hi} are the proportions of trials each is used, and L and H are conditional matching functions. If Lo is used only for very weak stimuli in the sense that $P_{Hi}(x)$ increases rapidly from zero to one, then one still predicts all of the *qualitative* properties accepted as evidence for the power law, namely, increasing, asymptotically linear, concave plots of Log f plotted against Log x.

This elaboration of projective theory has the curious implication that magnitude estimates need not be monotonic functions of physical intensity. Since L has steeper slope than H, there will be xs such that $L(x)$ is higher than $H(x)$. In an experiment of Ross and Di Lollo (1968) observers judged very many light weights. Without warning they were given a much heavier weight to judge. Suppose the observer retained a high slope, low point of view for moderately heavy weights, but with even heavier weights changed to a high point of view. Then one would expect precisely the pattern of nonmonotonicities reported.

Another empirical consideration suggesting this elaboration of projective theory is the observation occasionally reported by experimenters having a great deal of experience with direct measurement. It is reported that in the method of free magnitude estimation (in which the subject is not assigned a modulus) it is especially easy for the subject to give reliable estimates. It is not clear in relation theory that free magnitude estimation should be preferable to any other method. In the projective mapping theory there are many possibilities. In the free method the subject selects the point of view. In the variants, the experimenter influences the selection of the point of view. The unconstrained subject is free

(1) to select a point of view giving him an opportunity to use a range of numbers of great familiarity,

(2) to choose a point of view giving good resolution over the range of intensities into which most of the stimuli project,

(3) to choose a point of view that is easy to recall for successive trials, and

(4) to choose a point of view so that the number intensity curves will have approximately the same shape as the weight intensity curves.

The next step in developing this line of reasoning seems to be to obtain detailed individual data with deliberate manipulations to control the point of view. In the context of such experiments it may be profitable to design training procedures for giving the observer control of the point of view and statistical procedures for inferring the point of view from the fine structure of data.

4.8. On the Regression Effect

This application has been selected to illustrate arguments in which points of view that are not degenerate play an essential role. An intransitivity of matching called the regression effect and a generalization about the relation between uncertainty of judgment and intensity of stimulation called Ekman's law are considered briefly.

When an observer adjusts a bright light until it appears as intense as a loud tone, he consistently underestimates the adjusted continuum in the following sense: if light ℓ_1 is adjusted to match tone t_1 and then the tone is adjusted to t_2 matching ℓ_1, then t_2 is generally less physically intense than t_1. This phenomenon is called regression effect. It has been replicated and is valid for many pairs of continua (Stevens & Greenbaum, 1966). There is also a tendency for overestimation of faint adjusted stimuli.

In earlier mapping theories, regression was an artifact to be averaged away or made less apparent by complex experimental procedures. In the present theory magnitudes are only partially ordered, so the intransitivity of matches is not an embarrassment. Two interpretations of the phenomenon are offered below. In both cases the intransitivity follows from the manner in which the observer tries to equate magnitudes of different shape.

In the first interpretation the observer matches magnitudes of differing shape by computing an index of agreement. As a crude example, he might calculate the correlation coefficient between representative values on intensity curves and select stimuli to maximize it.

If the observer is maximizing an index of agreement between magnitudes, the intransitivities are predicted. For to find ℓ_1 he is adjusting lights to find a maximum in one set of index values. But to find t_2 he is searching for a maximum in a different set. The intransitivity occurs when the ℓ_1,t_2 pair has a higher index of agreement than the ℓ_1,t_1 pair. And since the ℓ_1,t_1 pair is included in the set of pairs available for the second match, the intransitivity is generally expected.

Some experimenters have attempted to obtain precise matches by instructing the observer to bracket. When bracketing the observer chooses a light ℓ_- just clearly less intense than tone t_1 and a light ℓ_+ just clearly more intense. Finally the observer ignores the tone and selects a light $\ell = \ell_1$ which is in some sense exactly[13] in between ℓ_- and ℓ_+.

[13] If the light ℓ is selected as exactly in between by bisecting an angle on the dial of an instrument used to control the lights, then the instrument can be adjusted so that by successive approximations a bias is introduced that is about the same size but in the opposite direction as the regression effect. In the analysis that follows it is assumed that the observer can learn to select the matching intermediate light without being influenced by angles on adjusting dials.

If the shape (skewness) of the intensity curve changes appropriately then the regression effect can also be deduced from this bracketing. One set of hypotheses sufficient to assure exactly the right changes in shape is this:

1. The light and tone continua are straight.
2. The light and tone continua are parallel to the intensity continua.
3. Psychological space is two dimensional.
4. Both the brightness and the loudness points of view are radially symmetric. (Here a point of view is called *radially symmetric* when there is one point of view such that every rotation of psychological space leaving this point of view fixed leaves the distribution defining the point of view fixed.)

To make the bracketing procedure explicit some notation is needed to express the fact that ℓ_- is just clearly less intense, ℓ_+ is just clearly more, and ℓ_1 is exactly in between. Suppose the bracketing lights are selected with reference to functions

$$f(x, m) = \text{the measure of intensities smaller than } x \text{ calculated with the magnitude of stimulus } m, \quad \text{and}$$

$$g(x, m) = \text{the measure of intensities larger than } x.$$

If brackets ℓ_-, ℓ_+ are chosen so that for small criteria $\epsilon < \delta$ there are intensities $x < y$ such that

$$\delta = f(x, \ell_-) = g(y, \ell_+) \quad \text{and}$$

$$\epsilon = f(x, t_1) = g(y, t_1)$$

and match $\ell = \ell_1$ such that for some $x < y$

$$f(x, \ell) = g(y, \ell) \quad \text{and}$$

$$g(x, \ell_-) = f(y, \ell_+),$$

then by elementary arguments one can deduce that there will be an apparent underestimation of very intense stimuli and an apparent overestimation of very feeble stimuli.

A further consequence of the unnecessarily strong simplifying assumptions used to deduce the regression effect is an approximation of Ekman's law (S. S. Stevens, 1966; also Ekman, 1961) on the uncertainty of judgment. For high levels of stimulation the ratio of a measure of scatter such as interquartile range to the modal intensity will be very nearly constant.

ACKNOWLEDGMENTS

Valuable criticism on this work has been received from Jean-Claude Falmagne, Peter Freyd, Francis Irwin, Dorothea Jameson, Ann King, R. Duncan

Luce, L. E. Marks, Jacob Nachmias, Klaus Riegel, Harris Savin, J. C. Stevens, and Thomas Stroud.

REFERENCES

Coxeter, H. M. S. *The real projective plane.* (2nd ed.) Cambridge: Cambridge University Press, 1955.

Curtis, D., & Rule, S. Evidence for a two-stage model of magnitude estimation. Paper read at Psychophysical Measurement Seminar of Mathematical Psychology Meetings, La Jolla, California, 1972.

Ekman, G. Some aspects of psychophysical research. In W. A. Rosenblith (Ed.), *Sensory communication.* Cambridge, Mass.: MIT Press, 1961.

Hanes, R. M. The construction of a subjective brightness scale from fractionation data. *Journal of Experimental Psychology*, 1949, **39,** 714–728.

Hurvich, L. M. Hering and the scientific establishment. *American Psychologist*, 1969, **24,** 497–514.

Krantz, D. H. A theory of magnitude estimation and cross-modality matching. *Journal of Mathematical Psychology*, 1972, **9,** 168–199.

Levine, M. V. Transformations that render curves parallel. *Journal of Mathematical Psychology*, 1970, **7,** 410–443.

Levine, M. V. Transforming curves into curves with the same shape. *Journal of Mathematical Psychology*, 1972, **9,** 1–16.

Levine, M. V. Nonadditive analogues of the basic mathematical results of additive measurement. *Journal of Mathematical Psychology*, 1974, in press. (a)

Levine, M. V. Exact measurement with functions defined on small sets. 1974, submitted to *Journal of Mathematical Psychology.* (b)

Levine, M. V. Geometric interpretations of some generalizations of direct measurement. 1974, in preparation. (c)

Marks, L. E. *Sensory processes: The new psychophysics.* New York: Academic Press, 1973.

Nielsen, T. I. Volition: A new experimental approach. *Scandinavian Journal of Psychology*, 1963, **4,** 225–230.

Restle, F. Rapid judgment of magnitudes: Half magnitudes. Mathematical Psychology Program Report Series, 69-5. Bloomington, Indiana: Indiana University, 1969. (a)

Restle, F. Speed of adding and comparing numbers. *Journal of Experimental Psychology*, 1969, **83,** 274–278. (b)

Ross, J. The task of magnitude estimation. *Abstract Guide* of XXth International Congress of Psychology, Tokyo, 1972.

Ross, J., & Di Lollo, V. A vector model for psychophysical judgment. *Journal of Experimental Psychology*, Monograph Supplement, 1968, **77,** 1–16.

Stevens, J. C. Convergence of psychophysical power functions. Written communication to Psychophysical Measurement Seminar of Mathematical Psychology Meetings, La Jolla, California, 1972.

Stevens, J. C., & Rubin, L. I. Psychophysical scales of apparent heaviness and the size-weight illusion. *Perception and Psychophysics*, 1970, **8,** 225–230.

Stevens, S. S. A metric for social consensus. *Science*, 1966, **151,** 530–541.

Stevens, S. S., & Greenbaum, H. B. Regression effect in psychophysical judgment. *Perception and Psychophysics*, 1966, **1**, 439–446.

Suppes, P., & Zinnes, J. L. Basic measurement theory. In R. D. Luce, R. R. Bush, and E. Galanter (Eds.), *Handbook of mathematical psychology*. Vol. 1. New York: Wiley, 1963.

Teghtsoonian, R. Range effects in psychophysical scaling and a revision of Stevens's law. *American Journal of Psychology*, 1973, **86,** 3–27.

Information Integration Theory: A Brief Survey

Norman H. Anderson
UNIVERSITY OF CALIFORNIA, SAN DIEGO

1. INTRODUCTION

1.1. Integration Theory

Information integration. A conception of the organism as an integrator of stimulus information is time honored in perception and judgment. Most judgments, if indeed not all, reflect several coacting stimuli that are combined or integrated to produce the response. Person perception is a good example. Our opinions about a person result from an integration of diverse pieces of information: personal interaction, direct observation, written records of diverse kinds, remarks of others. Person perception is not unique in this; evaluating a job offer, tasting a carbonated drink, catching a ball all require information integration.

This chapter surveys an attempt to develop a unified general theory of information integration (Anderson, 1968a, 1970a, 1971a, 1973a, 1973b). This theory has developed along with an extensive program of experimental research, and it has an exceptionally solid empirical foundation. Much of this work has been done within one experimental task, namely, person perception as described in Section 2.1. However, related work ranging from social attitudes to utility theory reveals the same principles in quite different areas as illustrated in Section 7.

Although integration processes pervade behavior and have been studied by numerous investigators, attempts at a unified theoretical system have been relatively limited. Systematic developments have been made by Garner (1962; see also Garner & Morton, 1969) using statistical information theory, by Edwards (e.g., 1968) using Bayesian theory of mathematical statistics, and by Brunswik (1956; see also Hammond, 1966) using multiple regression analysis. To characterize these developments in a few sentences is not possible.[1] However, all three share a predominant concern with normative analysis in which the mathematical model is used as a standard against which to measure the success or achievement of the organism rather than as a description of the psychological processes themselves. The present approach, in contrast, is primarily descriptive in nature.

Helson's (1964) theory should also be mentioned as it attempts a broad coverage. However, it is not primarily an integration theory even though the adaptation level itself is considered as an average of the various prevailing stimulus factors. The adaptation level is only a reference point relative to which the focal stimuli are judged. Helson gives little attention to the problem of how the various aspects of the focal stimulus are integrated to produce the judgment.

Two aspects of the present approach deserve comment. One is the use of simple algebraic models to describe various integration processes. The other is a theory of functional measurement to get the subjective or psychological values of the stimulus variables.

Algebraic models. The present theory makes systematic use of algebraic models of perception and judgment. A striking outcome of the investigations has been the repeated finding that these simple models can give a detailed, quantitative account of fairly complex cognitive activity.

Most of the models fall in one of two main classes. One class includes adding, subtracting, and averaging models. Adding and subtracting are formally similar, of course, but may be psychologically different. Adding and averaging are different, both psychologically and mathematically. Under certain circumstances, they make identical predictions and have a very simple analysis. With differential weighting, however, the averaging model becomes nonlinear and more difficult to handle (Sec. 5).

More recently, the methods of integration theory have been extended to handle multiplying models and dividing models. Multiplying models arise

---

[1] Slovic and Lichtenstein (1971) give an extended discussion of Brunswik's approach, the Bayesian development, the present theory, as well as the eclectic approach of the Oregonians.

in traditional utility theory, for example, and dividing models arise in comparative judgment. Some applications are summarized in Section 7.

Functional measurement. One aspect of these algebraic models requires special notice. They are expressed in subjective metrics, or psychological values, of the response and of the stimuli. Many investigators have ignored the need for subjective metrics and have employed handy, arbitrary scales. That may suffice for certain purposes, but it can lead to serious misinterpretations (Anderson, 1971a, 1972a). Without the psychological values, a completely adequate treatment of the models is not possible.

To get the subjective metrics requires a theory of measurement, one of the more contentious areas of psychology. The present approach includes a theory of functional measurement that yields the subjective metrics in a simple way. One example is discussed in connection with Figure 2 (Sec. 2), which illustrates the scaling of personality-trait adjectives for individual subjects. Another example is in Section 7.8, which illustrates the first general method for the long-standing problem of simultaneous measurement of subjective probability and utility. As these examples show, substantive theory and measurement theory are inseparable. The development of functional measurement theory has gone hand in hand with the development of a theory of information integration.

1.2. Valuation and Integration

Valuation and integration are two basic operations in integration theory. Integration refers to the processes whereby the stimuli are combined to determine an overall response; that is the main concern of this chapter. Valuation, which refers to the processes that determine the stimulus parameters, is no less important. A detailed illustration of the valuation problem is discussed in Section 7.9, and some general comments are given in Section 8. Two or three points need to be noted here.

Two stimulus parameters have special importance in the algebraic models: scale value and weight. The concept of scale value is used in its ordinary sense. It can be considered as the location of the stimulus on the dimension of judgment.

The concept of weight is less intuitive, and it may be helpful to view weight as the amount of information in the stimulus. Weight may thus be controlled experimentally by varying the number of equivalent stimuli in a given subset, or by manipulating source reliability of a given piece of information. Whether weight can always be considered as amount of information is unclear because various other factors can also influence weight. In some cases, weight and value of a stimulus are related: A more extreme stimulus will tend to be more diagnostic as well (e.g., Anderson, 1972a; Leon, Oden, & Anderson, 1973).

Weight and value both depend on the dimension of judgment. The same stimulus will have different relevance and different valence for different judgment tasks. Each task sets up a valuation operation by virtue of which the stimulus parameters are defined. For present purposes, the valuation operation will be taken for granted. That is not always feasible, of course. If the stimuli are inconsistent, for example, then valuation cannot be separated from integration, even in a formal way.

1.3. Averaging Model

Consider a set of N stimuli, S_i, with the value s_i and weight w_i. The averaging model specifies the response as a weighted average of the scale values:

$$R = C + \sum_{i=0}^{N} w_i s_i \Big/ \sum_{i=0}^{N} w_i + \epsilon. \tag{1}$$

The additive constant, C, allows for an arbitrary zero in the response scale, and will be assumed here to be zero. The error term, ϵ, is an additive random variable with zero mean that represents response variability. The error term is explicitly considered only in this section. In later sections, R may refer to either the 'observed' or the 'true' value of the response, but this will be clear from context.

The effective or *relative* weight of S_i is $w_i/\sum w_i$; the denominator in Equation 1 forces the relative weights to sum to one, the condition for an averaging model.

The sum in Equation 1 is over all relevant stimuli. It explicitly includes an internal, organismic state variable S_0 with weight w_0 and value s_0. In many cases, S_0 represents the integrated resultant of past stimuli, the current state in an ongoing task. However, S_0 may also represent response dispositions as well as motivational variables.

Most applications employ an independence assumption. Within any judgment task, the weight and value of each stimulus are assumed to be constant, regardless of what other stimuli it may be combined with. This is a stringent assumption since it apparently disallows assimilation, contrast, redundancy, or inconsistency effects. Fortunately, these effects are less prevalent than might be feared.

Relative weights will certainly depend on context. Since the weights must sum to one, adding a new relevant stimulus to a set will cause the relative weights of the old stimuli to decrease. Accordingly, each stimulus is considered to have an absolute weight, and the independence assumption applies to these absolute weights.

1.4. Testing Goodness of Fit

Factorial design. To test the averaging model, it is convenient, though not completely necessary, to impose two restrictions. The first is to construct the stimulus combinations from factorial designs. Two stimulus variables would define a two-way, row × column matrix whose elements are stimulus pairs. The response in cell *ij* of the design would then be

$$R_{ij} = \frac{w_0 s_0 + w_{Ri} s_{Ri} + w_{Cj} s_{Cj}}{w_0 + w_{Ri} + w_{Cj}} + \epsilon_{ij}, \tag{2}$$

where w_{Ri} and s_{Ri} are the weight and value of the stimulus in row i, w_{Cj} and s_{Cj} are the weight and value of the stimulus in column j, the ϵ_{ij} are random errors with constant mean, and R_{ij} is the response to the stimulus combination (S_{Ri}, S_{Cj}) in cell *ij* of the two-way matrix.

The second restriction calls for equal weighting of the stimuli within each factor. That is, it is required that the w_{Ri} all be equal, to w_R, say, and that the w_{Cj} all be equal, to w_C. Equation 2 then becomes

$$\begin{aligned} R_{ij} &= \frac{w_0 s_0 + w_R s_{Ri} + w_C s_{Cj}}{w_0 + w_R + w_C} + \epsilon_{ij} \\ &= \frac{w_0 s_0}{W} + \frac{w_R s_{Ri} + w_C s_{Cj}}{W} + \epsilon_{ij}, \end{aligned} \tag{3}$$

where $W = w_0 + w_R + w_C$ is a constant.

In Equation 3, the first term on the right can be set equal to 0, and W can be set equal to 1 without loss of generality. These merely define the zero and unit of the response scale. The averaging model then reduces to the simple additive form,

$$R_{ij} = w_R s_{Ri} + w_C s_{Cj} + \epsilon_{ij}. \tag{4}$$

If three stimulus variables were used, an exactly analogous form would be obtained:

$$R_{ijk} = w_R s_{Ri} + w_C s_{Cj} + w_L s_{Lk} + \epsilon_{ijk}, \tag{5}$$

where w_L and s_{Lk} are the value and weight of the stimuli in the third, 'layer' stimulus variable.

Parallelism prediction. Equation 4 implies that the data should exhibit a simple pattern of parallelism. If error variability is ignored, then rows 1 and 2 differ by the same amount in each column:

$$R_{1j} - R_{2j} = w_R(s_{R1} - s_{R2}). \tag{6}$$

Hence these two rows should plot as parallel lines. The same holds for any other two rows, so that the row curves as a group should exhibit parallelism.

Examples can be seen in Figures 1, 2, 4, 6, 7, and 8. These examples also illustrate that the parallelism property is independent of the spacing of the column stimuli on the horizontal axis.

Equation 5, for the three-factor design, leads to a more complex pattern of parallelism. Equation 6 still follows for any fixed k, and thus also for the average over k. Hence the row $\times$ column data table should exhibit parallelism. The same holds for the row $\times$ layer, and for the column $\times$ layer data tables. These three tests are independent in the sense that any two may show parallelism without constraining the third to do so.

Analysis of variance test. Although the parallelism test is extremely useful, it will never be exactly satisfied because of response variability. Some more objective assessment is usually needed, therefore, of whether the observed deviations from parallelism represent anything more than chance factors.

The ordinary analysis of variance can be used for this purpose because there is a direct relation between Equations 4 and 5 and the linear model used in the analysis of variance. The graphical prediction of parallelism from Equation 6 is equivalent to a zero row $\times$ column 'interaction.' With a factorial design, therefore, the averaging model implies that the observed row $\times$ column interactions should be statistically nonsignificant. The three-way design of Equation 5 leads to three such two-way interactions, plus a three-way interaction, all of which should be nonsignificant. If any interaction is found significant, that infirms the averaging model. The same holds for any linear model even if each stimulus is allowed its own weight parameter.

Analysis of variance is discussed in standard texts (e.g., Myers, 1972; Mandel, 1964; Winer, 1971) and is not considered here. Its use does depend on the further statistical assumptions that the ϵ_{ij} are independent random variables, normally distributed with constant variance. In the present applications, these statistical assumptions are no problem except that some precautions may be needed to ensure independence in repeated measurement designs. With such precautions, the analysis of variance allows simple, powerful tests even at the level of the single subject.

1.5. Functional Measurement

Interval response scales. The test of goodness of fit between model and data does two things at once: It provides a joint validation of the model itself and of the response scale. Unless the response measure was an 'equal interval' scale, it would produce discrepancies from parallelism for a correct model. Satisfactory fit testifies to the adequacy of the response measure. Validation of the model thus provides a functional scaling of the response.

In any one experiment, of course, defects in the model may just cancel defects in the response scale. No single experiment can be conclusive. The validity of the functional measurement approach rests on its coherence with an extended network of experimental analysis as illustrated below.

This point is important because the bulk of the experimental work on integration theory has used rating responses. A priori, there is no great reason to believe that ratings will be interval scales, and many workers have avoided them for just that reason. An important consequence of the present research program, beyond its substantive achievements, has been the cumulative support it has given to treating ratings as interval scales, at least if suitable precautions are observed (Sec. 2.1).

Interval stimulus scales. Functional scales of the stimuli can also be obtained if the model passes the test of fit. These are just the marginal means of the factorial design. From Equation 4, for example, the mean value of the response in row i is

$$\bar{R}_i = w_R s_{Ri} + w_C \bar{s}_C + \bar{\epsilon}_i, \tag{7}$$

where averages over columns are denoted by overbars and omission of the j subscript. Equation 7 implies that the row mean is a linear function of the row scale value, and vice versa. Thus the observed row means are estimates of the scale values of the row stimuli on an interval scale. In general, the zero and unit are arbitrary, which means that more than two row stimuli would be desired for scaling purposes. The model can be tested with only two stimuli in each factor, but then only between-factor comparisons of differences in marginal means are useful (see Sec. 6).

Much the same analysis holds for the multiplying model and the general multilinear models of Section 7.7. However, the analysis is more difficult for the differential weighting models of Section 5.

Ordinal response scales. Functional measurement includes a monotone response scaling feature that has basic theoretical importance. In general, it cannot be expected that the observed response will be on an interval scale. A general theory of measurement must provide for analysis of ordinal response data.

Loudness bisection illustrates this point quite well. Bisection can be represented as an averaging task in which the subject adjusts one sound to lie halfway between two other sounds in loudness. The subject operates on the subjective loudness scale, but the experimenter makes his measurements on the physical intensity scale. These two scales are known to be nonlinearly related. The measured response, therefore, cannot fit the model that describes the process underlying the behavior.

Nonlinear response scales must be expected in various situations, especially with animals and children. In many cases, however, the psychological and physical scales will be monotonically related, so that one can be obtained from the other by a monotone transformation. Function measurement explicitly allows for monotone transformation of the observed response into an interval scale (Anderson, 1962b).

General comments on monotone response scaling have been given in various articles (e.g., Anderson, 1962b, 1969a, 1970a, 1971a), and a useful transformation equation has been illustrated by Anderson and Jacobson (1968). Bogartz and Wackwitz (1971) have made an important development of the power series approach, and Weiss (1973) has constructed a computer program, FUNPOT, which implements this approach.

Rank-order data can also be used in functional measurement, as illustrated in Weiss and Anderson (1972). In general, however, rank-order data analyses do not seem too useful. Even if the observed response measure is not an interval scale, it will usually contain considerable metric information. Order analyses ignore this metric information which is inefficient. More seriously, that makes it extremely difficult to get a rigorous test of goodness of fit. The rapid progress that has been obtained with functional measurement results from using numerical response measures.

Traditional approaches to scaling have typically placed first emphasis on scaling the single stimuli. Functional measurement begins with the response to a set or combination of stimuli, and the stimulus values are often of no great interest. This approach may seem roundabout, but it turns out to be technically superior. It has greater precision since it avoids extraneous variation from estimated scale values. The analysis is simpler. And invalidity from possible shift of stimulus values between the scaling and the integration task is avoided. A brief comparison with two current methods may help illuminate the nature of the functional measurement approach.

Relation to magnitude estimation. Both functional measurement and magnitude estimation (Stevens, 1957, 1971) emphasize the use of numerical response measures. However, Stevens has repeatedly condemned ordinary rating scales, such as those used in functional measurement, as being nonlinear and invalid because they disagree with magnitude estimation on certain stimulus dimensions. The basis of Stevens' argument remains obscure. As is well known (e.g., Anderson, 1970a, p. 166; Garner & Creelman, 1967; Zinnes, 1969), Stevens does not provide a criterion against which to validate his response scale.

Such a criterion is central to functional measurement theory. As already seen, the validity of the response measure is assessed against the criterion imposed by thc algebraic judgment model. This approach leads naturally to the study of tasks involving stimulus integration. In contrast, most work with magnitude estimation has been judgments of single stimuli, and that is inadequate for testing an integration model.

Actually, magnitude estimation has been used in a few experiments with integration tasks (see Sec. 7.3; Weiss, 1972). This evidence points rather uniformly to a bias in the magnitude estimation response. It is as though the number continuum obeyed a Weber law so that the felt difference between 120 and 110 would be less than between 20 and 10. Such bias would invalidate the magnitude estimation procedure. It could, of course, be removed using the monotone scaling feature of functional measurement, though that requires an integration-type task.

Relation to paired comparisons. Thurstone's (1959) method of paired comparisons includes a validating criterion of internal consistency. Because it is possible to disprove Thurstone's model, it is possible to prove it. In contrast to magnitude estimation, Thurstone's approach constitutes a true theory of measurement.

But though paired comparisons has been the dominant scaling procedure in psychology, it has one serious limitation. It cannot measure scale values for individual persons for most verbal or symbolic stimuli (Bock & Jones, 1968, p. 4; Leon, Oden, & Anderson, 1973). The reason is that the paired comparisons technique requires imperfect discrimination between the members of a pair. That is no problem for psychophysical stimuli such as tones or lifted weights when the members of each pair are close together. For verbal or symbolic stimuli, however, each person will ordinarily have clear-cut preferences within each stimulus pair and that disallows scaling for any one person. In practice, the problem is ignored by pooling data over groups of subjects. If all subjects had the same rank order, that procedure would not work either. In practice, a group scale can be obtained, but it is manufactured out of real individual differences that are lost in the process. Such pooled, group scales can be useful in normative or sociological investigations that are often concerned with group preferences. However, group scales may not give an adequate representation for a single individual, especially for social, moral, or esthetic judgments that show large individual differences.

Functional measurement can assess values for the individual person (see Fig. 2) and that is an important advantage. Functional measurement does require establishing an integration model and, ordinarily at least, an interval response scale. It may not be as robust as paired comparisons, which requires only choice data. On the other hand, functional measurement provides scale values that are operative in the judgmental situation itself (e.g., Sec. 7.8, Table 6).

2. QUANTITATIVE EVIDENCE: THE PARALLELISM TEST

This section illustrates the parallelism test of the simple averaging model. For this purpose, the person perception task that has been used in many of these experiments is briefly described.

2.1. Person Perception

Judgmental task. In the typical experiment, the subject receives a set of personality-trait adjectives that describe a person (*efficient*, *scholarly*, *moody*), for example. His task is to form an impression of the person and to judge how likable the person would be. The adjectives are usually presented simultaneously, though serial presentation is used in some experiments (Secs. 4.4, 6.1).

This task was popularized by Asch (1946) and it has been used extensively in the research on integration theory. It has some notable advantages. The task is meaningful to the subjects and has great practical relevance. The stimuli are simple and directly meaningful. There are numerous trait words with considerable commonalty of meaning (Anderson, 1968b). Despite the simplicity of the task, there is a substantial cognitive component, and considerable potential for amplifying it.

The general result is that the judgments follow a simple quantitative model. Furthermore, the same principles that operate in the person perception task operate in other judgment tasks as well. Thus, the adjective task is probably as near ideal as could be hoped for.

Experimental procedure. In general, the quality of one's data is a direct function of experimental care. Three simple but important precautions deserve specific note.

First, the rating scale response is arbitrary to the subject. Some practice is ordinarily desirable to set up the frame of reference and to give the subject opportunity to stabilize both his judgmental processes and his usage of the response scale. A few minutes' practice seems adequate in most experiments.

Second, there is considerable evidence that rating scales are biased near their endpoints; for an example, see Section 4.1. It is customary, therefore, to use end-anchors, stimuli that are higher or lower in value than the regular experimental stimuli. These anchors define and tie down the ends of the rating scale. The data of interest then come from the interior of the scale. There is some evidence, incidentally, that graphic ratings may be superior to numerical ratings (e.g., Shanteau & Anderson, 1969).

Third, the standard instructions are intended to minimize effects of semantic inconsistency or redundancy among the adjectives in a given set. For example, each adjective is attributed to a different acquaintance of the person being described. Inconsistency is thus attributable to differences of view among the acquaintances, and such a statement is often worked into the instructions. Effects of redundancy are minimized because even repeating the same adjective will carry additional weight if it comes from a different source. Minimizing stimulus interaction helps ensure the independence assumption that the weight and value parameters of each adjective are the same in all combinations. Also, the instructions usually state that each adjective is equally important, and that equal attention should be paid to each. That is intended to help ensure the assumption of equal weighting, which is necessary for the parallelism prediction under the averaging model.

Stimulus interaction may itself be the main interest in certain studies (Sec. 8.4). In that case, of course, the task instructions would need corresponding changes (e.g., Anderson & Jacobson, 1965, p. 534).

2.2. Tests of the Integration Model

Figure 1 replots data from an early test of the integration model in the person perception task described in Section 2.1 (Anderson, 1962a). Subjects judged persons described by three personality-trait adjectives. The adjectives were combined in a 3^3 factorial design to obtain the 27 stimuli represented in the graph. If the model is correct, then the three curves in each panel should be parallel. Further, the three panels should be congruent, differing only by a constant vertical displacement. These parallelism properties hold rather well with no apparent systematic discrepancies. Tentatively, therefore, it would be concluded that both the integration model and the rating response scale have been validated in this graphical test.

A completely adequate assessment of the model requires analysis at the single subject level. Group data can be useful and efficient in initial exploration, but real individual discrepancies from the model may cancel out in the averaged group data.

This particular experiment was designed specifically to allow single subject analysis, and it is instructive to look at the data of the first two subjects, shown in Figure 2. For simplicity, the data are averaged over the third factor of the design to obtain a 3×3 plot and the actual adjective stimuli are listed in the figure. Thus, the top curve in each graph represents the likableness of persons who are all *level-headed* (row stimulus), but who differ in being *good-natured*, *bold*, or *humorless* (column stimulus). The two lower curves represent the same descriptions with *level-headed* replaced by *unsophisticated* and *ungrateful*.

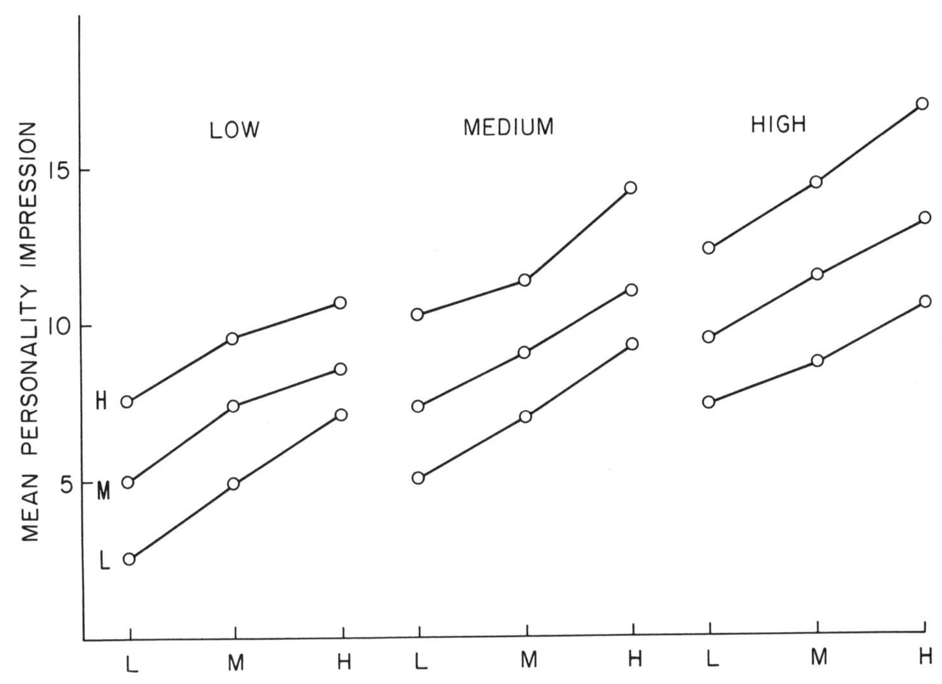

FIGURE 1.
Likableness of 27 persons described by different combinations of personality traits. Response on a 1–20, dislikable-likable rating scale. L, M, and H denote personality traits of low, medium, and high value. Each of the three panels represents a 3 × 3, row × column design, for the listed value of a third trait (low, medium, or high). The parallelism pattern verifies the theoretical prediction that the trait information is integrated by an averaging process. Data after Anderson (1962a).

As can be seen, parallelism holds quite well for both subjects. This was supported by the statistical analysis in which the interaction terms (Sec. 1.4) were nonsignificant for each subject. As a technical detail, the first two daily sessions were considered as practice, and the data of the last three days provided an estimate of error variability for each subject. Subsequent work has indicated that good results can be obtained with much smaller amounts of practice.

Figure 2 has special interest because it illustrates functional measurement on the stimulus side. Given the success of the model, the elevations of the three curves constitute an interval scale of the row adjectives. The two subjects evidently have different values. For Subject FF, *unsophisticated* is approximately halfway between *ungrateful* and *level-headed*, but for Subject RH, *unsophisticated* is nearly equal to *level-headed*. This illustrates the ability of functional measurement procedure to operate at the level of the single subject.

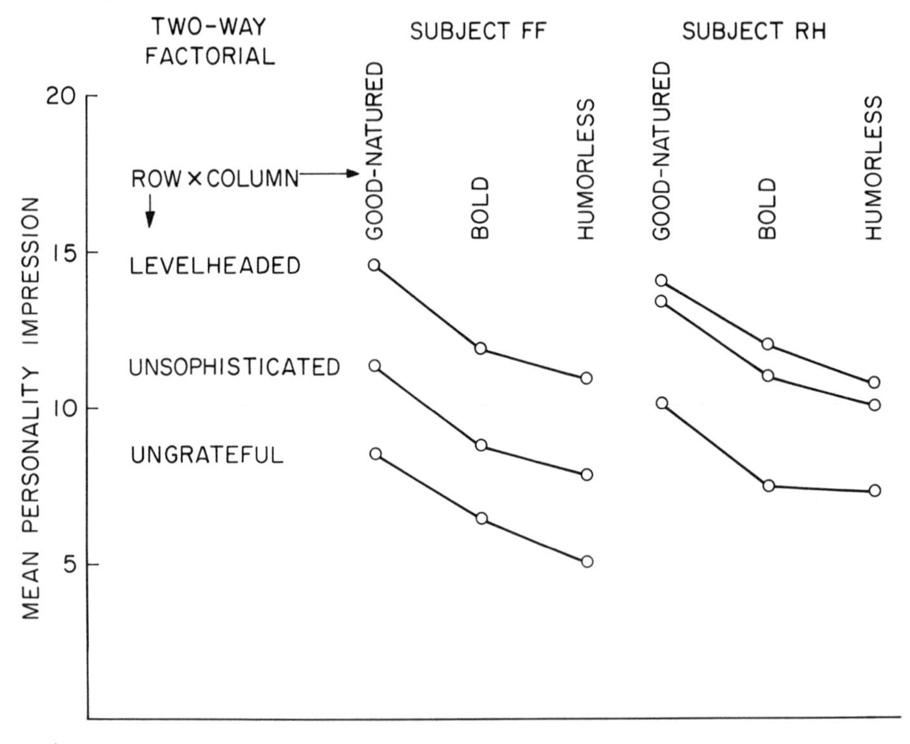

FIGURE 2.
Data for two subjects from experiment of Figure 1 support the parallelism prediction of the averaging model on a single-subject basis. Vertical spacing of the curves constitutes an equal-interval scale of the three listed row traits. Note that the stimulus values for the two subjects are nonlinearly related. Data after Anderson (1962a).

One experiment seldom proves a model, but the promise of this initial attempt has been confirmed by numerous subsequent experiments (see also Sec. 7). It should be explicitly noted that discrepancies from the parallelism prediction do appear. Three sources of such discrepancies may be noted.

First, a few subjects in almost any careful experiment will show some non-parallelism. Such discrepancies always need careful assessment but they are not necessarily serious. They may represent minor nonlinearities in response usage, number preferences, for example, or idiosyncratic judgments of a few stimulus combinations. Such effects must be expected, and of themselves would require little or no qualification of the main integration process.

Second, there are certain kinds of stimulus combinations to which the simple model is not expected to apply. Contradictory information and redundant information are examples. The averaging model might still hold, of course, but it would need to allow for a dependence of the weight parameter of each adjective on the other adjectives of the set. Interestingly enough, the evidence indicates that subjects are not overly sensitive to inconsistency and redundancy. There is no great difficulty in seeing a person as both *friendly* and *hostile*, or as *competitive* and *cooperative*.

Third, there is some evidence for a systematic deviation from parallelism that reflects differential weighting of the stimuli. This is entirely consistent with the averaging formulation and is discussed in Section 5.

2.3. Stimulus Scaling

Traditional measurement theories, such as the Thurstonian system, are primarily concerned with stimulus scaling per se. In functional measurement the emphasis is quite different. Primary concern is on algebraic models and substantive theory, and the scales themselves have secondary interest. As a consequence, many tests are made with 2×2 designs, for example, which give essentially no information on the stimulus scales. Even three levels of a factor, as in Figure 2, provide only one stimulus value since the zero and unit of the scale are arbitrary.

Functional scaling on the stimulus side is illustrated in Table 1. Each of the 20 row adjectives was paired with each of the 6 listed column adjectives and rated on a graphic scale. Judgments were made of social desirability rather than of likableness.

The marginal means of the factorial stimulus design are in the columns headed s. These are the estimated functional scale values. The data have the parallelism property, and the analysis of variance showed no significant interaction. This validates the stimulus scale. It may be concluded that the numbers listed in Table 1 are equivalent to the covert, subjective values of these adjectives in person perception.

3. QUALITATIVE EVIDENCE

3.1. Critical Tests

The parallelism observed in Figures 1 and 2 and in Table 1 argues that some simple process is at work. More than one model could account for these results, however. Both adding and averaging models are plausible candidates, and either one could account for the parallelism. Since these two models represent rather different psychological processes, it is desirable to find a way to distinguish between them experimentally.

One critical test between the adding and the averaging hypotheses is given in Table 2. Compare the response to a set of two very favorable adjectives,

TABLE 1
Functional scaling of social desirability

	Earnest (Intelligent)	Systematic (Shrewd)	Unproductive (Shallow)	s		Earnest (Intelligent)	Systematic (Shrewd)	Unproductive (Shallow)	s
Happy	157	132	88	126	Friendly	159	131	105	132
Cheerful	151	131	95	126	Helpful	152	132	96	127
Optimistic	154	135	94	128	Congenial	151	133	101	128
Light-hearted	149	121	87	119	Outgoing	159	130	102	130
Hopeful	150	124	84	119	Obliging	146	119	95	120
Preoccupied	114	87	44	82	Inoffensive	138	120	90	116
Nervous	108	77	45	77	Dependent	107	85	48	80
Unhappy	95	71	45	71	Possessive	100	80	48	76
Gloomy	96	65	38	66	Bossy	95	76	43	71
Depressed	94	69	38	66	Unfriendly	85	70	38	64

Note: Subjects rated social desirability of male college students described by two adjectives. Each of the twenty row adjectives was paired with each of the six column adjectives for a total of 120 sets. Each of 12 naive subjects judged each set twice on a 200-mm graphic rating scale in a single session. The functional scale values, in the columns headed *s*, are obtained as marginal means of the three columns. (Data from Anderson, 1973e).

TABLE 2
Test of averaging and adding hypotheses in person perception

Set Type	Response	Set Type	Response	Set Type	Response
HH	72.8	HHM^+M^+	71.1	HHHH	79.4
M^+M^+	57.6	—	—	$M^+M^+M^+M^+$	63.2
M^-M^-	42.2	—	—	$M^-M^-M^-M^-$	39.5
LL	23.7	LLM^-M^-	25.7	LLLL	17.6

Note: Entries are mean judgments of likableness on open-ended rating scale with neutral point at 50. H, M^+, M^-, and L stand for personality traits of very high, moderately high, moderately low, and very low value. Data from Anderson (1965a).

HH, with the response to the same two adjectives with two added mildly favorable adjectives, HHM^+M^+. A simple adding model implies a higher response to HHM^+M^+ than to HH; addition of favorable information, M^+M^+, should make the person more likable. In contrast, a simple averaging model can produce a lower response to HHM^+M^+ than to HH; addition of mildly favorable to highly favorable information decreases the average value of the stimulus information (see also Sec. 4.1; Anderson, 1965a, p. 399).

The two critical comparisons of Table 2 contradict the adding prediction: HHM^+M^+ yields a less favorable response than HH. Similarly, for negative information, LLM^-M^- yields a less unfavorable response than LL. This result has been verified repeatedly (Anderson, 1965a, 1968c; Anderson & Alexander, 1971; Hendrick, 1968; Hamilton & Huffman, 1971; Hewitt, 1972; Lampel & Anderson, 1968; Leon, Oden, & Anderson, 1973; Oden & Anderson, 1971) though exceptions have been obtained (Hamilton & Huffman, 1971; Takahashi, 1970).

A superior form of the critical test is shown in Figure 3. The data are date ratings of boys described by a photograph alone, or a photograph plus two personality-trait adjectives. The latter two curves, denoted by HM and HL, cross over the former curve. This crossover is critical for a simple adding model. If the HM adjectives add to the low photograph, they cannot subtract from the high photograph as the data would imply. The averaging hypothesis, of course, gives an immediate explanation of the crossover interaction. This form of the critical test has the advantage that it requires only that the added information be somewhere near neutral in value. The test in Table 2 required a separate verification that M^+M^+ was indeed positive, and that M^-M^- was indeed negative. See also Sections 7.1 and 7.6.

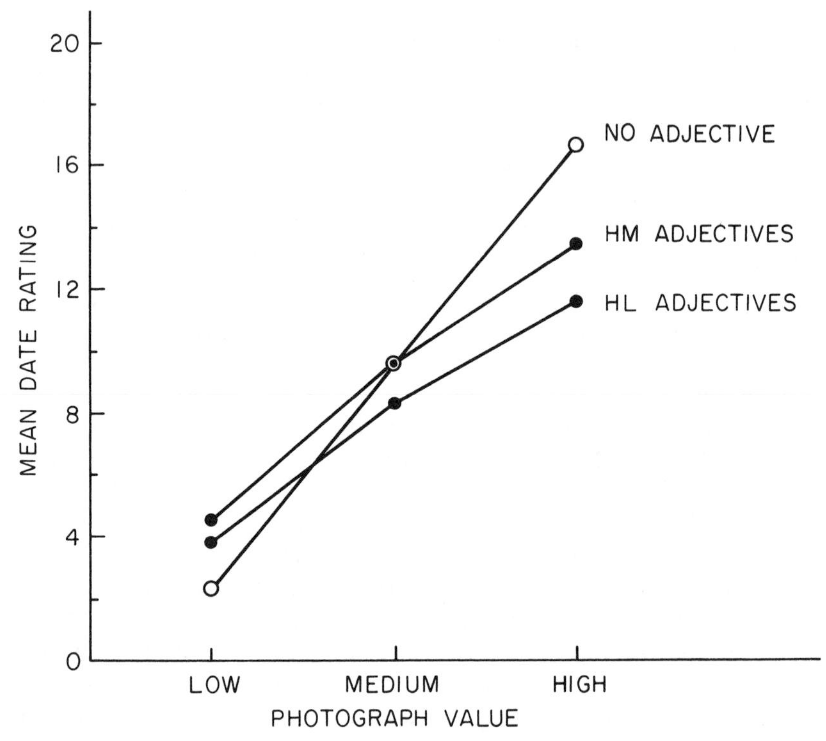

FIGURE 3.
Dateableness ratings of boys described by a photograph, plus personality-trait adjectives. HM denotes a high-medium pair of traits, HL denotes a high-low pair of traits. An adding model cannot account for the crossover exhibited by the curve for no adjectives (unless stimulus interaction is allowed; see Sec. 4.2). Data after Lampel and Anderson (1968).

3.2. Strategic Role of Qualitative Tests

It may seem odd to emphasize qualitative tests of quantitative models, but in fact they have played an extremely important role in the development of information integration theory. Since this matter appears to have general interest, it deserves brief discussion.

The advantages of the qualitative test arise because it rests on a difference in predicted direction. It is insensitive, therefore, to nonlinearity in the response scale (Anderson, 1965a). Indeed, the test can be made with rank order or choice data (Anderson & Alexander, 1971). This robustness of the qualitative test is quite advantageous.

But more important is that the qualitative test eliminates an entire class of models. For example, the set-size effect (Sec. 4.1) might have been interpreted in terms of an adding process together with a law of diminishing returns. Other attempts might also be made to retain the basic conception of an additive integration rule complicated by other integration processes, or by interactive valuation processes. These and other variations on a basic adding process are ruled out simultaneously.

As this brief discussion illustrates, an important role of qualitative tests is one of disproof. In the present case, their function is to rule out a general adding formulation. They do favor the averaging hypothesis, of course, but quantitative tests are necessary to establish the model. The validity of the averaging hypothesis rests in part on the success of the parallelism prediction discussed in the previous section, in part on further evidence discussed in Section 4.

3.3. Averaging Mechanisms

The work on the averaging hypothesis only allows the conclusion that the subject acts *as if* he were averaging. Several quite different mechanisms will produce averages, but little is known about which, if any, might underlie the response. It seems unlikely that the subject carries out the arithmetic steps indicated by Equation 1. However, two analog mechanisms are attractive possibilities.

One averaging mechanism is the example of point masses on a massless plank, familiar from elementary physics. The location of each mass is its scale value, and the physical mass equals the weight parameter. The mean is the balance point, center of gravity, or fulcrum. If all masses are equal, then the sum of distances to the fulcrum is zero. In this view, the subject is considered to weigh up the information and to strike a balance. Such a mechanism may apply to simple perceptual tasks such as averaging of number or position (Anderson, 1968d).

A rather different mechanism is obtained by considering each stimulus as a mixture containing positive, negative, and neutral elements. Scale value equals net concentration of charged elements, and weight equals volume. Integration corresponds to combining mixtures and it is easy to see that this mechanism yields a weighted average.

This representation of the stimulus as a set of elements, aspects, or features is traditional associationism. It deserves consideration as an approach to a more molecular semantic analysis of the integration processes (see also Secs. 8.1 and 8.4). As yet, not much is known beyond some unsystematic introspective reports. Recent work on semantic memory (e.g., Rumelhart, Lindsay, & Norman, 1971) may be helpful.

4. OBJECTIONS AND FURTHER EVIDENCE

The results of the two previous sections might make it appear that verification of the averaging hypothesis for stimulus integration was simple and straightforward. That was far from true. Three serious empirical objections still remain to be discussed. Also a variant of the adding formulation that allows for stimulus interaction needs to be considered. One additional problem, differential weighting, is taken up in the following section.

4.1. Set-Size Effect

One difficulty for the averaging model has already appeared in Table 2 of Section 3.1. For each adjective value, four adjectives show a more extreme response than two adjectives. The response of 79.4 to HHHH, for example, is more extreme than the response of 72.8 to HH. This 'set-size effect' occurs in all four comparisons.

Of itself, the set-size effect clearly argues for an adding model. Although the response is not a linear function of set size, it would be conceptually simple to introduce a law of diminishing returns. But this path is barred because the adding model has already been eliminated by the qualitative tests of the previous section.

On the other hand, a simple averaging model cannot account for the set-size effect. The mean value of four H adjectives is the same as the mean value of two H adjectives. A simple averaging model would predict the same response to both.

To account for the set-size effect, averaging theory makes use of the internal state variable, I_0, with value s_0 and weight w_0. This internal state is assumed to be averaged in along with the overt stimuli. For a set of k adjectives of equal value s_1 and equal weight w_1, the response is

$$R_k = \frac{w_0 s_0 + k w_1 s_1}{w_0 + k w_1}. \tag{8}$$

This set-size equation is a growth function of k with asymptote s_1. Thus it provides a qualitative account of the data of Table 2.

A quantitative test of the set-size equation was obtained in two experiments summarized in Table 3. Subjects judged descriptions containing 1, 2, 4, 6, or 9 adjectives, all high or all low in value. The value of s_0 was set at zero, the center of the rating scale in this experiment, and the scale values of the high and low adjectives, which are the asymptotes of Equation 8, were set equal to the endpoints of the rating scale.

Equation 8 may be normalized by setting $w_1 = w$, $w_0 = 1 - w$, so that

TABLE 3
Weight estimates as a function of set size

Experiment	Adjective Value	Number of Adjectives per Set					
		1	2	3	4	6	9
I	High	0.48	0.45	0.43	0.46	0.53	—
	Low	0.44	0.43	0.44	0.47	0.65	—
	Mean	0.46	0.44	0.44	0.46	0.59	—
II	High	0.44	0.37	0.36	0.36	0.37	0.49
	Low	0.40	0.38	0.45	0.39	0.42	0.58
	Mean	0.42	0.38	0.40	0.38	0.40	0.54

Note: Entries are normalized values of w value for a single adjective. The averaging model predicts w to be constant across set size. See text for interpretation of discrepancy at the largest set in each experiment. Data from Anderson (1967a).

w is the weight of a single adjective relative to the internal state. In this form, w is the only unknown and it may be estimated from the observed value of R_k for each k. If the set-size equation is correct, then this w-estimate should be the same for each value of k.

This test is exhibited in Table 3, which lists the w estimates averaged across subjects. These values are very nearly independent of set size, except for the big upward jump at the largest set in each experiment.

This discrepancy does not result from redundancy or from information overload that causes some information to be neglected. Either of these factors is plausible, but they would produce a decline in w, not the observed upward jump.

The explanation suggested by the first experiment was that the discrepancy reflected an end-effect response bias, a tendency to respond $+20$ or -20 when near the ends of the rating scale. The usual stimulus end-anchors, which function to nullify such response end-effects, were not used in this one experiment because it was desired to simplify the estimation by setting the scale values equal to the endpoints on the rating scale, that is, to the asymptote of Equation 8.

The second experiment tested this interpretation by adding sets of nine adjectives. If the discrepancy in the first experiment represents a breakdown of the model, the same discrepancy should reappear in the second experiment for sets of six adjectives. However, the hypothesized response bias implies that the locus of the discrepancy should shift from the set of six to the set of nine adjectives. The set of nine adjectives should be affected because, by virtue of the set-size effect, it has the response nearest the scale endpoints. That purifies the response to the set of six adjectives which should then yield the same w as the smaller sets.

As can be seen in Table 3, that is exactly what happened. Thus, the observed discrepancy seems only a minor technical complication. It is concluded that the averaging formulation can provide a quantitative account of the set-size effect.

4.2. Adding Model with Contrast

The adding model can be salvaged if stimulus interaction is incorporated. For example, a contrast effect could explain the critical tests of Section 2. In Table 2, the decrement produced by adding M^+M^+ to HH would be interpreted to mean that M^+M^+ was actually viewed as negative in the HH context. It is easy to see that all the qualitative tests of Section 2 could be accounted for within an adding formulation if the assumption of stimulus independence is not required.

The contrast hypothesis can be tested very simply by the method of component judgments. After forming the overall impression of the person, the subject is asked to judge one of the individual adjectives on how much he would like that particular trait of that particular person. Any contrast effect should appear directly in these component judgments.

Data from one such experiment are shown in Figure 4. Subjects judged a person described by three trait adjectives and then judged the likableness of one of the component traits of that person. The top curve gives the judgment of the same high adjective in four different sets that ranged from high to low in value, as listed on the horizontal axis. The downward slope of the curve means that the component judgments are displaced toward the value of other adjectives of the set. The same holds for the other three curves. Instead of contrast, a strong positive context effect (assimilation) is obtained. This result has been obtained repeatedly (Anderson, 1966a, 1971b; Anderson & Lampel, 1965; Kaplan, 1971a; Wyer & Dermer, 1968; Wyer & Watson, 1969; Takahashi, 1971b).

This positive context effect clearly rules out the adding model with contrast. However, it raises its own problem since it suggests that the stimulus values are changed by the other stimuli in the set. This question of stimulus nonindependence is taken up in the next two subsections.

4.3. Positive Context Effect

Two interpretations have been given for the positive context effect in Figure 4. One is that it represents a true assimilation effect: The scale values of each adjective shift toward the values of the other adjectives with which they happen to be combined.

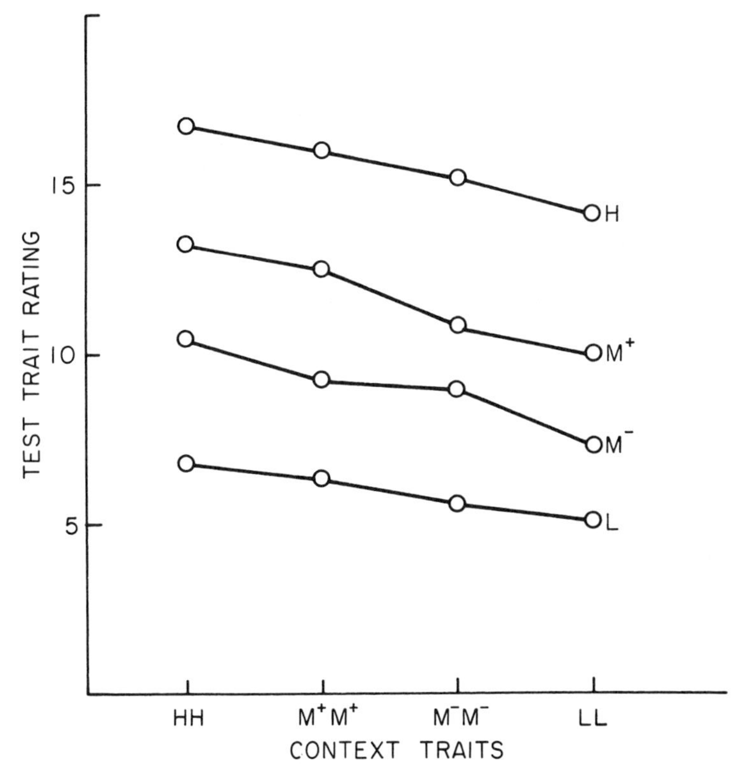

FIGURE 4.
Likableness ratings of component traits in person perception. Subjects first judge likableness of a person described by three traits, the context pair listed on the horizontal, and the test trait listed by each curve. They then rate the test trait on how much they would like 'that particular trait of that particular person.' The component rating of the same high adjective (top curve) shifts from about 16 to about 14 as a function of context value. Data after Anderson (1966a).

The other interpretation is that the positive context effect is only a generalized halo effect. Asked to judge the component, the subject gives a composite of the context-free value of the component and his overall impression of the person. Application of the averaging model to the component judgments yields

$$R = ws + (1 - w)\bar{s}, \tag{9}$$

where s is the context-free value of the component and $\bar{s}$ is the value of the person impression.

This halo effect model is complicated by the dependence of $\bar{s}$ on s. However, under the auxiliary assumption that $\bar{s}$ is a linear function of the components, Equation 9 leads to the parallelism prediction in a factorial design. This prediction is confirmed by the data of Figure 4. As an incidental comment, this model provides a method of stimulus scaling regardless of the nature of the positive context effect.

Of course, true assimilation or change in meaning could also obey the averaging model. Such potential stimulus interaction has great interest in its own right. It also has technical interest as an interaction that escapes detection by the analysis of variance test: if the stimulus values undergo linear assimilation and these changed values are then integrated by a linear model, then the overall impression will still follow the parallelism prediction.

Evidence against the change of meaning interpretation has been given by a comparison between no-paragraph and paragraph conditions (Anderson, 1971b). The no-paragraph condition was the same as in Figure 4; subjects judged the person, then one of the component adjectives. In the paragraph condition, subjects first wrote a paragraph describing the person in their own words.

This comparison has a straightforward rationale. If the adjectives interact and change meaning in the standard no-paragraph condition, they should do so even more in the paragraph condition because the subjects work over the information more thoroughly. But two large experiments found no difference between the two conditions, in accord with the halo effect interpretation. A third experiment, based on a different rationale, also supported the judgmental interpretation.

A different approach was employed by Kaplan (1971a). Adjectives were selected as being high or low in connotative variability but equal in scale value. The change-of-meaning hypothesis implies a greater shift in component judgments of the high variability adjectives. In contrast, the integration model predicts constant shift, regardless of connotative variability, and that is what was found. An earlier experiment by Wyer and Watson (1969) had claimed to find greater shift for high variability words. However, as Kaplan pointed out, this was true only in half of their data. The overall effect was actually in the opposite direction. See also Section 7.13.

These results, and those of the next subsection as well, give no support to the interpretation of the positive context effect as true change of meaning. Instead, it appears that the adjectives are integrated into the impression at their context-free value. Once integrated, they lose their separate existence and become part of the whole. The effect is analogous to the traditional halo effect, though it differs in asking for a judgment of a component from which the whole has been constructed.

The halo effect, incidentally, is usually considered an artifact, but it is a sensible strategy from an informational point of view. The subject is required to make a judgment and makes direct use of the two relevant pieces of information.

4.4. Primacy and Recency with Serial Presentation

If the adjectives are presented one at a time in succession, the order in which they are arranged will usually influence the final response. Such order effects are of theoretical concern since they mean that the scale values, or the weights, or both, depend on order of presentation. If the same adjectives in different orderings produce a different response, then the w and s parameters cannot be constant, as has been assumed in the integration model. It is important, therefore, to determine the existence and causes of order effects, for both theoretical and practical reasons.

The first investigation of this problem (Asch, 1946) obtained a primacy effect. Of two sequences of the same adjectives,

H H H L L L and L L L H H H,

the first yields the more favorable response. Asch speculated that this primacy effect was caused by a change in meaning of the later adjectives. A directed impression was assumed to be set up by the initial adjectives that constrained the interpretation of the later adjectives to produce an effective assimilation. The Ls in the first sequence would be raised in value, whereas the Hs in the second sequence would be lowered in value. Such value changes would produce primacy.

Asch's general view was based on the assumptions of stimulus interaction and change of meaning. Such a view is certainly plausible and no doubt true in some sense. However, solid evidence is elusive, and the only real evidence that Asch provided was the primacy effect. Since primacy has potential significance for the model analyses, as just noted, it was studied in an extended series of experiments by the writer and his associates (Anderson, 1965b, 1968c; Anderson & Barrios, 1961; Anderson & Hubert, 1963; Anderson & Jacobson, 1965; Anderson & Norman, 1964; Stewart, 1965). These reports all infirm the change-of-meaning hypothesis and support an alternative interpretation in terms of attention decrement, which results in a steady decrease in the weight parameter over successive serial positions (see Sec. 6). Since this evidence has been summarized elsewhere (Anderson, 1965b), only the two more recent reports of Tesser (1968) and Hendrick and Costantini (1970) are noted here.

Tesser read sequences of six adjectives, obtained the person impression, and then repeated the list, this time asking for component judgments. Since Tesser obtained a primacy effect in the overall impression, the change-of-meaning hypothesis implies that the component judgments of a given adjective should depend systematically on its serial position. Further, the dependence should be sizable since it is the presumed cause of the primacy. No such dependence was observed, which supports the integration model. The integration model predicts a very slight dependence, but it should be only a fraction of the order effect itself. Tesser also pointed out that the integration model predicts higher correlations between overall impression and component judgments for earlier words than for later words. Some support was obtained for that prediction.

Hendrick and Costantini (1970) made two important contributions to the problem of order effects. First, they showed that primacy depended on the affective difference between the early and late adjectives, but not on the degree of semantic inconsistency. Equal primacy was obtained from words that were opposite in both meaning and value, and from words that were opposite in value while lying on different semantic dimensions. This result clearly disagrees with the change-of-meaning hypothesis. Second, they showed that merely having subjects pronounce the words changed primacy to recency. Under the change-of-meaning hypothesis, pronunciation should if anything increase the primacy effect by emphasizing inconsistency between the earlier and later words. In contrast, the attention decrement hypothesis is in accord with the obtained result.

5. DIFFERENTIAL WEIGHTING

5.1. Qualitative Analysis

The simple parallelism prediction depends squarely on the condition of equal weighting. Weights may differ from one design factor to another, but within each factor the stimulus levels must have the same weight parameter.

In many tasks it will not be feasible to satisfy the equal weighting condition. That complicates the analysis. The averaging model then becomes nonlinear, and the parallelism prediction does not generally obtain.

An extreme case of unequal weighting has already appeared in the crossover interaction of Figure 3. The crossover shows that the net effect of the photographs, which is measured by the vertical extent of each single curve, is less when the photo is accompanied by trait adjectives than when alone. This vertical extent represents the effective weight of the photo. Since the weights must sum to one, inclusion of the trait information reduces the effective weight of the photo, and hence the vertical extent of the HM and HL curves. This example also provides an incidental illustration of the fact that the weight parameter cannot be absorbed into the stimulus scale by a response transformation.

A qualitative prediction of the pattern of nonparallelism can readily be obtained if a rough idea of the trend of the weight parameter is available (Anderson, 1971a, pp. 183–184). For example, if the scale values are all positive, and if the weight is a monotone-increasing function of scale value, then the curves should converge as scale value increases. If differential weighting is anticipated, therefore, it may be advisable to get a separate index of the weight, even if only in rank-order form.

The problem of differential weighting arises with special force when judgments along other personality dimensions are considered. Averaging theory should still apply, but differential weighting seemingly becomes the rule. For example, the traits *dependable* and *sociable* are about equally important in judging likableness. However, they carry different amounts of information for judgments of reliability, say, or of humorousness, and so would have unequal weight values for such judgments.

The likableness dimension is probably unique in facilitating equal weighting. Even in this case, there is evidence for some degree of differential weighting, especially of more extreme negative stimuli. The effect is not large, and when it was first obtained the possibility of response nonlinearity had no less plausibility than that of differential weighting (Anderson, 1965a). However, the effect is dependable (Anderson, 1968c; Anderson & Alexander, 1971) and is consistent with component judgments of the trait weights. Numerous other investigators have reported similar results (e.g., Cusumano & Richey, 1970; Hodges, 1974; Kanouse & Hanson, 1972; Lampel & Anderson, 1968).

Differential weighting is relevant to the averaging-versus-adding question discussed in Sections 2–4. The averaging model can account for certain kinds of nonparallelism directly in terms of the weight parameter. However, the standard adding models predict parallelism even with differential weighting. To account for nonparallelism would require abandoning the assumption of stimulus independence. That line of attack does not seem promising (Secs. 4.1 and 4.2).

5.2. Quantitative Analysis

When differential weighting affects only one factor of the design, an exact analysis can be obtained. That has great potential interest since it provides weight estimates directly from the marginal means. For a two-way design, the model can be written

$$R_{ij} = C + (w_0 s_0 + w_R s_{Ri} + w_{Cj} s_{Cj})/(w_0 + w_R + w_{Cj}), \qquad (10)$$

where equal weighting is assumed for rows, differential weighting for columns.

Equation 10 leads directly to a simple graphical test of fit. For fixed j, R_{ij} is a linear function of s_{Ri}. Consequently, the entries in each column should plot as a linear function of the row means. An exact statistical test based on bilinearity analysis is also available (Anderson, 1971a).

Equation 10 also leads to an interesting scheme for estimating weights. The difference between the first two rows, say, can be written as

$$R_{1j} - R_{2j} = w_R(s_{R1} - s_{R2})/(w_0 + w_R + w_{Cj}). \tag{11}$$

This may be solved to yield an interval scale for w_{Cj}:

$$w_{Cj} = c_1 + c_2/(R_{1j} - R_{2j}), \tag{12}$$

where c_1 and c_2 are constants for all j.

Equation 12 gives estimates of the weights that can only be obtained with equal weighting under special circumstances. It thus provides a possible means for measuring redundancy and inconsistency. It should be noted that the applications have had limited success, possibly because the difference score in the denominator causes statistical difficulties.

The general case of the averaging model, with separate weight and value parameters for each stimulus, does not allow a simple exact analysis. However, general purpose, computer estimation programs are now available that yield straightforward approximate analyses (e.g., Chandler, 1969). For example, in a rescaling of the Coombs-Thurstone social offenses using the functional measurement procedure (Leon, Oden, & Anderson, 1973), ten scale values and nine weights were estimated from 45 data points in about a minute per subject.

In many cases, the weight can be assumed to be a simple function of scale value. The program can then be used under the restriction that

$$w = 1 + as + bs^2. \tag{13}$$

Only two constants, a and b, would be needed for the weight estimates and that could markedly simplify the analysis. An application to clinical judgment is given in Anderson (1972a).

6. SERIAL INTEGRATION

6.1. Serial Averaging Model

In many tasks, information occurs and must be integrated in a temporal sequence. Order effects are typically found, the same information having different effects depending on the order in which it is presented. This violates the independence assumption that the stimulus parameters are the same in different combinations.

The averaging model can be extended to handle order effects by allowing the weight parameter to depend on serial position. Early stimuli might be important in crystalizing the impression, for example, or later stimuli might be fresher in memory. The model can provide estimates of the effect of each serial position to yield a complete, serial position curve.

For simplicity, it is assumed that the possible stimuli at any given serial position have the same natural weight. The response to a given sequence of N stimuli can then be written

$$R = w_0 s_0 + \sum_{i=1}^{N} w_i s_i, \tag{14}$$

where s_i and w_i are the value and relative weight of the ith stimulus in the sequence.

The key idea in analyzing this model is to treat the serial positions as factors and to construct the stimulus sequences from a factorial design. The parallelism prediction then applies, and the model predicts no statistical interaction among the serial factors.

One main result from this analysis is the serial curve of the weight parameter. To illustrate, suppose that just two stimuli, H_i and L_i, are presented at serial position i, and that the difference, $s_{Hi} - s_{Li}$, is constant for all i. Let R_{Hi} and R_{Li} be the mean response to all those sequences with H and L, respectively, at position i. Then it is easy to show that

$$R_{Hi} - R_{Li} = w_i (s_{Hi} - s_{Li}). \tag{15}$$

The other terms in Equation 14 cancel when the difference is taken because of the balance in the stimulus design.

Equation 15 is the desired result. The weight parameter is proportional to the observed difference on the left because $(s_{Hi} - s_{Li})$ is constant. Thus, the marginal means of the design provide direct estimates of the weight parameters on a ratio scale. This method fractionates the response at the end of the sequence into contributions from the individual serial positions.

To interpret the w_i as a serial position curve requires the further assumption that the stimuli have the same natural weight at every serial position. In general, w_i would be a product of two components, w'_i and p_i, where w'_i is the natural weight of the stimuli at serial position i, and p_i is the effect of position i itself. When the w'_i are equal, then the w_i are proportional to p_i, and hence they define a proper serial position curve.

This analysis has been used in several different areas of judgment with considerable though not complete success. The most directly relevant is impression formation (Anderson, 1973c) in which the serial position curve is interpreted in terms of the attention decrement hypothesis (Sec. 4.4). Other applications cover number averaging (Anderson, 1964a), psychophysical judgment (Anderson & Jacobson, 1968; Parducci, Thaler, & Anderson, 1968; Weiss & Anderson, 1969), attitude change (Anderson & Farkas, 1973), probability learning (Anderson, 1969b; Friedman, Carterette, & Anderson, 1968), and Bayesian decision-making (Shanteau, 1970a, 1972). A few brief comments on some of these applications are given in the next three subsections.

6.2. Length Averaging

In Weiss and Anderson (1969), each subject observed six lines projected one at a time for several seconds each. After the last line had disappeared, he adjusted a variable length to indicate his judgment of the average length of the six lines. The stimulus line at each serial position had one of two lengths, 16 or 24 centimeters, and the 64 possible sequences constituted a 2^6 factorial design. Each of eight subjects was run for nine sessions and judged each sequence four times.

The analysis of variance showed that the model fit the data quite well for seven of the eight subjects. The discrepancies for the remaining subject were apparently due to an idiosyncratic response to the one sequence with the 24-cm line in all positions. The report also gives various theoretical and methodological comments, as well as data from two related experiments.

Serial position curves were obtained using the procedures outlined above and these are shown in Figure 5. The plotted data are the differences, $R_{24} - R_{16}$, from Equation 15, and are proportional to the weight parameter. The upward trend of the curves represents a general recency effect, the later positions having a mildly greater influence on the response.

The recency in Figure 5 seems to be typical of psychophysical averaging tasks. Its cause is unknown. Two hypotheses, in terms of memory decay and confidence, were tested by Weiss and Anderson (1969) with limited success.

Success of the averaging model is not trivial even though subjects were instructed to 'average.' Each subject was free to choose his own measure of central tendency, and his averaging rule could even depend on the particular sequence of lengths. Indeed, the generality of the averaging model has been questioned by Birnbaum, Parducci, and Gifford (1971) who obtained a strong convergence interaction with simultaneous presentation of two lines. However, their data may reflect differential weighting; the shape of their interaction is what would be expected if the longer lines had greater natural weight (see also Sec. 7.4).

6.3. Intuitive Number Averaging

The first experiment to test the serial integration model used a number-averaging task (Anderson, 1964a). The subject saw a sequence of seven two-digit numbers, one at a time, and gave an 'intuitive' cumulative average as each new number appeared. The last number was an end-filler and is ignored here. A different pair of numbers was used at each serial position, but their difference was constant so that Equation 15 is applicable.

In this experiment, a serial position curve can be obtained from each

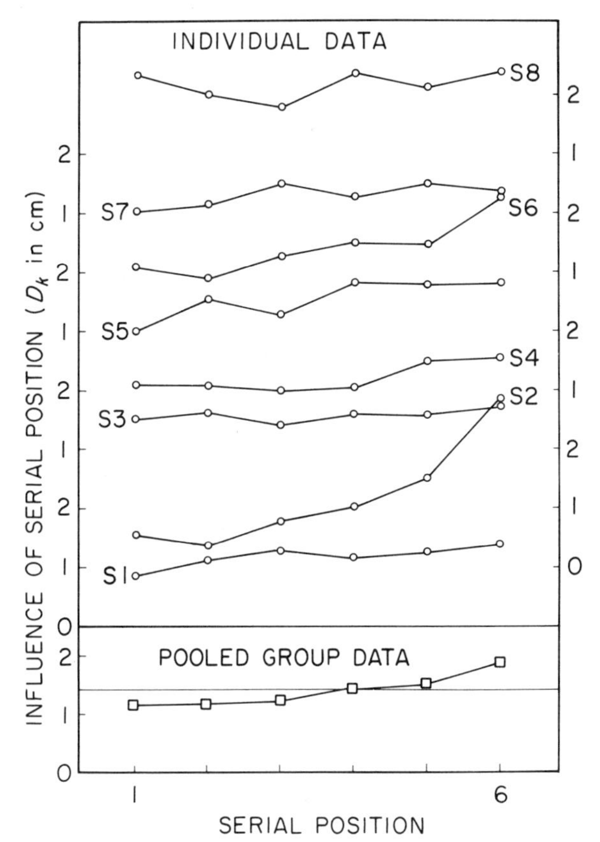

FIGURE 5.
Serial position curves for length averaging. Six lines were presented in sequence, and the subject gave a graphic judgment of their average length. The plotted data are proportional to the weight parameter at each serial position. Curves for individual subjects displaced vertically. Data from Weiss and Anderson (1969).

response. These are shown in Table 4 in the form of weight estimates with w_0 set at zero. In each curve, the most recent stimulus exerts a moderately strong effect, but much of this recency effect seems to disappear when further stimuli are added. Apparently there is a short-term salience analogous to that found recently for attitude change (Anderson & Farkas, 1973). For this and other reasons, the procedure of cumulative responding may not be as desirable as asking for a response only at the end of the sequence.

Neither number averaging nor length averaging has been much studied. The success of the model for these tasks suggests that they may be quite useful in studying memory processes involved in integration tasks. Different memory

TABLE 4
Serial position curves for number averaging

Response Position	Stimulus Serial Position					
	1	2	3	4	5	6
2	0.472	0.528				
3	0.305	0.314	0.380			
4	0.218	0.231	0.219	0.332		
5	0.143	0.177	0.177	0.195	0.308	
6	0.121	0.123	0.126	0.138	0.180	0.313

Note: Subjects respond with running intuitive average of sequence of seven two-digit numbers. Entries are weight parameters estimated from serial integration model, and show dependence of each response on the stimuli seen to that point. Data after Anderson (1964a).

systems appear to be required for the overall impression and for the individual stimuli, at least in the personality impression task (Anderson & Hubert, 1963).

6.4. Learning

Integration theory puts some traditional problems of learning in better perspective: Stimuli are conceptualized as informers, not as reinforcers. The organism is considered as an integrator of stimulus information, not as a conditioning machine. Learning itself is considered as serial information integration. Equation 15 above is a more general form of the sequential dependencies obtained from the customary stochastic learning models (Anderson, 1959b, 1964b, 1964c, 1966b; Atkinson & Estes, 1963). It is also more realistic since it allows for changes in learning rate (w) over trials.

Learning is thus viewed as a process of integrating information about some state of the world. From the informational view, therefore, Shanteau's work on decision-making (Sec. 7.2) would be considered as a study of learning. As this example indicates, the present view emphasizes a more cognitive class of learning tasks than is congenial to conditioning theory.

The informational view of learning is similar in spirit to the S-S approach favored by Tolman, and contrasts with the S-R approaches of Hull, Spence, and Skinner. Other recent developments (e.g., Atkinson & Wickens, 1969) have also emphasized informational concepts. Two important studies of probability learning using the serial integration model have been given by Friedman, Carterette, and Anderson (1968) and by Levin, Dulberg, Dooley, and Hinrichs (1972). Kaplan and Anderson (1973) include a brief discussion of motivation within the informational view.

6.5. Probability Learning

Much of the early work on mathematical learning models, especially in Estes' (1964) stimulus sampling theory, was centered on probability learning. In these experiments, the subject predicted which of two binary events would occur next in a random sequence. Surprisingly, the subjects tended to show 'matching behavior,' with the frequencies of their predictions being approximately equal to the frequencies of the two events. This behavior assumed great theoretical interest when it was shown that it was predicted by the conditioning models of stimulus sampling theory.

However, more detailed analysis showed that the conditioning models could not account for the sequential structure of the data. The theoretical analysis of sequential dependencies (Anderson, 1959b) showed that the effect of a given event should decay exponentially according to the function, $\theta(1 - \theta)^{k-1}$, where θ is the learning rate, and k is the number of succeeding trials. Data from noncontingent schedules departed radically from this exponential decay pattern. Anderson (1960) obtained the following results. Since each observed value is based on 16,000 observations, it is clear that the stimulus sampling model does not account for the data.

Previous Trial	1	2	3	4
Theoretical dependency	θ	$\theta(1 - \theta)$	$\theta(1 - \theta)^2$	$\theta(1 - \theta)^3$
Observed values	.22	.07	−.08	.01
Predicted values	—	.17	.14	.10

Despite these theoretical discrepancies, Estes (1964, p. 122) felt that ". . . the structure of the choice data can be accounted for quite well on the basis of an elementary learning process. . . ." However, similar deviations have been found in later work (see e.g., Anderson, 1964c, 1966b; Friedman, Carterette, & Anderson, 1968). It was this inability of the simple mathematical models of learning theory to account for the sequential dependencies that led to their abandonment. The averaging model of information integration theory does considerably better, although it too has difficulties with sequence patterning (Friedman, Carterette, & Anderson 1968).

7. GENERAL ALGEBRAIC JUDGMENT THEORY

The discussion so far has centered on one model, the averaging model for person perception, and one judgment task. However, the same ideas and methods have been applied to a wide range of other judgment tasks. This section gives a brief sketch of some of this work. Two characteristics of this approach deserve emphasis.

First, these tasks have the common property that they require information integration. Second, they rest on the use of simple algebraic models and on functional measurement theory. It is on this basis that a unified treatment can be given to judgments in such seemingly distinct areas as person perception, attitude change, psychophysical integration, perceptual illusions, and utility theory.

7.1. Attitude Change

Figure 6 reports judgments of U.S. presidents on a dimension of statesmanship. The main part of the design, represented by the solid lines, combined paragraphs that were low or high in value with paragraphs that were low,

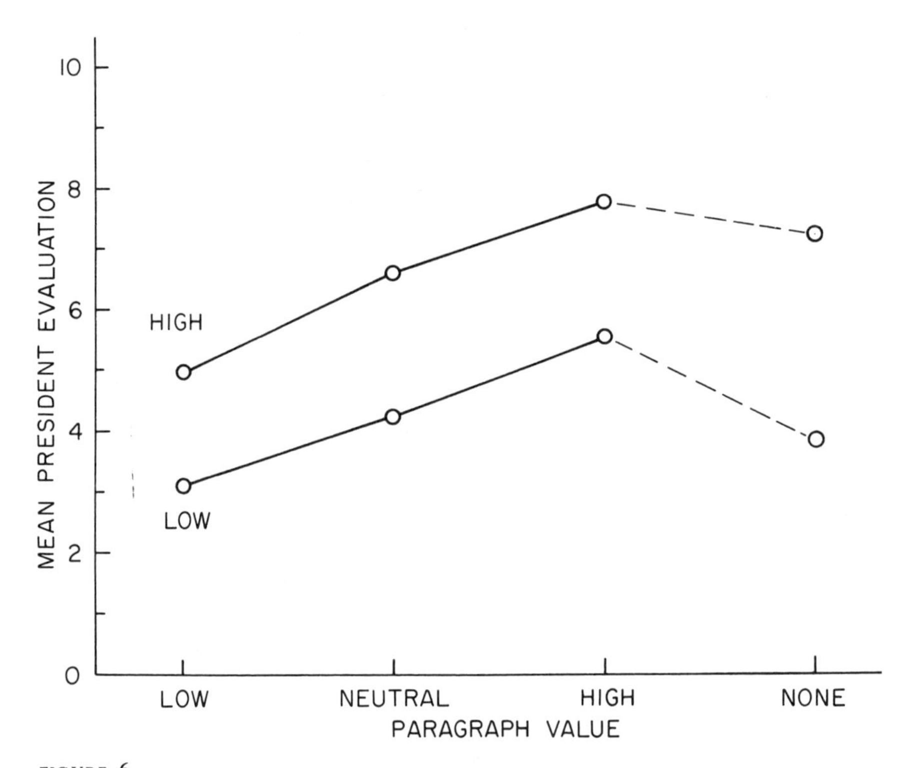

FIGURE 6.
U.S. presidents characterized by descriptive paragraphs were judged on statesmanship. The paragraphs formed a 2×3 design and the parallelism of the two solid curves supports the integration model. The two points connected by dashed lines provided a critical test between averaging and adding (see Sec. 3.1). Data after Anderson (1973d).

neutral, or high in value to form a 2 × 3 design. The integration model yields the usual parallelism prediction, and that is well supported in these data.

The two points connected by dashed lines provide a critical test between the averaging and adding models. The response to high is more extreme when it is alone than when it is paired with neutral, 7.2 vs 6.6. The same holds for the comparison of low with low + neutral, 3.8 vs 4.3. This infirms adding and supports averaging, according to the logic of Section 3. Related work is given in Anderson (1959a), Anderson and Farkas (1973), and Sawyers and Anderson (1971). A collection of 220 president paragraphs is given in Anderson, Sawyers, and Farkas (1972).

Two of the major directions in attitude theory have attempted to base attitudes on traditional verbal learning, or on a general consistency principle. The present direction is different, since it attempts to base attitudes on a principle of information integration.

In the verbal learning approach (see Insko, 1967; McGuire, 1964), the attitude or opinion is assumed to be mediated by the verbal material that is learned. That implies a strong, direct relationship between retention of attitudes and retention of the verbal materials. In fact, however, these correlations have generally been quite low.

Integration theory assumes two memory systems, an assumption based on direct experimental evidence (Anderson & Hubert, 1963). The verbal stimuli act as a carrier of the meaning, but once that is extracted and integrated into the current attitude, the verbal stimulus is no longer necessary and is stored in a different memory system. This conceptualization is consistent with the low correlations between retention of attitude and verbal material (e.g., McGuire, 1964, pp. 266–267). Further discussion is given in Anderson and Farkas (1973).

In the consistency principle approach, the subject is seen as striving to reduce imbalance, incongruity, dissonance, etc., among his cognitions. The difficulty with basing attitude theory on a consistency principle is that it is too narrow. Much if not most attitude change appears to be straightforward integration of information, in which felt inconsistency is either nonexistent or plays no effective role. A theoretical elaboration of this point is given in Anderson (1971a), together with comparisons between integration theory and various specific consistency theories (see also Sec. 8.4).

7.2. Bayesian Judgment Theory

Shanteau (1970a, 1972) has made a fundamental contribution in his careful comparison of integration theory and the Bayesian approach employed by Edwards (1968) and his associates. The standard Bayesian two-urn inference task was used. The subject's task was to judge the probability that one of two complementary binary urns was being sampled in a succession of draws.

Shanteau showed that a serial integration model (Sec. 6) gave an excellent account of the raw probability judgments. However, the log-odds response used in much of the Bayesian work was nonadditive, contrary to the Bayesian model. Two other comparisons also favored the integration model. The serial curves showed mild long-term recency whereas the Bayesian model requires that they be flat. Also the effect of diagnosticity (proportion of white beads in the urn) was much less than required by the Bayesian model.

This brief comparison is unfair to the Bayesian model in one important respect. The Bayesian model derives directly from mathematical statistics. Although it attempts to allow for personal probability, it is a normative model that prescribes the statistically optimal criterion against which the subject's response is to be compared. In contrast, the integration model is descriptive and attempts to delineate the processes that underlie the behavior. That it fits the data better is due in part to its inclusion of parameters to describe memory effects, for example, that the Bayesian model disallows.

Moreover, it is well known (Edwards, 1968) that subjects do not obey the Bayesian model. Instead, they typically show marked 'conservatism'; their responses are less extreme than the statistically correct response. Conservatism has been the focal topic of the Bayesian research, as Slovic and Lichtenstein (1971) point out. In many practical applications, man-machine systems, for example, accuracy and optimality are all important. For such purposes, the Bayesian model may be entirely appropriate.

Unfortunately, there is an inevitable tendency to use the Bayesian model inappropriately. Considerable effort has been made to 'explain' conservatism in terms of various psychological processes. But from the present view, conservatism is a noneffect. It has a purely nominal existence, by reference to a model that has no psychological content. It is such inappropriate usage that is subject to criticism.

This criticism is underscored by the way that data are presented in typical Bayesian reports. The raw data are seldom given, but instead are transformed into statistical quantities, accuracy ratios or inferred log likelihood ratios, for example, by applying the Bayesian model to the data. Such data may have meaning within the Bayesian framework, but they are difficult to interpret in any other framework.

The importance of a descriptive, process orientation is further emphasized by Shanteau's (1970a) finding that subjects in the Bayesian task seemed to be making estimations, basing their judgments directly on the proportion of sample beads. Estimation is a qualitatively different kind of behavior from the inferences assumed in the Bayesian model. Kahneman and Tversky (1972) provide further evidence for Shanteau's conclusion about the prepotence of the sample proportion.

7.3. Preference and Difference Judgments

Judgments of the difference in value of two stimuli should follow a subtracting model. If the stimuli are presented in a factorial design, the subtracting model can be written,

$$R_{ij} = s_{Ri} - s_{Cj}. \tag{16}$$

This linear form implies parallelism in the graphs, and no interaction in the analysis of variance. This prediction was supported by Shanteau and Anderson (1969) for preference judgments between drinks, sandwiches, and lunches as shown in Figure 7. Further support was obtained by Birnbaum (1972b) for the personality impression task.

A number of experiments have studied judgments of the difference between two psychophysical stimuli (see Curtis, 1970; Dawson, 1971). These results give mixed support to the subtracting model. The series of papers by Curtis

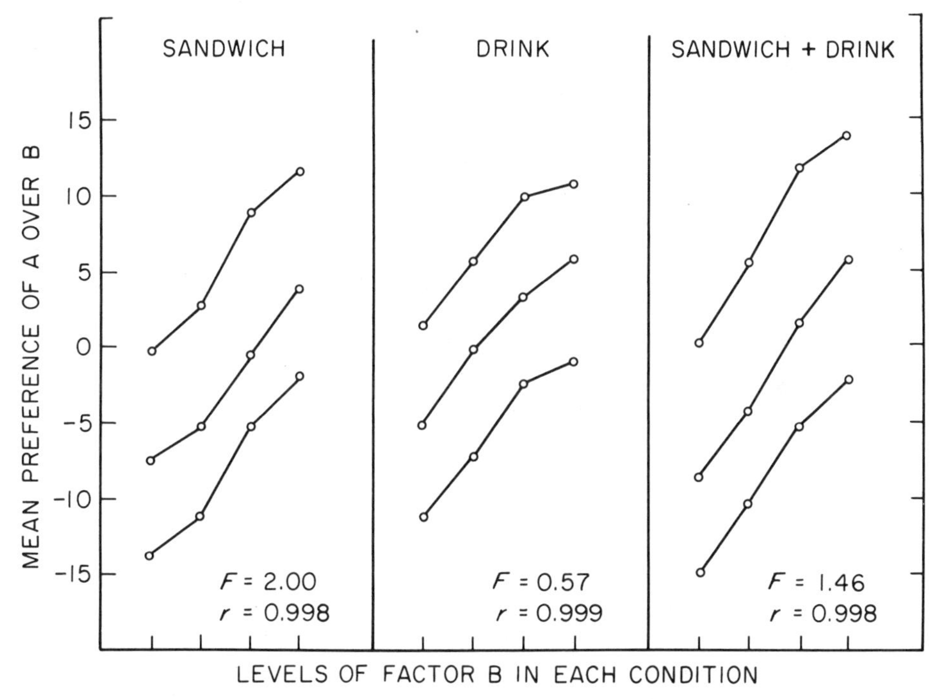

FIGURE 7.
Subtracting model for preference judgments predicts parallel curves. Subjects judge degree of preference between two sandwiches, two drinks, or two sandwich-drink combinations. The listed F ratios are the analysis of variance interaction test of parallelism (Sec. 1.4). The r values are correlations between predicted and observed. Data from Shanteau and Anderson (1969).

and his associates tend to support the model if a category rating response is used, but to show discrepancies with magnitude estimation.

An interesting note to this question can be seen in Figure 4 of Stevens (1971) who replots difference judgments of loudness from Beck and Shaw (1967) and Dawson (1971). Although Stevens does not remark on the point, the near parallelism of the fitted curves can be taken as support for the subtracting model. But though magnitude estimation was used, the response is actually plotted on a logarithmic scale. From the present view, Stevens' figure suggests that magnitude estimation yields a biased response, and that something like a log transformation is needed to remove the bias.

Existing data do not allow a firm evaluation of the subtracting model for psychophysical stimuli. Much of the work suffers from use of a stimulus scaling procedure that assumes a power function relation between the physical and psychological scale values. In addition, the quantitative analyses of the model have not generally been conclusive (see also Sec. 7.12).

Functional measurement procedure should be useful in this question since it does not require any assumptions about the stimulus scale. It also provides a rigorous and powerful test of discrepancies from the model. A recent report by Birnbaum and Veit (1973b) has used this approach to good effect in judgments of lifted weights. Further references are in Anderson (1973a).

7.4. Psychophysical Averaging

Several experiments have studied judgments of the average value of a set of psychophysical stimuli. Averaging judgments, like the difference judgments noted just above, present a mixed picture. Most experiments designed to study the serial position curve have supported the averaging model (Anderson, 1967b; Anderson & Jacobson, 1968; Weiss & Anderson, 1969). In addition, a report on angle averaging (Weiss & Anderson, 1972) has interest since it illustrates the use of rank-order data in functional measurement.

However, a number of papers have reported deviations from the model (e.g., Parducci, Thaler, & Anderson, 1968; Birnbaum, Parducci, & Gifford, 1971). These have the same form in almost every case, namely, a convergence interaction with increasing stimulus intensity. The simplest interpretation is that the more intense stimuli have greater weight as well as greater scale value. Differential weighting in an averaging model would then imply deviation from parallelism in the observed direction. Unfortunately, most of the published reports that have obtained the nonparallelism have failed to use the experimental precautions with their rating scales that are standard in the present research program. End-effects in the response scale, in particular, could act to produce the observed nonparallelism and it seems likely that that is part of the cause. However, some nonparallelism seems to remain even with a production response for averaging of motor movements (Levin, Craft, & Norman, 1971, Fig. 1).

7.5. Size-Weight Illusion

A linear integration model for the size-weight illusion can be written as

$$R_{ij} = w_R s_{Ri} - w_C s^*_{Cj}, \qquad (17)$$

where s_{Ri} and s^*_{Cj} are the scale values of the kinesthetic and visual cues, respectively. Thus, s_{Ri} may be considered as the felt heaviness of the object when lifted with closed eyes, whereas s^*_{Cj} represents an expectancy based on the visual size. The negative sign of w_C, the weight of the size cue, emphasizes the contrast effect in this illusion.

The parallelism prediction of this model holds up rather well in the two experiments of Figure 8. The slope of each curve, proportional to w_C, reflects

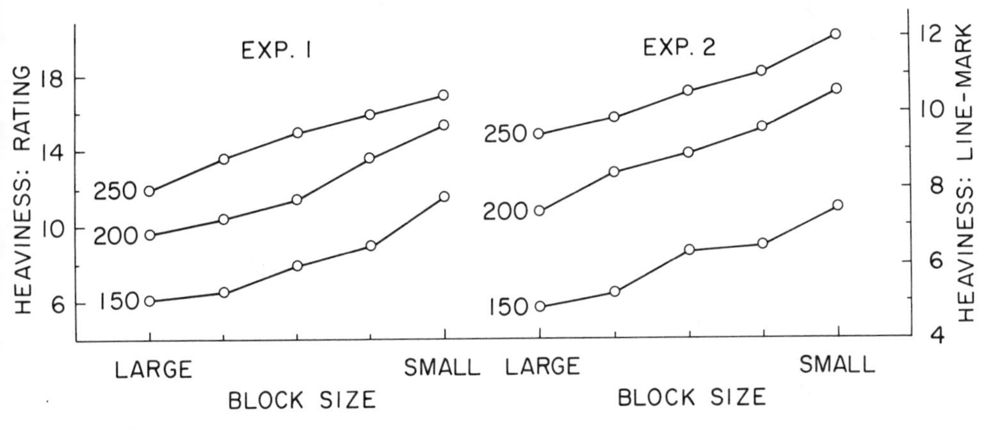

FIGURE 8.
Heaviness of cubes varied in weight (150, 200, or 250 grams) and size (large to small) in a two-way factorial design. The parallelism of the curves implies that the two cues, size and weight, are integrated according to the linear model of Equation 17. The upward slope of the curves indicates that the scale value of the size is an approximately linear function of the cube root of the volume. From functional measurement theory, vertical spacing of the curves constitutes an interval scale of subjective weight. The psychophysical function shows a slight negative acceleration. Data after Anderson (1970b).

the illusion that the same physical weight is judged lighter when it is larger. As Scripture (1897) commented, a pound of lead *is* heavier than a pound of feathers.

The most popular explanation of the illusion has been that people really judge density. That would imply a divergence interaction in the curves

(Anderson, 1970b). Data that would appear to support a density interpretation (Stevens & Rubin, 1970) may really indicate a bias in magnitude estimation (Sec. 7.3). An application of a model similar to Equation 17 for the Ebbinghaus illusion gave good results using a size production response (Massaro & Anderson, 1971). An interesting extension of Equation 17 to a size-numerosity illusion has been given by Birnbaum and Veit (1973a). Further support for the size-weight model is in Anderson (1972b), which also shows a cross-task validation of functional measurement.

7.6. Meals

Meals are a useful stimulus class, interesting in their own right, and also by contrast with the personality adjective task. Both involve an integrative judgment, but the semantic interrelations among the adjectives have no direct analog for the foods composing a meal. As a consequence, it seems implausible that foods would change meaning as a function of context (Anderson & Norman, 1964). In addition, an adding-type model might be expected to apply to meals which lends this task further theoretical interest.

That some linear-type law applies to judgments of meals is indicated in the left panel of Figure 9. Subjects judged likableness of meals composed of main course and vegetable. The parallelism of the three vegetable curves implies that the two components obey a linear integration rule. The closeness of the three curves reflects the small weight of vegetable relative to main course.

That an averaging model is required is shown by the crossover interaction in the right panel of Figure 9. The N and NN curves represent one and two vegetables of neutral value, respectively. The addition of a neutral vegetable thus increases palatability for a disliked meal, and decreases palatability of a liked meal, in accord with the averaging hypothesis. Similar results were obtained for judgments of naval officers and of criminal groups in this same experiment (Oden & Anderson, 1971). An adding formulation cannot account for these results unless some kind of stimulus interaction is postulated (see also Sec. 4.2).

7.7. Multiplying Models and Multilinear Models

In the next several applications, the stimulus variables combine by multiplying, or by joint adding and multiplying rules. The application of functional measurement theory to such models has been discussed elsewhere (Anderson, 1970a, 1971a; Anderson & Shanteau, 1970; Anderson & Weiss, 1971; Shanteau & Anderson, 1972). Since the essential ideas are simple, only a brief sketch is given here.

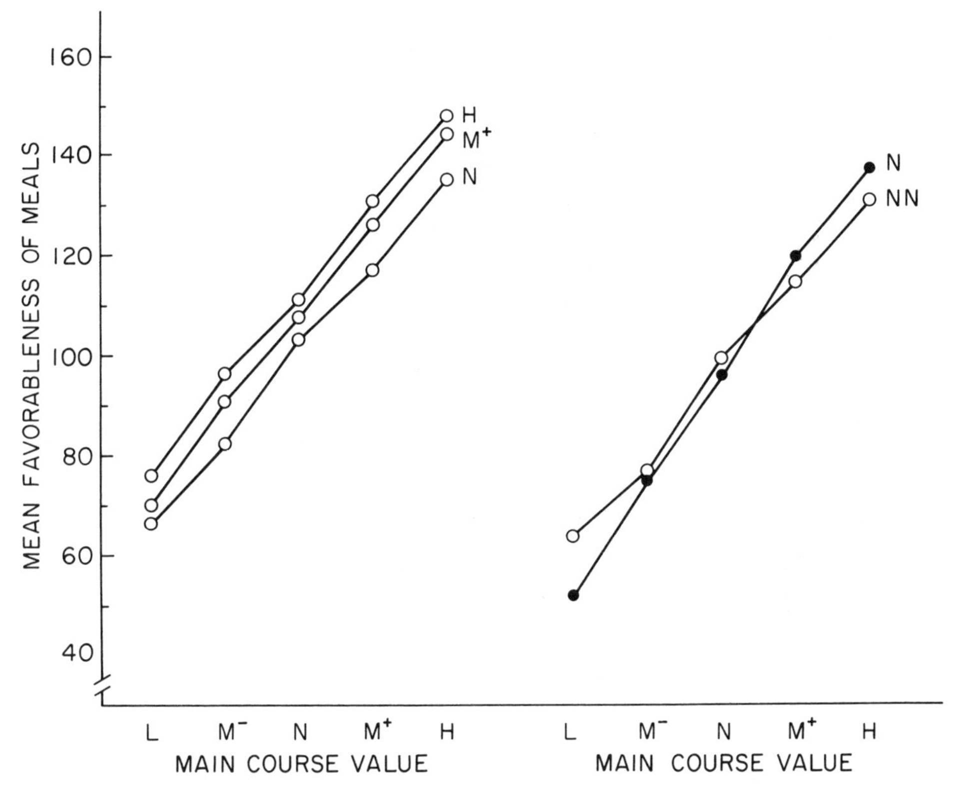

FIGURE 9.
Integration model predicts parallelism for likableness of meals described by the two-way, vegetable × main course design in the left panel. The H, M^+, and N curves correspond to vegetables of high, mildly favorable, and neutral value. Averaging model predicts the crossover in the right panel of the curves for one and two neutral vegetables; see text. Data after Oden and Anderson (1971).

With two stimulus variables in a factorial design, the multiplying model may be written,

$$R_{ij} = C + w_{Ri}s_{Cj}. \tag{18}$$

This equation may be combined with the expression for the column means,

$$\overline{R}_{.j} = C + \overline{w}_R s_{Cj}, \tag{19}$$

to yield

$$R_{ij} = C + (w_{Ri}/\overline{w}_R)\overline{R}_{.j} - (w_{Ri}/\overline{w}_R)C. \tag{20}$$

Equation 20 yields a simple graphical test of fit. If the observed column means are used as the scale values of the column stimuli, then each row of data should plot as a diverging fan of straight lines. Deviations from this bilinear form would infirm the model, and a rigorous statistical test based on analysis of variance is available. The multiplying model implies nonzero interaction, of course, but it should all be concentrated in the linear $\times$ linear component, and the residual interaction should be nonsignificant.

This approach may be extended directly to multilinear models in which several stimulus variables combine by joint adding-multiplying rules. If A, B, and C represent stimulus variables, the model $AB + C$ implies nonsignificant $A \times C$, $B \times C$, and $A \times B \times C$ interactions in the analysis of variance test. The bilinear component of the $A \times B$ interaction should be significant, and its residual should not. This same approach extends directly to more complex multilinear models.

Functional scales of the stimulus variables are also available if the model passes the test of goodness of fit. As illustrated by Equation 20, the marginal means of the factorial design estimate the subjective stimulus values on an interval scale.

One precaution is needed in estimating scale values from marginal means. If both positive and negative values are present, they may largely cancel in the direct average, yielding unreliable estimates. This difficulty can usually be avoided by complementing the curves with negative slope. An illustration is given in Figure 11 discussed in Section 7.9.

7.8. Utility Theory

A central problem in utility theory has been that of measuring the subjective values of probability and goods. Functional measurement provides the first method for solving this problem that is practically effective and generally applicable.[2]

---

[2] The reviewer and one of the editors have suggested that Davidson, Suppes, and Siegel (1957), and Tversky (1967a, 1967b) might have prior claim to a general method. Both these approaches seem rather specialized, however. The techniques of Davidson, Suppes, and Siegel are ingenious but too complicated to summarize briefly. They might perhaps be considered as a general method in principle, but a satisfactory error theory is lacking, and in any case the method did not seem to work very well. An extensive discussion is given in Luce and Suppes (1965, Secs. 4.2 and 4.3, Table 2).

Tversky's two papers present considerable evidence to support a simple multiplying model for gambles like the single positive bets of Figure 10. Tversky attempted an ordinal, nonmetric analysis based on conjoint measurement theory, but that failed to provide a satisfactory test of fit; correlation coefficients are not generally appropriate for testing fit (Anderson, 1972a; Birnbaum, 1973). The difficulty with ordinal analysis is lack of an error theory (Weiss & Anderson, 1972).

To get a proper test of fit, Tversky took logarithms of the raw scores and applied analysis of variance, testing for significant interaction, much as in functional measurement procedure. From Equation 18, it can be seen that taking logs will convert the two-factor multiplying model to an adding model if $C = 0$. In that case, the parallelism test of Section 2 is directly applicable.

Tversky's log-response method is limited to special cases since it requires $C = 0$ and

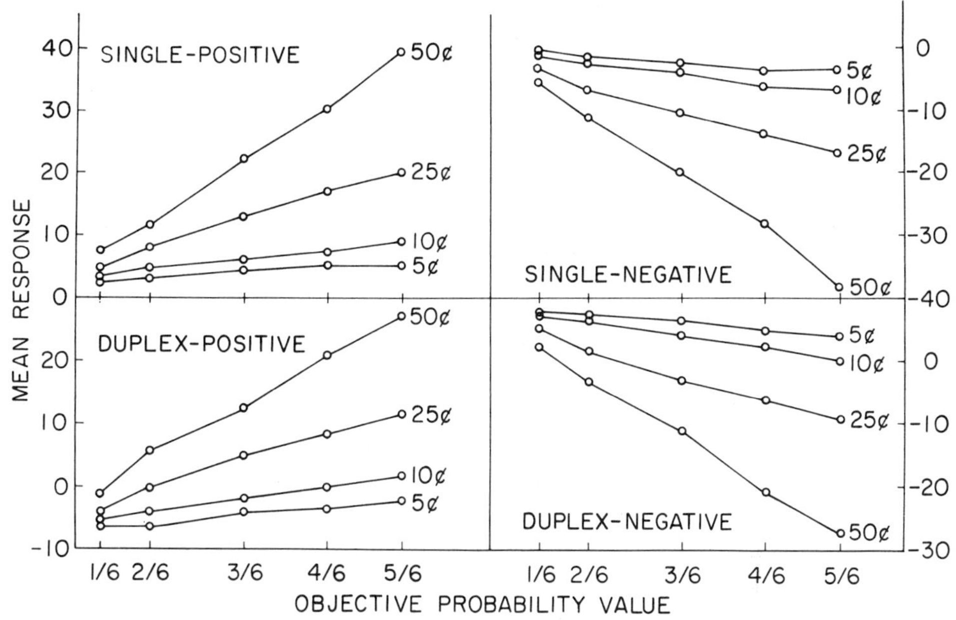

FIGURE 10.
Judged worth of bets consisting of a certain chance to win certain sums of money. Integration theory implies that each panel should be a diverging fan of straight lines when plotted with subjective probability on the horizontal (see text). This figure illustrates the first general method for simultaneous measurement of subjective probability and utility. From Anderson and Shanteau (1970). Copyright 1970 by the American Psychological Association. Reprinted by permission.

Illustrative data are given in Figure 10. The upper two panels plot the judged worth of single bets consisting of a certain probability to win or lose a certain sum of money. The data evidently follow the bilinear form quite well, suggesting that the subjects do act as if they multiply the two stimulus variables, probability and money.

These graphs require the subjective probability values, obtained following the procedure of Section 7.7. The probability stimuli were specified by a die, but these need to be spaced on the horizontal axis according to their subjective values to get the bilinear form. These subjective values were obtained directly from the marginal means of the raw data tables. The unequal spacing

---

disallows negative numbers (losses). The log-response trick also could not handle the duplex bets of Figure 10 which require a model of the form, $A \times B + C \times D$.

On the other hand, present methods will apply to Tversky's set-IV data (Tversky, 1967a, Tables 3 and 4) which were thought to follow the model, $A(B + C)$. Indeed, if these data are graphed, they appear to show the parallelism and bilinearity forms that are implied by the given model. Tversky's two papers thus agree with the present emphasis on the use of numerical response measures.

on the horizontal means that the functional stimulus values are a nonlinear function of their objective values.

Table 5 compares the subjective and objective values for probability and

TABLE 5
Comparison of objective and subjective values of probability and money stimuli

Objective Values	Subjective Values
Money	Worth
5¢	5
10¢	10
25¢	28.4
50¢	58.3
Probability	Likelihood
1/6	1
2/6	2
3/6	3.92
4/6	5.63
5/6	7.19

Note: From Anderson and Shanteau (1970). Copyright 1970 by the American Psychological Association. Reprinted by permission.

money. Since the estimates are only on an interval scale, the two lowest values of each subjective scale were set numerically equal to the corresponding objective values. Statistical analysis then showed that the remaining subjective values were significantly different from their objective counterparts. In other words, both subjective scales are nonlinear in the objective values. This difference was quite reliable for probability, but marginal for money.

The two lower panels of Figure 10 plot the judged worth of pairs of bets. The duplex-positive panel represents duplex bets that included the single-positive bets, plus a lose component, analogously varied. The model for the duplex bets can be written schematically as a multilinear model,

$$\text{Worth} = P_W \times \$_W + P_L \times \$_L,$$

where P_W is the probability of winning $\$_W$, and P_L is the probability of losing $\$_L$ (Slovic & Lichtenstein, 1968). The bilinear form of the two lower panels means that these bets also are integrated by a multiplying operation.

However, the adding operation in the model did not fare as well. Small but reliable deviations from parallelism were obtained, unfortunately of no clear pattern. The possibility that these deviations from additivity reflect response nonlinearity can perhaps be ruled out because the success of the multiplying model suggests an interval response scale.

Related work by Shanteau (1970b) used symbolic stimuli (e.g., "Better than Even" chance to win "Sandals") rather than the numerical stimuli of Figure 10. This work also supported the multiplying model, and again found deviations from the adding model. Shanteau was able to show that these deviations had a subadditive form, possibly interpretable as a law of diminishing return. He also found evidence for subadditivity in analyses of previous papers on utility theory that had been taken to support additivity.

An interesting facet of Shanteau's results is in Table 6, which compares the

TABLE 6
Subjective values of probability phrases obtained by direct ratings and by functional scaling

Probability Phrase	Direct Rating	Functional Scale Value
No chance	0	0
Unlikely	0.22	0.18
Somewhat unlikely	0.30	0.27
Not quite even	0.44	0.50
Tossup	0.50	0.54
Better than even	0.56	0.66
Fairly likely	0.66	0.69
Highly probable	0.90	0.84
Sure thing	1	1

Note: Data from Shanteau (1970b).

direct ratings of the probability phrases used in his experiment with the functional scale values. The differences between the two sets of values are moderate in size and statistically reliable. Evidently the probability phrases have a different value when they stand alone than when they function in the judgment task. In contrast, there was good agreement between the direct ratings of the goods and their functional scale values.

7.9. Adverb-Adjective Phrases

Cliff (1959) has suggested that adverbs act as multipliers on certain classes of adjectives, and this multiplying model has been supported in a number of experiments (e.g., Howe, 1966; Lilly, 1968). However, this work has relied on Thurstonian scales for the stimuli. Tests of the model have been somewhat crude, therefore, and only apply to pooled group data (Sec. 1.5).

Functional measurement. Illustrative data from a functional measurement analysis are shown in Figure 11. Two adjectives, *sincere* and *insincere*, were combined with the nine listed adverbs in a two-way factorial design. Forty subjects judged social desirability of these person descriptions on a graphic rating scale. Marginal means of the data in Figure 11 were used to obtain the functional adverb scale on the horizontal. For this purpose, incidentally, the data for *insincere* were reflected about the center point of scale. The mean of the two curves is approximately horizontal which yields unreliable estimates. The reflection procedure is justified since *sincere* and *insincere* are known to have opposite values.

The multiplying model implies that the two curves should be diverging straight lines. Actually, a mild curvilinearity can be seen, and the same was

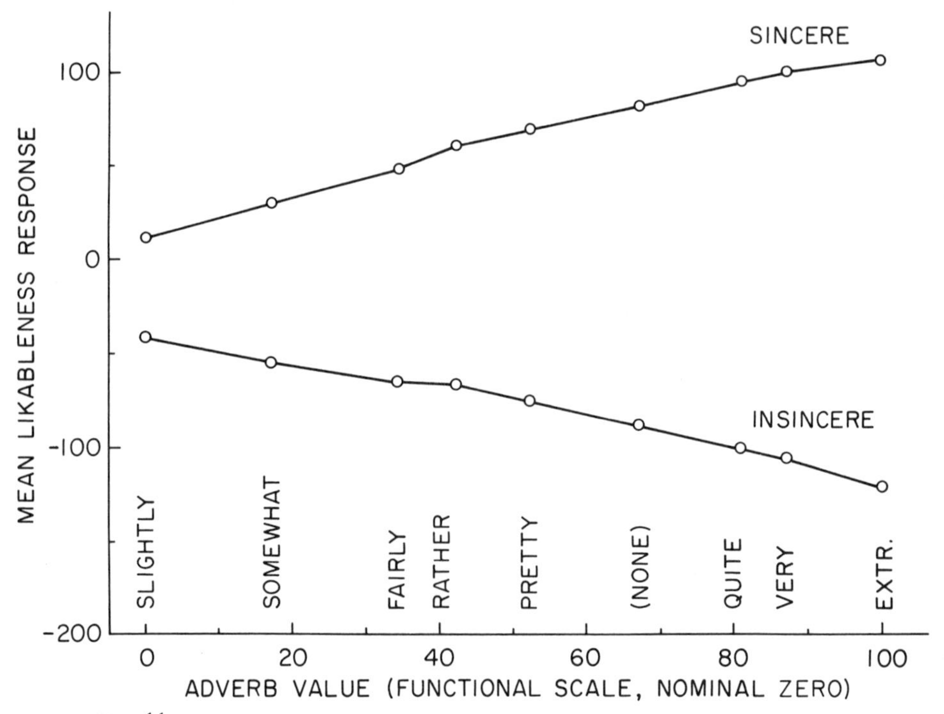

FIGURE 11.
Functional measurement analysis of multiplying model for adverb-adjective combinations. Adverbs spaced on horizontal axis according to their functional scale value. Multiplying model implies that the two curves should be two diverging straight lines. Data from unpublished experiment (Anderson, 1970).

found with other pairs of opposite adjectives in this experiment. It is tempting to attribute this discrepancy to mild nonlinearity in the response scale. However, the generally good behavior of these rating scales suggests that the discrepancy may be real.

Integration theory analysis. The multiplying model for adverb-adjective phrases has special interest since it illustrates the distinction between form and process in algebraic models. Also, it illustrates a valuation operation (Sec. 1.2). A closer analysis suggests that the subjects do not really multiply, that the model has only an 'as if' status, and indeed that it should hold only for a limited class of words. Since this issue has general interest for algebraic judgment theory, the conceptual analysis of adverb-adjective phrases is briefly summarized here.

Each adjective defines its own dimension of judgment. The adverbs are quantifiers, quite like numbers, and are represented as points on the adjective dimension. It is assumed that they have the same proportionate locations along different adjective dimensions. The adverb locations for two different adjectives should thus plot as linear functions of each other; if the adjectives had the same intensity range, this line would have unit slope.

Of course, the judgments of social desirability are not on the adjective dimension itself. A valuation operation is required, which is a function from the adjective dimension to the actual dimension of judgment.

The valuation function is defined by two reference points and a slope parameter. The two reference points are a zero point, and an optimum point analogous to Coombs' (1964) concept of an ideal. The valuation function is zero at the zero point, maximal at the ideal, and decreases linearly on either side of the ideal. The slope of this function is the relevance of the adjective dimension to the judgment dimension.

This analysis is conceptually different from Cliff's formulation, and leads to several opposed predictions. First, the multiplying model will not hold when the ideal is interior to the range since a sawtooth is then predicted. *Cautious* is an example; *slightly cautious* and *extremely cautious* would both be less desirable than *moderately cautious*. Second, the relative location of each adverbless adjective along its own dimension need not be a fixed constant, but is allowed to depend on linguistic usage. Third, the zero point need not have a fixed relative location. Thus *slightly* could be negative for some positive adjectives, and positive for others.

Of no less theoretical interest is that the multiplying model is only formal in the present framework. The adverbs do not multiply the adjectives in any psychological sense, but merely define a location on the adjective dimension. The psychological meaning of the multiplying idea is that the adverbs have proportionate locations across different adjective dimensions, and that the valuation function is linear. To the extent that the multiplying model appears to hold, it reflects a linguistic fact of quantification, and a linear valuation operation.

7.10. Information Purchasing

A particularly interesting application of functional measurement to a multilinear model was given by Shanteau and Anderson (1972) in an experiment on purchasing of information in a decision task. The subject's ostensible task was to guess the color of a bead drawn from a binomial urn of known composition, P, for a money prize, M. To aid their guess, they could pay for the roll of a die that would tell them the true color with probability D, otherwise nothing. The judgment studied was the judged worth of the die information.

The need for information is maximal when $P = 0.5$, and decreases as P approaches 0 or 1. The value of the information increases with M, and its diagnosticity is a direct function of D. By analogy to the rational expected value model, it was hypothesized that the behavior would also follow a three-factor multiplying model. This can be written formally as

$$R = MPD, \tag{21}$$

but where the symbols are now to be interpreted in terms of the subjective values of the stimuli. In particular, P represents need for information.

The group data followed the model very nicely as can be seen in Figure 12. Each panel shows the graphic bilinear form, and the panels as a group show the trilinear form. Some discrepancies, centered on the $P \times D$ interaction, did appear in the single-subject analyses. Apparently there was some problem in integrating these two pieces of information although none of the other interactions caused problems.

More interesting is that some subjects followed alternative models as detected in the detailed analysis. For example, the integration rule,

$$R = MP + MD = M(P + D), \tag{22}$$

was detected by the following pattern of interactions: significant bilinear components for the $M \times P$ and $M \times D$ interactions; nonsignificant bilinear component of the $P \times D$ interaction; nonsignificant trilinear component of the $M \times P \times D$ interaction; and nonsignificant residuals for all four interactions. This kind of deviation from the three-factor multiplying model will not necessarily show up in the graphical test of the group data.

It is a reasonable conjecture that the subjects who followed Equation 22 were reducing the task to a simpler cognitive form. That such cognitive simplification does occur has been suggested by numerous writers, but not much

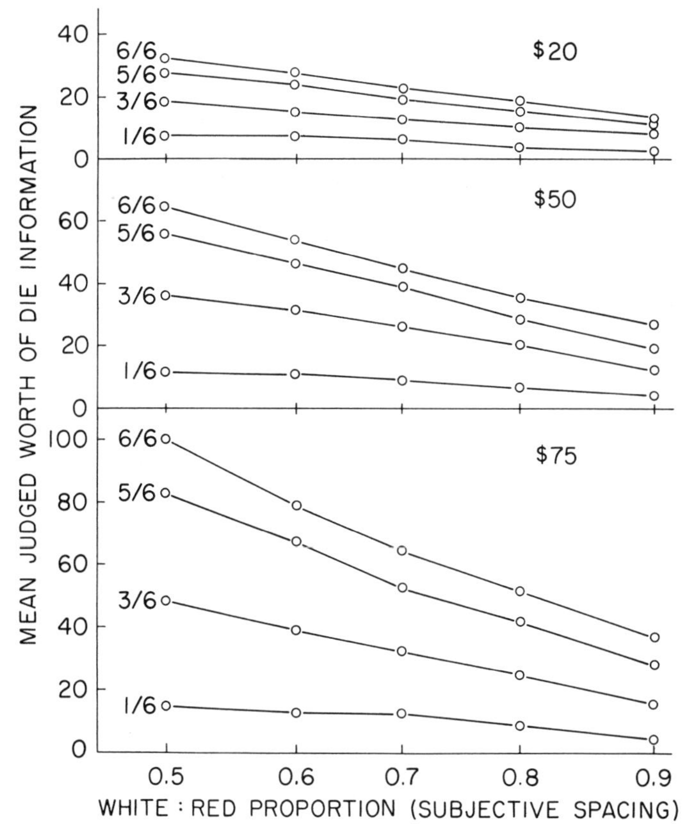

FIGURE 12.
Judged value of added information in risky decision making. The three-factor integration model implies that each panel should form a fan of diverging straight lines as a function of the subjective value of need for information plotted on the horizontal. From Shanteau and Anderson (1972). Copyright 1972 by the American Psychological Association. Reprinted by permission.

is known about the matter. The methods of integration theory may be useful for detecting such strategies.

The final phase of this experiment bears on the relative usefulness of the descriptive approach of integration theory compared with the normative Bayesian model. In this truth-or-lie condition, the roll of the die told the truth with probability D, otherwise a lie. Then, if $P = 0.9$, for example, a rational man would not pay anything for $D = 4/6$ or $5/6$ as that cannot increase his a priori certainty.

Figure 13 shows the data from this truth-or-lie condition. The data points

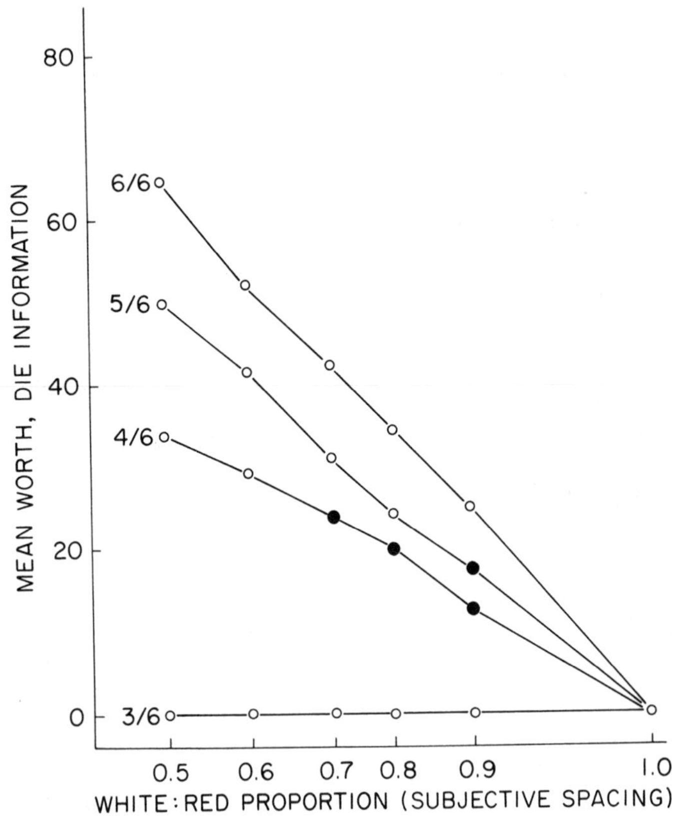

FIGURE 13.
Data from a subsidiary condition of the information purchase experiment of Figure 12. Integration theory implies the observed fanlike divergence. In contrast, a normative Bayesian model would imply that the four filled circles should be down at zero. From Shanteau and Anderson (1972). Copyright 1972 by the American Psychological Association. Reprinted by permission.

clearly follow the bilinear form, in agreement with integration theory. In contrast, the normative model implies that the four filled circles should lie on the horizontal curve of zero value.

7.11. Comparative Judgment

The judged magnitude of a stimulus will, in certain tasks, depend on both absolute and relative factors. Contextual stimuli in the neighborhood of the focal stimulus may have an effective yardstick function. If S denotes the stimulus to be judged, and Y_1 and Y_2 are two contextual stimuli, the simplest form of the model for comparative judgment would be

$$R = F(S) + F(S)/G_1(Y_1) + F(S)/G_2(Y_2), \tag{23}$$

where F, G_1, and G_2 are arbitrary functions (Anderson, 1970b).

Equation 23 is a multilinear model. If S, Y_1, and Y_2 can be independently varied in a factorial design, then Y_1 and Y_2 should follow the parallelism prediction, whereas S and Y_1 and S and Y_2 should both have the bilinear form. It may be emphasized that the analysis can be made directly on the raw response. Nothing need be known about the functions F, G_1, or G_2.

Support for this model can be seen in the data of Helson and Kozaki (1968) in which subjects judged numerosity of plates of dots in the context of a nonjudged anchor plate. The judgments for the four anchor conditions formed four diverging straight-line functions of actual numerosity. This bilinear form agrees with the present model of comparative judgment.

Surprisingly, although Helson and Kozaki considered their study to be an "experimentum crucis" supporting adaptation-level theory, their data actually appear to contradict their theory. In this experiment, they do not give the quantitative predictions, but presumably the "reformulated Fechner law" (Helson, 1964, Eq. 49) should be applicable. It reads

$$\begin{aligned} J_i &= C_0 + K \log (S_i/AL) \\ &= C_0 + K[\log S_i - \log AL], \end{aligned} \tag{24}$$

where J_i is the judged value of the physical stimulus S_i, AL is the adaptation level, and C_0 and K are constants.

Helson's equation implies parallelism. The four conditions can be represented as a two-way, stimulus $\times$ anchor design. The anchors affect only AL. The stimuli to be judged, the S_i, have a common effect on AL. Accordingly, Equation 24 can be treated as a simple subtracting model, and the actual responses should plot as parallel lines. The marked bilinearity visible in Figure 1 of Helson and Kozaki thus appears to contradict adaptation-level theory.

It should be emphasized that the test of Equation 24 is made directly on the raw response. Use of the physical values of S_i and AL is not only unnecessary, but generally inadvisable. The log transformation in Equation 24 reflects Helson's assumption about the relations between the subjective and physical stimulus values. Since that assumption is at best approximate, the present technique is preferable to the usual procedure in adaptation-level theory. It should be added that Helson often uses a different equation in fitting data (Helson, 1964, Eq. 41) though he does not comment on the inconsistency of these two equations.

Although adaptation-level theory has had widespread applications to categorical judgment, it compares unfavorably with Parducci's (1965) range-frequency theory (see Parducci & Perrett, 1971). Other applications, such as Restle's work on illusions (see Restle & Greeno, 1970), may still be valid. Restle works directly with the *AL* itself for which a multilinear model is postulated. Functional measurement procedure (Sec. 7.7) could also be used to advantage in that case also to obtain a more general and more rigorous analysis. Two relevant applications of integration theory to illusions are given by Massaro and Anderson (1971) and by Birnbaum and Veit (1973a).

7.12. Heteromodal Psychophysical Integration

Numerous investigators have tried to solve the measurement problem in psychophysics by using some kind of integration task, bisection, or loudness summation, for example (see Anderson, 1970a). These attempts have not been generally successful although recent studies by Fagot, Curtis, and their associates (see Curtis, 1970; Fagot & Stewart, 1969) have shown greater promise. The difficulties may reflect a basically incorrect integration model, but in most cases the issue remains unclear because previous methods of testing the model have not been entirely satisfactory.

A report by Feldman and Baird (1971) has double interest in this regard. On the positive side, it illustrates a potentially important integration task. On the negative side, it illustrates the weaknesses of traditional methods of analyzing integration models.

Feldman and Baird obtained judgments of 'total intensity' of a sound-light pair. Their second experiment employed a 5×5, sound $\times$ light design with 25 stimulus pairs. One hypothesis they wished to test was whether the 'total intensity' is an additive or linear function of the components. They actually tested the more restrictive model,

$$R = a(S_1^{n_1} + S_2^{n_2}) + b, \tag{25}$$

where R is the predicted response, S_1 and S_2 are the measured intensities of the physical stimuli, n_1 and n_2 are power law exponents, and a and b are constants.

Feldman and Baird followed the traditional approach in testing their model. They carefully measured values of S_1 and of S_2 on the physical scales; a power law was assumed so that $s_1 = S_1^{n_1}$, $s_2 = S_2^{n_2}$, where s and S denote subjective and objective values; n_1 and n_2 were estimated from judgments of the stimuli presented singly in the first experiment; a and b were estimated by least squares to maximize the fit between the predicted values of R and the actual judgments. This model gave a better graphical fit than an alternative model, but no test of goodness of fit was possible.

It is striking how much simpler the analysis is with functional measurement. Every step in Feldman and Baird's test can be bypassed, and the resultant test is more general and much more powerful.

All that is needed is to plot the raw responses as a two-way data table, in the manner of Figure 2, for example. The linear model predicts parallelism; it is as simple as that. No parameters need estimation, and the power law assumption is unnecessary.

It should be emphasized that this discussion is not meant to be critical of Feldman and Baird. Quite the contrary. Although their method of analysis was of the traditional type, they did employ a factorial design that is not too frequent in this type of work. Most important, if the heteromodal integration task does turn out to follow a simple model, it would be a very important tool in psychophysical measurement. [Since this was written, an experimental study of heteromodal integration (Anderson, 1974) has been completed with very promising results.]

7.13. Other Applications

Applications of information integration theory have been made to a variety of other situations. Some of these are cited briefly.

An important series of experiments by Kaplan (1971c, 1972) has studied the problem of response predisposition in person perception. Predisposition, which can be measured by an association test, is identified with the internal state variable, S_0. Theoretically, S_0 should be averaged in with the external stimuli, and Kaplan's data give substantial support to that conclusion. An especially interesting result was that subjects tended to discount inconsistent information in line with their predispositions. Kaplan's papers thus help to operationalize the concept of initial impression in person perception.

Several problems in person perception and attitude change have been studied by Himmelfarb. One set of experiments (Himmelfarb, 1973) shows that the 'resistance' to persuasion conferred by a neutral message reflects a simple increase in weight, not an active discounting process. Another paper (Himmelfarb, 1972), also on weighting, relates integration theory and attribution theory in person perception.

The interpretation of source credibility in terms of weight parameters has been studied by Rosenbaum and Levin (1968, 1969). This work is discussed by Kaplan and Anderson (1973) as it relates to interpersonal attraction.

Hendrick's work on the averaging hypothesis and on order effects has been cited in previous sections. He has also contributed to the study of inconsistency resolution in papers that show the importance of salience in the resolution process (Hendrick, 1972; Hendrick & Costantini, 1970). An interesting result of this work is that inconsistency in a person description seems to operate as a negative piece of information in its own right.

The same stimuli can be used in different tasks to which different models would apply. Thus, one task might call for the difference between two stimuli, another for their average. A basic question is whether the same stimulus scales will be obtained across the different tasks as in Anderson (1972b, 1974). Important work on this question has been done by Birnbaum (1972a, 1972b) for moral judgments and person perception, and by Birnbaum and Veit (1973a, 1973b) for psychophysical judgments.

Several interesting experiments on person perception, especially on the positive context effect of Section 4.3, have been reported by Wyer (e.g., Wyer & Dermer, 1968; Wyer & Watson, 1969) and by Takahashi (1971a, 1971b). These investigators have criticized the halo effect interpretation and have argued for a change-of-meaning formulation. However, many of their results can be readily interpreted within the present approach. For example, Takahashi's (1971a) manipulation of interrelatedness among the adjectives of a set would correspond to a manipulation of the w parameter of Equation 9. It should, therefore, act as a multiplier of the scale value of the component. These two variables should then obey the bilinear form of Section 7.7, and Takahashi's Figure 1 agrees roughly with that prediction.

An experiment by Butzin and Anderson (1973) studied information integration using toys as stimuli with children. The data supported the parallelism prediction, and a critical test (Sec. 3) ruled out adding in favor of averaging. These data suggest that children as young as five years may be able to use ratings as an interval response measure.

Judgments of groups have been studied by Leon, Oden, and Anderson (1973) and by Anderson, Lindner, and Lopes (1973). In this task, the group members constituted the pieces of information to be integrated to form a judgment of group attractiveness. The first study was primarily concerned with applying functional measurement to scale social offenses, and the group task was used mainly as a scaling tool. The second study was mainly concerned with group attractiveness in its own right and obtained results quite favorable to integration theory. Related work has been done by Lindner (1970, 1971) in an extensive experimental comparison between integration theory and various consistency theories.

Problems of information integration arise in every area of psychology, and they have been studied by innumerable investigators. This chapter is written around one theoretical formulation, and it is not possible to give adequate attention to the contributions made by others. However, the work of the Oregonians deserves special mention. This includes a wide range of work on judgment and decision-making by Slovic, Lichtenstein, Dawes, Goldberg, Hoffman, and others (see Slovic & Lichtenstein, 1971), and on perception with special reference to algebraic models (e.g., Curtis, Attneave, & Harrington, 1968; Curtis, 1970; Fagot & Stewart, 1969; Hyman & Well, 1968). This work is eclectic, not tied to any one theoretical formulation, and its collective value is impressive.

8. ASSOCIATED PROBLEMS

The work summarized in this chapter has given a broad and solid base for information integration theory. The theory gives a good account of the data, both conceptually and in exact, quantitative form. It provides a unified treatment of diverse areas of psychology that ordinarily have had little interrelation. It may, therefore, be the beginning of a unified general theory of behavior.

Many problems still remain that have not been adequately treated. Some of these are noted in Slovic and Lichtenstein (1971, see especially Table 2, p. 676) and in Rapoport and Wallsten (1972). This section touches on a number of these problems, more to indicate their existence than to suggest solutions. It is worth noting that all of these problems have to do with valuation.

8.1. Valuation

The problem of value is as fundamental as the problem of integration, but it has been relatively neglected. Most of the work has been primarily directed at establishing integration models. Scaling and the study of the stimulus parameters in their own right have taken second place.

Valuation and integration are inseparable, and it deserves emphasis that the integration models do contribute to the study of valuation. They contribute to the definition of structure, as well as to measurement. For example, the weight-scale value distinction, which is essential to the averaging models (Sec. 1.2), represents a structural property of the stimulus. This conceptualization is no less important than the methods for scaling the two parameters.

The problem of value goes beyond mere measurement to a deeper conceptual analysis of the valuation operations. One aspect of this issue, already discussed, is that the stimulus parameters will depend on the situation. For example, stubbornness has its virtues in the laboratory, but not in the classroom. More general situational determinants of value are discussed in Sections 8.3 and 8.4.

Another aspect of valuation is that stimulus value will often be derivative from some chain of operations instead of given directly. In person perception, the likableness values of a descriptive adjective, such as *tactful*, have probably been learned and so act as immediate givens. But in judgments of occupational proficiency, of *tactful nurse*, for example, determination of the scale value and weight or relevance of the same adjective may require a chain of thought. A computer analogy may accentuate this distinction. Values of some mathematical function may be obtained directly by lookup in a stored table, or by a subroutine that evaluates the function as needed. As is well known, it is often more efficient to employ the subroutine even though each single evaluation may require more computation than the table lookup.

In this analogy, the valuation chains correspond to the subroutines. Little is known about them. One example is given in the discussion of adverb-adjective phrases of Section 7.9. A similar, as yet unsolved problem, is given by adjective-noun combinations. For example, it is known that *irresponsible mother* is more negative than *irresponsible person*, and similarly that *responsible mother* is more positive than *responsible person*. However, a clear conceptual analysis of such phrases is still lacking.

One class of valuation processes is sufficiently common to deserve specific comment. To illustrate, suppose a person is to evaluate and choose among several possible courses of action. This requires envisaging for each action some set of consequences. Each consequence will have its own scale value and weight, the latter reflecting both the importance and the subjective probability of the consequence. The overall value of each action will then be the weighted average of its envisaged consequences.

In practice, of course, such processing is usually carried out only till a clear decision is reached. The more important consequences present themselves initially and, only if these leave a close choice, are the lesser consequences searched for. In practice, most decisions are routine and hardly noticed. The close decisions become the center of attention, and their difficulty arises in part from the largely subjective, conjectural nature of the envisaged consequences.

The study of such valuation processes can get complicated, primarily because the consequences are not controlled experimentally, but instead must be developed by the person. Certain consequences may enter into the valuation without the person's being able to make them explicit. Those that he can make explicit may be difficult to measure, especially in terms of redundancy. Nevertheless, such valuation processes are extremely important in thought and judgment and deserve more attention.

Finally, it should be kept in mind that the present models have been one dimensional and may not be adequate for all purposes (Anderson, 1971a). Even with number averaging, the response might be better conceptualized as confidence distribution than as a single point. In person perceptions, several dimensions are involved (Edwards & Ostrom, 1971; Rosenberg, Nelson, &

Vivekananthan, 1968). The present approach could be directly generalized to a vector model, of course, but even that might not be a completely adequate representation. As yet, there are only scattered studies on the detailed structure of the stimuli, and of the response.

8.2. The Psychophysical Law

The so-called psychophysical law is the function relating the subjective and objective values of light, sound, and other physical stimuli. In present notation, it is just the function relating s and S. The main problem is to measure the subjective values, s, since the physical values, S, present no basic difficulty. This problem, which was raised by Fechner in 1860, has remained a source of controversy to this day (Sec. 1.5).

The psychophysical law is thus a problem in valuation. The functional measurement approach is simple. Any of several possible integration models (e.g., Secs. 7.3, 7.4, 7.5, and 7.12) could provide functional scales of s. Thus, the psychophysical law is a by-product of the psychological law for the integration task. This idea is not novel, and it has been employed by numerous investigators of psychophysical judgment (see Anderson, 1970a, 1973a). The results have not, however, commanded general agreement. Present methods may be more effective and more widely applicable (e.g., see Sec. 7.12).[3]

As yet, it is true, the functional measurement approach has not been tested adequately in psychophysics. Its applicability depends on establishing a substantive, psychological law, and though results to date have been quite promising, some exceptions have appeared (see Anderson, 1970a, 1973a). Moreover, much of this work has been primarily concerned with the integration process and so has used only a few stimulus values.

It should also be noted that various investigators (e.g., Parducci, 1965; Poulton, 1968; Treisman, 1964) have argued that the s scale depends specifically on the judgmental task so that it is not meaningful to speak of 'the' psychophysical law. From the present view, this argument is largely an empirical question on which there is not much evidence. However, two recent reports (Anderson, 1972b; Birnbaum & Veit, 1973b) suggest that the same scales function across different tasks.

---

[3] An empirical relation between s and S does not, of course, lead to a unique mathematical function. With a finite set of somewhat unreliable data points, any number of different mathematical functions could provide an adequate fit. Further, the scale for s may not have a known zero and that might become important in the mathematical function, much as in the physical laws involving temperature.

8.3. Learning and Motivation

Learning involves integration processes at two different levels. One is serial integration (Sec. 6.4), in which the information on successive trials is integrated to yield continually revised weight and value parameters of each stimulus cue. What is learned is knowledge, in particular, the stimulus parameters, w and s. These correspond to subjective validity and expectance, respectively. At this level, the integration is effectively a valuational process.

The second level of integration is concerned with the response at any trial. Here the traditional learning-performance distinction parallels the present valuation-integration distinction. Learning itself is valuational as already noted. Performance depends on integration of prevailing information and involves decision-making processes. Inclusion of cost-reward matrices is natural in the present approach, and illustrates its flexibility relative to traditional conditioning-reinforcement views.

The informational and expectancy basis of the present approach is similar to that of Tolman. Integration learning theory has advantages in precision and testability (see Hilgard, 1956, p. 208) though as yet it is relatively undeveloped. Indeed, Slovic and Lichtenstein (1971, Table 2, p. 676) list learning as an unexplored area within integration theory.

A few studies have been done in probability learning. A basic paper by Friedman, Carterette, and Anderson (1968) showed that the information integration model was markedly superior to models based on stimulus sampling theory (see also Anderson, 1966b). Related work has been done by Anderson (1969b) for continuous response tasks, and by Levin et al. (1972) who used several concurrent prediction tasks. In these applications, the reinforcing stimuli are reinterpreted as informational stimuli, or informers.

Cue compounding problems offer an ideal field of application for integration models. In such problems, the cue parameters, weight and value, are established in preliminary training. Data are then obtained from test trials with various combinations or compounds of cues (e.g., Anderson, 1969c). The subject's task, of course, is to integrate the individual cues into a single response. Theoretically, the response to any compound should be a weighted average of the cues in that compound. Work by Himmelfarb (1964, 1970) and by Friedman, Rollins, and Padilla (1968) and Robbins and Medin (1971) has given some support to weighted-average type formulations.

One difficulty in applying the averaging model to cue compounding is that weights tend to covary with scale values, bringing in problems of differential weighting (Sec. 5) that complicate parameter estimation. These problems could be considerably lessened by using tasks in which both cues and response are numerical rather than discrete.

Considerable work on learning tasks with numerical cues has been done in the Brunswikian framework (e.g., Azuma & Cronbach, 1966; Brehmer, 1969; Hammond & Summers, 1965; see also Slovic & Lichtenstein, 1971), a framework in which cue-integration paradigms have basic importance. Virtually all of this work has employed multiple regression analysis, however, and is difficult to relate to the present formulation. Functional measurement procedure may have some advantages in such tasks since, in contrast to the usual regression analyses, it allows for subjective cue values and for differential weighting within a given cue dimension (Anderson, 1972a).

A brief discussion of drive and motivation within integration learning theory is given by Kaplan and Anderson (1973). These problems are largely unexplored. Drive and motivation are valuational processes, in that their effect is on the w and s parameters. Changes in drive will evidently affect w, but whether they will also affect s is not entirely certain. A choice between bread and water will depend on need state; that their taste will covary is plausible but not definite.

Need state, it should be noted, will particularly affect the internal state variable, S_0. In certain tasks, that variable will be averaged with other prevailing stimuli to determine the response. Kaplan and Anderson (1973) note two or three reports that have a simple interpretation on this basis.

In Hull-Spence learning theory, a debated question is whether drive and incentive combine by adding or multiplying. The symbols E, H, D, and K are used to denote response strength, habit strength, drive, and incentive motivation, respectively. In Hull's formulation,

$$E = HDK,$$

whereas in Spence's formulation,

$$E = H(D + K).$$

Both models are multilinear models and can be handled by the methods noted in Section 7.7. Thus, both models imply that the $H \times D$ and $H \times K$ interactions are concentrated in their bilinear component, with zero residual. The models differ in two predictions. Spence's model predicts no $D \times K$ interaction and no $H \times D \times K$ interaction; Hull's model predicts that the former should be concentrated in its bilinear component, and that the latter should be concentrated in its trilinear component.

The experimental studies of this question have not been conclusive, mainly because a validated interval scale for response measurement has been lacking. Typical experiments have studied rats in a straight runway, or humans in eyelid conditioning, and the customary response measures in such studies may be only ordinal. But then an interaction term in the analysis of variance may be psychologically meaningless (Anderson, 1961, Fig. 1; 1962b).

A method for solving the response scaling problem was given by Anderson

(1962b) in connection with Lewin-Miller conflict theory, especially as interpreted within the Hull-Spence approach. The leverage for response scaling is an algebraic model that specifies how the underlying quantities are combined or integrated to produce the response (Anderson, 1962b, p. 412). Possibilities for scaling stimulus variables such as drive were also noted. The key idea is the use of a monotone transformation within the constraints of a factorial design. Bogartz and Wackwitz (1971) give a systematic statistical development of this approach, and a general computer routine has been developed by Weiss (1973).

Functional measurement procedure may also be useful in operant work. The near parallelism of Guttman's (1954) sucrose and glucose curves raises the attractive possibilities both that bar-press rate may yield an interval response measure, and that incentives may combine by adding or averaging (see also Young & Christensen, 1962; Christensen, 1962). Incentive, and presumably drive as well, could thus be scaled by using an incentive × incentive or a drive × incentive design. The latter, which presumably would follow a multiplying model, would have special interest as it would provide a basis for comparing drives induced by different operations.

Operant work with two concurrent schedules has indicated that birds distribute their time on the two keys in proportion to the relative reinforcement (e.g., Baum & Rachlin, 1969; Herrnstein, 1970; Staddon, 1968). If T_1 and T_2 are the total times spent in the two keys, and r_1 and r_2 are the two respective rates of reinforcement, the results appear to be consistent with the model,

$$\frac{T_1}{T_2} = \frac{r_1}{r_2}. \tag{26}$$

As it stands, this 'matching law' is conceptually simple to test because all four quantities are observables.

However, Baum and Rachlin noted that other factors such as amount and immediacy of reinforcement would also affect the time distribution. They suggested that time is distributed over the two keys in proportion to their 'values', so that

$$\frac{T_1}{T_2} = \frac{s_1}{s_2}. \tag{27}$$

Although Equations 26 and 27 look alike, they are quite different because, in general, the values s_1 and s_2 will be unknown. It is true that physical measures of amount and immediacy of reinforcement could be used, but they do not seem overly plausible, and failure of the model might then simply reflect an invalid stimulus scale. Furthermore, other reinforcement variables, such as type of grain, have no simple physical measure.

Functional measurement procedure allows a simple treatment of the general matching law. Equation 27 is a simple dividing model to which the methods of Section 7.7 apply. The experimental requirement is to vary the values in a $\text{key}_1 \times \text{key}_2$ factorial design. The observed response measure, (T_1/T_2), should then follow the bilinear form. Further, if the model passes the test, then functional scales of the stimulus values can be simply obtained for the single organism.

8.4. Stimulus Interaction

Stimulus interaction is one of the most important areas in integration theory. Most of this chapter has rested on an assumption of independence, of no interaction among the stimuli in a given combination. This assumption is justified in many cases, as has been seen, but the exceptions are numerous and of great interest. Some aspects of this problem have been touched on in Section 4, and in two other reports (Anderson, 1971a, pp. 198–201; 1972a). A few comments are made here to illustrate some facets of a complex and ramified problem area.

Perhaps the most attractive hypothesis about stimulus interaction is that it changes the scale value. Contrast and assimilation are familiar examples, and such effects are certainly found in some perceptual tasks. Even then, they may still be interpretable in terms of an algebraic judgment model, as with the size-weight illusion and the Ebbinghaus illusion (Sec. 7.5).

For verbal and symbolic stimuli, on the other hand, there is little solid evidence for assimilation or contrast. The positive context effect and the primacy effect (Secs. 4.3 and 4.4) may possibly represent true assimilation, but that seems pretty doubtful on current evidence. Contrast and assimilation were once widely used as explanatory notions in social psychology. Now it is generally recognized that much of what has looked like contrast is merely a shift in response scale usage rather than perceptual change in stimulus value. Despite very considerable efforts, evidence for true contrast is meager (see Anderson, 1971a; Upshaw, 1969).

Stimulus interaction does affect the weight parameter, causing certain stimuli to be discounted from their natural importance. Redundancy discounting is an obvious example. Completely redundant information should carry zero weight, though its scale value would be unaffected. Partially redundant information would be handled by partial discounting. Analyses of redundancy within integration theory have been given by Schmidt (1969) and Kaplan (1971b).

Inconsistency among stimuli is one of the most interesting causes of interaction, but not much is known about how subjects do handle inconsistent information. Discounting seems to be primarily involved (Anderson & Jacobson, 1965; Schümer & Cohen, 1968) rather than change in scale value, though the latter process may play a part in certain judgment tasks (Sidowski &

Anderson, 1967). The 'instrumentalization' and 'elimination' strategies discussed by Pepitone and Hayden (1955) would also represent discounting. In addition, some evidence suggests that inconsistency of itself may constitute an effective negative stimulus (Hendrick, 1972), or evoke an additional dimension of judgment (Cohen & Schümer, 1968), even though the integration otherwise follows a simple averaging model. An interesting application of an expectancy-discrepancy model to attribution theory has been made by Lopes (1972).

Several theories in social psychology have attempted to base themselves on a consistency principle. Most of them suffer from a nominal definition of inconsistency which, though rational in an armchair sense, is not psychologically valid. In practice, people seem to be much less troubled by inconsistency than might have been expected (Anderson & Jacobson, 1965; Hendrick & Costantini, 1970), and they average the information as given without changing the value or weight of the stimuli. Unfortunately, much of the work on the consistency problem has assumed that the subjects were resolving inconsistency without recognizing that that assumption stands in need of empirical proof. This topic is discussed further in Anderson (1968a, 1971a, pp. 185ff), and an important experimental investigation has been given by Lindner (1971).

A somewhat different form of stimulus interaction can be termed cognitive simplification. Various investigators have suggested that subjects will simplify a complex task by ignoring some of the elements (see Shepard, 1964). That can be considered as discounting, of an attentional kind, since ignoring a stimulus element corresponds to giving it zero weight. Two experimental applications of integration theory have been made in decision-making (Shanteau & Anderson, 1972) and moral judgment (Leon, Oden, & Anderson, 1973).

It is also worth noting that averaging is itself a form of stimulus interaction. The effective weights are the relative weights, and they depend on the weights of all the stimuli in the combination. This reflects the gestalt character of the averaging rule, in which the role of each part depends on the whole.

Stimulus interaction introduces a difficult methodological problem. When the weight and value parameters of a given stimulus shift from one stimulus combination to another, the simple predictions used for testing the noninteractive models no longer apply. Present methods may still be useful in two ways, however. With careful design, deviations from prediction can be used to demonstrate the existence of interaction. For example, two interacting stimuli may be included in an otherwise noninteractive design. The parallelism in the noninteractive part of the design then serves as a baseline against which to interpret the predicted discrepancy. As another possibility, interactive stimulus sets may be included as units in a noninteractive design. Thus, subjects might be asked to form impressions of groups whose members are described by interactive sets of personality adjectives. If the group judgment is itself noninteractive, functional measurement may then be applied to scale the group members. This last design exemplifies a two-stage integration model discussed in more detail elsewhere (Anderson, 1973a).

The most attractive approach to the analysis of stimulus interaction is the method of component judgments. Subjects simply rate the value and importance of each stimulus in their overall impression. If this could be made to work, it would have the greatest importance. Unfortunately, present evidence is none too hopeful. Component value judgments seem to be subject to severe context effects (Sec. 4.3). Present evidence on weight judgments is mixed (Anderson, 1972a) and further work is most desirable.

ACKNOWLEDGMENTS

This work was supported by National Science Foundation Grant GB-21028, and facilitated by NIMH grants to the Center for Human Information Processing, University of California, San Diego. I wish to thank Robyn Dawes for a detailed review of a previous draft. I am also indebted for various comments to Michael Birnbaum, Reid Hastie, Clyde Hendrick, Martin Kaplan, Irwin Levin, Dominic Massaro, Burton Rodin, James Shanteau, Raymond Smith, and Gordon Stanley.

REFERENCES

Anderson, N. H. Test of a model for opinion change. *Journal of Abnormal and Social Psychology*, 1959, **59,** 371–381. (a)

Anderson, N. H. An analysis of sequential dependencies. In R. R. Bush and W. K. Estes (Eds.), *Studies in mathematical learning theory*. Stanford: Stanford University Press, 1959. (b)

Anderson, N. H. Effect of first-order conditional probability in a two-choice learning situation. *Journal of Experimental Psychology*, 1960, **59,** 73–93.

Anderson, N. H. Scales and statistics: Parametric and nonparametric. *Psychological Bulletin*, 1961, **58,** 305–316.

Anderson, N. H. Application of an additive model to impression formation. *Science*, 1962, **138,** 817–818. (a)

Anderson, N. H. On the quantification of Miller's conflict theory. *Psychological Review*, 1962, **69,** 400–414. (b)

Anderson, N. H. Test of a model for number-averaging behavior. *Psychonomic Science*, 1964, **1,** 191–192. (a)

Anderson, N. H. Linear models for responses measured on a continuous scale. *Journal of Mathematical Psychology*, 1964, **1,** 121–142. (b)

Anderson, N. H. An evaluation of stimulus sampling theory: Comments on Prof. Estes' paper. In A. W. Melton (Ed.), *Categories of human learning*. New York: Academic Press, 1964. (c)

Anderson, N. H. Averaging versus adding as a stimulus-combination rule in impression formation. *Journal of Experimental Psychology*, 1965, **70,** 394–400. (a)
Anderson, N. H. Primacy effects in personality impression formation using a generalized order effect paradigm. *Journal of Personality and Social Psychology*, 1965, **2,** 1–9. (b)
Anderson, N. H. Component ratings in impression formation. *Psychonomic Science*, 1966, **6,** 279–280. (a)
Anderson, N. H. Test of a prediction of stimulus sampling theory in probability learning. *Journal of Experimental Psychology*, 1966, **71,** 499–510. (b)
Anderson, N. H. Averaging model analysis of set size effect in impression formation. *Journal of Experimental Psychology*, 1967, **75,** 158–165. (a)
Anderson, N. H. Application of a weighted average model to a psychophysical averaging task. *Psychonomic Science*, 1967, **8,** 227–228. (b)
Anderson, N. H. A simple model for information integration. In R. P. Abelson, E. Aronson, W. J. McGuire, T. M. Newcomb, M. J. Rosenberg, and P. H. Tannenbaum (Eds.), *Theories of cognitive consistency: A sourcebook*. Chicago: Rand McNally, 1968. (a)
Anderson, N. H. Likableness ratings of 555 personality-trait words. *Journal of Personality and Social Psychology*, 1968, **9,** 272–279. (b)
Anderson, N. H. Application of a linear-serial model to a personality-impression task using serial presentation. *Journal of Personality and Social Psychology*, 1968, **10,** 354–362. (c)
Anderson, N. H. Averaging of space and number stimuli with simultaneous presentation. *Journal of Experimental Psychology*, 1968, **77,** 383–392. (d)
Anderson, N. H. Comment on "An analysis-of-variance model for the assessment of configural cue utilization in clinical judgment." *Psychological Bulletin*, 1969, **72,** 63–65. (a)
Anderson, N. H. Application of a model for numerical response to a probability learning situation. *Journal of Experimental Psychology*, 1969, **80,** 19–27. (b)
Anderson, N. H. Effects of choice and verbal feedback on preference values. *Journal of Experimental Psychology*, 1969, **79,** 77–84. (c)
Anderson, N. H. Functional measurement and psychophysical judgment. *Psychological Review*, 1970, **77,** 153–170. (a)
Anderson, N. H. Averaging model applied to the size-weight illusion. *Perception and Psychophysics*, 1970, **8,** 1–4. (b)
Anderson, N. H. Integration theory and attitude change. *Psychological Review*, 1971, **78,** 171–206. (a)
Anderson, N. H. Two more tests against change of meaning in adjective combinations. *Journal of Verbal Learning and Verbal Behavior*, 1971, **10,** 75–85. (b)
Anderson, N. H. Looking for configurality in clinical judgment. *Psychological Bulletin*, 1972, **78,** 93–102. (a)
Anderson, N. H. Cross-task validation of functional measurement. *Perception and Psychophysics*, 1972, **12,** 389–395. (b)
Anderson, N. H. Algebraic models in perception. In E. C. Carterette and M. P. Friedman (Eds.), *Handbook of perception*. New York: Academic Press, 1973. (a)
Anderson, N. H. Cognition algebra: Integration theory applied to social attribu-

tion. In L. Berkowitz (Ed.), *Advances in experimental social psychology*. Vol. 7. New York: Academic Press, 1973. (b)

Anderson, N. H. Serial position curves in impression formation. *Journal of Experimental Psychology*, 1973, **97,** 8–12. (c)

Anderson, N. H. Integration theory applied to attitudes about U.S. presidents. *Journal of Educational Psychology*, 1973, **64,** 1–8. (d)

Anderson, N. H. Functional measurement of social desirability. *Sociometry*, 1973, **36,** 89–98. (e)

Anderson, N. H. Cross-task validation of functional measurement using judgments of total magnitude. *Journal of Experimental Psychology*, 1974, in press.

Anderson, N. H., & Alexander, G. R. Choice test of the averaging hypothesis for information integration. *Cognitive Psychology*, 1971, **2,** 313–324.

Anderson, N. H., & Barrios, A. A. Primacy effects in personality impression formation. *Journal of Abnormal and Social Psychology*, 1961, **63,** 346–350.

Anderson, N. H., & Farkas, A. J. New light on order effects in attitude change. *Journal of Personality and Social Psychology*, 1973, **28,** 88–93.

Anderson, N. H., & Hubert, S. Effects of concomitant verbal recall on order effects in personality impression formation. *Journal of Verbal Learning and Verbal Behavior*, 1963, **2,** 379–391.

Anderson, N. H., & Jacobson, A. Effect of stimulus inconsistency and discounting instructions in personality impression formation. *Journal of Personality and Social Psychology*, 1965, **2,** 531–539.

Anderson, N. H., & Jacobson, A. Further data on a weighted average model for judgment in a lifted weight task. *Perception and Psychophysics*, 1968, **4,** 81–84.

Anderson, N. H., & Lampel, A. K. Effect of context on ratings of personality traits. *Psychonomic Science*, 1965, **3,** 433–434.

Anderson, N. H., Lindner, R., & Lopes, L. L. Integration theory applied to judgments of groups. *Journal of Personality and Social Psychology*, 1973, **26,** 400–408.

Anderson, N. H., & Norman, A. Order effects in impression formation in four classes of stimuli. *Journal of Abnormal and Social Psychology*, 1964, **69,** 467–471.

Anderson, N. H., Sawyers, B. K., & Farkas, A. J. President paragraphs. *Behavior Research Methods and Instrumentation*, 1972, **4,** 177–192.

Anderson, N. H., & Shanteau, J. C. Information integration in risky decision making. *Journal of Experimental Psychology*, 1970, **84,** 441–451.

Anderson, N. H., & Weiss, D. J. Test of a multiplying model for estimated area of rectangles. *American Journal of Psychology*, 1971, **84,** 543–548.

Asch, S. E. Forming impressions of personality. *Journal of Abnormal and Social Psychology*, 1946, **41,** 258–290.

Atkinson, R. C., & Estes, W. K. Stimulus sampling theory. In R. D. Luce, R. R. Bush, and E. Galanter (Eds.), *Handbook of mathematical psychology*. Vol. 2. New York: Wiley, 1963.

Atkinson, R. C., & Wickens, T. D. *Human memory and the concept of reinforcement*. Technical Report No. 145. Stanford, Calif.: Institute for Mathematical Studies in the Social Sciences, Stanford University, 1969.

Azuma, H., & Cronbach, L. J. Cue-response correlations in the attainment of a scalar concept. *American Journal of Psychology*, 1966, **79,** 38–49.

Baum, W. M., & Rachlin, H. C. Choice as time allocation. *Journal of the Experimental Analysis of Behavior*, 1969, **12,** 861–874.

Beck, J., & Shaw, W. A. Ratio-estimations of loudness-intervals. *American Journal of Psychology*, 1967, **80,** 59–65.

Birnbaum, M. H. The nonadditivity of impressions. Unpublished doctoral dissertation, University of California, Los Angeles, 1972. (a)

Birnbaum, M. H. Morality judgments: Tests of an averaging model. *Journal of Experimental Psychology*, 1972, **93,** 35–42. (b)

Birnbaum, M. H. The Devil rides again: Correlation as an index of fit. *Psychological Bulletin*, 1973, **79,** 239–242.

Birnbaum, M. H., Parducci, A., & Gifford, R. K. Contextual effects in information integration. *Journal of Experimental Psychology*, 1971, **88,** 158–170.

Birnbaum, M. H., & Veit, C. T. Judgmental illusion produced by contrast with expectancy. *Perception and Psychophysics*, 1973, **13,** 149–152. (a)

Birnbaum, M. H., & Veit, C. T. Psychophysical measurement: Information integration with difference, ratio, and averaging tasks. *Perception and Psychophysics*, 1973, in press. (b)

Bock, R. D., & Jones, L. V. *The measurement and prediction of judgment and choice.* San Francisco: Holden-Day, 1968.

Bogartz, R. S., & Wackwitz, J. H. Polynomial response scaling and functional measurement. *Journal of Mathematical Psychology*, 1971, **8,** 418–443.

Brehmer, B. Cognitive dependence on additive and configural cue-criterion relations. *American Journal of Psychology*, 1969, **82,** 490–503.

Brunswik, E. *Perception and the representative design of psychological experiments.* Berkeley: University of California Press, 1956.

Butzin, C. A., & Anderson, N. H. Functional measurement of children's judgments. *Child Development*, 1973, **44,** 529–537.

Chandler, J. P. STEPIT—Finds local minima of a smooth function of several parameters. *Behavioral Science*, 1969, **14,** 81–82.

Christensen, K. R. Isohedonic contours in the sucrose-sodium chloride area of gustatory stimulation. *Journal of Comparative and Physiological Psychology*, 1962, **55,** 337–341.

Cliff, N. Adverbs as multipliers. *Psychological Review*, 1959, **66,** 27–44.

Cohen, R., & Schümer, R. Eine Untersuchung zur sozialen Urteilsbildung. I. Die Verarbeitung von Informationen unterschiedlicher Konsonanz. *Archiv für die gesamte Psychologie*, 1968, **120,** 151–179.

Coombs, C. H. *A theory of data.* New York: Wiley, 1964.

Curtis, D. W. Magnitude estimations and category judgments of brightness and brightness intervals. *Journal of Experimental Psychology*, 1970, **83,** 201–208.

Curtis, D. W., Attneave, F., & Harrington, T. L. A test of a two-stage model of magnitude estimation. *Perception and Psychophysics*, 1968, **3,** 25–31.

Cusumano, D. R., & Richey, M. H. Negative salience in impressions of character: Effects of extremeness of stimulus information. *Psychonomic Science*, 1970, **20,** 81–83.

Davidson, D., Suppes, P., & Siegel, S. *Decision-making: An experimental approach.* Stanford: Stanford University Press, 1957.

Dawson, W. E. Magnitude estimation of apparent sums and differences. *Perception and Psychophysics*, 1971, **9,** 368–374.
Edwards, J. D., & Ostrom, T. M. Cognitive structure of neutral attitudes. *Journal of Experimental Social Psychology*, 1971, **7,** 36–47.
Edwards, W. Conservatism in human information processing. In B. Kleinmuntz (Ed.), *Formal representation of human judgment*. New York: Wiley, 1968.
Estes, W. K. Probability learning. In A. W. Melton (Ed.), *Categories of human learning*. New York: Academic Press, 1964.
Fagot, R. F., & Stewart, M. Tests of product and additive scaling axioms. *Perception and Psychophysics*, 1969, **5,** 117–123.
Feldman, J., & Baird, J. C. Magnitude estimation of multidimensional stimuli. *Perception and Psychophysics*, 1971, **10,** 418–422.
Friedman, M. P., Carterette, E. C., & Anderson, N. H. Long-term probability learning with a random schedule of reinforcement. *Journal of Experimental Psychology*, 1968, **78,** 442–455.
Friedman, M. P., Rollins, H., & Padilla, G. The role of cue validity in stimulus compounding. *Journal of Mathematical Psychology*, 1968, **5,** 300–310.
Garner, W. R. *Uncertainty and structure as psychological concepts*. New York: Wiley, 1962.
Garner, W. R., & Creelman, C. D. Problems and methods of psychological scaling. In H. Helson and W. Bevan (Eds.), *Contemporary approaches to psychology*. Princeton, N.J.: Van Nostrand, 1967.
Garner, W. R., & Morton, J. Perceptual independence: Definitions, models, and experimental paradigms. *Psychological Bulletin*, 1969, **72,** 233–259.
Guttman, N. Equal-reinforcement values for sucrose and glucose solutions compared with equal-sweetness values. *Journal of Comparative and Physiological Psychology*, 1954, **47,** 358–361.
Hamilton, D. L., & Huffman, L. J. Generality of impression-formation processes for evaluative and nonevaluative judgments. *Journal of Personality and Social Psychology*, 1971, **20,** 200–207.
Hammond, K. R. (Ed.). *The psychology of Egon Brunswik*. New York: Holt, Rinehart & Winston, 1966.
Hammond, K. R., & Summers, D. A. Cognitive dependence on linear and nonlinear cues. *Psychological Review*, 1965, **72,** 215–224.
Helson, H. *Adaptation-level theory*. New York: Harper & Row, 1964.
Helson, H., & Kozaki, A. Anchor effects using numerical estimates of simple dot patterns. *Perception and Psychophysics*, 1968, **4,** 163–164.
Hendrick, C. Averaging vs. summation in impression formation. *Perceptual and Motor Skills*, 1968, **27,** 1295–1302.
Hendrick, C. Effects of salience of stimulus inconsistency on impression formation. *Journal of Personality and Social Psychology*, 1972, **22,** 219–222.
Hendrick, C., & Costantini, A. F. Effects of varying trait inconsistency and response requirements on the primacy effect in impression formation. *Journal of Personality and Social Psychology*, 1970, **15,** 158–164.
Herrnstein, R. J. On the law of effect. *Journai of the Experimental Analysis of Behavior*, 1970, **13,** 243–266.

Hewitt, J. Integration of information about others. *Psychological Reports*, 1972, **30,** 1007–1010.

Hilgard, E. R. *Theories of learning*. (2nd ed.). New York: Appleton-Century-Crofts, 1956.

Himmelfarb, S. The utilization and combination of cues in the perception of persons. Unpublished doctoral dissertation, University of California, Los Angeles, 1964.

Himmelfarb, S. Effects of cue validity differences in weighting information. *Journal of Mathematical Psychology*, 1970, **7,** 531–539.

Himmelfarb, S. Integration and attribution theories in personality impression formation. *Journal of Personality and Social Psychology*, 1972, **23,** 309–313.

Himmelfarb, S. "Resistance" to persuasion induced by information integration. In S. Himmelfarb and A. H. Eagly (Eds.), *Readings in attitude change*. New York: Wiley, 1973.

Hodges, B. H. The effect of valence on relative weighting in impression formation. *Journal of Personality and Social Psychology*, 1974, in press.

Howe, E. S. Verb tense, negatives, and other determinants of the intensity of evaluative meaning. *Journal of Verbal Learning and Verbal Behavior*, 1966, **5,** 147–155.

Hyman, R., & Well, A. Perceptual separability and spatial models. *Perception and Psychophysics*, 1968, **3,** 161–165.

Insko, C. A. *Theories of attitude change*. New York: Appleton-Century-Crofts, 1967.

Kahneman, D., & Tversky, A. Subjective probability: A judgment of representativeness. *Cognitive Psychology*, 1972, **3,** 430–454.

Kanouse, D. E., & Hanson, L. R. *Negativity in evaluations*. New York: General Learning Press, 1972.

Kaplan, M. F. Context effects in impression formation: The weighted average versus the meaning change formulation. *Journal of Personality and Social Psychology*, 1971, **19,** 92–99. (a)

Kaplan, M. F. The determination of trait redundancy in personality impression formation. *Psychonomic Science*, 1971, **23,** 280–282. (b)

Kaplan, M. F. The effect of judgmental dispositions on forming impressions of personality. *Canadian Journal of Behavioural Science*, 1971, **3,** 259–267. (c)

Kaplan, M. F. The modifying effect of stimulus information on the consistency of individual differences in impression formation. *Journal of Experimental Research on Personality*, 1972, **6,** 213–219.

Kaplan, M. F., & Anderson, N. H. Information integration theory and reinforcement theory as approaches to interpersonal attraction. *Journal of Personality and Social Psychology*, 1973, **28,** 301–312.

Lampel, A. K., & Anderson, N. H. Combining visual and verbal information in an impression-formation task. *Journal of Personality and Social Psychology*, 1968, **9,** 1–6.

Leon, M., Oden, G. C., & Anderson, N. H. Functional measurement of social values. *Journal of Personality and Social Psychology*, 1973, **27,** 301–310.

Levin, I. P., Craft, J. L., & Norman, K. L. Averaging of motor movements: Tests of an additive model. *Journal of Experimental Psychology*, 1971, **91,** 287–294.

Levin, I. P., Dulberg, C. S., Dooley, J. F., & Hinrichs, J. V. Sequential dependencies in single-item and multiple-item probability learning. *Journal of Experimental Psychology*, 1972, **93,** 262–267.

Lilly, R. S. The qualification of evaluative adjectives by frequency adverbs. *Journal of Verbal Learning and Verbal Behavior*, 1968, **7,** 333–336.

Lindner, R. Associative and dissociative bonds in personality impression formation. In R. G. Smith, *Speech-communication: Theory and models*. New York: Harper & Row, 1970.

Lindner, R. Congruity, balance, and information integration in personality impressions. Unpublished doctoral dissertation, Indiana University, 1971.

Lopes, L. L. A unified integration model for "Prior expectancy and behavioral extremity as determinants of attitude attribution." *Journal of Experimental Social Psychology*, 1972, **8,** 156–160.

Luce, R. D., & Suppes, P. Preference, utility, and subjective probability. In R. D. Luce, R. R. Bush, and E. Galanter (Eds.), *Handbook of mathematical psychology*. Vol. III. New York: Wiley, 1965.

McGuire, W. J. The nature of attitudes and attitude change. In G. Lindzey and E. Aronson (Eds.), *The handbook of social psychology*. Vol. 3. (2nd ed.). Reading, Mass.: Addison-Wesley, 1964.

Mandel, J. *The statistical analysis of experimental data*. New York: Wiley, 1964.

Massaro, D. W., & Anderson, N. H. Judgmental model of the Ebbinghaus illusion. *Journal of Experimental Psychology*, 1971, **89,** 147–151.

Myers, J. L. *Fundamentals of experimental design*. Boston: Allyn and Bacon, 1972.

Oden, G. C., & Anderson, N. H. Differential weighting in integration theory. *Journal of Experimental Psychology*, 1971, **89,** 152–161.

Parducci, A. Category judgment: A range-frequency model. *Psychological Review*, 1965, **72,** 407–418.

Parducci, A., & Perrett, L. F. Category rating scales. *Journal of Experimental Psychology*, 1971, **89,** 427–452.

Parducci, A., Thaler, L., & Anderson, N. H. Stimulus averaging and the context for judgment. *Perception and Psychophysics*, 1968, **3,** 145–150.

Pepitone, A., & Hayden, R. Some evidence for conflict resolution in impression formation. *Journal of Abnormal and Social Psychology*, 1955, **51,** 302–307.

Poulton, E. C. The new psychophysics: Six models for magnitude estimation. *Psychological Bulletin*, 1968, **69,** 1–19.

Rapoport, A., & Wallsten, T. S. Individual decision behavior. *Annual Review of Psychology*, 1972, **23,** 131–176.

Restle, F., & Greeno, J. G. *Introduction to mathematical psychology*. Reading, Mass.: Addison-Wesley, 1970.

Robbins, D., & Medin, D. L. Cue selection after multiple-cue probability training. *Journal of Experimental Psychology*, 1971, **91,** 333–335.

Rosenbaum, M. E., & Levin, I. P. Impression formation as a function of source credibility and order of presentation of contradictory information. *Journal of Personality and Social Psychology*, 1968, **10,** 167–174.

Rosenbaum, M. E., & Levin, I. P. Impression formation as a function of source credibility and the polarity of information. *Journal of Personality and Social Psychology*, 1969, **12,** 34–37.

Rosenberg, S., Nelson, C., & Vivekananthan, P. S. A multidimensional approach to the structure of personality impressions. *Journal of Personality and Social Psychology*, 1968, **9,** 283–294.

Rumelhart, D. E., Lindsay, P. H., & Norman, D. A. *A process model for long-term memory*. Technical Report CHIP 17. San Diego, Calif.: Center for Human Information Processing, University of California, 1971.

Sawyers, B. K., & Anderson, N. H. Test of integration theory in attitude change. *Journal of Personality and Social Psychology*, 1971, **18,** 230–233.

Schmidt, C. F. Personality impression formation as a function of relatedness of information and length of set. *Journal of Personality and Social Psychology*, 1969, **12,** 6–11.

Schümer, R., & Cohen, R. Eine Untersuchung zur sozialen Urteilsbildung. II. Bemerkungen zur verschiedenen konkurrierenden Modellen der Urteilsbildung. *Archiv für die gesamte Psychologie*, 1968, **120,** 180–202.

Scripture, E. W. *The new psychology*. New York: Scribner's, 1897.

Shanteau, J. C. An additive model for sequential decision making. *Journal of Experimental Psychology*, 1970, **85,** 181–191. (a)

Shanteau, J. C. Component processes in risky decision judgments. Unpublished doctoral dissertation, University of California, San Diego, 1970. (b)

Shanteau, J. C. Descriptive versus normative models of sequential inference judgment. *Journal of Experimental Psychology*, 1972, **93,** 63–68.

Shanteau, J. C., & Anderson, N. H. Test of a conflict model for preference judgment. *Journal of Mathematical Psychology*, 1969, **6,** 312–325.

Shanteau, J. C., & Anderson, N. H. Integration theory applied to judgments of the value of information. *Journal of Experimental Psychology*, 1972, **92,** 266–275.

Shepard, R. N. On subjectively optimum selections among multi-attribute alterna. tives. In M. W. Shelly and G. L. Bryan (Eds.), *Human judgments and optimality*. New York: Wiley, 1964.

Sidowski, J. B., & Anderson, N. H. Judgments of city-occupation combinations. *Psychonomic Science*, 1967, **7,** 279–280.

Slovic, P., & Lichtenstein, S. Relative importance of probabilities and payoffs in risk taking. *Journal of Experimental Psychology*, 1968, **78,** 1–18.

Slovic, P., & Lichtenstein, S. Comparison of Bayesian and regression approaches to the study of information processing in judgment. *Organizational Behavior and Human Performance*, 1971, **6,** 649–744.

Staddon, J. E. R. Spaced responding and choice: A preliminary analysis. *Journal of the Experimental Analysis of Behavior*, 1968, **11,** 669–682.

Stevens, J. C., & Rubin, L. L. Psychophysical scales of apparent heaviness and the size-weight illusion. *Perception and Psychophysics*, 1970, **8,** 225–230.

Stevens, S. S. On the psychophysical law. *Psychological Review*, 1957, **64,** 153–181.

Stevens, S. S. Issues in psychophysical measurement. *Psychological Review*, 1971, **78,** 426–450.

Stewart, R. H. Effect of continuous responding on the order effect in personality impression formation. *Journal of Personality and Social Psychology*, 1965, **1,** 161–165.

Takahashi, S. Analysis of weighted averaging model on integration of information in personality impression formation. *Japanese Psychological Research*, 1970, **12,** 154–162.

Takahashi, S. Effect of inter-relatedness of information on context effect in personality impression formation. *Japanese Psychological Research*, 1971, **13,** 167–175. (a)

Takahashi, S. Effect of the context upon personality-impression formation. *Japanese Journal of Psychology*, 1971, **41,** 307–313. (b)

Tesser, A. Differential weighting and directed meaning as explanations of primacy in impression formation. *Psychonomic Science*, 1968, **11,** 299–300.

Thurstone, L. L. *The measurement of values*. Chicago: University of Chicago Press, 1959.

Treisman, M. Sensory scaling and the psychophysical law. *The Quarterly Journal of Experimental Psychology*, 1964, **16,** 11–22.

Tversky, A. Additivity, utility, and subjective probability. *Journal of Mathematical Psychology*, 1967, **4,** 175–201. (a)

Tversky, A. Utility theory and additivity analysis of risky choices. *Journal of Experimental Psychology*, 1967, **75,** 27–36. (b)

Upshaw, H. S. The personal reference scale: An approach to social judgment. In L. Berkowitz (Ed.), *Advances in experimental social psychology*. Vol. 4. New York: Academic Press, 1969.

Weiss, D. J. Averaging: An empirical validity criterion for magnitude estimation. *Perception and Psychophysics*, 1972, **12,** 385–388.

Weiss, D. J. FUNPOT, a FORTRAN program for finding a polynomial transformation to reduce any source of variance in a factorial design. *Behavioral Science*, 1973, **18,** 150.

Weiss, D. J., & Anderson, N. H. Subjective averaging of length with serial presentation. *Journal of Experimental Psychology*, 1969, **82,** 52–63.

Weiss, D. J., & Anderson, N. H. Use of rank order data in functional measurement. *Psychological Bulletin*, 1972, **78,** 64–69.

Winer, B. J. *Statistical principles in experimental design*. (2nd ed.). New York: McGraw-Hill, 1971.

Wyer, R. S., & Dermer, M. Effect of context and instructional set upon evaluation of personality-trait adjectives. *Journal of Personality and Social Psychology*, 1968, **9,** 7–14.

Wyer, R. S., & Watson, S. F. Context effects in impression formation. *Journal of Personality and Social Psychology*, 1969, **12,** 22–33.

Young, P. T., & Christensen, K. R. Algebraic summation of hedonic processes. *Journal of Comparative and Physiological Psychology*, 1962, **55,** 332–336.

Zinnes, J. L. Scaling. *Annual Review of Psychology*, 1969, **20,** 447–478.

Quantitative Theories of the Integrative Action of the Retina

Norma Graham
COLUMBIA UNIVERSITY

Floyd Ratliff
THE ROCKEFELLER UNIVERSITY

1. INTRODUCTION

One of the great advantages of the use of mathematics in science is that it encompasses equally well both the concrete and the abstract. The mere counting of a number of tangible objects along a single dimension is as much a proper part of the mathematical world as is the formulation of an abstract multidimensional concept that may exist only in thought. Thus, since the purely mathematical aspects of both the real and the imaginary are of one and the same kind, there is no barrier to prevent transitions from the one to the other. The major purpose of this survey of quantitative studies of neurophysiology of vision is to illustrate how easy transitions from empirical data to abstract ideas and back again, via mathematics, provide a most important tool for the investigator—a ready means for the continual interplay of evolving theory and experimental advances.

This survey focuses on the retina, which—as its name implies—is a network. The functional units in it are interconnected and interact with one another; the neural activity generated by illumination of any one photoreceptor may influence, or be influenced by, the activity generated by many others. Consequently, many visual phenomena may depend as much upon the properties of the network of interconnected retinal neurons, acting as a whole, as they do upon the properties of the individual components, acting alone.

This vital and organic activity of the retina may be summarized briefly as follows: Light absorbed by photoreceptors stimulates them to generate neural activity in the form of graded local excitatory and inhibitory influences. These opposed influences are then integrated, over both space and time, by the neural networks in the retina. By means of this simple calculus the retina abstracts information about significant features of the spatial and temporal pattern of light and shade on the receptor mosaic. Ultimately, these abstractions are expressed in a neural code signaled by a pattern of discrete identical nerve impulses that are transmitted along the optic nerve to the brain. This set of abstractions and transformations may, quite properly, be called the 'logic' of the retina. The fundamental process involved—the integration (summation) of opposed excitatory (positive) and inhibitory (negative) influences—is amenable to a rigorous mathematical treatment. And it is the purpose of this chapter to review the quantitative theories of the integrative action of the retina that have been advanced. The review is confined to the retinas of the three animals on which most of the quantitative experimental and theoretical work has been done—the horseshoe crab (*Limulus*), the cat, and the goldfish.

For the relatively simple eye of *Limulus* the underlying physiological processes are fairly well understood and the theory well developed; accurate predictions can now be made of the *Limulus* optic nerve responses to almost any spatial and temporal pattern of illumination on the retina. For the more complex eyes of the cat and goldfish, the underlying processes are less well known and the theories less well developed, but a more complete understanding of the physiological mechanisms and more general theories are now beginning to emerge.

1.1. Historical Background

The basic idea that the retina has a certain 'logic' of its own based on the interplay of opposed excitatory and inhibitory influences is not new. Truly scientific work on the subject began well over 150 years ago, and the quantitative approach to the problem is now more than 100 years old. A glimpse of this early work is provided by the following brief mention of the contributions of four giants in the field: the German poet and author Johann Wolfgang von Goethe (1749–1832), the French chemist Michel Eugene Chevreul (1786–1889), the German physician and physiologist Ewald Hering (1834–1918), and the Austrian physicist-philosopher-psychologist Ernst Mach (1838–1916).[1]

[1] This historical review was adapted from Ratliff (1973).

Goethe's *Outline of a Color Theory*, published in 1808, seemed to contradict at every point the well-established and widely accepted Newtonian theory of light and, being a mere poet, he was severely criticized by scientists from all sides. But much of the confusion about and criticism of Goethe's work resulted from the failure of scientists, both then and now, to understand and to appreciate the distinction between the physical and the psychophysiological aspects of color. Many of Goethe's conceptions about what he called "physiological colors" were well founded on psychophysical experiments on brightness contrast and color contrast, and he was perhaps the first to state explicitly the fundamental ideas of the interaction of opposed influences in the retina that are to be elaborated in this chapter. He wrote: "When darkness is presented to the eye it demands brightness, and *vice versa;* it shows its vital energy, its fitness to receive the impression of the object, precisely by spontaneously tending to an opposite state." And, "If a colored object impinges on one part of the retina, the remaining portion has a tendency to produce the compensatory color. . . . Thus yellow demands purple; orange, blue; red, green; and *vice versa:* Thus again all intermediate gradations reciprocally evoke each other; the simpler color demanding the compound, and *vice versa.*"

No matter how far from the mark Goethe may have been in his views on the physical aspects of light, it is now evident that his concept of a simultaneous reciprocal interaction of opposed or complementary influences among neighboring parts of the retina is basic to the understanding of the psychophysiology of the visual system. For a translation of and commentary on Goethe's work on color see Matthaei (1971).

Chevreul, whose work on fats and organic analysis opened up two major fields of research in biochemistry, also made important early contributions to the psychophysiology of vision. When he was called upon, in 1824, to become director of the dye plant at the Gobelin Tapestry Works in Paris there had been numerous complaints about the quality of certain pigments prepared there. These led him to undertake a study of the psychology of color. He soon discovered that many of the complaints about the pigments themselves were unfounded; some of the supposed defects were entirely due to subjective contrast effects. The evident simultaneity and reciprocity of the contrast of contiguous colors and contiguous shades led Chevreul to attempt to represent the phenomena in quasi-mathematical terms—rather like a set of simultaneous equations—which would express all the facts succinctly. These findings were published in 1839 in his great work on *The principles of harmony and contrast of colors and their application to the arts.* So important is this work for artists and artisans that it is still a standard reference—reprinted as recently as 1967.

To Hering belongs much of the credit for recognizing that reciprocal interactions in the retina underly our normal everyday visual experience as well as the so-called illusions and distortions such as border contrast and color contrast. As he put it, "The most important consequences of reciprocal interactions are not at all those expressed in contrast phenomena, that is, in the alleged false seeing of the 'real' colors of objects. On the contrary, it is precisely the so-called correct seeing of these colors that depends in its very essence on such reciprocal interactions. . . . A closer familiarity with the consequences of this reciprocal interaction is essential for understanding the nature of our vision. . . ."

Hering's extremely important hypothesis about the role of reciprocal interactions in normal color vision, although long supported by psychophysical evidence, has only recently been vindicated by direct electrophysiological experiments. The classical Young-Helmholtz trichromatic theory must be extended, as Hering's work indicated, to include excitatory and inhibitory interactions among various pairs of the three basic cone types with their three different sensitivities (maxima in the red, green, and blue) to the visible spectrum. For a recent translation of his major work see Hering (1964).

Mach, while experimenting with rotating discs used to produce various spatial distributions of light and shade, found light and dark bands appearing where, according to physical calculations, none were expected. Formerly, such contrast phenomena as these Mach bands—as they are now called—had generally been attributed to 'unconscious inferences' or 'errors of judgment.' But for Mach these were merely various ways of expressing the still unexplained facts. He sought an explanation in terms of the interdependence of neighboring points on the retina. As he expressed it:

> Since every retinal point perceives itself, so to speak, as above or below the average of its neighbors, there results a characteristic type of perception. Whatever is near the mean of the surroundings becomes effaced, whatever is above or below is disproportionately brought into prominence. One could say that the retina schematizes and caricatures. The teleological significance of this process is clear in itself. It is an analog of abstraction and of the formation of concepts.

Mach's quantitative analysis of the Mach bands and related contrast phenomena (for translations see Ratliff, 1965) was probably the first attempt to express the integrative action of the nervous system in precise mathematical terms. But his application of mathematical modes of thought to the study of the nervous system (in 1865!) was so far ahead of the times that his papers on the subject attracted little attention when first published. Indeed, the general concept of a simultaneous reciprocal interaction of opposing or complementary influences among neighboring parts of the retina, formulated more or less independently by Goethe, Chevreul, Hering, and Mach, was not widely accepted in their own time. And for many years thereafter psychophysiologists generally regarded the retina as a mere passive and mechanical transducer rather than the vital integrative organ that it is.

1.2. Modern Unit Analyses of Retinal Activity

C. S. Sherrington's researches into the physiology of the central nervous system laid the foundation of modern neurophysiology. His *Integrative Action of the Nervous System* (1906), although based largely on reflex action, put forth in a systematic way practically all of the basic concepts of the functional organization of the nervous system that we know today. In particular, his carefully worked out concepts of central excitatory states and central inhibitory states are basic to modern interpretations. It was fundamental in Sherrington's teaching that these excitatory and inhibitory states, both capable of gradation, were simply of opposite sign and would add up algebraically. It is interesting to note that some of these ideas of Sherrington's about interaction in the nervous system came in part from his own little known psychophysical studies entitled "On reciprocal action in the retina as studied by means of some rotating discs" (1897) and in part indirectly from Mach by way of McDougall's (1903) studies on perceptual contrast phenomena.

But the most important single factor that advanced the theory of the integrative action of the retina, and of the visual system in general, was the relatively recent development of techniques for directly measuring the activity of the visual neurons. Less than half a century ago Adrian and Matthews (1927a, 1927b) reported the first successful recording of the electrical activity of the whole optic nerve. Even in this whole nerve (of the conger eel), consisting of a very large number of axons of retinal ganglion cells, they could discern some evidence of excitatory and inhibitory interactions within the retina (Adrian & Matthews, 1928). Precise quantitative studies became possible soon thereafter when Hartline and Graham (1932) dissected single axons apart from the optic nerve of the marine arthropod, *Limulus* (the horsehoe crab), and recorded their unitary activity. Within a few years these unit analyses were extended to the retinas of vertebrates (Hartline, 1938a) and molluscs (Hartline, 1938b). Since Hartline's initial studies, literally thousands of qualitative and quantitative experiments have been carried out on single neurons in a wide variety of animals and at all levels of the visual system—from the photoreceptors to the brain. Since the optic nerve forms a 'bottleneck' through which all visual information must pass as it is transmitted from the retina to the brain, any meaningful analysis of retinal function must concern itself with the patterns of nerve impulses conducted by it.

The most striking feature of the activity of the individual axons that make up an optic nerve is that it appears in the form of discrete electrochemical pulses, each one very much like all the others. Thus the primary data in experiments on these neurons are the *time intervals* between impulses. Once measured, these data may be converted into other forms, typically into *rate*. For individual neurons in the optic nerve of *Limulus*, where the interimpulse variability is not great, a useful measure—especially in studies of the dynamics—is the *instantaneous rate*, the computation of which is illustrated in Figure 1. For any particular interval, the reciprocal of the interval (i.e., the impulse rate over that period) is assigned to all the time between the beginning and the end of the interval. The instantaneous rate thus forms a piecewise continuous function with the mathematical advantage that it can be manipulated in the same way as the other continuous functions (light stimuli, underlying cellular processes of excitation and inhibition) to which it is related. For preparations in which interimpulse variability is large and therefore the instantaneous rate fluctuates widely, as in the cat and goldfish, a useful substitute for instantaneous rate is the *impulse histogram*. The histogram is constructed by counting impulses within successive equal time intervals (Fig. 1). For steady-state responses, in any preparation, simple rate measures over a long period of any arbitrary length are commonly used.

2. EXCITATION AND INHIBITION IN THE EYE OF LIMULUS

2.1. The Steady State

The earliest records of optic nerve activity in the compound lateral eye of *Limulus* showed that, in general, the more intense the illumination, the higher the instantaneous rate. There are no excitatory interactions in this eye. Activity can be elicited from an optic nerve fiber only by illuminating the receptor unit (ommatidium) from which it arises. However, during the course of the experiments, it was casually observed that turning on the room lights often *diminished* the activity in an optic nerve fiber that was being prepared for study (Hartline, 1949). Once this observation was considered seriously, it was easy to demonstrate that shading regions of the eye neighboring the receptor unit under study restored the receptor's activity. Evidently activating one region of the eye inhibited the activity of neighboring regions (for review, see Hartline & Ratliff, 1972). Qualitatively, experiments showed that the amount of inhibition depended on several variables: The inhibition was found to be greater, the greater the intensity of illumination on the neighbors, the larger the area of the neighboring region activated, or the smaller the distance between the activated neighbors and the receptor unit under study (Hartline, Wagner, & Ratliff, 1956). But first attempts at expressing the relationship between the amount of inhibition and these variables in some simple quantitative way were not successful. For example, plotting decrement in a test receptor's firing rate as a function of the intensity of illumination on the neighbors yields a complicated curvilinear function.

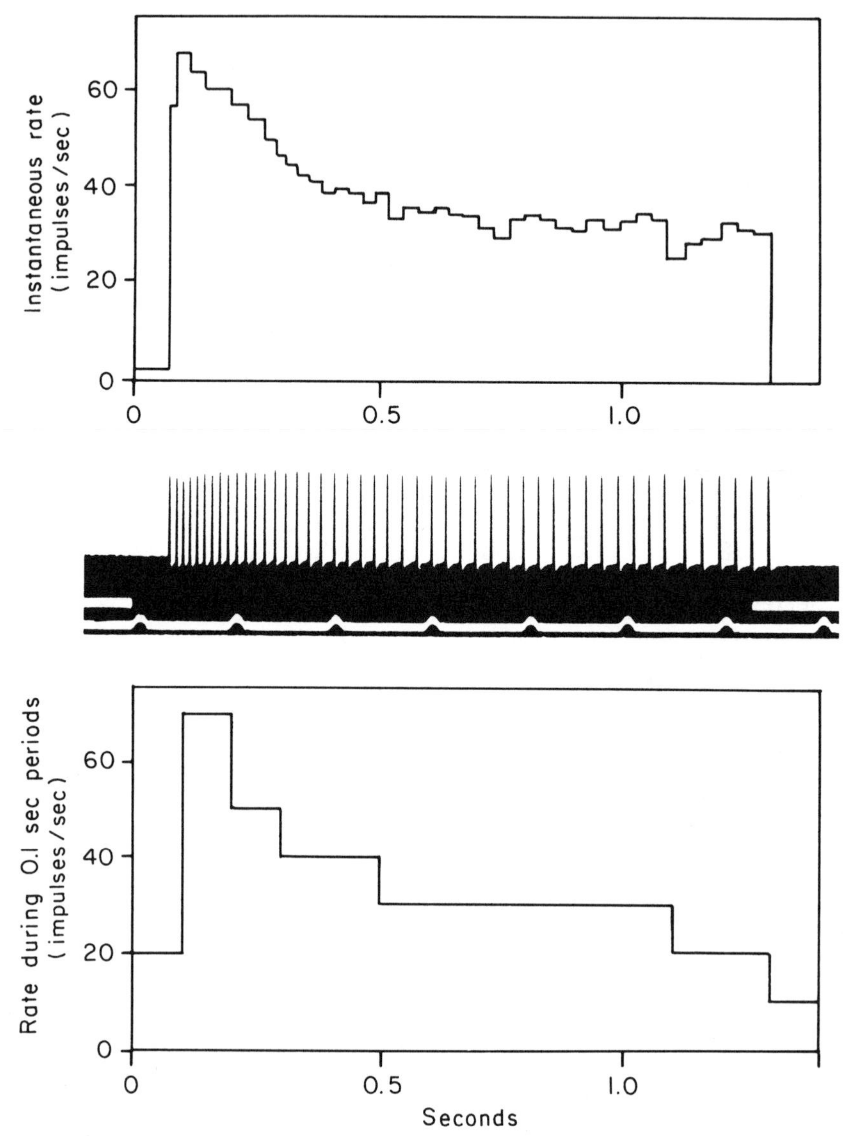

FIGURE 1.
Center: An oscillographic record of a train of impulses discharged in a single optic nerve fiber arising from a receptor unit (ommatidium) in the compound eye of *Limulus*. Time marked in 1/5 sec. Period of illumination marked by blackening of white line above time marks. *Above:* Plot of 'instantaneous' rate of discharge constructed by plotting the reciprocal of each time interval between impulses over that same interval. The latent period (interval between onset of illumination and discharge of first impulse) is plotted similarly. *Below:* Impulse rate histogram constructed by counting the number of impulses within successive equal time periods (each 0.1 sec) and plotting the rate over the corresponding interval. For short counting periods or for low rates, fractional intervals between impulses should be considered. (See Lange, Hartline, & Ratliff, 1966b.)

An important and useful simplification resulted when it was considered that the inhibition might be recurrent—that is, the amount of inhibition exerted on the receptor under study by a particular neighbor might depend on the firing rate of that neighbor rather than on the illumination stimulating that neighbor. (This was suggested by an apparently imperfect summation of inhibition and by the phenomenon of disinhibition—to be discussed later.) This possibility was directly tested by recording simultaneously from two interacting receptor units (see schema in Fig. 2) as the intensity stimulating

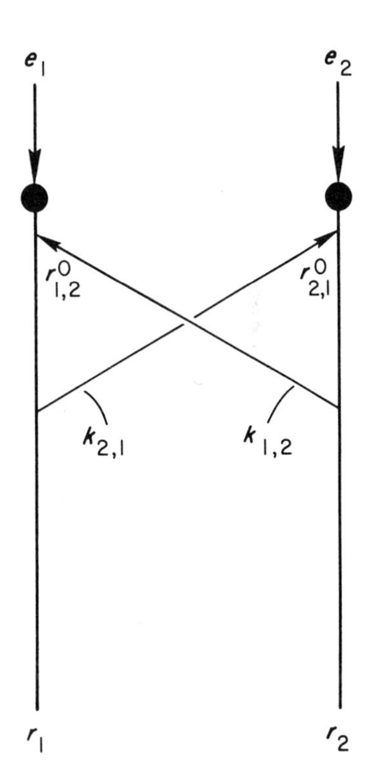

FIGURE 2.
Schema of mutual inhibitory interaction of two neighboring receptor units in the compound eye of *Limulus*. The excitation e is measured by the response r when one unit alone is illuminated. The threshold frequency that must be exceeded before a receptor unit can exert any inhibition on its neighbor is represented by r^0. It and the inhibitory coefficient k are labeled to indicate the direction of the action: $r^0_{1,2}$ is the frequency of receptor 2 at which it begins to inhibit receptor 1; $r^0_{2,1}$ is the reverse. In the same way, $k_{1,2}$ is the coefficient of the inhibitory action of receptor 2 on 1; $k_{2,1}$ the reverse. It is necessary to specify the direction of the action because the mutual influences are seldom identical, although generally they are similar. Note that the inhibition is recurrent, that is, each element exerts inhibition back on the site of impulse generation in the other; inhibition of one receptor unit depends on the response r of the other rather than on the excitation e. In the actual retina of *Limulus* the interactions are not confined to adjacent neighbors, the inhibitory influences extend over a considerable field. (See Fig. 3.) (Redrawn after Ratliff, 1968.)

them was varied over a wide range (Hartline & Ratliff, 1957). The decrement in the steady-state firing rate of one of the receptors turned out to depend linearly on the steady-state *firing* rate of the other, once a threshold had been reached. This may be expressed in terms of a pair of simultaneous piecewise linear equations:

$$
\begin{aligned}
r_1 &= e_1 - k_{1,2}(r_2 - r^0_{1,2}), \\
r_2 &= e_2 - k_{2,1}(r_1 - r^0_{2,1}),
\end{aligned}
\tag{1}
$$

where r_1 is the response (rate of discharge of impulses) of receptor unit 1, e_1 is the 'excitation' of receptor unit 1 expressed as discharge rate in absence of inhibition from neighbors. In the first equation $k_{1,2}$ is the coefficient of the inhibitory action of unit 2 on 1, and $r^0_{1,2}$ the threshold of that inhibition. In the second equation, the inhibitory action of 1 on 2 is expressed similarly. Note that there are certain limitations to these equations: there can be no negative rates of discharge, there is some upper bound to the rate, and whenever the discharge rate of the neighbor (r_2 in the first equation) is less than the threshold of inhibition ($r^0_{1,2}$) the term in parentheses is taken to be equal to zero.

To extend the quantitative description to large numbers of receptor units, we must know how the inhibitory influences combine with one another. It has been shown that the inhibitory influences from two groups of receptor units, which are widely separated so that the two groups do not interact with one another, combine by simple addition when acting simultaneously on an intermediate receptor (Hartline & Ratliff, 1958). Consequently, the activity of n interacting receptors may be described by a set of simultaneous piecewise linear equations, each with n inhibitory terms combined by simple addition

$$r_p = e_p - \sum_{j=1}^{n} k_{p,j}(r_j - r^0_{p,j}). \tag{2}$$

Notice that there is a term where $p = j$ or, in short, where the receptor unit p is inhibiting itself (see Stevens, 1964). This self-inhibition term can be disregarded in the present discussion of the steady state, but it will turn out to be important when we come to the underlying physiology of the *Limulus* receptor unit and to the dynamics of the inhibition.

The above equations contain no explicit terms for distance. No special terms are required; the effects of distance are already implicit in the equations for they are expressed in the concomitant changes in threshold (r^0) and inhibitory coefficient (k). In general, the threshold increases with distance and the inhibitory coefficient decreases. The exact form of the distribution of the inhibitory coefficients, which has been determined by recording from some 20 to 30 fibers, can best be described as a broad Gaussian curve (a broad normal distribution) from which a concentric narrow Gaussian curve (a narrow normal distribution) has been subtracted (Fig. 3). That is, it rises to a maximum some distance away from the point of excitation and then gradually diminishes to zero (Barlow, 1967). This dependence of the mutual inhibition among receptors on distance and the recurrent nature of the inhibition are, of course, most significant factors in determining the responses to various configurations of illumination. Following are two examples.

The inhibitory effect produced by the combined influences of several groups of receptors is sometimes less than the effect produced by one group—an apparent contradiction of the law of spatial summation expressed in Equation 2 above. But this results only because the inhibitory influence exerted by a particular unit depends upon its own activity r rather than directly upon the

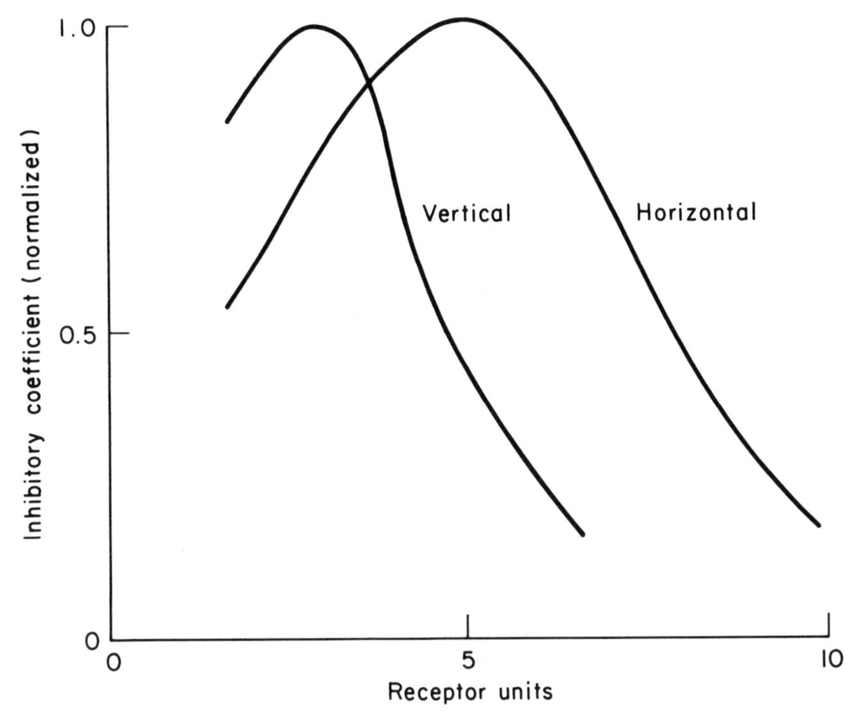

FIGURE 3.
The dependence of the magnitude of the inhibitory effect on the separation of the ommatidia (receptor units) in the retinal mosaic. The curves represent mean values obtained in several different experiments. Since the absolute magnitude of the effect varies from one preparation to the next, the data were normalized by assigning the maximum a value of 1.0 and adjusting other coefficients proportionately. The effect falls off more rapidly in the vertical direction than in the horizontal, thus the field as a whole has an elliptical shape. Note that the maximum effect is some distance away from the source of the inhibition. (Redrawn after Barlow, 1969.)

light stimulus incident upon it—that is, because the inhibition in *Limulus* is *recurrent* (feeding *back* upon neighbors) rather than *nonrecurrent* (feeding *forward* upon neighbors). Since the amount of activity in each of several groups illuminated together may be less than when illuminated separately (due to mutual recurrent inhibition and inhibitory thresholds), their combined simultaneous influences may not necessarily equal the sum of their separate influences.

The responses to one particular configuration of illumination illustrate this clearly and demonstrate, at the same time, the phenomenon of disinhibition (Hartline & Ratliff, 1957). When, in the vicinity of a mutually inhibiting pair of receptors (*A* and *B*), additional receptors located so that they inhibit only *B* are illuminated, the firing rate of *A* increases. This apparent facilitation of *A* actually results from inhibition of inhibition—the additional illuminated receptors inhibit *B* which therefore inhibits *A* less than before. (A supposed facilitation that was observed in the spinal cord has since been shown to be another example of disinhibition. See Wilson, Diecke, & Talbot, 1960; Wilson & Burgess, 1961.)

So-called border contrast effects and the Mach bands, well known in human vision, also appear to depend upon the spatial distribution of lateral inhibition. One can see the Mach bands at the borders of a half-shadow cast by an object in good sunlight, such as the shadow of one's own head and shoulders on a sidewalk. The objective distribution of illumination is high and uniform in the full sunlight, more or less uniformly graded in the half shadow, and uniformly low in the full shadow. Objectively there are no maxima and minima, yet there appears to be a narrow dark band at the dark edge of the half shadow and a narrow bright band at the bright edge. Analogous effects have been observed in the lateral eye of *Limulus* (Ratliff & Hartline, 1959), and are implicit in Equation 2. Because the inhibition diminishes with distance, the difference between the responses of neighboring receptor units is greatest at or near the boundaries of differently illuminated regions (Fig. 4). A unit within a dimly illuminated region, but near the boundary, is inhibited not only by its dimly illuminated neighbors, but also by some brightly illuminated ones. The total inhibition on this receptor unit, therefore, will be greater than on the other dimly illuminated units farther from the boundary. Consequently, its response will be less than theirs. Conversely, a unit within the brightly illuminated area, but also near the boundary, will have a higher frequency of discharge than other brightly illuminated units located far from the boundary, since they are receiving strong inhibition from brightly illuminated neighbors all around, while the one near the boundary receives some strong inhibition from brightly illuminated neighbors and some weak inhibition from dimly illuminated neighbors. Consequently, its response is the greatest of all. Thus maxima and minima appear in the neural response, even though there are none in the spatial distribution of illumination.

2.2. Physiological Mechanisms of Excitation and Inhibition

Up to this point the activity of the *Limulus* eye has been described with few references to the physiological processes intervening between the light stimulus and the discharge of impulses in the optic nerve. Neither the original formulation nor the description of the abstract model embodied in Equation 2 required knowledge of these processes. Now, however, the physiology is discussed in some detail, for not only is it interesting in itself but the investigation

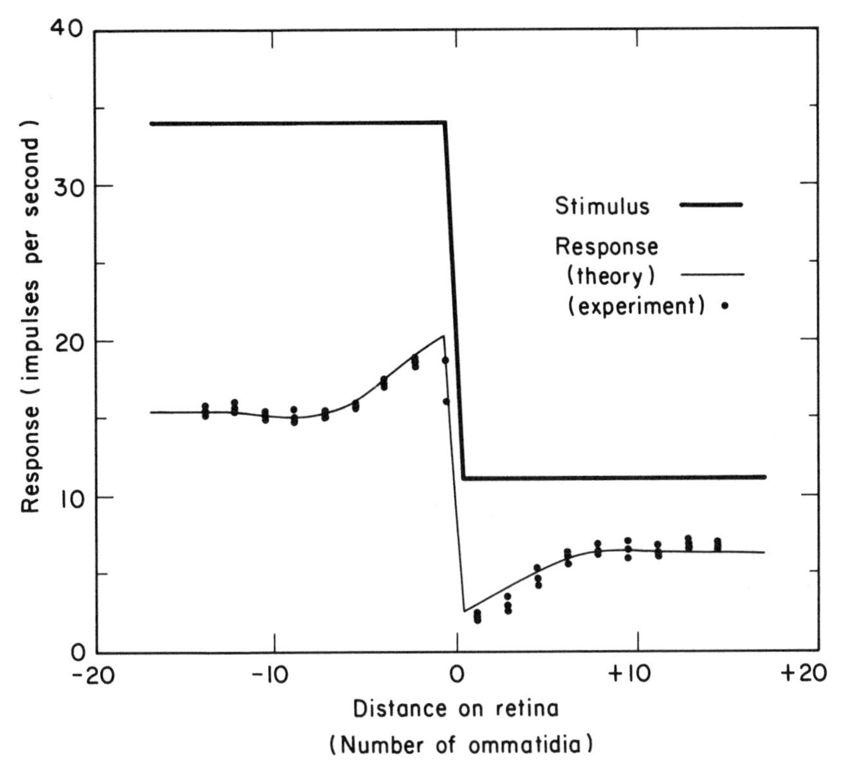

FIGURE 4.
Response of an optic nerve fiber in the compound eye of *Limulus* to a step pattern of illumination. The solid thick line represents the stimulus as measured by the response that would have been obtained if there had been no lateral inhibition. The solid thin line represents the response, with lateral inhibition, calculated using Equation 2 modified to include certain effects of the two different levels of excitation (see discussion at the end of Sec. 2.2). The points show actual responses measured as the stimulus was moved across the eye in successive stages. The three data points at each stage are the results of three different passages of the stimulus pattern across the eye. (Redrawn after Barlow & Quarles, personal communication, 1972.)

of it led to the development of a more exact and more general model of the *Limulus* eye encompassing both the steady state and the dynamics.

When Purple and Dodge (Purple, 1964; Purple & Dodge, 1965) set out to study the cellular mechanisms of excitation and inhibition in the *Limulus* eye, they could form working hypotheses based on both the large body of knowledge accumulated about the cellular mechanisms of similar processes in other kinds of neurons and the information already known about the *Limulus* receptor unit (ommatidium).

It was well known that the membrane of a nerve cell plays a crucial part in all its activity. (See Katz, 1966, for a thorough presentation of the material that is briefly summarized in the following paragraphs.) When the nerve is 'resting' there is a constant voltage difference between the outside and the inside of the cell—the 'resting potential'—which occurs because the membrane is selectively permeable to different ions. In general, excitatory influences affect the cell by increasing the permeability of the membrane to certain ions and thereby drive the potential difference across the cell membrane toward zero (that is, excitatory influences depolarize the cell). A large enough depolarization of the membrane will trigger an impulse and, if the depolarization is maintained, a train of impulses. Inhibitory influences affect the cell similarly by increasing the permeability of the membrane to some ions, but in this case the ions involved cause the potential difference across the cell membrane to be driven farther from zero (that is, inhibitory influences hyperpolarize the cell). Thus, inhibitory influences hinder the triggering of a nerve impulse.

The action of excitatory and inhibitory influences can be modeled as the 'equivalent electric circuit' shown in Figure 5. E_r is the resting potential, E_e is the potential towards which excitatory influences push the cell, and E_i is the potential towards which inhibitory influences push the cell. E is the resulting potential across the membrane (the so-called membrane potential). V, the difference between this resulting potential E and the resting potential E_r, is called the 'generator potential.' Several resistances represent the membrane's selective permeability to various ions. R_e, a variable resistance, reflects the amount of excitatory influence. The greater the excitatory influence, the more permeable the membrane is to those ions associated with excitation, and thus the lower the resistance R_e and the higher the excitatory conductance $g_e = 1/R_e$. Similarly, the greater the inhibitory influence, the more permeable the membrane is to those ions associated with inhibition, and thus the lower the corresponding resistance R_i and the higher the inhibitory conductance $g_i = 1/R_i$. R_r is fixed and equals the resistance of the membrane in its resting state. The capacitance of the membrane will be ignored because of its relatively short time constant (see Purple, 1964, p. 91). The equations underneath the diagram in Figure 5 can be derived from Kirchhoff's rules for electric circuits.

The general model of Figure 5 can be recast in a form specific to the *Limulus* receptor unit by making the following assumptions. Light absorbed by the photoreceptor causes excitatory conductance changes leading to depolarization of the receptor nerve cell. It has been suggested that the total excitatory conductance and potential change is the result of the superposition of 'quantum bumps,' small changes each due perhaps to the absorption of one photon of light and the subsequent release of transmitter substance (these quantum bumps are considered further in the section on dynamics). But exactly how light energy on the front of the ommatidium leads to the changes in the membrane permeability and consequent changes in membrane potential of the

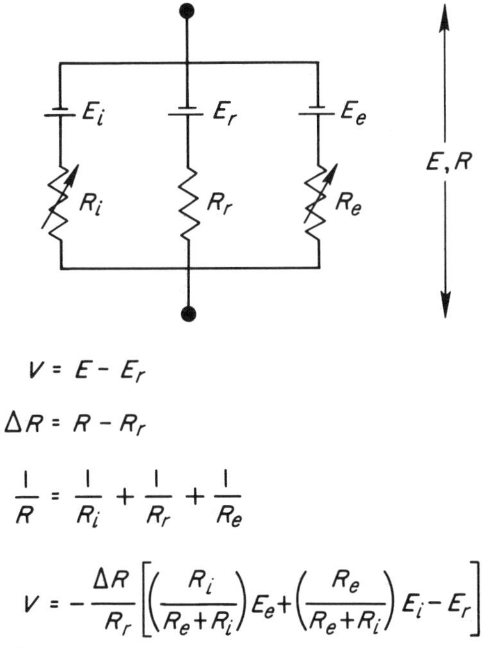

FIGURE 5.
An equivalent circuit for the integration of excitatory and inhibitory influences. E_r is the electromotive force of the resting potential, and R_r is the resting resistance value. E_e is the electromotive force of excitation and R_e is the variable resistance associated with excitation. E_i is the electromotive force of inhibition and R_i is the variable resistance associated with inhibition. E and R are the resulting potential and resistance measured across the cell membrane. The departures of the membrane potential and resistance from the resting values are given by V and ΔR. The expression for V may be derived from Kirchhoff's rules for electric circuits.

eccentric cell is still unknown. Nerve impulses coming either from neighboring receptor units (producing 'lateral inhibition') or from the receptor unit in question (producing 'self-inhibition') cause inhibitory conductance changes leading to hyperpolarization. These inhibitory conductance changes might well result from the usual kind of synaptic process as it is reasonable to suppose, on the basis of anatomical evidence, that axon collaterals both from neighboring receptor units and doubling back from the axon of the receptor unit in question form synapses with that receptor unit's axon (Gur, Purple, & Whitehead, 1972). Each inhibitory impulse, arriving at such a synapse, would cause a given quantity of neural transmitter to be released; this transmitter, after diffusing across the synapse, will cause a given conductance change in the postsynaptic membrane. Thus, in the model, every impulse from a particular receptor unit is assumed to cause an identical change (both in time-course and magnitude) in the inhibitory conductance $g_i = 1/R_i$. Direct measurements of the membrane potential and conductance of the *Limulus* receptor nerve cell do show changes in the appropriate directions both when the receptor unit is stimulated by light and after nerve impulses from its neighbors or itself (MacNichol, 1956; Fuortes, 1959; Purple, 1964; Purple & Dodge, 1965). However, quantitative consideration of the measurements reveals certain defects in this simple equivalent circuit model of the *Limulus* receptor unit. By taking further into account the known anatomy of the *Limulus* eye, Purple and Dodge were able to modify the model and overcome its defects.

Each ommatidium (Fig. 6A) contains one bipolar neuron—the eccentric cell—that initiates impulse potentials that proceed down its axon into the central nervous system. The dendrite of the eccentric cell is surrounded by 10 to 15 retinular cells arrayed like the segments of an orange. Since the retinular cells contain the photopigment, excitatory influences probably act on the eccentric cell at the junction of its dendrite with the retinular cells. Inhibitory influences, on the other hand, appear to affect the eccentric cell at sites on the other side of the cell body in the lateral plexus, a region of many interconnections between fine branches of the axons of all the ommatidia. This separation of excitatory and inhibitory sites has been handled by using Rall's extension of the classical 'cable' theory of electronic spread in uniform cylindrical axons to the more complicated geometry of branching dendrites, and cell bodies (Rall, 1962). Calculations (based on properties of the eccentric cell estimated from anatomical and electrophysiological data) using Rall's theory suggested that the dendrite that contains the excitatory site could be treated as if it were part of the cell body but that the lateral plexus containing the inhibitory site could not. Accordingly, the model of Figure 5 was expanded into the model of Figure 6B by assuming (a) there is a length of 'uniform axon' X between the cell body and the site of inhibitory conductance change, (b) there is a very long length of uniform axon Y on the other side of the inhibitory site, and (c) the point of impulse generation is at or beyond the inhibitory site. The lengths of uniform axons, shown as cylinders in the diagram, are assumed to have the properties of a passive cylindrical cable.[2]

A theorem, proved by Bruce Knight (see Purple, 1964, p. 91), made the above assumptions very easy to use. He showed that properties of an arbitrary length of passive axon can be represented as a three-resistor network. Further, a semi-infinite length of axon can be represented as its resistance to ground. These simplifications allow the model in Figure 6B to be reduced

[2] This model is still a simplification of the actual situation in which there are many-branched axon collaterals each with different electrical properties instead of passive lengths of uniform axons and inhibition acts at many inhibitory synapses distributed among the collaterals instead of acting at one point on the axon.

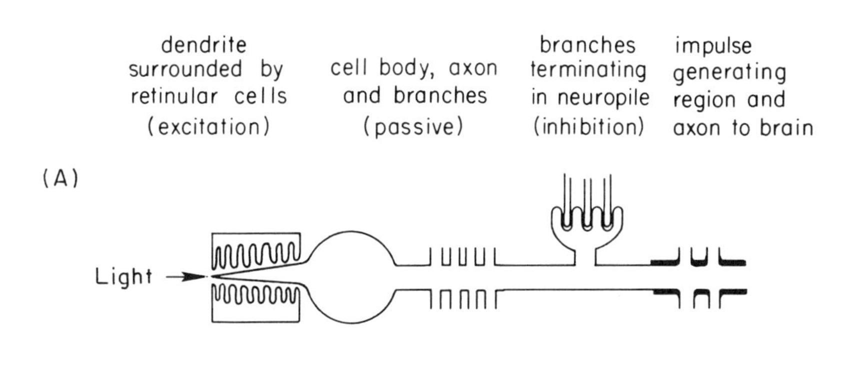

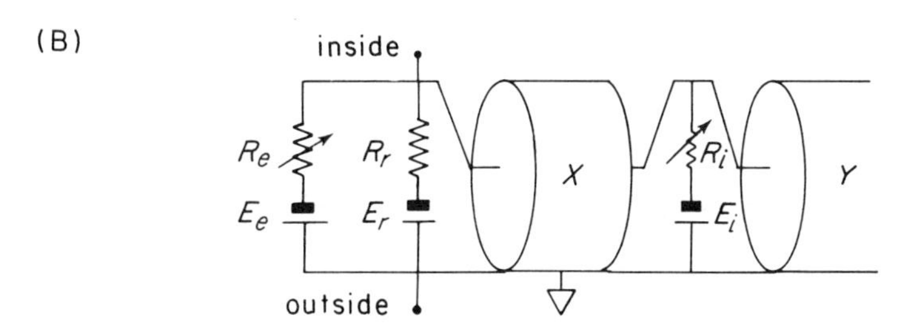

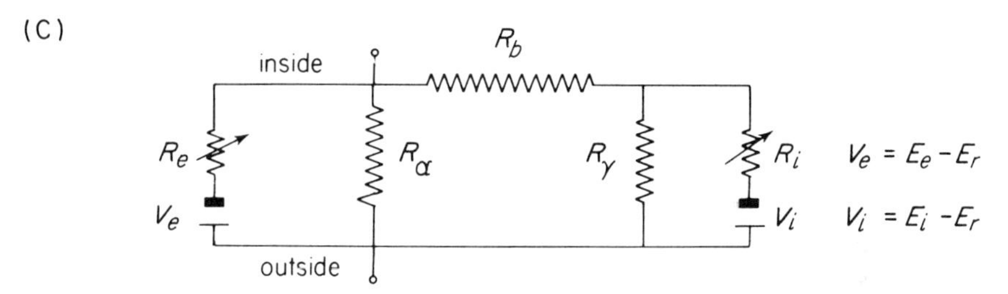

FIGURE 6.
(A) Schematic diagram of a single ommatidium. (B) Equivalent circuit of Figure 5 modified to include the separation of excitatory and inhibitory sites by a length of cablelike axon. (C) Equivalent circuit for the cable model of part B. The resistances R_α and R_γ incorporate the properties of the cablelike stretches of axon. (Adapted from Purple, 1964; Purple & Dodge, 1965.)

to the model in Figure 6C. The resistance R_α and R_γ incorporate the information about the properties, including length, of the stretches of axon X and Y. This model is in turn simple enough that Kirchhoff's rules for circuits can easily be used to derive formulas that describe the network. From computer solutions of these formulas calculated for various sets of model parameters (various lengths of axon X for example), predictions can be made about the membrane resistance and potential measured at the cell body in response to electrical and light stimulation. As it turns out, with appropriate choice of parameters this model predicts the observed conductance and potential measurements well (Purple, 1964; Purple & Dodge, 1965).

By adding one more property to the model—that the frequency of impulses is directly proportional to the depolarization (the generator potential) at the site of impulse initiation—the model may be compared with the more abstract Equation 2. Figure 7 summarizes the predictions of this final version of the equivalent circuit model. It shows the decrement in the firing rate of a test ommatidium resulting from six different levels of inhibitory conductance plotted as a function of excitation on the test ommatidium. Since, according to the model, firing rate is a linear function of the membrane potential and change in membrane potential is a hyperbolic function of inhibitory conductance (as is easier to show for the simple model of Fig. 5 but also true in this case), the predicted decrement in firing rate due to inhibition is a hyperbolic function of the inhibitory conductance. Further, since a hyperbolic function is quite linear for part of its range, the predicted decrement in firing rate will be approximately proportional to inhibitory conductance for small values but will begin to saturate gradually as inhibitory conductance reaches higher values. In other words, in Figure 7 the curves for different levels of

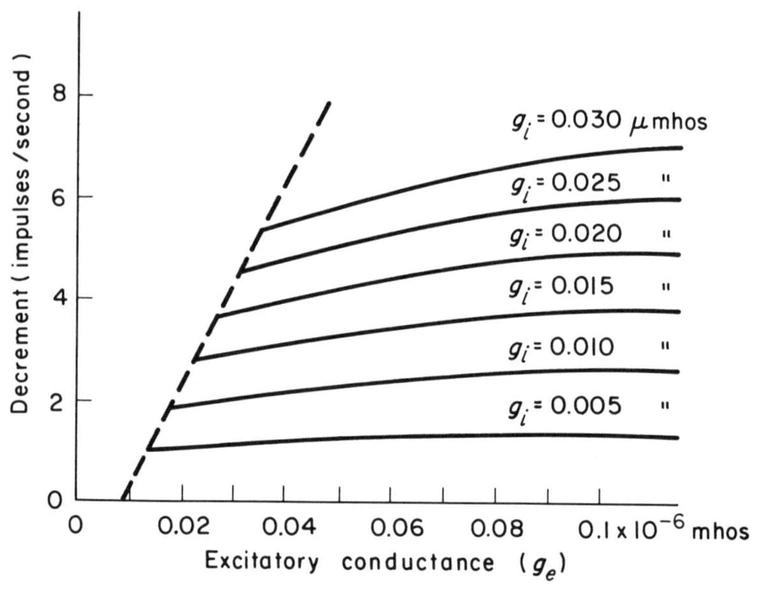

FIGURE 7.
Prediction by the model, shown in Figure 6C, of the decrement in a test ommatidium's firing rate due to inhibition. The abscissa represents the amount of excitation on the test ommatidium expressed in terms of the excitatory conductance (g_e). Decrements resulting from six levels of inhibitory conductance (g_i) are shown. (Adapted from Purple & Dodge, 1965.)

inhibitory conductance are approximately equally spaced, although at higher levels the curves are slightly closer together than at lower levels. This approximate proportionality between decrement in firing rate and inhibitory conductance means that the equivalent circuit model agrees almost exactly with Equation 2 in two important properties.

1. 'Linear recurrent inhibition'—the property that the decrement in a test ommatidium's firing rate is a linear function of a neighbor's firing rate above inhibitory threshold (in the case of lateral inhibition) or of its own firing rate (in the case of self-inhibition). That is, $e_p - r_p = k_{pj}(r_j - r_{pj}^0)$ for the case of two ommatidia p and j (see Eq. 2). The equivalent circuit model predicts this relationship because the total inhibitory conductance is the sum of the small conductance changes caused by the individual impulses. Since all the impulses coming from a particular neighbor (at frequencies above the threshold r_{pj}^0) cause equal conductance increments, the summed inhibitory conductance due to that neighbor will be proportional to the suprathreshold firing rate of that neighbor. (The constant of proportionality may vary from neighbor to neighbor as is reflected in the different values of k_{pj}.) Since decrement in firing rate is approximately proportional to total inhibitory conductance, the property of (approximate) linear recurrent inhibition is predicted. (The cellular mechanism for threshold is as yet unknown. However, direct intracellular recordings of the postsynaptic lateral inhibitory potential show that at least two closely spaced antidromic impulses are necessary to produce any postsynaptic potential at all (Knight, Toyoda, & Dodge, 1970). Thus in the model it is simply assumed that below a certain frequency, impulses produce no inhibitory conductance change.)

2. 'Spatial summation'—the property that the total decrement in a test ommatidium's firing rate due to inhibition by several neighboring ommatidia is the sum of the decrements due to each individually. Since the total inhibitory conductance due to the inhibition from several neighbors is exactly the sum of the conductances due to each alone, and since the decrement in the test ommatidium's firing rate is approximately proportional to inhibitory conductance, the property of (approximate) spatial summation is predicted.

The above discussion of agreement between the abstract Equation 2 and the equivalent circuit model implicitly assumed a constant excitatory level in the test ommatidium. What happens as excitation is changed? The equation has the following property in this situation.

3. 'Inhibition independent of excitation'—the decrement in firing rate of the test ommatidium resulting from any given pattern of firing by its neighbors is independent of the amount of excitation in the test ommatidium—that is, for a given $\sum_j k_{pj}(r_j - r_{pj}^0)$, $e_p - r_p$ is the same for all values of e_p.

Figure 7 shows that the equivalent circuit model predicts approximate independence—the decrement in firing rate does not change very much as the excitation on the test ommatidium is changed.

In summary, the equivalent circuit model (Purple, 1964; Purple & Dodge, 1965) that explains both lateral and self-inhibition in terms of known cellular mechanisms agrees well with Equation 2 in three fundamental properties: the inhibition is recurrent and linear above threshold, there is spatial summation of inhibition, and the decrement in firing rate caused by a given amount of inhibition is independent of excitation on the test ommatidium. Hence, one can reasonably say that the cellular mechanisms incorporated into the Purple and Dodge model 'explain' the equations that were based entirely on input-output relations.

Those details on which the cellular (equivalent circuit) model and Equation 2 disagree with one another or with direct experimental observations are interesting and can lead to fúrther refinements of both theory and experiment. Barlow and Quarles (personal communication, 1974) found that the magnitude of the response to a 'Mach band' pattern measured in the *Limulus* eye differed somewhat from the magnitude of the response predicted by the abstract model of Equation 2. At that time there was already some experimental evidence (Lange, 1965) that the inhibitory decrement in the firing rate of a test ommatidium depends on the level of excitation of the test ommatidium (as it does somewhat in the equivalent circuit model). Further experiments (Barlow & Lange, 1974) show this to be a substantial effect, especially at higher levels of excitation much above those generally used in routine experiments. Taking this dependence into account, Barlow and Quarles then found excellent agreement between experimental data and theoretical prediction (see Fig. 4).

2.3. Dynamics

To extend the steady-state model (either the abstract version in Equation 2 or the physiological equivalent circuit) to include the dynamics requires the inclusion of functions representing the time-courses of each of the three component processes: excitation, lateral inhibition, and self-inhibition. Within the last few years, two complementary methods have been successfully used to determine the time-courses of these processes. In one method, the time-courses were deduced by Fourier analysis from measurements of the response of the ommatidium, both membrane potential and firing rate, to stimulation sinusoidally varying in time. In a linear system or within small stimulus ranges in any system, the response of the system to a sinusoidal stimulus is itself a sinusoid of the same frequency differing from the stimulus only in phase and amplitude.[3]

[3] A linear model is, roughly, an adding and subtracting model—a model in which the response to the sum of two inputs is the sum of the responses to each input alone. More formally, a linear model is a model in which the rule that assigns responses to inputs is a

Thus, to specify the response of the system to all sinusoids one need only specify the 'transfer function,' a function that gives the amplitude (difference between peak and trough) and phase of the sinusoidal response as a function of the frequency of the sinusoidal stimulus. Further, the response of the system to any stimulus can be calculated from the response to sinusoidal stimuli. And, as a last advantage, the transfer function of a system containing several parts can easily be specified in terms of the transfer functions of each part separately. (See Cornsweet, 1970, for an introduction to the use of transfer functions in the study of vision. A number of textbooks describe 'Fourier analysis' or 'linear analysis.')

The second method for studying the time-courses was to measure directly the temporal variations in the membrane potential and conductance occurring during excitation and after inhibitory impulses. These experiments agree with the deductions from the measured transfer functions.

Let us now consider in turn the temporal characteristics revealed by these two methods for the processes underlying excitation, self-inhibition, and lateral inhibition, and some of the consequences for the dynamics of the retinal network.

Excitation. The excitatory component of the generator potential appears to be the sum of numerous 'quantum bumps,' discrete miniature potentials in the depolarizing direction produced by light excitation of the receptors (Yeandle, 1957; Fuortes & Yeandle, 1964; Adolph, 1964; Dodge, Knight, & Toyoda, 1968). The dynamics of excitation, therefore, are determined by the time-course of these quantum bumps. As shown in Figure 8, each quantum bump has a moderately abrupt onset and a slower exponential decay. But the amplitude and duration of these quantum bumps are not always the same. The amplitude and duration of the quantum bumps are markedly dependent on—that is, become adapted to—the mean level of steady illumination (Fig. 9). As the light becomes brighter, the amplitude of each individual quantum bump becomes smaller and the duration becomes shorter. Thus, although the rate of occurrence of these quantum bumps increases in proportion to light intensity, the concomitant decreases in their amplitude and duration are such that in the steady state there is an approximately logarithmic relation between light intensity and the sum of the quantum bumps—the generator potential (solid line of Fig. 9). Further, at low enough frequencies of sinusoidal stimulation, where there is sufficient time during each cycle for some adaptation of the quantum bumps to occur, the generator potential response

---

linear transformation. For example, the model expressed by Equation 2 gives the responses r_j as a linear transformation of the excitations e_j as long as the r_j are above the thresholds, r^0_{pj}. Any nonlinear model can be approximated by a linear model when the possible inputs are constrained to a small range.

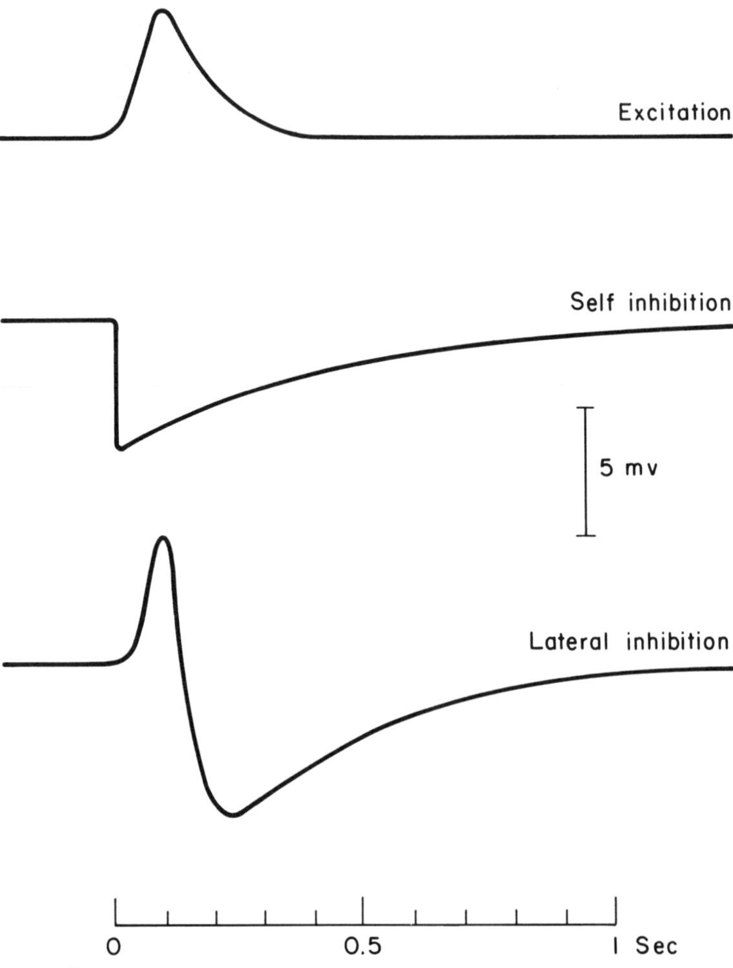

FIGURE 8.
Time-courses of the electrical events underlying excitation, self-inhibition, and lateral inhibition. A discrete miniature excitatory potential (quantum bump) may result from absorption of a single photon of light. A self-inhibitory potential in a given ommatidium follows the discharge of each impulse by that ommatidium. A lateral inhibitory potential follows the discharge of each impulse by neighboring ommatidia. (From Ratliff et al., 1974.)

at each moment approaches that given by the steady-state function (solid line). The change in generator potential resulting from a rapid change in intensity, however, is too fast for significant adaptation of the quantum bumps to take place and thus depends only on rate of occurrence of the quantum bumps. Therefore, at high frequencies of sinusoidal stimulation, the generator potential (the sum of the quantum bumps) at each moment is

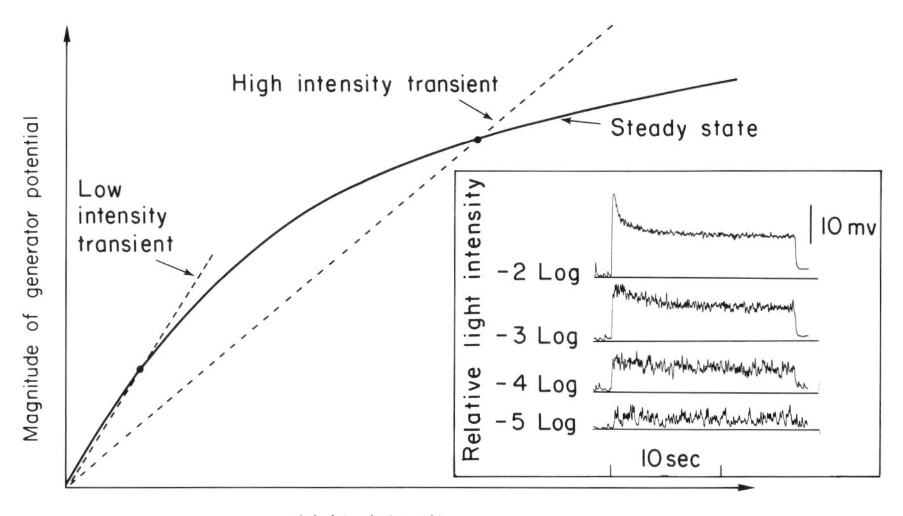

FIGURE 9.
Magnitude of generator potential as a function of light intensity. The solid line shows the approximately logarithmic stimulus-response function obtained for steady-state stimulation. The two dashed lines show the linear stimulus-response function obtained for transient changes in intensity level around a high mean intensity level (right-hand dashed line) and around a low mean intensity level (left-hand dashed line). The intensity levels at which the dashed lines cross the solid curve correspond to the mean intensity levels about which the transient variations occur. *Inset:* Generator potential records for long duration light stimuli of different intensities. Nerve impulses were blocked by tetrodotoxin. The generator potential appears to be the sum of small discrete changes in potential, called quantum bumps. At the lowest intensity, (−5 log), the bumps are infrequent and of maximum amplitude. Thus the generator potential, the sum of the individual bumps, is very irregular, and one can easily distinguish individual bumps in the generator potential record. As intensity increases, the frequency of the bumps increases in direct proportion but their amplitude and duration decrease somewhat. Thus, as intensity increases, the records of the generator potential become much smoother. (Graph redrawn from Dodge, 1969; records from Dodge, Knight, & Toyoda, 1968.)

directly proportional to light intensity (dashed curves). The constant of proportionality depends on the mean level of the sinusoidal stimulus, being smaller for high mean intensity stimuli where the individual quantum bumps are smaller (right-hand dashed curve) and larger for low mean intensity stimuli where the individual quantum bumps are larger (left-hand dashed curve). Consequently, the transfer functions of the excitatory process will have different shapes for high and low mean intensity stimuli (Fig. 10). At high mean levels of illumination, the (peak-trough) amplitude of the response to low frequencies will be smaller than the amplitude of the response to higher frequencies because the slope of the steady-state stimulus-response function (solid line in Fig. 9) is smaller than the slope of the high-frequency stimulus-response function (dashed line in Fig. 9). In other words, the ampli-

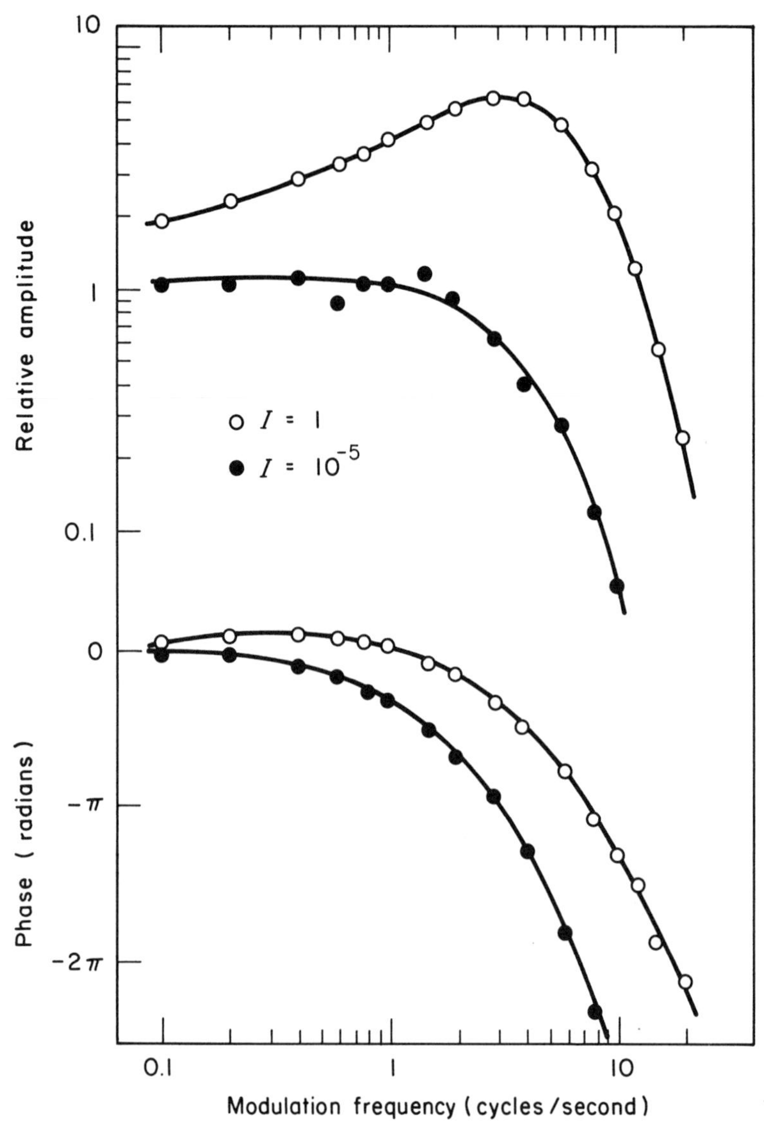

FIGURE 10.
Temporal transfer function for the excitatory process. The peak-trough amplitude and phase of the generator potential in response to light that was sinusoidally varying at various frequencies for two mean luminance levels. (Impulse generation was blocked by tetrodotoxin so the lateral and self-inhibitory processes could not occur.) (From Dodge, 1969.)

tude characteristic of the transfer function shows a 'low-frequency decline.' At low mean levels of illumination, however, there is no low-frequency decline because the two stimulus-response curves do not differ.

At all mean intensities, the duration of the quantum bumps becomes an important factor as the frequency of the sinusoidal stimulus gets so high that its period is not much longer than the duration of individual quantum bumps. For these high frequencies, the relatively long duration of the individual quantum bumps causes 'smearing' or 'integrating' of the responses initiated at different times in the stimulus cycle. Thus the peak-trough amplitude in the response to a high-frequency stimulus is diminished or, in other words, there is a 'high-frequency' cutoff in the transfer function.

In addition to determining the temporal transfer functions for the excitatory process, the properties of the quantum bumps determine another dynamic feature of the ommatidium's response—the small, but significant, variability in interimpulse intervals. That the variability in the interimpulse interval decreased with increased light intensity much the way that variability in the generator potential did suggested that the variability among intervals was coupled to the variability of the generator potential (Ratliff, Hartline, & Lange, 1968). Using 'spectral analysis,' mathematical methods related to the Fourier analysis methods discussed earlier, this hypothesis was confirmed. By considering two sources of variability—that variability inherent in the generator potential of the test ommatidium itself (resulting from the quantum bumps, see Fig. 9) and that variability transmitted to the test ommatidium via inhibition from the generator potentials of neighboring ommatidia—one can completely account for the variability in the train of impulses produced by the test ommatidium (Shapley, 1971a, 1971b).

Self-inhibition. The low-frequency decline observed in the transfer function for light intensity to impulse rate (see Fig. 11) depends not only on the adaptation of quantum bumps but also on self-inhibition—the inhibition that each impulse discharged exerts back on its own generator potential. The self-inhibitory potential after each impulse is hyperpolarizing and has an abrupt onset and an exponential decay with a relatively long time constant of about 0.5 sec (Fig. 8). Self-inhibition therefore depresses the steady response to steady-state stimulation and also reduces the peak-trough amplitude of the response to low frequencies of modulation. However, due to its long time constant, it cannot follow high frequencies of modulation and thus does not affect the peak-trough amplitude of the response to high-frequency stimuli (Knight, Toyoda, & Dodge, 1970; Biederman-Thorson & Thorson, 1971). Unlike adaptation of quantum bumps, self-inhibition produces a low-frequency decline at *all* mean luminances. The time constant of self-inhibition—indeed, the concept of self-inhibition itself—was first deduced by mathematical analyses of changes in firing rate (see Fig. 1) resulting from step increments and decrements in the stimulus (Stevens, 1964). Direct electrophysiological observations (Purple & Dodge, 1965) later confirmed this deduction. Although simple in principle, the direct linkage of the self-inhibitory potential to the occurrence of an impulse makes the theoretical analysis of the self-inhibitory feedback complex. For details, see Knight, Toyoda, and Dodge (1970).

Lateral inhibition. Because of the characteristics of the quantum bumps and the self-inhibition, a low-frequency decline is an inherent property of the transfer function for light intensity to impulse rate of any individual ommatidium, and the decline is more pronounced at higher mean luminances. And when neighboring ommatidia are also illuminated, the lateral inhibitory potentials that they produce contribute still more to the low-frequency decline because they, like self-inhibition, are of relatively long duration compared to the quantum bumps (Knight, Toyoda, & Dodge, 1970; Biederman-Thorson & Thorson, 1971). Thus they can depress the responses to low frequencies but not the responses to high frequencies. The lateral inhibitory potential, however, has a more complex time-course than does the self-inhibition (Fig. 8). There is first a brief excitatory depolarization followed by a long lasting hyperpolarization. The form of the lateral inhibitory potential was, like the time-course of the self-inhibitory potential, first deduced by mathematical methods—it was determined by taking the Fourier transform of the lateral inhibitory transfer function (modulation of impulse rate in neighbors to modulation of generator potential in the test ommatidium). In subsequent experiments the time-course of the lateral inhibitory potential following each impulse discharged by neighbors was directly observed by electrophysiological means and the mathematical deductions confirmed. Although the amplitude of the inhibitory potential varies greatly with distance between test and inhibiting ommatidia, the time-course evidently does not change with distance; lateral inhibitory transfer functions measured at different distances show large changes in the magnitude of the amplitude characteristic but no significant changes in phase characteristics (Ratliff, Knight, Dodge, & Hartline, 1974). The above time-course for lateral inhibition applies only to inhibition above threshold. The dynamics of the threshold for inhibition are only beginning to be explored (Lange, Hartline, & Ratliff, 1966a; Ratliff, Hartline, & Lange, 1966; Graham, Ratliff, & Hartline, 1973).

Composite spatiotemporal properties. Once the dynamic properties of the three component processes—excitation, self-inhibition, and lateral inhibition—have been expressed in terms of their time-course (Fig. 8) or in terms of their transfer functions, either form can be used to extend Equation 2 to include the dynamic properties. The transfer function form, used because of its greater simplicity and its closer relationship to the majority of the experimental data, gives the following dynamic equations:

$$r_p(f) = g(f)I(f) - k_{pp}T_s(f) - T_L(f) \sum_{p \neq j} k_{pj} r_j(f), \qquad (3)$$

where $r_p(f)$ is the peak-trough amplitude in the firing rate of ommatidium p in response to a sinusoidal light stimulus of frequency f, $I(f)$ is the peak-trough amplitude of the light stimulating p, $g(f)$ is the transfer function of the excitatory process, $T_s(f)$ is the transfer function of the self-inhibitory process, $T_L(f)$ is the transfer function of the lateral inhibitory process, and the k_{pj} and k_{pp} are the lateral and self-inhibitory coefficients already introduced. ($T_s(f)$ and $T_L(f)$ are normalized so that in the steady state these equations will reduce to Equation 2. Because the time-course of lateral inhibition is independent of distance, the same transfer function $T_L(f)$ can apply to all pairs of ommatidia.) Notice that these equations apply only for suprathreshold stimuli, and introduce a minor error by ignoring the direct linkage between the lateral and self-inhibitory potentials and the individual impulses that produce them.

Equation 3 can be used to predict fully the behavior of the *Limulus* retina, including responses to simultaneous variations of stimuli in both space and time. Figure 11 shows theoretical predictions based on these equations and an experimental confirmation of them (Knight, Toyoda, & Dodge, 1970). The predicted light-to-impulse-rate transfer function when a spot of light is confined to the test ommatidium is shown by the solid lines and the corresponding measured transfer function by the filled circles. The predicted transfer function and data points when the spot of light is enlarged to illuminate not only the test receptor but also about 20 of its neighbors are shown by the dashed lines and open circles. Notice that in both situations there is a high-frequency cutoff resulting from limitations of the generator mechanism and a low-frequency decline due to adaptation of the quantum bumps and self-inhibition as already discussed. When the neighbors are illuminated, the low-frequency decline is further accentuated by lateral inhibition from them.

In both cases the high-frequency cutoff and the low-frequency decline 'tune' the system (network) to best transmit the intermediate frequencies. Notice, however, that the peak-to-trough amplitude of the response at the best tuned frequency is greater with lateral inhibition (large spot) than it is without (small spot) (Ratliff, Knight, Toyoda, & Hartline, 1967). This 'amplification' is a direct consequence of the long delay (about 150 msec) to the peak of the inhibitory potential (Ratliff, Knight, & Graham, 1969). Because of this delay, for stimulation at the best tuned frequency (period about 300 msec), the inhibition will be maximal at the trough of the response and minimal at the peak, thus tending to produce the greatest possible peak-to-trough amplitude.

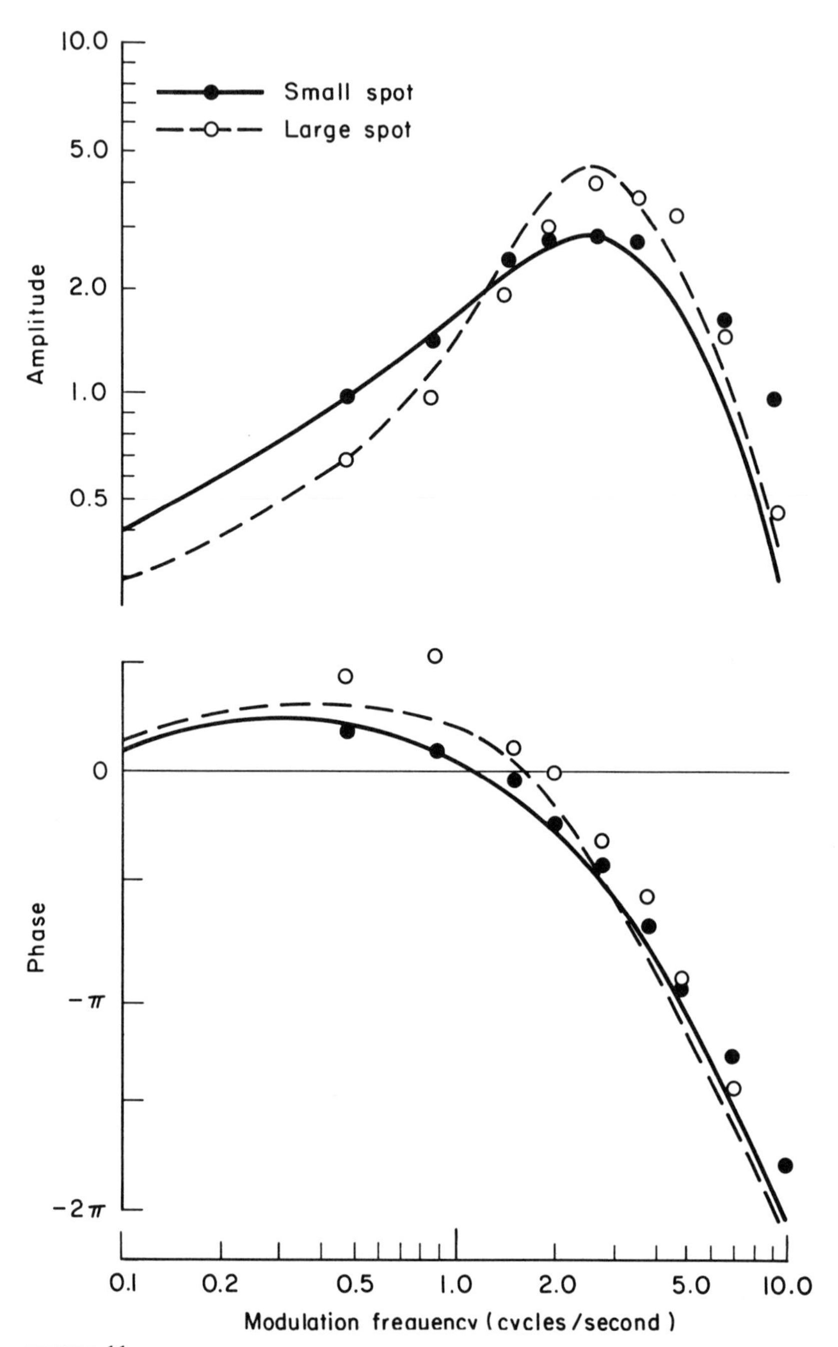

FIGURE 11.
Theoretical and observed transfer functions for light-to-impulse rate. Filled circles show peak-trough amplitude and phase of impulse rate for a single ommatidium illuminated alone by a small spot of light. Open circles show amplitude and phase for that same ommatidium when the spot of light centered on it was enlarged to include 18 neighboring ommatidia. Amplitude is given in sec^{-1} peak-to-peak. Mean rate was 25/sec for solid circles and 20/sec for open circles. (Adapted from Knight, Toyoda, & Dodge, 1970.)

There is no question that these effects are due to the delayed lateral inhibition. It is easy to demonstrate this by introducing further artificial delays. This can be done simply by delaying the stimulus to the neighbors that are exerting the inhibition on the test receptor (Ratliff, Knight, & Milkman, 1970). As the delay is increased the peak of the transfer function becomes considerably amplified and shifts to lower and lower frequencies. As predicted by the theory, second-order maxima and minima also appear as the delay is lengthened.

It is easy to transfer to the spatial domain the basic concepts of Fourier analysis that were discussed above for the temporal domain. Instead of sinusoidal variations in time, one simply considers sinusoidal variations in space—and, instead of the temporal distributions of excitation and inhibition, the spatial distribution. The resulting spatial transfer function shows a pronounced maximum at intermediate spatial frequencies. The broad extent of the inhibitory field (Fig. 3) produces the low-spatial-frequency decline and the limitations imposed by the optical apparatus and dimensions of the receptor units account for the high-spatial-frequency cutoff. Theoretical predictions even show an amplification at intermediate spatial frequencies (because of the distance to the point of maximum lateral inhibition as shown in Fig. 3) analogous to the amplification at intermediate temporal frequencies. This analogy between the spatial and temporal frequency domains has not yet been demonstrated by experiment.

A most important property utilized by Fourier methods of analysis is that any function may be expressed as the sum of sinusoids of various frequencies, amplitudes, and phases. Thus in a linear system or over a small range in any system the response to any stimulus may be calculated as the sum of the responses to the component sinusoids. An example of such a calculated response along with the response measured directly is illustrated in Figure 12. A group of receptors was illuminated with temporal square-wave modulation (a train of abrupt steps above and below the mean luminance) while a neighboring unit was illuminated steadily. From the previously measured responses of the group of receptors to sinusoidal stimulation (excitatory transfer function) and the concomitant inhibitory effect of these sinusoidal responses on the steadily illuminated neighbor (lateral inhibitory transfer function), the predicted responses to the square-wave modulation were calculated using the theory of Equation 3. The actual responses to square-wave modulation, both the directly measured excitatory responses in the group and the indirect inhibitory effects that were produced in the neighbor, agree very well with the calculations (Ratliff et al., 1974).

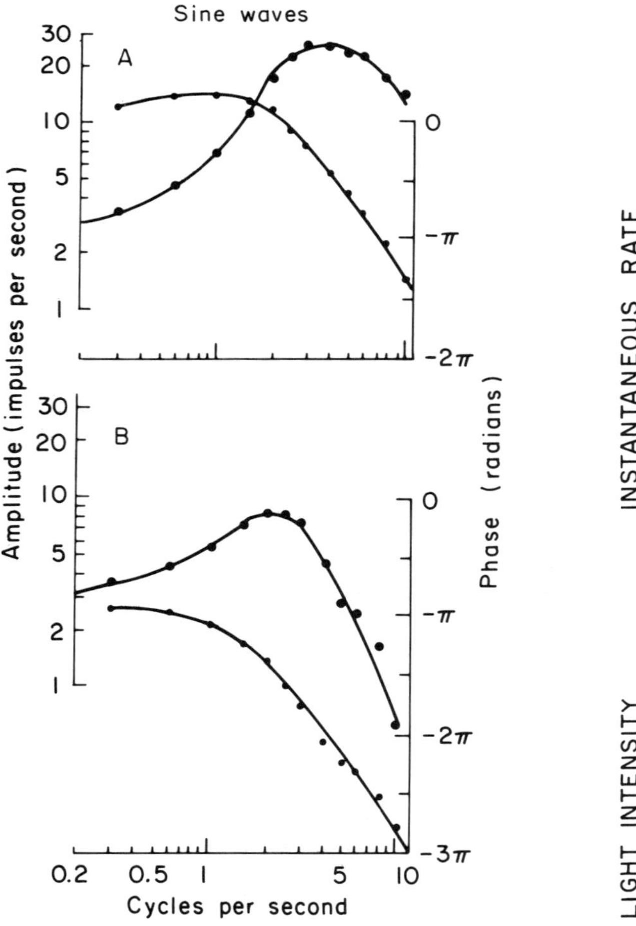
Sine waves
A
B
Amplitude (impulses per second)
Phase (radians)
0
$-\pi$
-2π
-3π
30
20
10
5
2
1
0.2
0.5
1
5
10
Cycles per second

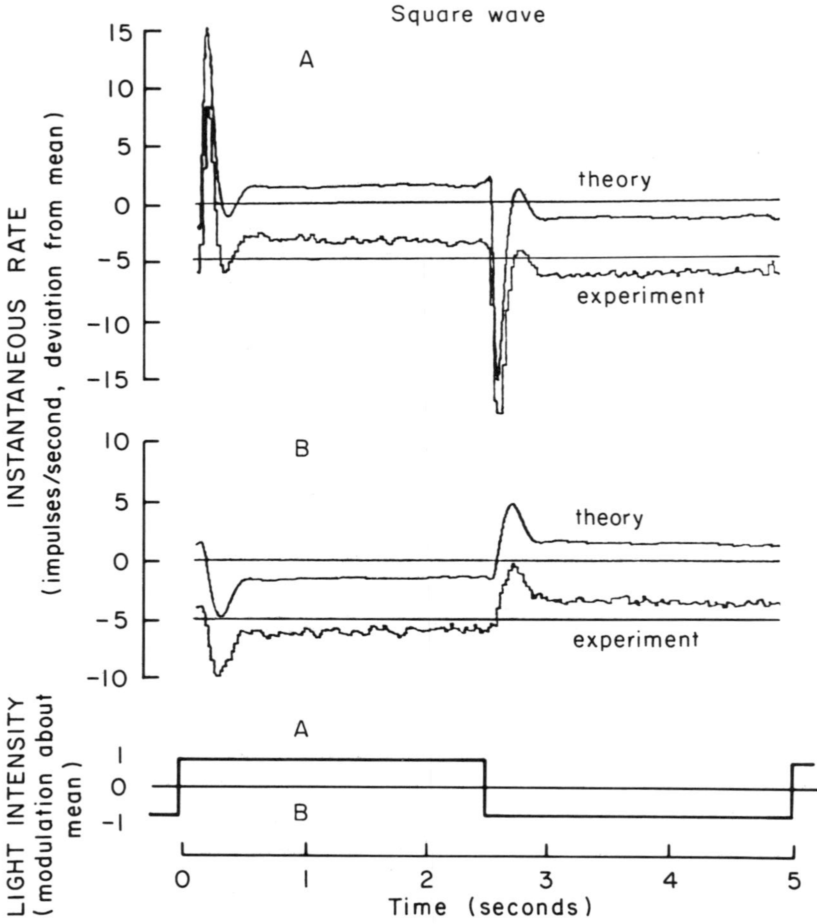
Square wave
A
B
INSTANTANEOUS RATE
(impulses/second, deviation from mean)
15
10
5
0
-5
-10
-15
theory
experiment
LIGHT INTENSITY
(modulation about mean)
1
0
-1
0
1
2
3
4
5
Time (seconds)

2.4. Summary

The responses of the *Limulus* retina to various arbitrary spatial and temporal patterns of illumination have been predicted accurately using a set of n simultaneous equations—one equation for each of the n receptor units and in each of which the three basic processes of excitation, self-inhibition, and lateral inhibition are represented as separate spatiotemporal transfer functions.

Excitation of a receptor unit in *Limulus* can only be produced by illumination falling directly on it. Each absorbed quantum of light seems to produce a brief miniature excitatory potential—a quantum bump—and these quantum bumps add together to form the excitatory component of the generator potential. The nonlinearities found in the response of the *Limulus* receptor unit as light intensity is raised (decreased sensitivity and changes in temporal response characteristics) are primarily due to the properties of this excitatory process: The amplitude and duration of the individual quantum bumps become smaller as the number of recent photon absorptions becomes larger. The variability in the response of the receptor unit is also due to the excitatory process: The variability inherent in the quantal nature of light is apparently reflected in the occurrence of quantum bumps and is thus transmitted to the train of impulses discharged by the ommatidia.

Self-inhibition also involves only a single receptor unit. Each impulse discharged down a receptor unit's axon produces, in that same receptor unit, a relatively long duration (about 0.5 sec) inhibitory potential that subtracts from the excitatory potential caused by light. This self-inhibition is an important factor in many of the temporal features of a receptor unit's response, including the transient peak in the response to an increment of light.

The lateral influence of neighboring receptor units in the *Limulus* retina on one another is inhibitory. Each impulse discharged down one unit's axon

---

FIGURE 12.
Theoretical predictions and experimental observations of concurrent excitatory and inhibitory responses in neighboring ommatidia. *Left-hand graph:* The excitatory transfer function (A) shows the peak-trough amplitude (left ordinate) and phase shift (right ordinate) of the rate of discharge of one member of a group of ommatidia stimulated by sinusoidally modulated illumination. The inhibitory transfer function (B) is the peak-trough amplitude and phase shift of the rate of discharge of a neighboring ommatidium, which though illuminated steadily was inhibited by the group of ommatidia responding to sinusoidal illumination. *Right-hand graph:* The excitatory response (A) is the discharge rate of one member of the group of ommatidia stimulated by temporal square wave illumination. The inhibitory response (B) is the discharge rate of the one neighboring ommatidium, which though illuminated steadily was inhibited by the ommatidia responding to the square wave illumination. The theoretical curves are predicted from the responses (left-hand graph) to the sinusoidal stimuli. The experimental observations are displaced 5 impulses/sec downward from the theoretical predictions. (From Ratliff et al., 1974.)

will produce, in another unit, an inhibitory potential of a duration (0.15 sec) that is shorter than the duration of the self-inhibitory potential. (The lateral inhibitory potential is always preceded by a small brief lateral excitatory potential.) The magnitude of the lateral inhibition depends on the distance between the two receptor units, being largest in general when the two units are three or four units apart although never nearly as large as the self-inhibitory potential. Due to the differences in their spatial and temporal characteristics, the interaction of lateral inhibition with self-inhibition and excitation can produce many complicated spatial and temporal effects, of which the best known are probably the 'Mach bands' or 'border contrast' effects and the 'tuning' to relatively narrow bands of temporal frequencies.

3. INTEGRATIVE ACTION OF RETINAL GANGLION CELLS IN CAT AND GOLDFISH

3.1. General Model

Some characteristics of retinal ganglion cells. The Sherringtonian concept of the *receptive field* is basic to an understanding of the integrative action of the vertebrate retina. It was first applied by Hartline (1938a) to the study of vertebrate retinal ganglion cells (those cells whose axons form the optic nerve that extends from the eye to the central nervous system). In Hartline's words (1941): "The retinal region occupied by visual sense cells whose connections converge upon a given retinal ganglion cell shall be termed the receptive field of that ganglion cell. . . . Not only do excitatory influences converge upon each ganglion cell from different parts of its receptive field, but . . . inhibitory influences converge as well."

Hartline observed three basic types of ganglion cells in the frog retina: 'on,' 'off,' and 'on-off'—so named because they responded most actively at the onset of light, the offset of light, or at both onset and offset. He noted the existence of intermediate types, however, and it has since become evident that there are many others that are highly specialized in other respects (sensitivity to motion, orientation, color, etc.). Since vertebrate retinas are far more complex than the retina of *Limulus* and the recording and analysis of unit activity in them more recent, quantitative theories of their integrative action are less complete. However, in the cat and goldfish the retinal ganglion cells have been studied extensively in recent years, and a quantitative model of their discharge of impulses in response to illumination is beginning to emerge. And despite the ages of evolution separating these vertebrates from *Limulus* and despite the great differences between the anatomy of vertebrate retinas and that of *Limulus* (compare Figs. 2 and 13), the behavior of the cat and

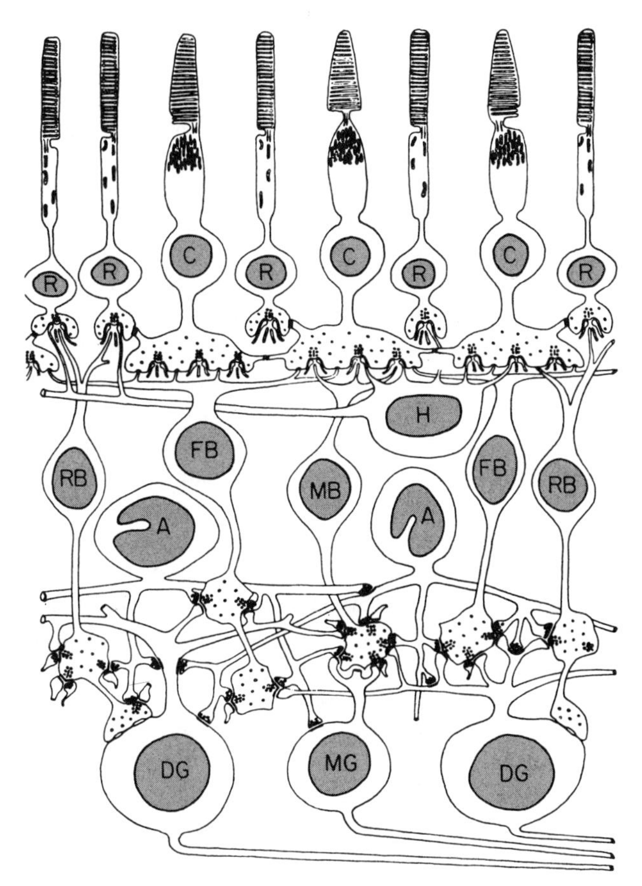

FIGURE 13.
Schematic diagram of a vertebrate retina. At the top are the photoreceptors—the rods (R) and the cones (C). Linking them with the ganglion cells (DG, MG) are the bipolars (RB, MB, FB). More extensive lateral connections are made by the horizontal cells (H) and amacrine cells (A). (Adapted from Dowling & Boycott, 1966.)

goldfish retinal ganglion cells is turning out to be remarkably similar to the behavior of the *Limulus* eccentric cell in many respects.

The spatial separation of opposed excitatory and inhibitory processes that was found in the *Limulus* retina is found also in the cat and goldfish retinas.[4]

[4] The primary emphasis of the following is on the retinal ganglion cells themselves, rather than on the relationship of these neurons to the functioning of the whole visual system. Thus facts that may be very important in understanding significant phenomena of visual perception but that are not important in discussing the working of these individual neurons

Such a spatial separation in a vertebrate retina was first discovered in the cat by Kuffler (1953) and later in the goldfish by Wagner, MacNichol, and Wolbarsht (1963). From a given retinal ganglion cell, light shining in one area of the retina elicits an 'on-type' response characterized by an increase in firing rate at the onset of a light and a decrease in firing at the offset of light, and light shining in another area elicits an 'off-type' response characterized by a decrease in firing rate at the onset of light and an increase in firing at the offset of light. These two opposing types of response are illustrated in Figure 14 both as photographs of the actual nerve potentials displayed on an oscilloscope face and as poststimulus-time-histograms (counts of the number of nerve impulses occurring in each short time period). For this goldfish cell, responses of the off-type are elicited from a central circular area; responses of the on-type are elicited from a surrounding annular area. Therefore the cell is said to have an off-center on-surround receptive field for light of 710 nms. About an equal number of cells of the opposite organization (on-center, off-surround) are also found. In the cat, response type is (almost) independent of the wavelength of light, but in the goldfish a cell may have an on-center off-surround organization for some wavelengths and an off-center on-surround organization for others. (Spekreijse, Wagner, & Wolbarsht, 1972, present in detail the various wavelength dependencies found in goldfish retinal ganglion cells.)

In any of these various types of cells, two spots of light that individually elicit opposite types of response tend to cancel each other's effect when simultaneously presented. Thus, for example, for the cell of Figure 14 both the 'off' response to a spot of light in the center of the cell's receptive field and the 'on' response to an annulus of light in the surround of the receptive field will be smaller when both spot and annulus are presented simultaneously than when each is presented alone. As is expected from such an antagonistic spatial organization (see Fig. 4 and discussion following Fig. 11), effects analogous to Mach bands are observed in the responses of retinal ganglion cells to edge patterns (Baumgartner, 1961; Enroth-Cugell & Robson, 1966), and the spatial transfer function of these cells typically shows a peak sensitivity at medium spatial frequencies (Enroth-Cugell & Robson, 1966).

Temporal characteristics of the responses of cat and goldfish retinal ganglion cells are also similar to the characteristics of *Limulus* responses. The

---

are mentioned only briefly. An example of one such fact is the variability of receptive field size. In the cat and primate (but not goldfish), the size of receptive fields is not uniform even for receptive fields in one part of the retina (Wiesel, 1960; Hubel & Wiesel, 1960; Enroth-Cugell & Robson, 1966). This variation in size may have important consequences for the perception of patterns containing units of different size (Enroth-Cugell & Robson, 1966; Campbell & Robson, 1968; Thomas, 1970; Enroth-Cugell & Shapley, 1973b), but such psychophysiological relations cannot be treated in detail here—they are beyond the scope of this chapter.

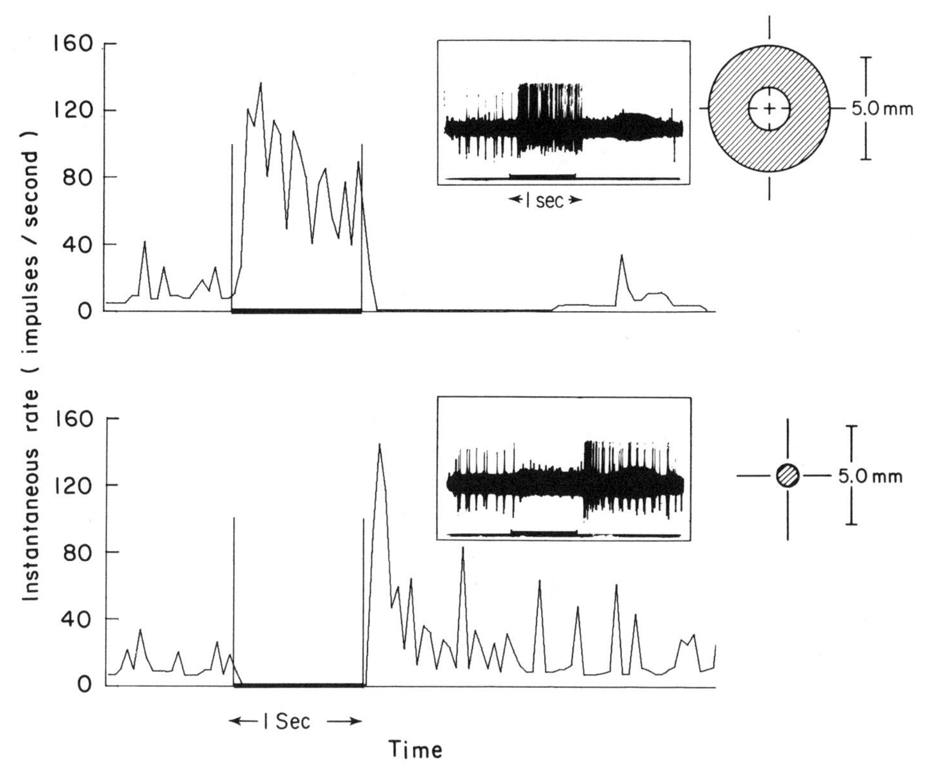

FIGURE 14.
Responses of a goldfish retinal ganglion cell to 710 nm light shining in various places in its receptive field. The responses are shown both as photographs of the nerve potentials displayed on an oscilloscope face (insets) and as impulse histograms—the number of spikes occurring per 40 msec bin averaged over 10 stimulus presentations. The 1-second period during which the stimulus was on is indicated by a heavy dark bar. The stimulus for each record is illustrated on the right. The upper record shows the on-response to illumination of the receptive field surround; the bottom record shows the off-response to illumination of the receptive field center. (Adapted from Levine, 1972.)

transients in the responses to steps of illumination (Fig. 14) resemble those transients in the *Limulus* responses caused by 'self-inhibition' and adaptation of quantum bumps (Figs. 1 and 9). And the response of the surround mechanism of a retinal ganglion cell may well be delayed with respect to the response of the center mechanism just as the lateral inhibition in the *Limulus* is delayed with respect to the excitation.

Linear model. Rodieck and Stone (1965a, 1965b; Rodieck, 1965) proposed and tested a linear model of the cat retinal ganglion cells that incorporated the features described above and strongly resembled the model of the *Limulus* eye. Figure 15 is a diagram of their model. Compare the responses

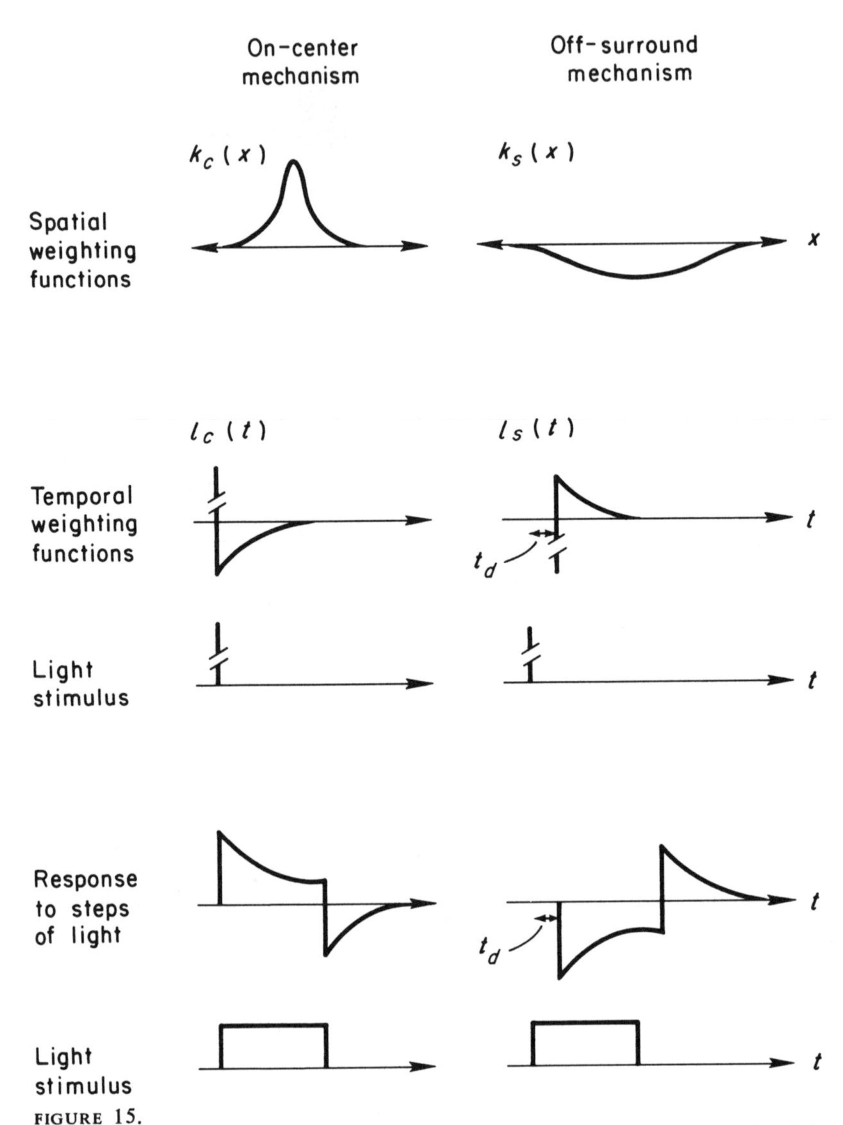

FIGURE 15.
Graphical representation of on-center and off-surround mechanisms postulated by Rodieck (1965) and Rodieck and Stone (1965a, 1965b). The top graphs show the spatial distribution of responsiveness for each mechanism. The middle graphs show the temporal weighting functions (the time-courses of the response of each mechanism to a brief pulse of light). The bottom graphs show the time-course of the response of each mechanism to steps of light. See text for definition of symbols.

computed with their model for the on-center and off-surround mechanisms of a cat retinal ganglion cell with those computed using Equation 3 for the excitatory responses and inhibitory effects in *Limulus* (Fig. 12). For an on-center off-surround retinal ganglion cell, whose receptive field is centered at point x, the response at time t, called $r(t, x)$ is postulated to be the sum of a positive input from a 'center mechanism' and a negative input from a 'surround mechanism.' For an off-center cell, the signs are reversed. Each mechanism's contribution to the ganglion cell response is the sum (or integral when writing the functions in continuous form) of the luminances at each point on the retina weighted according to that mechanism's spatial distribution and time-course. Formally

$$r(t, x) = \iint I(y, \tau)k_c(x - y)l_c(t - \tau)\, d\tau\, dy + \iint I(y, \tau)k_s(x - y)l_s(t - \tau)\, d\tau\, dy, \tag{4}$$

where $r(t, x)$ is the response of the ganglion cell, $I(y, \tau)$ is the luminance at point y and time τ, $k_c(x)$ and $k_s(x)$ are the spatial weighting functions of the center and surround mechanisms, respectively, and $l_c(t)$ and $l_s(t)$ are the temporal weighting functions (the response to an instantaneous pulse of light) of the center and surround mechanisms, respectively. The spatial and temporal weighting functions that Rodieck and Stone chose are shown in Figure 15. For the center spatial distribution $k_c(x)$ they used a narrow Gaussian function, and for the surround spatial distribution $k_s(x)$, a wide Gaussian. (Notice that both mechanisms are sensitive to light shining in the central portion of the field but that the center mechanism is much more sensitive.) The center temporal weighting function $l_c(t)$ was an instantaneous pulse of excitation (a Dirac delta function) followed by an exponentially decaying inhibitory pulse (formally analogous to 'self-inhibition'). The surround temporal weighting function $l_s(t)$ was a delay of length t_d followed by the negative of the center temporal weighting function. To provide an alternate picture of the temporal assumptions of the model, the temporal responses of each mechanism to steps of light are also shown in Figure 15. Although Equation 4 is unlike the *Limulus* model in having an 'excitatory center' that spreads over some distance and in having nonrecurrent inhibition rather than recurrent (the 'inhibitory' surround's response depends on the luminance at each point, not on the response), the resemblance to the *Limulus* model is clear—summation of two opposing mechanisms, spatial summation within a mechanism, temporal transients of an exponentially decaying form, and different time-courses for the two mechanisms.

Rodieck and Stone tested this linear model by computing its responses to various moving patterns and comparing the model's behavior to the behavior of cat retinal ganglion cells (Fig. 16). Qualitatively, the model's predictions

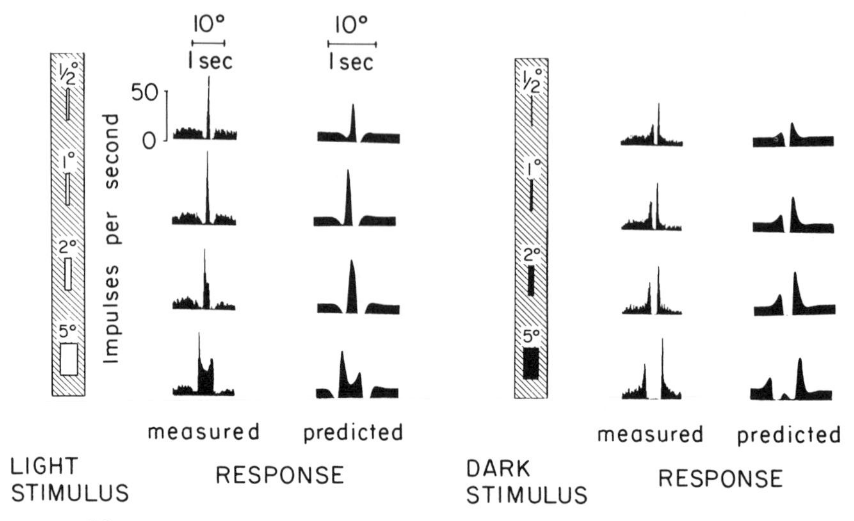

FIGURE 16.
Responses of an on-center cat retinal ganglion cell as light and dark rectangular bars of various widths were moved through its receptive field and the predictions of these responses from Rodieck and Stone's model. All bars were 10° high and moved horizontally at a constant speed of 10°/sec. The measured responses are shown as impulse histograms with 250 bins of 16 msec duration averaged over 30 repetitions of the stimulus. Maximum rate (ordinate) about 75 impulses/sec. The vertical scale of the predicted responses is arbitrary, and the horizontal scales of the predicted and measured responses are slightly different as indicated above the records on the left. (Adapted from Rodieck, 1965; Rodieck & Stone, 1965a.)

were good. However, it is easy to show that no linear model will completely describe the cat or goldfish retinal ganglion cells.

Nonlinearities. The response of a retinal ganglion cell to a given increment (a sudden increase or 'step up' in luminance) becomes smaller as the steady background on which the increment is superimposed is made more intense. Equivalently, the intensity of an increment of light necessary to evoke a response of given magnitude increases as the background intensity increases. Figure 17, adapted from Shapley et al. (1972), illustrates this latter change with background level for the response of cat retinal ganglion cells. Background intensity is plotted on the horizontal axis and the intensity of the increment that produced a criterion response magnitude is plotted on the vertical axis. (Response magnitude was measured as maximum firing rate in response to the increment minus steady firing rate to background. The criterion response magnitude was 30 impulses/sec.) The vertical axis is also labeled in units of response magnitude divided by stimulus magnitude, a ratio which is a common definition of the 'gain' of a system. (Although other definitions are sometimes used, here 'gain' will always be used this way.) As can be seen in Figure 17, the incremental intensity necessary to evoke a constant response (the gain) is approximately constant at low-intensity backgrounds. And then incremental intensity gradually increases (gain decreases) until at higher backgrounds, incremental intensity increases (gain decreases) in proportion to background intensity (producing a slope of one on the log-log plot). Similar functions have been found by other investigators (Barlow & Levick, 1969a, 1969b; Cleland & Enroth-Cugell, 1968; Maffei, Fiorentini, & Cervetto, 1971; Yoon, 1972). In a linear system, the response to a given increment of light is always the same, or in other words, gain is the same at all background intensities. Consequently, a major 'gain-changing' nonlinearity must be introduced into the retinal ganglion cell model to explain results like those of Figure 17. Further, this is not the only type of result showing nonlinearity in cat and goldfish retinal ganglion cell responses. Evidence to be considered later suggests the existence of several different nonlinear processes, a reasonable conclusion considering the anatomical complexity of the vertebrate retina.

General model. To provide a framework within which to discuss these complex processes, a diagram of a general model for cat and goldfish retinal ganglion cells is presented in Figure 18. For the goldfish retinal cell, the model to be complete should be elaborated to include the opposed color-coded processes. But, whenever, as is the case in most experiments to be discussed here, light selected to affect only one of the color-coded processes is used, the model of Figure 18 will be adequate. In the general model, there are three stages: early local processing at each retinal point; combination (separately for the center and surround mechanisms) of the early local signals originating at different places on the retina; and lastly combination of the signal from the center mechanism with the opposed signal from the surround mechanism. The model of Figure 18 certainly does not include all possible models of retinal ganglion cells. For example, one might imagine that at the first combination points, signals coming from part of the surround might be combined with those coming from part of the center, and then at the final combination points these partial center-surround combinations might themselves combine. However, the model of Figure 18 is sufficiently general to serve as a framework within which to discuss the known results from the cat and goldfish retinal ganglion cells.

What is known about the processing at each stage will be examined, starting at the final combination point and working backwards. In particular, the nonlinear or 'gain-changing' effects of varying light intensity will be discussed, but other types of nonlinear processing will also be considered. Throughout most of this discussion temporal factors will be ignored and the responses of the cells talked about as if they were in a 'steady-state.' At the end of the discussion, however, the temporal properties of cat and goldfish retinal gan-

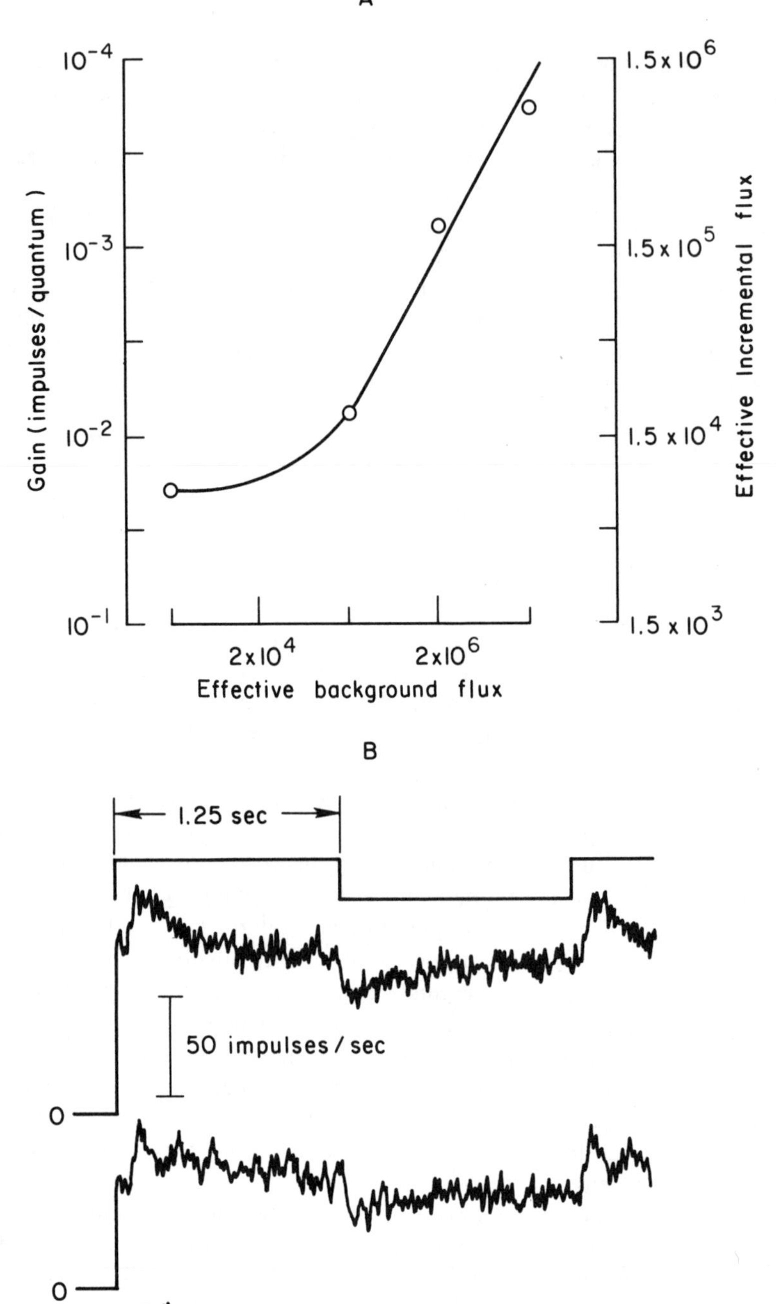
A
10^{-4}
10^{-3}
10^{-2}
10^{-1}
Gain (impulses / quantum)
1.5×10^6
1.5×10^5
1.5×10^4
1.5×10^3
Effective Incremental flux
2×10^4
2×10^6
Effective background flux
B
1.25 sec
50 impulses / sec
0
0
0

glion cells will be examined briefly. Finally, the recent division of the cat's retinal ganglion cells into two classes, *X* (sustained) and *Y* (transient), will be discussed (Enroth-Cugell & Robson, 1966). The *Y* cells exhibit more nonlinear behavior than do the *X* cells and it would be most satisfying to discuss the processing at each stage of the general model for these two classes separately. This cannot be done systematically, however, because most studies have not distinguished the two classes.

3.2. Combinations of Signals from Center and Surround

At the final combination point, the signal from the center appears simply to add to the signal from the surround. This has been well demonstrated for the cat retinal ganglion cell by Enroth-Cugell and Pinto (1970; 1972). They used an annulus flashing on and off in the periphery to elicit a response from the surround mechanism, a spot flashing on and off in the center to elicit a response from the center mechanism, and then both stimuli flashing together (in phase or antiphase). The stimuli and responses, as well as a computed sum are shown in Figure 19. The response of the cell to the two stimuli flashing together was indistinguishable from the sum of the responses to the individual stimuli as added by computer. This demonstration is particularly convincing because of certain precautions taken in the experimental procedure: (a) An artificial pupil was used and the distribution of light on the retina actually measured. (The optics of the cat eye are not good (Wässle, 1971; Bonds, Enroth-Cugell, & Pinto, 1972) and consequently the distribution of light on the retina may be very unlike that that the experimenter is trying to produce); (b) Various criteria were used to decide whether a given response was a pure-center response or a pure-surround response or mixed. Obtaining pure responses was important in this experiment because if both spot and annulus had stimulated the same mechanism (if either had elicited an impure response), early nonlinearities described in the next section would have con-

---

FIGURE 17.
(A) The 'gain' function for the response of an on-center cat retinal ganglion cell to a small spot of light in the receptive field center superimposed on a large steady background. At each intensity level of the background, the intensity of the small spot stimulus was adjusted until the magnitude of the response to the illumination of the spot was exactly 30 impulses/sec (measured as the difference between the peak of the response to the illumination of the spot and the steady response just before the spot was illuminated). Effective background flux (intensity times area of overlap between the stimulus and the receptive field center) is plotted on the horizontal axis and the incremental flux producing the criterion response magnitude is plotted on the right-hand vertical axis. 'Gain' (the criterion response magnitude divided by the incremental flux) is plotted on the left-hand vertical axis. (B) The response of the cell to the incremental stimulus at the three lower background levels. (Figure adapted from Shapley et al., 1972.)

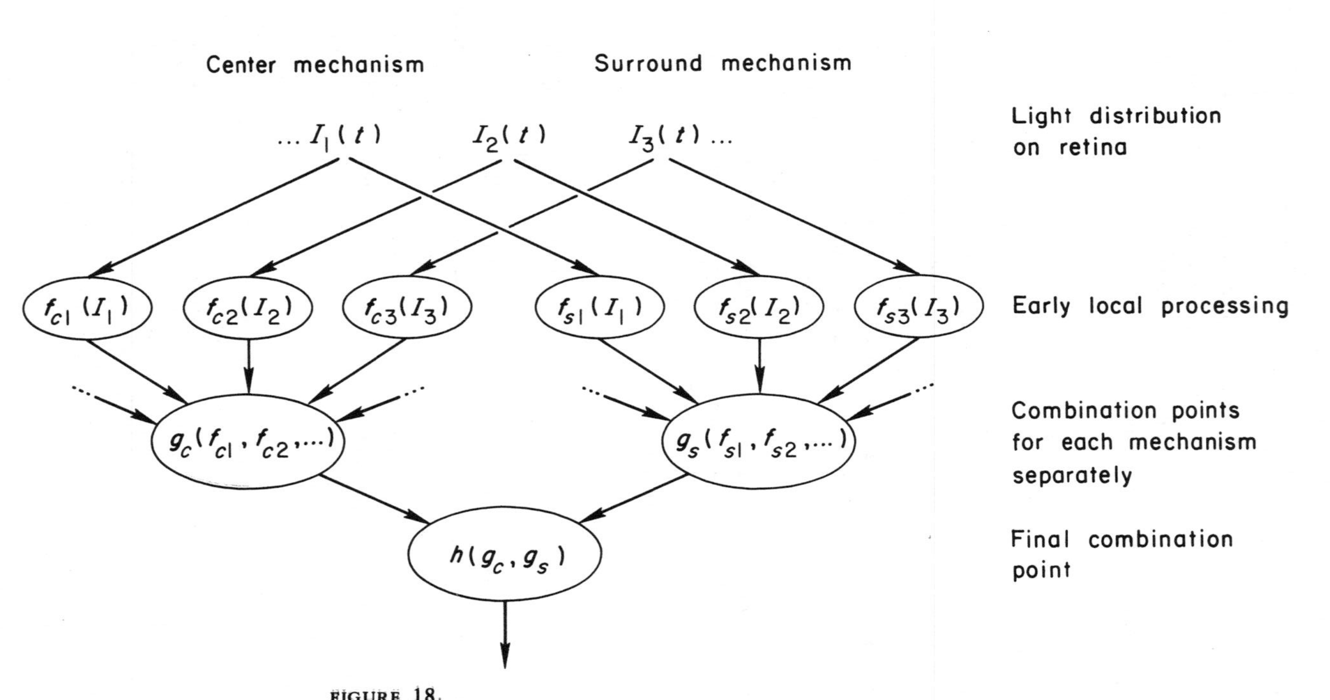

FIGURE 18.
Diagram of general model for cat and goldfish retinal ganglion cells.

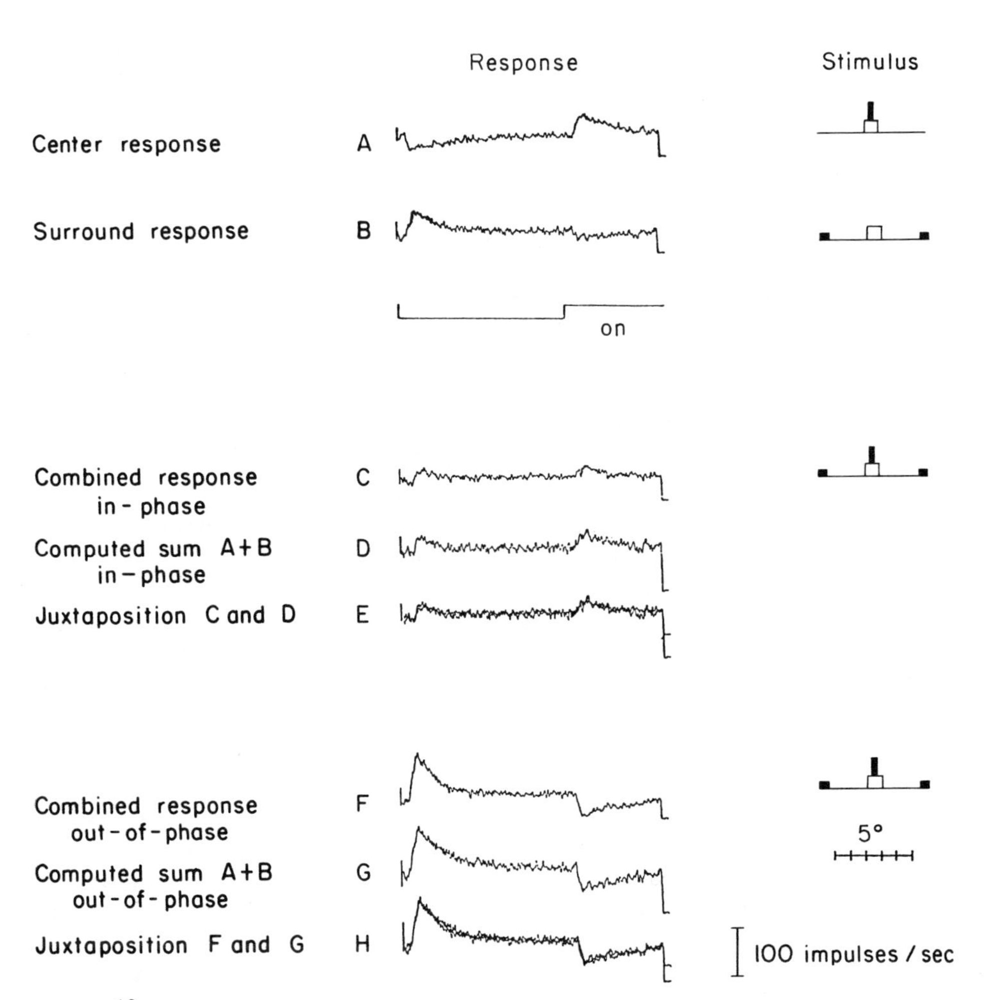

FIGURE 19.
Linear combination of responses from the center and surround mechanisms of an on-center cat retinal ganglion cell. (A) Pure center response to a spot flashing on and off in the receptive field center (indicated by filled stimulus profile on right) which was superimposed on a steadily illuminated spot (open stimulus profile). (B) Pure surround response to an annulus flashing on and off in the receptive field surround in the presence of the steadily illuminated center spot. (C) The cell's response to both the central spot and surround annulus flashing in phase (in presence of steadily illuminated center spot). (D) Computed sum of responses A and B in phase. (E) Juxtaposition of (C) and (D). (F) The cell's response to both the central spot and surround annulus flashing out of phase (in presence of steadily illuminated center spot). (G) Computed sum of responses (A) and (B) out of phase. (H) Juxtaposition of (F) and (G). Stimuli were flashing on and off at 0.4 cps. (Adapted from Enroth-Cugell & Pinto, 1972.)

cealed the linearity at the final combination point. There is evidence that a class of cells exists from which one can never get pure responses, particularly not pure surround responses (Enroth-Cugell & Pinto, 1972). If that is the case, the conclusions from this experiment cannot extend to those cells.

Another experiment (Maffei, Cervetto, & Fiorentini, 1970) indicating linearity at the final combination point for the cat retinal ganglion cell showed an interesting result of such combination. A spot in the receptive field center and a spot in the receptive field surround were flickered sinusoidally, either individually or together. The response to both was equal to the sum of the responses to each alone (Fig. 20). The differences between the transfer functions for the center and surround spots, however, lead to a complicated combination transfer function showing 'amplification' (similar to that in *Limulus*, see Fig. 11) and two peaks (also similar to *Limulus* when lateral inhibition is artificially delayed, see Ratliff, Knight, & Milkman, 1970).

Notice that linear combination of the signals from center and surround mechanisms does not guarantee that the magnitude of the combined response will be a monotonic function of light intensity, even if the magnitudes of both the center and surround mechanisms' individual responses are, for nonlinearities within the separate mechanisms may make the surround responses relatively more important in one part of the light intensity range than in another. Indeed, for on-center cells, the magnitude of the combined response (both initial transient and maintained discharge) to light of different intensities first increases as light intensity is raised but then decreases with further increases in light intensity. This decrease at higher light intensities is attributed to an increase in the relative contribution from the surround, that is, the positive signal from the center dominates at low intensities and the negative signal from the surround dominates at high intensities (Barlow & Levick, 1969b; Winters & Walters, 1970; Maffei, Fiorentini, & Cervetto, 1971). Similar results have been found for off-center cells, although there is some indication that off-center cells may differ from on-center cells in this regard.

After the basic addition of the signals from center and surround there is, at least in some situations, a 'rectification nonlinearity,' a zeroing of all responses below some fixed level. Formally,

$$h(t) = \begin{cases} g_c(t) + g_s(t), & \text{if } g_c(t) + g_s(t) \text{ greater than some fixed level,} \\ 0, & \text{otherwise,} \end{cases} \tag{5}$$

where $h(t)$ is the response of the retinal ganglion cell and $g_c(t)$ and $g_s(t)$ are the responses of the center and surround mechanisms individually. A rectification nonlinearity was suggested for the goldfish retinal ganglion cell by Spekreijse (1969) to explain the truncated appearance of responses to flickering light. The rectification nonlinearity's instantaneous nature (its dependence only on the input at the present moment, not on past inputs) has

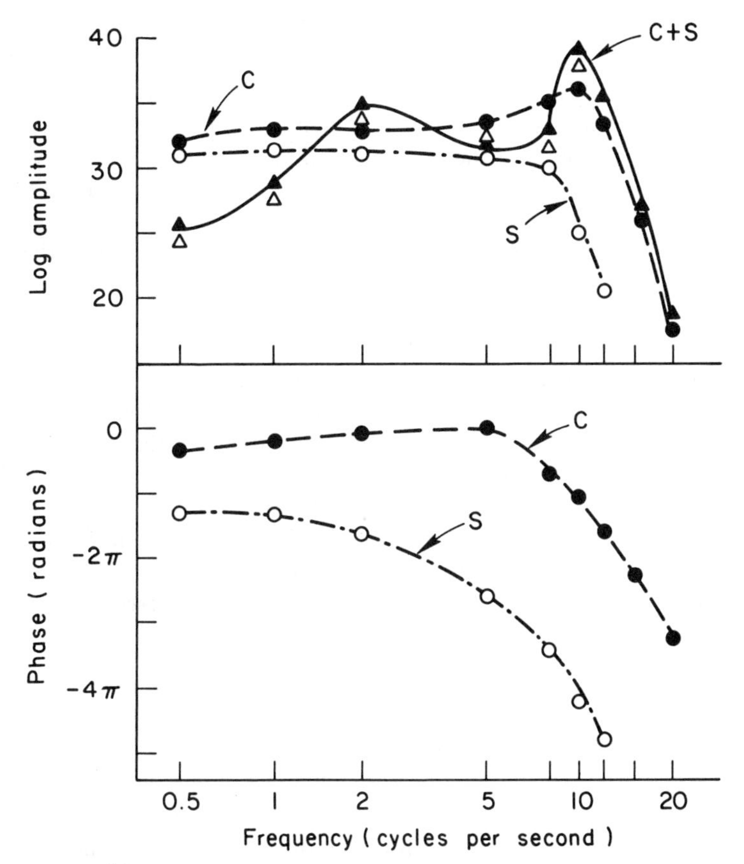

FIGURE 20.
Transfer functions for the center mechanism, the surround mechanism, and both mechanisms together of a cat on-center retinal ganglion cell. The transfer functions were obtained from the responses to sinusoidally varying illumination of a spot in the receptive field center (filled circles), of a spot in the receptive field surround (open circles), and of both spots together in phase (filled triangles). The transfer function for both spots together computed from the linear sum of the responses to the central spot alone and peripheral spot alone (considering phase as well as amplitude) is shown by the open triangles. (Adapted from Maffei, Cervetto, & Fiorentini, 1970.)

received some confirmation since the responses can be 'relinearized' by introduction of auxiliary noise signals (Spekreijse, 1969) and since the responses to white-noise stimuli and to sinusoidal flicker yield the same transfer functions as is expected if the nonlinearity is instantaneous (Schellart & Spekreijse, 1972). Spekreijse and van den Berg (1971) showed that the rectification must occur at the final combination point of center and surround after the signals from center and surround have been added. For the cat eye, Grüsser and his co-workers (Büttner, Büttner, & Grüsser, 1971, for example) report such a rectification of sine-wave responses, although several other investigators have not (Cleland & Enroth-Cugell, 1966; Hughes & Maffei, 1966, for example). This discrepancy may be due both to the depths of modulation used in the stimuli and to differences in the physiological states of the preparations.

As Enroth-Cugell and Pinto (1972) discuss in some detail, it is hard to rule out the possibility that some type of nonlinear combination is imitating linear addition at this or any other point in the model. However, if that is the case, the imitation is good.

Simple and interaction nonlinearities. The rectification nonlinearity $h(t)$ is an example of what will be called a *simple nonlinearity*—it is equivalent to a linear combination of signals coming from different sources followed by some nonlinear function of that sum. Nonlinear transformations that are not simple will be called *interaction nonlinearities*. Examples of interaction nonlinearities are (a) division of the signal from the center mechanism by the signal from the surround mechanism at the final combination point, (b) 'shunting' of the center signal by the surround signal at the final combination point (Furman, 1965; Sperling, 1970), and (c) pairwise interaction of the early local signals at the center or surround combination point such as

$$g_c(f_{c1}, f_{c2}, f_{c3} \ldots) = \sum_i a_i f_{ci} + \sum_i \sum_j b_{ij} f_{ci} f_{cj}.$$

Since interaction nonlinearities have not been considered extensively as models of retinal ganglion cell data, they will not be dealt with in detail here (but see Sperling, 1970). However, it should be remembered that, although interaction nonlinearities have not been considered extensively, neither have they been entirely ruled out.

3.3. Separate Combination Points for Center and Surround

Since the final combination of center and surround signals appears to be linear except for a rectification, the gain-changing effect of changing light intensity that is apparent in the cell's response (for example, Fig. 17) must take place prior to the final combination point. In fact, there appear to be such nonlinear effects at both the early local stage and at the separate combination points for the center and the surround mechanisms. Cleland and Enroth-Cugell (1968) provided a particularly clear and quantitative demonstration that there is a change in gain resulting from the transformation at the center combination point.

Four stimulus arrangements were used in their experiments: (a) To measure the distribution of sensitivity across the receptive field center—a small test spot of *variable* intensity at *various* locations on the retina, superimposed on a very large *constant* background stimulus; (b) To compare the threshold for several small spots with that for one spot—*variable* number and intensity of small test spots superimposed on a *constant* large background; (c) To measure the trade-off between area and intensity for the test stimulus—a circular test stimulus of *variable* area and intensity superimposed on a concentric circular background stimulus of *constant* area and intensity; (d) To measure the trade-off between area and·intensity for the adapting or 'gain-changing' background stimulus—a circular test stimulus of *constant* area and intensity superimposed on a concentric circular background stimulus of *variable* area and intensity. In each arrangement, the variable quantities were adjusted until the response to the intermittent test stimulus was of just threshold magnitude. (Thus, even if there were nonlinearities at the final combination point, they would not matter since the response $h(t)$ and thus the input $g_c(t)$ to the final combination point were being held constant.)

The results of the experiments demonstrated that the response of the center mechanism of a retinal ganglion cell to a test stimulus superimposed on a background stimulus depended only on two properties of the stimulus situation—the 'center effective flux' of the test stimulus and the 'center effective flux' of the background stimulus. Center effective flux is defined as the sum (or, in general, the integral) of the intensities at each point in the stimulus weighted by the sensitivities at each point of the center mechanism. For a spatially homogeneous stimulus and a center mechanism having approximately uniform sensitivity over the whole center, center effective flux is defined as $I \times A_0$, where I is the intensity of the stimulus and A_0 is the area of overlap between the stimulus and the center of the receptive field. In other words, Cleland and Enroth-Cugell (1968) showed that as long as the center effective fluxes of the test and background stimuli were held constant, the spatial distribution of stimulation could be varied without changing the response.

Of course, all possible spatial distributions of stimulation have not been tested so the above statement may not be true for all stimuli. Suppose, however, that it is. What would this dependence solely on center effective flux imply for the general model? First, it suggests that in this experimental situation the early local transformations for the center mechanism are linear and identical for all points within the center of the field (and of course are zero for points outside the center). Notice that combining intensities of light falling on separate retinal points to form the center effective flux $I \times A_0$ is equivalent to having all light falling within the center illuminate one retinal point. If the early local transformations were not linear or differed substantially in temporal properties, the effects of light illuminating different retinal points would have been distorted relative to one another by early nonlinearities or different time-courses and thus the light could not have acted as if it were all illuminating one point. (Although evidently linear in these experiments, the early local processing is sometimes found to be nonlinear as is discussed in detail below.)

Second, the importance of center effective flux implies that the gain-changing nonlinearity cannot be postponed until the third stage of the model (except in the trivial way of postponing some nonlinearity acting on the center signal alone until the final combination point). The experiment in which the adapting background was varied (stimulus arrangement d) shows that the responses to a test stimulus superimposed on an adapting background depend only on that part of the background illuminating the center mechanism and not at all on light illuminating the surround mechanism.

Finally, the dependence of the response on center effective flux implies that the nonlinearity acting at the center combination point must be a 'simple' nonlinearity as defined earlier. It must be equivalent to a linear addition of the early local signals coming from different points in the retina followed by some nonlinear function of that sum. For, since the early local transformations are linear and identical in this situation, the sum of the inputs arriving at the center combination $\left(\sum_i f_{ci}\right)$ is proportional to the center effective flux. Thus, the experimental results can be rephrased as follows: Regardless of the spatial distribution of the stimulus, as long as the sum of inputs to the center combination point is constant, the response of the center mechanism is constant. If the nonlinearity at the center combination point were not simple, i.e., were not equivalent to a nonlinear function acting after the sum of inputs had been formed, a constant sum of inputs would not insure a constant center mechanism response.

Cleland and Enroth-Cugell (1970) made further measurements and showed that, for the range of light intensities in which they were working, the magnitude of the response of the center mechanism $\bar{g}_c$ to a test stimulus of effective center flux ΔF superimposed on a background stimulus of effective center flux F can be expressed as

$$\bar{g}_c = \frac{k \cdot \Delta F}{(\Delta F + F)^{0.9}}. \tag{6}$$

Although these experiments could not be quantitatively repeated on the surround mechanism due to difficulties in isolating pure surround responses, Enroth-Cugell and Pinto (1972) found some evidence that the surround mechanism acts like the center mechanism: Its response to a test stimulus and the adaptation of such a response that is caused by a steady background both seem to depend on the 'surround effective flux' (intensity weighted by the spatial distribution of the surround's responsiveness) and not on other spatial aspects of the stimulus. However, in a large class of cells ('surround-concealing' cells) it was impossible to isolate a pure-surround response to any stimulus, and Enroth-Cugell and Pinto did not study these cells. Further, Cleland, Levick, and Sanderson (1973) present results suggesting that for Y cells there is *no* gain-changing nonlinearity at the surround combination point: An adapting background on one part of a Y cell's receptive field surround did not affect the response to a test stimulus on another part. For X cells, the results were mixed: Some cells were like the Y cells in this regard but others did show evidence of a gain-changing nonlinearity at the surround combination point (an adapting background on one part of the surround did affect the response to a test stimulus on another part). Since the relationship between the surround-concealing versus surround-revealing distinction and the X versus Y distinction is unclear, and since Cleland, Levick, and Sanderson did not use 'pure surround' responses, it is difficult at this time to compare the results of these two studies directly.

Further evidence of a gain-changing process at the center combination point of the cat retinal ganglion cells comes from experiments in which separate areas of the center of the receptive field are illuminated either individually or together (Büttner & Grüsser, 1968; Stone & Fabian, 1968). If the transformation at the center combination point were a linear summation of early local signals coming from different retinal points (remembering that the transformation at the final combination point is linear except for rectification), the response to the two small areas illuminated together would be equal to the sum of the responses to each alone. However, the response when they are illuminated together is a good deal smaller than the sum of the responses to each alone. Similar evidence comes from studies in which both the intensity and area of a centered stimulus were varied independently and the responses measured (Winters & Walters, 1970; Creutzfeldt, Sakmann, Scheich, & Korn, 1970).

To express this 'compressive' nonlinear summation, formulas of the following type have been suggested (Büttner & Grüsser, 1968; Grüsser, Schaible, & Vierkant-Glath, 1970; Creutzfeldt et al., 1970; Fischer & Freund, 1970; Grüsser, 1971).

$$\bar{g}_c = \frac{\sum_i \bar{f}_{ci}}{a + b \sum_i \bar{f}_{ci}}, \tag{7}$$

where $\bar{g}_c$ and the $\bar{f}_{ci}$ are measures of the magnitudes of the center mechanism's response and the early local signals, and a and b are constants to be estimated. Such formulas apply to stimuli turned on simultaneously and do not consider the effect of a steady background. Nevertheless the similarity of the two formulations is clear: As the input to the center mechanism (ΔF in Eq. 6 and $\sum_i \bar{f}_{ci}$ in Eq. 7) increases, the response of the center mechanism does not increase in direct proportion but increases at a progressively slower rate as the quantities in the denominators of the expressions become larger.

The evidence so far presented demonstrating a gain-changing nonlinearity at the center (and perhaps surround) combination points has all come from cat retinal ganglion cells. Corroborating evidence indicating nonlinearity at the center combination point has been found for the goldfish retinal ganglion cell: Certain temporal properties of responses to stimuli in the center of the receptive field, which are closely related to gain-change as defined earlier, depend on stimulus intensity summed over area rather than just on intensity (Schellart & Spekreijse, 1972); also, the responses to two small spots in the receptive field center indicate that there must be a nonlinearity at or after the point at which the signals from two spots combine (Levine, 1972; Abramov & Levine, 1974).

3.4. Early Processing

The evidence for some nonlinear early local processing in addition to the nonlinearity at the center combination point is quite extensive in the case of the goldfish retinal ganglion cell. Easter (1968a) demonstrated that the response to two small spots at a given intensity (where the spots illuminate retinal locations of equal sensitivity) is larger than the response to one spot at twice the given intensity. This result cannot be explained by a simple nonlinearity at the center combination point after linear early local processing, for, in that case, the sum of the early local signals arriving at the center combination point would be the same in both the one-spot and two-spot situations, and thus the responses would be the same. This result can be explained by a nonlinear early local transformation that compresses responses so that doubling the stimulus intensity at one retinal location does not double the early local signal from that location. Thus the sum of the early local signals from two spots of a given intensity will be greater than the early local signal from one spot of twice that intensity. Easter suggested that the nonlinear early local transformation was a square-root function of light intensity and this suggestion has been supported, at least above threshold, by Levine (1972). (As Easter discussed in some detail, these results could also be explained by an interaction nonlinearity at the center combination point, although some possible interaction nonlinearities could be ruled out by another experiment.)

This same two-spot experiment done in the receptive field centers of cat retinal ganglion cells has yielded results similar to those described above for the goldfish (Büttner & Grüsser, 1968; Grüsser, 1971; Stone & Fabian, 1968). And another kind of experiment using single stimuli of varying area also provided evidence that, for cat retinal ganglion cells, distributing center effective flux over a wide area was more effective than concentrating it in a small area (Creutzfeldt et al., 1970). These results can be explained, as for the goldfish, by a compressive early local nonlinearity (which can be followed by a simple nonlinearity at the center combination point). Thus there is a discrepancy between these studies and those discussed earlier (Cleland & Enroth-Cugell, 1968) in which as long as center effective flux was constant, whatever the spatial arrangement, the response was constant. The cause of the discrepancy is not clear. However, it may be due to differences in the range of luminances and responses in the various studies: The early local transformation may be linear for low values although nonlinear for high. Indeed, as is particularly clear in the studies by Creutzfeldt et al. (1970) on the cat and Levine (1972) on the goldfish, the evidence for early local nonlinearity is greatly reduced as the response decreases toward threshold.

In connection with the early local nonlinear (square-root) transformation shown in the center mechanism of the goldfish retinal ganglion cell, an odd coincidence appears to exist: The compressive nonlinearity is exactly balanced out by the spatial distribution of sensitivity thereby producing apparently linear spatial summation. Levine (1972) and Abramov and Levine (1974) have shown not only that the early local transformations are nonlinear, but also that they are not identical across the center of the receptive field. Unlike the cat center mechanism (Cleland & Enroth-Cugell, 1968), the goldfish center mechanism is more responsive to light shining near the midpoint of the receptive field center than to light shining further away from the midpoint[5] (Levine, 1972; Abramov & Levine, 1972; Easter, 1968a, as reinterpreted by Levine, 1972). This spatial distribution of responsiveness exactly balances out the early local square-root transformation for the case of circular stimuli centered in the receptive field, so that for these stimuli, $I \times A_0$ (intensity times area of overlap between the stimulus and the receptive field center) is exactly proportional to the summed inputs to the central combination point $\sum f_{ci}$. (In the case of an arbitrary stimulus, the summed inputs $\sum f_{ci}$ must be calculated using the fact that the f_{ci} are proportional to the square root of intensity and that the constant of proportionality varies for different retinal locations i). And so, by this odd coincidence, whenever $I \times A_0$ is held constant for centered circular stimuli, the response of the center mechanism should be constant (Ricco's law). That it is constant has been found by Easter (1968a) and Levine (1972).

Early local nonlinearity for a surround mechanism has been demonstrated for at least one kind of cell: the *Y* type cat retinal ganglion cell. For five *Y* cells studied, Cleland, Levick, and Sanderson (1973) found that a steady

[5] Receptive field centers of almost all goldfish retinal ganglion cells are about the same size, 1 mm in diameter or about 12° of visual angle.

adapting light shining on one part of the receptive field surround decreased the magnitude of the response to a test stimulus illuminating that same part much more than it decreased the response to a test stimulus illuminating some other part of the surround. A similar result was found for some, but not all X cells.

In addition to the nonlinear gain-changing effects, introduced by raising the light intensity, that seem to occur at both the early local and separate combination point stages, several other kinds of evidence suggest further complications in the mechanism of the retinal ganglion cell. Since these complications may well be due to processing prior to the separate combination points of center and surround, two of them are discussed briefly here: local adaptation pools, and early center-surround interaction.

Local adaptation pools. A major modification of the general model of Figure 18 is probably required to describe the goldfish retinal ganglion cells. In these cells (Easter, 1968b), steady background light on one retinal spot decreases the cell's response to an increment of light at that spot. It also decreases the response at neighboring spots, but more for nearby ones than for distant ones (when all spots are within the center and equated for sensitivity). This result cannot be attributed to an early local nonlinearity because the background light affects some neighboring spots and not just the spot it directly shines on. Nor can this result be attributed to the simple nonlinearity at the center combination point because the effect of the background light does not extend uniformly to all points within the center of the receptive field. This result could be explained by 'local adaptation pools' such as those shown in Figure 21. In this modification, the early local signal that originates at one point on the retina is affected (before arriving at the center combination point) by a signal A_i from a 'local adaptation pool.' The signal A_i is determined by the light in an area around the point producing the early local signal—an area usually smaller than the center of the receptive field. (Note that it is possible that the square-root transformation assigned to the early local path is actually a manifestation of the local adaptation pool.) Further information relevant to the characteristics of the local adaptation pools in the goldfish can be found in Easter (1968b).[6]

Are there 'local adaptation pools' in the cat? The results of Cleland and Enroth-Cugell's (1970) experiment varying the area and intensity of a background stimulus suggest not. However, adaptation experiments using separate small spots at various distances have not yet been done in the cat and

---

[6] Similar local adaptation pools have been found in the frog by Burkhardt and Berntson (1972). These authors have also suggested a function for the local adaptation pools—the pools may account for the extraordinary sensitivity of certain cells (quite common in the frog) to moving as compared to stationary stimuli.

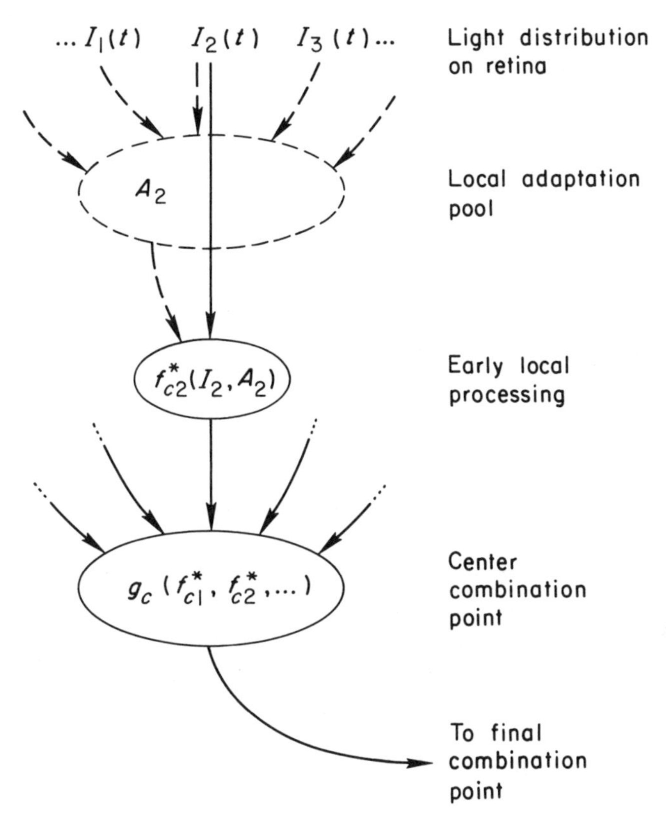

FIGURE 21.
One early local pathway in the center mechanism of the ganglion cell model modified to include early local adaptation pools.

the temporal parameters used in the cat experiments generally have been quite different from those used in the goldfish.

Center-surround interactions. Maffei and his coworkers have suggested that light stimulating the center mechanism of the cat retinal ganglion cell may affect the gain or adaptation level of the surround mechanism, and vice versa (Maffei, 1968; Maffei & Cervetto, 1968; Maffei, Cervetto, & Fiorentini, 1970; Maffei, Fiorentini, & Cervetto, 1971). In the case of influence of the surround mechanism on the gain of the center mechanism, this suggestion is not supported by the results of Cleland and Enroth-Cugell (1968). And some of Maffei and his co-workers' results might be explainable without such interaction if the responses that they attributed to only one mechanism were actually mixed responses coming from both mechanisms. However, it seems difficult to explain all their findings in that fashion, particularly those suggesting that the responsiveness of the surround mechanism is affected by activity in the center mechanism (Maffei, Fiorentini, & Cervetto, 1971).

3.5. Temporal Properties

Since each stage in the retinal ganglion cell mechanism has its own temporal characteristics, the overall temporal properties of the cell may be quite complicated. Figure 20 has already illustrated the complexities in the responses to time-varying stimuli that can arise just from linear addition of center and surround responses. The earlier stages appear to add even greater complexity.

One major experimental result is that the temporal characteristics of the retinal ganglion cell responses change as adaptation level is changed. The latency of the response to a flash or an increment decreases as background intensity is raised (Cleland & Enroth-Cugell, 1970; Levick & Zacks, 1970; Bicking, 1965). At low luminance levels, the response to an increment of luminance in the receptive field center rises to a maximum shortly after the increment and stays there, but at high luminance levels, the response rises to a peak and then decays exponentially to a maintained steady state (Shapley et al., 1972; Stone & Fabian, 1968; Yoon, 1972). See Figure 17. As might be expected from these responses to increments, the transfer function relating peak-trough amplitude in the cell's response to temporal frequency of sinusoidal flicker (for stimuli either in the center or in the surround of the field) changes with mean luminance level, exhibiting a low-frequency decline at high mean luminances (the peak-trough amplitude in the response to low frequencies is smaller than that in the response to medium frequencies) but no low-frequency decline at low mean luminances (Schellart & Spekreijse, 1972; Enroth-Cugell & Shapley, 1973a). Actually, rather than using stimuli flickering sinusoidally, Schellart and Spekreijse (1972) introduced a new method of determining the temporal transfer function indirectly by measuring the cell's response to a visual white-noise stimulus. The transfer functions at various mean luminances for cat and goldfish retinal ganglion cells look remarkably like the comparable *Limulus* electrophysiological and human psychophysical transfer functions.

Shapley et al. (1972), Enroth-Cugell and Shapley (1973b), and Schellart and Spekreijse (1972) all present evidence that these changes in the temporal characteristics of the responses depend on luminance summed appropriately over area (on the quantity theoretically equivalent to the summed inputs to the center combination point) rather than on any other property of the stimulus. This suggests that much of the time-course of the response is determined by the gain-changing nonlinearity at the central combination point rather than by the early local stages (although the early local stage will also influence the time-course). Enroth-Cugell and Shapley (1973a) have accordingly proposed a model for the center combination point of the cat retinal ganglion cell, which explains simultaneously the gain-change (the different magnitudes of response for an incremental stimulus superimposed on different background levels) and the different temporal characteristics of the response for different background levels. They suggested the following equations for the response of the center mechanism $g_c(t)$:

$$g_c(t) = \frac{A(t)}{\exp(B(t))};$$

$$A(t) = \int F_c(t - \tau)\left(\frac{\tau}{\tau_F}\right)^n e^{-\tau/\tau_F}\, d\tau; \tag{8}$$

$$B(t) = \int k g_c(t - \tau)\frac{e^{-\tau/\tau_H}}{\tau_H}\, d\tau.$$

$A(t)$ is the summed input to the center combination point calculated from $F_c(t)$, the center effective flux at time t, where n and τ_F are constants. $B(t)$ is a feedback signal which is equal to the center mechanism's response averaged over past time with an exponential weighting function specified by the constants k and τ_H. A limitation of this model is that it does not predict the result for the cat retinal ganglion cell that is embodied in Equation 6, which shows that the response to an increment is exactly proportional to the incremental flux (ΔF) as long as incremental plus background flux ($\Delta F + F$) is held constant. Nevertheless, this model is an important advance because it describes both the temporal characteristics and the nonlinear characteristics of the ganglion cell responses simultaneously, and does successfully predict the changes in the magnitude and time-course of the response to an incremental stimulus as background luminance is changed and also predicts some of the changes in the transfer functions at different mean luminances.

The extent of this linkage between gain-changing nonlinearity and temporal properties may be a characteristic distinguishing X cells from Y cells (see next section): according to Enroth-Cugell and Shapley (1973a), the linkage may be more dramatic for Y cells (which are the type of cells from which they present results) than for X cells.

A question often raised is whether the temporal properties of the center and the surround mechanism are the same. For stimulus spots of similar size and luminance, the center mechanism of the cat retinal ganglion cell can follow flicker out to higher temporal frequencies than can the surround mechanism (Maffei, Cervetto, & Fiorentini, 1970). If the luminance and areas of the stimuli are chosen appropriately to compensate for the different sensitivities of the two mechanisms, however, this difference between the two mechanisms is not found, at least not in the goldfish. Rather the temporal transfer functions for center and surround mechanisms can always be made identical except for a phase difference indicating some delay of the surround response with respect to the center response (Schellart & Spekreijse, 1972). Thus the temporal properties of the center and surround mechanisms are the same, except for this delay, as long as the mechanisms are operating at the same adaptation level.

Similarly, the temporal transfer functions for the two opposing color-coded processes found in the center mechanism of the goldfish retinal ganglion cell can be made equal except for a phase difference indicating a delay of the 'green' process (the process with maximum sensitivity to the middle wavelengths) with respect to the 'red' process (the process with maximum sensitivity to the long wavelengths) (Schellart & Spekreijse, 1972).[7]

3.6. *X* (Sustained) and *Y* (Transient) Cells

Recently two types of cells have been discovered in the cat retina (Enroth-Cugell & Robson, 1966; Cleland, Dubin, & Levick, 1971; Fukada, 1971; Fukada & Saito, 1971) and perhaps also in the primate retina (Gouras, 1968, 1969). It has been proposed that the existence of these two types of cells may reveal a basic division of visual function: *X* cells (also called sustained because they respond throughout a maintained stimulus) being specialized for spatial information, and *Y* cells (also called transient because they tend to respond only at the beginning of a maintained stimulus) specialized for temporal information (Cleland, Dubin, & Levick, 1971; Fukada & Saito, 1971).

An important difference between these cell types is revealed in their responses to the onset and offset of a stationary sinusoidal grating (when the grating is not present, the retina is illuminated uniformly at the average luminance of the grating). Such a sinusoidal grating is pictured in Figure 22, along with *X* and *Y* cell responses to four placements of the grating. Consider what happens when the grating is so placed with respect to the cell's receptive field that as much area receives an increase in illumination as receives a decrease in illumination when the grating is turned on and off (90° and 270° phase positions in Fig. 22, situations in which the center effective flux and surround effective flux remain constant). *X* cells did not respond at either the onset or the offset of the grating. *Y* cells, however, showed a tran-

[7] Bicking (1965) proposed a very interesting model of goldfish retinal ganglion cells. It is quite different from the models we have been discussing in that it makes complex assumptions about the actual mechanism for generating impulses from an underlying continuously varying slow potential (rather than just assuming that impulse frequency is proportional to the magnitude of some underlying continuously varying response function h). It predicts quite well certain temporal characteristics, a number of changes with stimulus intensity level, and some aspects of the variability among responses to the same stimulus.

sient increase in firing at both onset and offset (Enroth-Cugell & Robson, 1966). An analogous result using two small spots flickering in antiphase has been demonstrated (Büttner, Büttner, & Grüsser, 1971). A further experiment yielding similar results was reported by Cleland, Levick, and Sanderson (1973) who used radially symmetric patterns of alternating black and white sectors (like windmills) centered on cells' receptive fields. No matter how such a pattern is rotated around its center, the effective flux stimulating the center of the receptive field and that stimulating the surround remains constant (assuming a circular receptive field). Consistent with the results of Figure 22, *X* cells' discharge rates were never affected by rotation of such a pattern but *Y* cells always responded to small jerky movements of the pattern with brief but definite bursts of impulses.

This behavior of the *X* cells in response to gratings and windmill patterns can be predicted by a model with linear early local transformations (all having the same time-course) followed either by a linear transformation or by a simple nonlinear transformation at the center and surround combination points (and with relative responsiveness of both the center and surround mechanisms assumed spatially symmetric as always appears to be the case). For, as a result of the linear early local stage, all increases in inputs to the separate combination points would be balanced out by exactly compensating decreases at both the onset and the offset of the stationary grating (at 90° and 270° phase positions) in the experiment of Figure 22 or whenever the radially symmetric windmill pattern was rotated. Hence this model with linear early local processing predicts that the responses of both the center and surround mechanisms and consequently the response of the cell should remain constant during the onset and offset of the appropriately placed stationary grating and during any rotation of the radially symmetric pattern as does the *X* cell response. The compressive early local nonlinearity suggested by two-spot summation experiments might not have shown up in these experiments because the luminance was being varied within a relatively small range.

The response of *Y* cells to the gratings and windmill patterns can be predicted by a model with early local transformations that are nonlinear even for small ranges of luminance variation (followed either by linear or by simple nonlinear transformations at the separate combination point). The *Y* cells' response might also be predicted if the early local transformations were linear, but their time-courses were not identical and not symmetrically distributed around the center of the field, for in that case the time-course of the response to the decreases in illumination would be different from the time-course of the response to the increases in illumination. (As mentioned in connection with the *X* cells, the compressive early local nonlinearity suggested by the two-spot summation experiments might not show up in this situation because luminance is being varied within a relatively small range.)

It appears unlikely, however, that either of these explanations involving

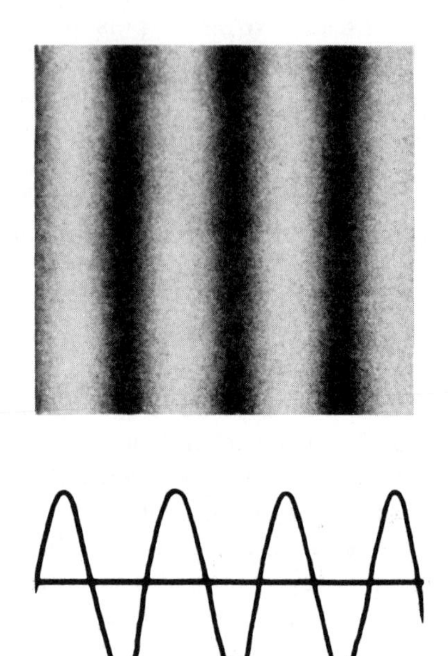

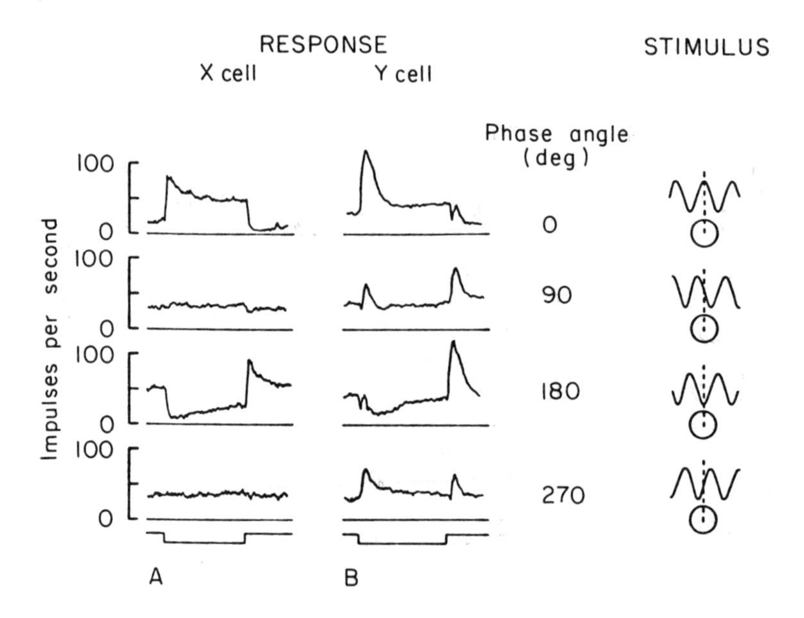
RESPONSE
STIMULUS
X cell
Y cell
Phase angle (deg)
100
0
Impulses per second
0
90
180
270
A
B

early local transformations is a complete description of the processes differentiating *Y* cells from *X* cells. For example, when the spatial frequency of a moving grating is varied but the temporal frequency is held constant (the number of bars per second passing each point is held constant), the average discharge rate of *X* cells is the same for all spatial frequencies but the average discharge rate of *Y* cells depends heavily on the spatial frequency (Enroth-Cugell & Robson, 1966; Cleland, Dubin, & Levick, 1971). The difference between *X* cells' and *Y* cells' behavior as spatial frequency is changed suggests that some variations in the processes of spatial summation (the processes at the combination points in terms of the general model) are also involved in differentiating the two types.

Whatever the fundamental differences between *X* cells and *Y* cells, the differences may well involve the surround and not the center mechanism, since *Y* cells responses to moving radially symmetric patterns are not changed very much if the center five degrees of the pattern is covered (Cleland, Levick, & Sanderson, 1973). And, as mentioned before, the effect of a steady adapting background on the receptive field surround is always localized for *Y* cells but can be either localized or not for *X* cells. Further, *Y* cells but not *X* cells exhibit the periphery effect: stimuli moving in a part of the visual field far outside the usual limits of the cell's receptive field cause an increase in the firing rate of the cell (Cleland, Dubin, & Levick, 1971).

There are many further differences between the two types of cells. *X* cells tend to have smaller receptive fields than *Y* cells and their receptive fields tend to be closer to the center of the eye (Enroth-Cugell & Robson, 1966; Cleland, Dubin, & Levick, 1971). *X* cells tend to have higher maintained discharge rates than *Y* cells (Enroth-Cugell & Robson, 1966; Cleland, Levick, & Sanderson, 1973). For *X* cells, the thresholds for different-sized spots of light and for annuli with different inner diameters (centered on the receptive field) show clear evidence of opposing center and surround mechanisms, but for *Y* cells, the thresholds do not (Cleland, Levick, & Sanderson, 1973). *Y* cells can respond to objects moving much faster than those *X* cells can respond to (Cleland, Dubin, & Levick, 1971). As mentioned earlier, the dynamics of the *Y* cells may be more dependent on the level of light adaptation than those of

---

FIGURE 22.
On the left is a photograph of a sinusoidal grating pattern and a sketch of its luminance profile with its mean intensity indicated by a horizontal line. On the right are shown responses of an off-center *X* (sustained) and an off-center *Y* (transient) cat retinal ganglion cell to the onset and offset of a stationary sinusoidal grating pattern. Downward deflection of the lowest tracings indicates offset of the grating and upward deflection indicates onset. The grating was turned on and off at 0.45 cps. (When it was off, the field was uniformly illuminated at the mean intensity of the grating.) The position of the grating relative to the midpoint of the receptive field center was varied as sketched at the right of the figure. (Adapted from Enroth-Cugell & Robson, 1966.)

X cells (Enroth-Cugell & Shapley, 1973a). The conduction velocity of impulses traveling up the axons of the retinal ganglion cells is faster for *Y* cells than for *X* cells (Cleland, Dubin, & Levick, 1971; Fukada, 1971). Lastly, this division of cells into *X* and *Y* types is also found in the lateral geniculate nucleus; the *X* type lateral geniculate neurons receive all their inputs from *X* type retinal ganglion cells whereas the *Y* type lateral geniculate neurons receive all their inputs from *Y* type retinal ganglion cells (Cleland, Dubin, & Levick, 1971).

3.7. Summary

Many features of the responses of cat and goldfish retinal ganglion cells can be quantitatively described using a specific version of the general model shown in Figure 18. The signals from the center and surround mechanisms are linearly combined (and then perhaps rectified) at the final point. The signals from different points on the retina are combined by a 'simple' nonlinear transformation at the combination point for the center mechanism and perhaps also by a similar transformation at the combination point for the surround mechanism—these nonlinear transformations reduce the gain or sensitivity of the cell as light intensity is raised. The early local transformation is a compressive nonlinear function that also reduces the sensitivity of the cell as light intensity is raised.

This version of the model of Figure 18 must certainly be elaborated, however, in order to explain a number of other characteristics of the responses of these retinal ganglion cells. The differences between the *X* cells and *Y* cells must be explained. Temporal characteristics, only partially specified in the above, must be fully described, including the variability of the responses (which has been quantitatively studied although not described here, see Barlow & Levick, 1969a, 1969b). Wavelength selectivity (particularly in the goldfish; see Spekreijse, Wagner, & Wolbarsht, 1972, for example) must also be incorporated. Further elaborations might include local adaptation pools and an influence of the center mechanism on the surround mechanism's responsiveness.

4. CONCLUDING REMARKS

The optic nerve is a bottleneck through which must pass all visual information that reaches the central nervous system. Retinal ganglion cells (eccentric cells in *Limulus*) are thus located at a critical point in the visual system. On the one hand their behavior represents the outcome of the integration of the activity of many individual intraretinal components, and on the other hand this very same behavior represents the individual inputs into an even more complex integrative system—the brain.

In elucidating the intraretinal mechanism, knowledge of the retinal ganglion cell's behavior indicates the kinds of mechanisms to be looked for within the retina. Also, any presumed understanding of how these intraretinal processes form a chain of events leading from light to ganglion cell response can be tested by the extent of its agreement with the known behavior of the retinal ganglion cell. In *Limulus* the discovery of the underlying physiological processes represented in the equivalent circuit model was greatly facilitated by knowing the characteristics of optic nerve axons' responses, which suggested, for example, the existence of self-inhibition and indicated the time-course it must have. In cat and goldfish the correspondence between the actual intraretinal processes and the various processes represented abstractly in the general model of retinal ganglion cells has not yet been determined. There are, by a conservative count, four kinds of neurons in the vertebrate retina that are intermediary between light and retinal ganglion cells (receptors, horizontals, bipolars, amacrines). In the last few years, with the development of appropriate microelectrodes, micromanipulators, and staining techniques, it has become possible to record from these cells and to discover many of their individual properties. Although no overall picture of the whole network feeding into a particular ganglion cell has yet emerged, work along these lines is progressing rapidly.

Knowing what happens at the retina may also simplify the task of understanding what happens higher in the visual system. First, it enables the input to these higher cells to be specified in a form that is closer to their actual input than is the original light stimulus. Second, basic processes that have been isolated and studied at the retinal level may be identifiable as subunits in the mechanism for higher order cells that are more complicated in their behavior and therefore more difficult to analyze. For example, cells responding maximally to moving lines of particular orientation (where the lines can be anywhere in a wide part of the visual field) are found in the cortex of the cat. Although not yet quantitatively understood, it appears that these cells as well as other higher order cells integrate opposed excitatory and inhibitory influences in a manner similar to that of the simpler retinal ganglion cells. Moreover, it is likely that the mechanisms of these higher order cells in the cat are not unlike those of the complicated retinal ganglion cells found in many other vertebrate species. In fact, the retinal ganglion cells of cat and goldfish are simple compared with the retinal ganglion cells in many lower vertebrate species (frog, rabbit, pigeon), although apparently quite similar to those in most primates.

Visual perception depends not on the activity of any one cell alone, but on the integrated activity of the whole visual system. Under proper stimulus conditions, however, the action of one type of cell may be emphasized relative to that of the others. Various perceptual phenomena are known that seem indeed to reflect the characteristics of particular types of visual neurons. This correspondence has encouraged many investigators to suggest explanations for perceptual phenomena based on visual neurophysiology. The more quantitative the description of both the perceptual phenomena and the visual neurophysiology, the more likely the success in determining the actual relation of one to the other.

ACKNOWLEDGMENTS

The preparation of this chapter was supported in part by NSF Grant GB-6540 (P2B1569) and NIH Grants EY-188 and GM-1789. We are grateful to Donald C. Hood, R. Allison Ryan, and Robert M. Shapley for their careful reading and helpful criticism of early drafts of this chapter.

REFERENCES

Abramov, I., & Levine, M. W. Spatial summation in the receptive fields of goldfish ganglion cells, 1974, in preparation.

Adolph, A. R. Spontaneous slow potential fluctuations in the *Limulus* photoreceptor. *J. Gen. Physiol.*, 1964, **48,** 297–322.

Adrian, E. D., & Matthews, R. The action of light on the eye. Part I. The discharge of impulses in the optic nerve and its relation to the electric change in the retina. *J. Physiol.*, 1927, **63,** 378–414. (a)

Adrian, E. D., & Matthews, R. The action of light on the eye. Part II. The processes involved in retinal excitation. *J. Physiol.*, 1927, **64,** 279–301. (b)

Adrian, E. D., & Matthews, R. The action of light on the eye. Part III. The interaction of retinal neurones. *J. Physiol.*, 1928, **65,** 273–298.

Barlow, H. B., & Levick, W. R. Three factors limiting the reliable detection of light by retinal ganglion cells of the cat. *J. Physiol.*, 1969, **200,** 1–24. (a)

Barlow, H. B., & Levick, W. R. Changes in the maintained discharge with adaptation level in the cat retina. *J. Physiol.*, 1969, **202,** 699–718. (b)

Barlow, R. B., Jr. Inhibitory fields in the *Limulus* lateral eye. Unpublished doctoral dissertation, The Rockefeller University, 1967.

Barlow, R. B., Jr. Inhibitory fields in the *Limulus* lateral eye. *J. Gen. Physiol.*, 1969, **54,** 383–396.

Barlow, R. B., Jr., & Lange, G. D. A nonlinearity in the inhibitory interactions in the lateral eye of *Limulus*. *J. Gen. Physiol.*, 1974, in press.

Baumgartner, G. Kontraslichteffekte an retinalen Ganglienzellen: Ableitungen vom tractus opticus der Katze. In R. Jung and H. Kornhuber (Eds.), *Neurophysiologie und Psychophysik des visuellen Systems*. Berlin: Springer, 1961.

Bicking, L. A. Some quantitative studies on retinal ganglion cells. Unpublished doctoral dissertation, Johns Hopkins University, 1965.

Biederman-Thorson, M., & Thorson, J. Dynamics of excitation and inhibition in the light-adapted *Limulus* eye in situ. *J. Gen. Physiol.*, 1971, **58,** 1–19.

Bonds, A. B., Enroth-Cugell, C., & Pinto, L. H. Image quality of the cat eye measured during retinal ganglion cell experiments. *J. Physiol.*, 1972, **220,** 383–402.

Burkhardt, D. A., & Berntson, G. C. Light adaptation and excitation: Lateral spread of signals within the frog retina. *Vision Res.*, 1972, **12,** 1095–1112.

Büttner, C., Büttner, U., & Grüsser, O. J. Interaction of excitation and direct inhibition in the receptive field center of retinal neurons. *Pflügers Arch.*, 1971, **322,** 1–21.

Büttner, U., & Grüsser, O. J. Quantitativ Untersuchungen der räumlichen Erregungssummation im rezeptiven Feld retinaler Neurone der Katze. *Kybernetik*, 1968, **4,** 81–94.

Campbell, F. W., & Robson, J. G. Application of Fourier analysis to the visibility of gratings. *J. Physiol., Lond.*, 1968, **197**, 551–566.

Chevreul, M. E. *The principles of harmony and contrast of colors and their application to the arts.* New York: Rheinhold, 1967.

Cleland, B. G., Dubin, M. W., & Levick, W. R. Sustained and transient neurons in the cat's retina and lateral geniculate nucleus. *J. Physiol.*, 1971, **217,** 473–496.

Cleland, B. G., & Enroth-Cugell, C. Cat retinal ganglion cell responses to changing light intensities: Sinusoidal modulation in the time domain. *Acta Physiol. Scand.*, 1966, **68,** 365–381.

Cleland, B. G., & Enroth-Cugell, C. Quantitative aspects of sensitivity and summation in the cat retina. *J. Physiol., Lond.*, 1968, **198,** 17–38.

Cleland, B. G., & Enroth-Cugell, C. Quantitative aspects of gain and latency in the cat retina. *J. Physiol., Lond.*, 1970, **206,** 73–82.

Cleland, B. G., Levick, W. R., & Sanderson, K. J. Properties of sustained and transient ganglion cells in the cat retina. *J. Physiol.*, 1973, **288,** 649–680.

Cornsweet, T. N. *Visual perception.* New York: Academic Press, 1970.

Creutzfeldt, O. D., Sakmann, B., Scheich, H., & Korn, A. Sensitivity distribution and spatial summation within receptive-field center of retinal on-center ganglion cells and transfer function of the retina. *J. Neurophysiol.*, 1970, **33,** 654–671.

Dodge, F. A. Inhibition and excitation in the *Limulus* eye. In W. Reichardt (Ed.), *Processing of optical data by organisms and by machines.* Proceedings of the International School of Physics "Enrico Fermi," 1969, **43,** 341–365.

Dodge, F. A., Knight, B. W., & Toyoda, J. I. Voltage noise in *Limulus* visual cells. *Science*, 1968, **160,** 88–90.

Dowling, J., & Boycott, B. B. Organization of the primate retina: Electron microscopy. *Proc. Roy. Soc., Lond.*, 1966, **166B,** 80–111.

Easter, S. S. Excitation in the goldfish retina: Evidence for a non-linear intensity code. *J. Physiol., Lond.*, 1968, **195,** 253–271. (a)

Easter, S. S. Adaptation in the goldfish retina. *J. Physiol., Lond.*, 1968, **195,** 273–281. (b)

Enroth-Cugell, C., & Pinto, L. Algebraic summation of centre and surround inputs to retinal ganglion cells of the cat. *Nature*, 1970, **226,** 458–459.

Enroth-Cugell, C., & Pinto, L. Properties of the surround response mechanism of cat retinal ganglion cells and centre-surround interaction. *J. Physiol., Lond.*, 1972, **220,** 403–441.

Enroth-Cugell, C., & Robson, J. G. The contrast sensitivity of retinal ganglion cells of the cat. *J. Physiol.*, 1966, **187**, 517–552.
Enroth-Cugell, C., & Shapley, R. M. Adaptation and dynamics of cat retinal ganglion cells. *J. Physiol.*, 1973, **233,** 271–309. (a)
Enroth-Cugell, C., & Shapley, R. M. Flux, not illumination, is what cat retinal ganglion cells really care about. *J. Physiol.*, 1973, **233,** 311–326. (b)
Fischer, B., & Freund, H.-J. Eine mathematische Formulierung für Reiz-Reaktionsbeziehungen retinaler Ganglienzellen. *Kybernetik*, 1970, **7,** 160–166.
Fukada, Y. Receptive field organization of cat optic nerve fibers with special reference to conduction velocity. *Vision Res.*, 1971, **11,** 209–226.
Fukada, Y., & Saito, H.-A. The relationship between response characteristics to flicker stimulation and receptive field organization in the cat's optic nerve fibers. *Vision Res.*, 1971, **11,** 227–240.
Fuortes, M. G. F. Initiation of impulses in visual cells of *Limulus. J. Physiol.*, 1959, **148,** 14–28.
Fuortes, M. G. F., & Yeandle, S. Probability of occurrence of discrete potential waves in the eye of *Limulus. J. Gen. Physiol.*, 1964, **47,** 443–463.
Furman, G. C. Comparison of models for subtractive and shunting lateral-inhibition in receptor-neuron fields. *Kybernetik*, 1965, **2,** 257.
Gouras, P. An identification of cone mechanisms in monkey ganglion cells. *J. Physiol., Lond.*, 1968, **199,** 533–547.
Gouras, P. Antidromic responses of orthodromically identified ganglion cells in monkey retina. *J. Physiol., Lond.*, 1969, **204,** 407–414.
Graham, N., Ratliff, F., & Hartline, H. K. Facilitation of inhibition in the compound lateral eye of *Limulus. Proc. Nat. Acad. Sci.*, 1973, **70,** 894–898.
Grüsser, O. J. A quantitative analysis of spatial summation of excitation and inhibition within the receptive field of retinal ganglion cells of cats. *Vision Res.*, 1971, Suppl. 3, 103–127.
Grüsser, O. J., Schaible, D., & Vierkant-Glathe, J. A quantitative analysis of the spatial summation of excitation with the receptive field centers of retinal neurons. *Pflügers Arch.*, 1970, **319,** 101–121.
Gur, M., Purple, R. L., & Whitehead, R. Ultrastructure within the lateral plexus of the *Limulus* eye. *J. Gen. Physiol.*, 1972, **59,** 285–304.
Hartline, H. K. The response of single optic nerve fibers of the vertebrate eye to illumination of the retina. *Amer. J. Physiol.*, 1938, **121,** 400–415. (a)
Hartline, H. K. The discharge of impulses in the optic nerve of *Pecten* in response to illumination of the eye. *J. Cell. and Comp. Physiol.*, 1938, **11,** 465–478. (b)
Hartline, H. K. The neural mechanisms of vision. *The Harvey Lectures*, 1941–42, Series 37, 39–68.
Hartline, H. K. Inhibition of activity of visual receptors by illuminating nearby retinal elements in the *Limulus* eye. *Fed. Proc.*, 1949, **8,** 69.
Hartline, H. K., & Graham, C. H. Nerve impulses from single receptors in the eye. *J. Cell. and Comp. Physiol.*, 1932, **1,** 277–295.
Hartline, H. K., & Ratliff, F. Inhibitory interaction of receptor units in the eye of *Limulus. J. Gen. Physiol.*, 1957, **40,** 357–376.
Hartline, H. K., & Ratliff, F. Spatial summation of inhibitory influences in the eye

of *Limulus*, and the mutual interaction of receptor units. *J. Gen. Physiol.*, 1958, **41,** 1049–1066.

Hartline, H. K., & Ratliff, F. Inhibitory interactions in the retina of *Limulus*. In, *Handbook of sensory physiology*. Vol. 7. Berlin: Springer, 1972.

Hartline, H. K., Wagner, H. G., & Ratliff, F. Inhibition in the eye of *Limulus*. *J. Gen. Physiol.*, 1956, **39,** 651–673.

Hering, E. *Outlines of a theory of the light sense*. (Translated by L. M. Hurvich and D. Jameson) Cambridge, Mass.: Harvard University Press, 1964.

Hubel, D. H., & Wiesel, T. N. Receptive fields of optic nerve fibers in the spider monkey. *J. Physiol.*, 1960, **154,** 572–580.

Hughes, G. W., & Maffei, L. Retinal ganglion cell response to sinusoidal light stimulation. *J. Neurophysiol.*, 1966, **29,** 333–352.

Katz, B. *Nerve, muscle, and synapse*. New York: McGraw-Hill, 1966.

Knight, B. W., Toyoda, J. I., & Dodge, F. A. A quantitative description of the dynamics of excitation and inhibition in the eye of *Limulus*. *J. Gen. Physiol.*, 1970, **56,** 421–437.

Kuffler, S. W. Discharge patterns and functional organization of mammalian retina. *J. Neurophysiol.*, 1953, **16,** 37–68.

Lange, D. Dynamics of inhibitory interactions in the eye of *Limulus:* Experimental and theoretical studies. Unpublished doctoral dissertation, The Rockefeller University, 1965.

Lange, D., Hartline, H. K., & Ratliff, F. The dynamics of lateral inhibition in the compound eye of *Limulus* II. In C. G. Bernhard (Ed.), *The functional organization of the compound eye*. New York: Pergamon Press, 1966. (a)

Lange, D., Hartline, H. K., & Ratliff, F. Inhibitory interaction in the retina: Techniques of experimental and theoretical analysis. *Ann. N.Y. Acad. Sci.*, 1966, **128,** 955–971. (b)

Levick, W. R., & Zacks, J. L. Responses of cat retinal ganglion cells to brief flashes of light. *J. Physiol.*, 1970, **206,** 677–700.

Levine, M. W. An analysis of spatial summation in the receptive fields of goldfish ganglion cells. Unpublished doctoral dissertation, The Rockefeller University, 1972.

MacNichol, E. F., Jr. Visual receptors as biological transducers. In R. G. Grenell and L. J. Mullins (Eds.), *Molecular structure and functional activity of nerve cells*. Washington, D.C.: Amer. Inst. Biol. Sci., 1956.

Maffei, L. Inhibitory and facilitatory spatial interactions in retinal receptive fields. *Vision Res.*, 1968, **8,** 1187–1194.

Maffei, L., & Cervetto, L. Dynamic interactions in retinal receptive fields. *Vision Res.*, 1968, **8,** 1299–1303.

Maffei, L., Cervetto, L., & Fiorentini, A. Transfer characteristics of excitation and inhibition in cat retinal ganglion cells. *J. Neurophysiol.*, 1970, **33,** 276–284.

Maffei, L., Fiorentini, A., & Cervetto, L. Homeostasis in retinal receptive fields. *J. Neurophysiol.*, 1971, **34,** 579–587.

Matthaei, R. (Ed.) *Goethe's color theory*. New York: Van Nostrand Reinholdt, 1971.

McDougall, W. Intensification of visual sensation by smoothly graded contrast. *Proc. Physiol. Soc.*, 1903, **1,** 19–21.

Purple, R. L. The integration of excitatory and inhibitory influences in the eccentric cell in the eye of *Limulus*. Unpublished doctoral dissertation, The Rockefeller Institute, 1964.

Purple, R. L., & Dodge, F. A. Interaction of excitation and inhibition in the eccentric cell in the eye of *Limulus*. *Cold Spring Harbor Symposia*, 1965, **30,** 529–537.

Rall, W. Theory of physiological properties of dendrites. *Ann. N.Y. Acad. Sci.*, 1962, **96,** 1071–1092.

Ratliff, F. *Mach bands: Quantitative studies on neural networks in the retina*. San Francisco: Holden-Day, 1965.

Ratliff, F. On fields of inhibitory influence in a neural network. In E. R. Caianello (Ed.), *Neural networks*. New York: Springer, 1968.

Ratliff, F. The logic of the retina. In M. Marois (Ed.), *Theoretical physics to biology*. Basel, Switzerland: Karger, 1973.

Ratliff, F., & Hartline, H. K. The responses of *Limulus* optic nerve fibers to patterns of illumination on the receptor mosaic. *J. Gen. Physiol.*, 1959, **42,** 1241–1255.

Ratliff, F., Hartline, H. K., & Lange, D. The dynamics of lateral inhibition in the compound eye of *Limulus* I. In C. G. Bernhard (Ed.), *The functional organization of the compound eye*. New York: Pergamon Press, 1966.

Ratliff, F., Hartline, H. K., & Lange, D. Variability of interspike intervals in optic nerve fibers of *Limulus:* Effect of light and dark adaptation. *Proc. Nat. Acad. Sci.*, 1968, **60,** 464–469.

Ratliff, F., Knight, B. W., Dodge, F. A., & Hartline, H. K. Fourier analysis of dynamics of excitation and inhibition of eye of *Limulus:* Amplitude, phase, and distance. *Vision Res.*, 1974, in press.

Ratliff, F., Knight, B. W., & Graham, N. On tuning and amplification by lateral inhibition. *Proc. Nat. Acad. Sci.*, 1969, **62,** 733–740.

Ratliff, F., Knight, B. W., & Milkman, N. Superposition of excitatory and inhibitory influences in the retina of *Limulus:* Effect of delayed inhibition. *Proc. Nat. Acad. Sci.*, 1970, **67,** 1558–1564.

Ratliff, F., Knight, B. W., Toyoda, J. I., & Hartline, H. K. Enhancement of flicker by lateral inhibition. *Science*, 1967, **158,** 392–393.

Rodieck, R. W. Quantitative analysis of cat retinal ganglion cell responses to visual stimuli. *Vision Res.*, 1965, **5,** 583–601.

Rodieck, R. W., & Stone, J. Response of cat retinal ganglion cells to moving visual patterns. *J. Neurophysiol.*, 1965, **28,** 819–832. (a)

Rodieck, R. W., & Stone, J. Analysis of receptive fields of cat retinal ganglion cells. *J. Neurophysiol.*, 1965, **28,** 833–849. (b)

Schellart, N., & Spekreijse, H. Dynamic characteristics of retinal ganglion cell responses in goldfish. *J. Gen. Physiol.*, 1972, **59,** 1–21.

Shapley, R. M. Fluctuations of the impulse rate in *Limulus* eccentric cells. *J. Gen. Physiol.*, 1971, **57,** 539–556. (a)

Shapley, R. M. Effects of lateral inhibition on fluctuations of the impulse rate. *J. Gen. Physiol.*, 1971, **57,** 557–575. (b)

Shapley, R. M., Enroth-Cugell, C., Bonds, A. B., & Kirby, A. The automatic gain control of the retina and retinal dynamics. *Nature*, 1972, **236,** 352–353.

Sherrington, C. S. On reciprocal action in the retina as studied by means of some rotating discs. *J. Physiol.*, 1897, **21,** 33–54.

Sherrington, C. S. *The integrative action of the nervous system.* Silliman Lectures. London: Constable, 1906.

Spekreijse, H. Rectification in the goldfish retina: Analysis by sinusoidal and auxiliary stimulation. *Vision Res.*, 1969, **9,** 1461–1472.

Spekreijse, H., & van den Berg, T. J. T. P. Interaction between colour and spatial coded processes converging to retinal ganglion cells in goldfish. *J. Physiol.*, 1971, **215,** 679–693.

Spekreijse, H., Wagner, H., & Wolbarsht, M. L. Spectral and spatial coding of ganglion cell responses in goldfish retina. *J. Neurophysiol.*, 1972, **35,** 73–86.

Sperling, G. Model of visual adaptation and contrast detection. *Perception and Psychophysics*, 1970, **8,** 143–157.

Stevens, C. F. A quantitative theory of neural interactions: Theoretical and experimental investigations. Unpublished doctoral dissertation, The Rockefeller Institute, 1964.

Stone, J., & Fabian, M. Summing properties of the cat's retinal ganglion cell. *Vision Res.*, 1968, **8,** 1023–1040.

Thomas, J. P. Model of the function of receptive fields in human vision. *Psych. Review*, 1970, **77,** 121–134.

Wagner, H. G., MacNichol, E. F., Jr., & Wolbarsht, M. L. Functional basis for "on"-center and "off"-center receptive fields in the retina. *J. Opt. Soc. Amer.*, 1963, **53,** 66–70.

Wässle, H. Optical quality of the cat eye. *Vision Res.*, 1971, **11,** 995–1006.

Wiesel, T. N. Receptive fields of ganglion cells in the cat retina. *J. Physiol.*, 1960, **153,** 583–594.

Wilson, V. J., & Burgess, P. R. Changes in the membrane during recurrent disinhibition of spinal motoneurons. *Nature*, 1961, **191,** 918–919.

Wilson, V. J., Diecke, F. P., & Talbot, W. H. Action of tetanus toxin on conditioning of spinal motoneurons. *J. Neurophysiol.*, 1960, **23,** 659–666.

Winters, R. W., & Walters, J. W. Transient and steady state stimulus-response relations for cat retinal ganglion cells. *Vision Res.*, 1970, **10,** 461–477.

Yeandle, S. Studies on the slow potential and the effects of cations on the electrical responses of the *Limulus* ommatidium (with an appendix on the quantal nature of the slow potential). Unpublished doctoral dissertation, Johns Hopkins University, 1957.

Yoon, M. Influence of adaptation level on response pattern and sensitivity of ganglion cells in the cat's retina. *J. Physiol.*, 1972, **221,** 93–104.

Counting and Timing Mechanisms in Auditory Discrimination and Reaction Time

David M. Green
HARVARD UNIVERSITY

R. Duncan Luce
UNIVERSITY OF CALIFORNIA, IRVINE

1. INTRODUCTION

Models of behavior can be constructed in at least three different ways.

1. One can begin with a few empirical generalizations which, taken as postulates or axioms, lead deductively to a variety of testable predictions.
2. One can postulate some internal mechanism as mediating the behavior and, after estimating the parameters of that mechanism, then predict a variety of other behaviors.
3. One can actually investigate the internal workings of the organism, describe these mechanisms axiomatically, measure the needed parameters, and deduce the behaviors.

If the behavior in question is psychophysical, then the first two approaches call only for psychophysical data, whereas the third requires physiological data of some sort. A pure example of the last approach is rare; usually a physiological model is blended to some degree with a hypothetical model. One reason is that, even with animals, our clearest and most detailed information comes only from the peripheral nervous system, and so we are forced to speculate how that information is processed by higher centers. This chapter presents an example of this compromise approach for auditory psychophysics.

Examples of the other approaches can be found in this volume: (a) Falmagne; (b) Levine, Krantz; and (c) Graham and Ratliff.

Perhaps the single most pervasive characteristic of psychophysical data is the inconsistency of subjects when answering most questions we ask them about simple stimuli. Somewhere, between the stimulus and the response, randomness enters. One hundred years of research, the most careful of methodological practices, and the best signal sources provided by modern technology have not reduced below 6 percent the separation in intensity needed for two 1000 Hz tones to be discriminated 75 percent of the time. For this reason, many theorists believe that a general theory of psychophysics can hardly avoid an explicit formulation of this randomness. Other theorists, most notably Stevens (1957, 1971), have argued that this 'noise,' although pervasive, is completely incidental to the main effects in psychophysics, and so it is best averaged away. The former group, and we are in it, feel that the interlock between the global and local aspects of psychophysics is much more profound, although not as simple as some earlier theorists (Fechner, 1860; Luce, 1959; Thurstone, 1927) have suggested.

Variability can intrude itself at five places: (a) in the physical signal itself; (b) in its transduction from physical energy into the pulsed 'language' of the nervous system; (c) in the various transformations imposed as the peripheral neural information wends its way through the central nervous system; (d) in the decision process which converts the available information into an answer to whatever question has been asked about the nature of the signal; and (e) in the processes that lead to the execution of a response. Different theorists have focused on particular sources of randomness, attempting to show that one of these accounts for most of the overall variability. For example, Hecht, Schlaer, and Pirenne (1942), in a classic study of visual thresholds, held that the quantum variability of a threshold light source coupled with quantal losses prior to the retina were sufficient to account for the observed psychophysical variability. Later (Sec. 10) we argue that in a simple reaction-time experiment to intense signals the variability of observed times is dominated by conduction times and synaptic delays and that essentially no measurable variability is added to it by the sensory or decision aspects of the process. A number of authors (including Thurstone, 1927; Tanner & Swets, 1954; the whole resulting school of signal detectability; Green & Swets, 1966; Luce & Green, 1972) have taken the view that in many situations it is unnecessary to partition the variability due to the first three sources—the signal, its transduction into a neural language, and its transmission in the nervous system up to the point where a decision is made—and for a number of experimental tasks and measures, usually those involving some aspect of discrimination, the total variability associated with these three sources is all that need be considered. For other tasks and measures, however, the variability introduced by the decision process itself or that added by the remainder of the response process may play a significant role, as we discuss in Sections 4, 5, and 10.

From this point of view, a key initial question is how the information is encoded when it reaches the decision center. Sections 2 and 4 of the chapter deal with this topic. Section 2 summarizes some peripheral physiological evidence concerning the coding of auditory information. Section 3 treats what an ideal device, making optimum use of this information, could do. Section 4 advances some hypotheses, admittedly speculative, about the form that the sensory information takes as it is presented to the decision center. These hypotheses are supported to some degree by comparing their predictions with psychophysical data from a detection experiment with response deadlines. Another detailed comparison of these predictions is carried out in Section 5, which discusses the speed-accuracy trade off. Sections 6 through 8 provide an account of the classical data on the discrimination of changes in intensity or frequency of a pure tone signal. Section 9 briefly discusses how still other sources of variability can influence psychophysical data, especially those data in which a number of response categories are employed. Section 10 continues with a discussion of how various parameters of the model can be measured from reaction-time data and an explanation of how another source of variability can be estimated.

2. THE PERIPHERAL NEURAL REPRESENTATION OF AUDITORY SIGNALS

What we say here, and so in the rest of the chapter, pertains only to the representation of auditory signals. No comparable data for other modalities have yet been collected. We suspect that certain features of this auditory representation will be found in other senses, but many of the details will undoubtedly differ in important ways.

An auditory input signal is simply a continuous function of time, for example, it is pressure as a function of time. When one measures electrical activity in individual nerve fibers of the peripheral auditory nervous system, one does not see anything directly analogous to that function. Rather, each fiber conducts a train of electrical pulses of very brief duration and of approximately the same voltage. At first sight, these pulse trains are highly irregular; sometimes they are obviously affected by changes in the signal; at other times, apparently, they are not. A good deal of very careful work over three decades, especially the last one, has led to some understanding of the exact nature of the encoding involved, although one still cannot predict from a limited set of observations on an individual fiber how it will respond to an arbitrary input signal. For our purposes here, however, it will suffice to have a reasonably detailed description of the neural response to a very limited class of signals, namely, pure tones.

It is important to realize that the pulses themselves do not directly carry information about the signal. For example, their size does not change systematically with either signal intensity or frequency. Thus information about the signal must be carried either by the occurrence or by the absence of individual pulses, or by some aspect of their temporal pattern, or by the spatial pattern of activity over the whole auditory fiber bundle. Wever (1949, p. 128) summarized it well:

> The nerve impulse seems to be everywhere the same, regardless of the type of nerve in which it appears. The modes of variation of nerve transmission therefore are strictly limited. The following dimensions are generally regarded as exhausting the possibilities of representation by nerves of the physical characteristics of the stimulus: (a) the particular fiber or fibers set in operation, (b) the number of fibers excited at any one time, (c) the frequency of impulses in each fiber, (d) the duration of the train of impulses, and (e) the time relations of the separate impulses passing through different fibers. The problem of auditory theory is to show how these variables represent the properties of the stimulus and determine the nuances of auditory experience.

Without going deeply into the details of the arguments, no one today believes that the mere occurrence of a pulse contains any information whatsoever. One reason is that without any signal present, fibers fire spontaneously, sometimes at rather high rates. A second is that there is no sign of synchronization in the peripheral system which would make the absence of a pulse clear. So we turn to questions of temporal and spatial patterns.

For temporal patterns, the first question to resolve is which aspect of a pulse train corresponds to intensity. The well-known fact that reaction time to signal onset decreases with signal intensity (Chocholle, 1940; McGill, 1960) strongly suggests that there must be some deep interplay between intensity and time in the nervous system. One obvious conjecture is that pulse rate increases as signal intensity is increased and everything else is held constant. Peripheral data on the cat (Galambos & Davis, 1943; Kiang, 1965, 1968) and on the squirrel monkey (Rose, Brugge, Anderson, & Hind, 1967) confirm that something of this sort is true, although it is not quite so simple. If we restrict our attention to pure tone signals, the following seems to summarize the situation. Each fiber has a characteristic signal frequency to which it is most, but not exclusively, responsive. At this frequency, there is a lower and upper threshold. Below the lower threshold, it fires at its spontaneous rate; between the two, the rate increases by a factor of from 2 to 10, reaching a maximum rate at the upper threshold; for more intense signals the rate is either maintained or drops somewhat. As the frequency deviates from the characteristic one, both thresholds rise and the maximum firing rate remains about the same. Looked at another way, a pure tone of sufficient intensity activates a particular set of fibers in the sense that it drives their firing rates above their spontaneous rates. Changing the frequency causes some fibers to drop from the active category and others to enter it; increasing the intensity adds fibers to the active category. Thus, frequency change involves the substitution of fibers (metathetic continuum); intensity change, either the addition or the subtraction of fibers (prothetic continuum). This distinction was discussed by Stevens (1957) and is described fully by Wever (1949).

It is clear, therefore, that both frequency and intensity are represented spatially. In addition, of course, intensity is represented temporally as a firing rate. The question remains what, if any, additional information about frequency may exist in the temporal representation. The only way to answer this is to examine the detailed statistical structure of the pulse trains in the presence of steady, pure tone signals. Galambos and Davis (1943) were the first to do so with the care needed, and improved techniques have been employed by Kiang (1965) and his colleagues and by Rose et al. (1967). Kiang's group mostly used clicks, that is brief pressure pulses, as their stimuli. They concluded that to a first approximation the peripheral neural pulses form a renewal process in which the times between successive pulses are independent of one another and have the same distribution when the signal is constant. The times between neural bursts cluster at the reciprocal of the characteristic frequency of the fiber, indicating that the fibers are most likely to fire at only one phase of the essentially sinusoidal response produced by the brief stimulus. With no stimulus input the process appears to be approximately Poisson; i.e., a renewal process in which the distribution of interarrival times (IATs) between successive neural pulses is exponential. A deviation from the exponential occurs at 0 because very short (less than ½ msec) IATs appear to be lacking, presumably because of absolute refractoriness in the nerve. Increasingly, as various experimental artifacts have been removed, the data appear to be very well approximated by a Poisson process except for very brief times.

Frequency information about a pure tone signal, at least for frequencies below 2000 Hz (which includes most of the relevant musical range), is also encoded in the pulse train, as has been demonstrated most clearly by Rose and his collaborators. This can be seen by looking at IAT distributions, of which Figures 1 and 2 are typical, provided the measurements are sufficiently precise (at least to 100μsec). The distributions are startlingly multimodal, with one mode at about 0.5 to 1 msec and the others lying at multiples of the period of the input signal. We refer to the former as the 'sputtering' mode and to the others as 'normal' modes. Thus, for a 1000 Hz signal, the normal modes are 1 msec apart, for a 500 Hz signal they are 2 msec apart, etc. Moreover, the ratio of the heights of the successive normal modes are roughly constant suggesting a geometric distribution having some constant probability p of firing at each successive mode. The probability of the neuron firing exactly i periods after the last firing is $p(1 - p)^{i-1}$. The smearing of these geometric modes may reflect the randomness in the spontaneous Poisson

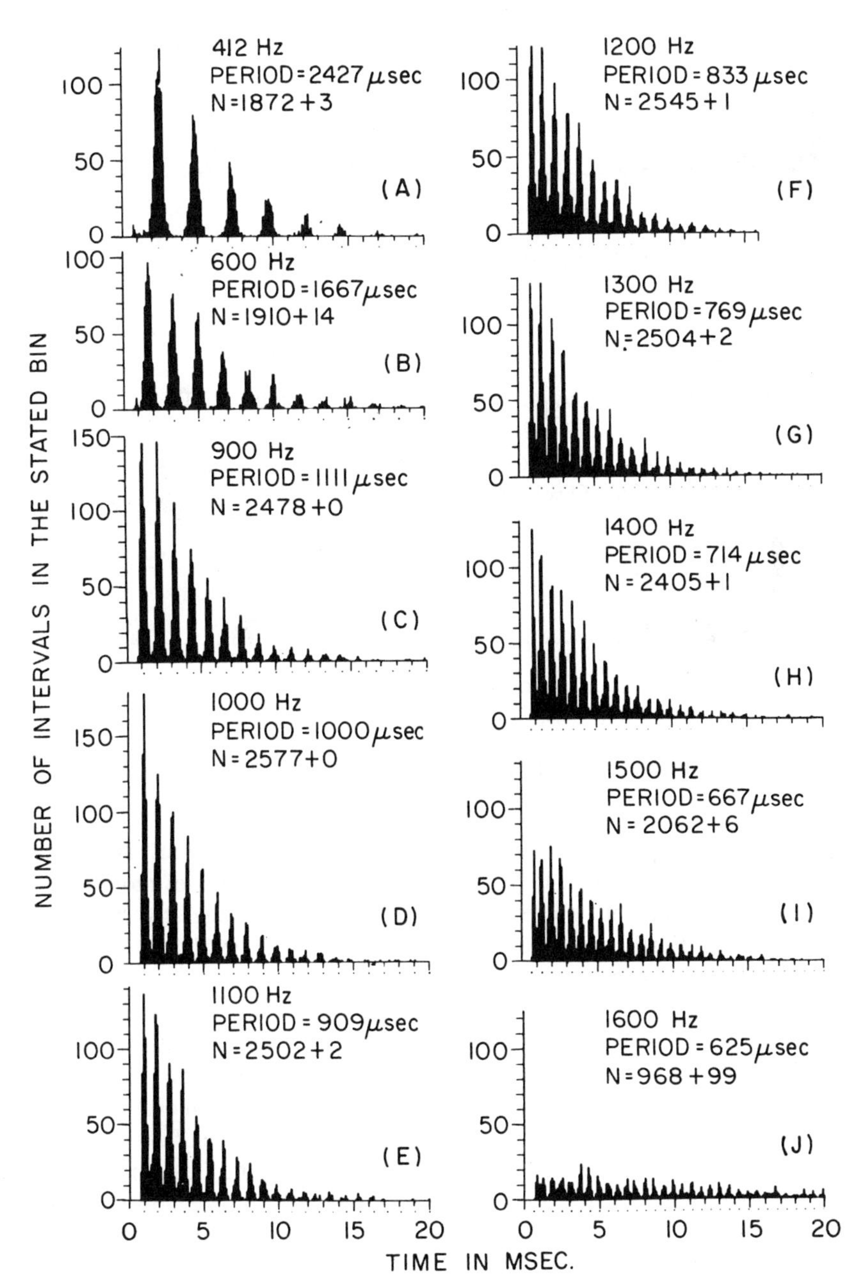

FIGURE 1.
Distribution of interspike times on a single cochlear fiber of a squirrel monkey when the acoustic signal is a pure tone of the intensities and frequencies shown. (Rose et al., 1967, Fig. 1.)

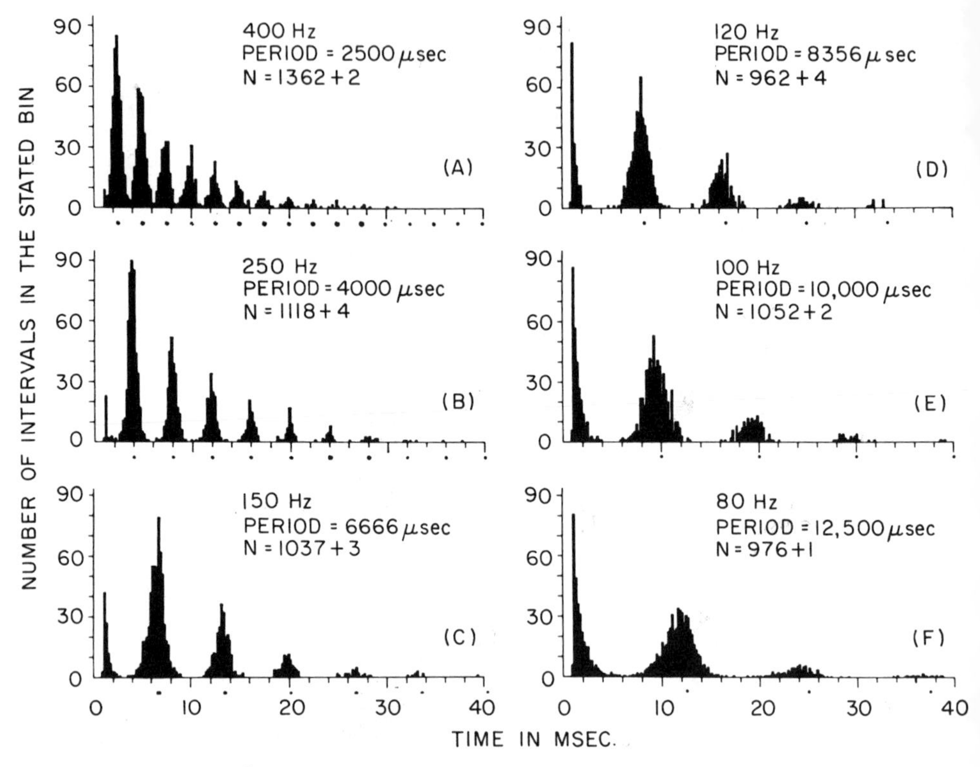

FIGURE 2.
Same as Figure 1, but for lower frequencies. (Rose et al., 1967, Fig. 4.)

process which has been modulated by the sinusoidal signal input. Note that, contrary to common belief, there is no sign of refractoriness of more than a fraction of a msec. This is even shorter than the value suggested by Kiang (1965, p. 101). Further, the geometric parameter p is clearly an increasing function of intensity, at least over some limited range.

For some purposes having to do with stimulus intensity, it is reasonable to smear the IAT distribution and to approximate it by an exponential. But one must be careful since the most probable IAT under an exponential density is 0 whereas it is $1/f$ in Figure 1. If the nervous system ever computes 1/IAT, this error of approximation can loom large (see Sec. 7).

We may conclude that at the periphery, the intensity and frequency of pure tones of low frequency are both represented spatially and temporally in the nervous system. Discrimination mechanisms that draw only on the spatial representation are usually called place mechanisms; those that draw only on the temporal representation, periodicity mechanisms. An obvious question is whether discriminative behavior is based on one or both of these mechanisms.

The fact that reaction time is a strong function of intensity suggests that longer time samples are required with weak signals (slow rates) than with intense ones, which in turn suggests that some temporal sampling is involved. Furthermore, as we have verified empirically, the reaction time to either intensity or frequency discrimination of the same quality (say, $d' = 1$) is the same, suggesting that comparable samples are taken for the two tasks. Other data below will reinforce our belief that much discriminative behavior depends on temporal mechanisms. It is much less clear whether a place mechanism plays any role whatsoever for low-frequency signals. It is entirely possible that the increase in total activity with intensity serves merely to increase the total sample size available. Wever (1949) developed a theory in which periodicity mechanisms dominate at low frequencies and place ones at high frequencies. Our calculations in Sections 6 and 8 can be thought of as elaborating the behavioral predictions of his periodicity mechanisms.

The kind of peripheral encoding of information just described sets firm limits on what the nervous system can possibly do. In whatever way it may ultimately transform this information, it can never decrease the amount of noise that is inherent in the stochastic representation of the signals. In particular, it can never undo the temporal smearing of instantaneous values of intensity. One approach, then, is to formalize the peripheral representation—this we do at the end of this section—and then ask what is the best that an ideal machine, using both the temporal and spatial representations, could do with that information. We discuss this approach in the next section, but it appears to be difficult to work it through to a definitive conclusion.

A more adventurous approach, which we undertake in Sections 4 through 8, is to guess what the rest of the (relevant) nervous system does when making decisions in terms of information encoded temporally, ignoring the spatial representation. This approach is both more specific, and so entails more detailed predictions, and far more speculative, and so probably is wrong. The question, of course, is not whether it is wrong in detail, which it is almost bound to be, but whether it is wrong in spirit and so in qualitative character.

We next formulate two models for the spike train on a single neuron. Each of these models describes imperfectly the data of which Figures 1 and 2 are typical. The defects are discussed as the models are presented.

In the first we assume that the pulse train is generated by a nonstationary Poisson process whose hazard function waxes and wanes with the signal amplitude. The particular hazard function postulated by Siebert (1970) is of the form

$$\mu(t; I, f) = \mu_0 \exp \{G[AH(f/f_0)] \cos 2\pi f t\}, \tag{1}$$

where t is time, f_0 is the characteristic frequency of the fiber, $I \propto A^2$ is the intensity of the pressure wave, and f is the frequency; G and H are functions that must be specified, although we will not do so here. Such a process is obviously phase locked to the sinusoidal signal; it is not a renewal process because successive IATs are not independent; but it is multimodal with modes at 0 and at multiples of the period of the signal. Thus, it accounts for the data almost exactly except, according to the model, the sputtering mode should be at or near 0 instead of ½ to 1 msec. The reason for this discrepancy is simply that the model does not provide for any refractoriness whatsoever.

In the second model we assume that the process is a renewal one in which the IAT random variable is of the form

$$\mathbf{IAT} = \frac{\mathbf{I}}{f} + \mathbf{X}, \tag{2}$$

where $\mathbf{I}$ is geometrically distributed, i.e.,

$$P(\mathbf{I} = i) = p(1 - p)^{i-1}, \quad i = 1, 2, \ldots, \tag{3}$$

with p some unknown function of intensity I and frequency f. The data suggest that the distribution of $\mathbf{X}$ is symmetric and bounded by an interval of the form $\left(-\frac{\epsilon}{f}, \frac{\epsilon}{f}\right)$, $0 < \epsilon < \frac{1}{2}$. In fact, it is reasonable to suppose that there is a symmetric density function g on $(-1, 1)$ with the property that

$$g(1) = g'(1) = 0,$$

and for a signal of frequency f that

$$P(\mathbf{X} = x) = \frac{f}{\epsilon} g\left(\frac{f}{\epsilon} x\right), \quad 0 < \epsilon < \frac{1}{2}. \tag{4}$$

Observe that by symmetry

$$E(\mathbf{X}) = 0 \tag{5}$$

and

$$V(\mathbf{X}) = \theta^2/f^2, \tag{6}$$

where

$$\theta^2 = \epsilon^2 \int_{-1}^{1} x^2 g(x)\, dx.$$

Combining Equations 2 and 3, we obtain for the density of IATs,

$$\psi(x) = P(\mathbf{IAT} = x) = \begin{cases} p(1 - p)^{i-1} \dfrac{f}{\epsilon} g\left[\left(x - \dfrac{i}{f}\right)\dfrac{f}{\epsilon}\right], & i = 1, 2, \ldots, \quad \left|x - \dfrac{i}{f}\right| \le \dfrac{\epsilon}{f}, \\ 0, & \text{otherwise.} \end{cases} \tag{7}$$

This renewal model has two major drawbacks. First, it has no 'sputtering' mode at all, that is, no IATs appreciably shorter than $1/f$. So when we study discrimination in terms of it we are implicitly assuming that the nervous system is able to filter out the Poisson noise that shows up as the 'sputtering' mode and that it carries out all of its computations on the normal modes. Exactly how this filtering is done we do not attempt to say. One plausible possibility is that some higher order neurons have a refractoriness that is slightly less than the period of the sinusoid. This amounts to a mild form of a place mechanism at some higher order center. Second, by assuming a renewal process (independent IATs) we have no mechanism to maintain the phase locking of the process to the signal. This is certainly wrong, but whether it matters is another question. If the nervous system actually bases its decisions on the IATs, as we shall assume, rather than paying attention to coincidences of pulses over different fibers, then the phase locking is an incidental matter of minor importance.

3. IDEAL DISCRIMINATION OF INTENSITY AND FREQUENCY

Siebert (1968, 1970) has pointed out that the Cramér-Rao inequality (see below) is the important key to determining the possible limits of discriminability. It establishes the limit on the discriminability of a small change of intensity, ΔI, or of frequency, Δf, in terms of the inherent statistical variability of the neural representation of the signal. Of course, we do not know a priori how well the organism actually approaches these theoretical limits. Still, the calculation could be useful in establishing an upper bound on the performance and, depending on how close actual performance is to this bound, it might establish constraints on the possible hypotheses we may entertain concerning the actual detection process.

The form of the Cramér-Rao inequality used by Siebert (1970) is rather more general than the most general statement given in Cramér (1946).[1] Suppose n processes are each sampled, the ith yielding a vector of κ_i observations (where κ_i may be a random variable), say $x_i = (x_{i1}, \ldots, x_{i\kappa_i})$. Combining these, we may use the abbreviation

$$(\boldsymbol{x}, \boldsymbol{\kappa}) = (x_1, \ldots, x_n, \kappa_1, \ldots, \kappa_n).$$

We assume that their joint probability density $\psi(\boldsymbol{x}, \boldsymbol{\kappa}; \xi)$ exists for each value of some parameter ξ (which we will take to be either I or f). Assuming that

[1] Our first draft of this section included a number of misinterpretations of Siebert's arguments, and we are deeply grateful to him for showing us exactly how the argument proceeds. The proofs included in Notes 2 and 3 are taken from his letters to us.

the density function is sufficiently regular (precise conditions are well known) and that $\hat{\xi}(\boldsymbol{x}, \boldsymbol{\kappa})$ is an unbiased estimator of ξ, then the variance of that estimate must satisfy[2]

$$\sigma_{\hat{\xi}}^2 \geq \frac{1}{\sum_{\kappa} \int_{x} \left(\frac{\partial \ln \psi}{\partial \xi}\right)^2 \psi \, dx}, \tag{8}$$

where

$$\sum_{\kappa} = \sum_{\kappa_1=0}^{\infty} \cdots \sum_{\kappa_n=0}^{\infty} \quad \text{and} \quad \int_{x} = \int_{x_1} \cdots \int_{x_n}.$$

Any estimator for which the equality holds in Equation 8 is called *efficient*, and any such discrimination mechanism is called *ideal.*

If, further, the n processes are independent in the sense that for each ξ,

$$\psi(\boldsymbol{x}, \boldsymbol{\kappa}; \xi) = \prod_{i=1}^{n} \psi_i(x_i, \kappa_i; \xi),$$

then one can show[3] that Equation 8 can be replaced by

---

[2] By definition of an unbiased estimator,

$$\sum \int \psi(\boldsymbol{x}, \boldsymbol{\kappa}; \xi)[\hat{\xi}(\boldsymbol{x}, \boldsymbol{\kappa}) - \xi] \, dx = 0.$$

Differentiate this with respect to ξ which, when ψ is sufficiently smooth, can be carried out to the right of $\sum \int$,

$$\begin{aligned} 0 &= \sum \int \left[\frac{\partial \psi}{\partial \xi}(\hat{\xi} - \xi) - \psi\right] dx \\ &= \sum \int \frac{\partial \ln \psi}{\partial \xi} \psi(\hat{\xi} - \xi) \, dx - 1 \\ &= \sum \int \left[\frac{\partial \ln \psi}{\partial \xi} \psi^{1/2}\right] [\psi^{1/2}(\hat{\xi} - \xi)] \, dx - 1. \end{aligned}$$

Assuming that ψ is sufficiently well behaved so that Schwarz' inequality holds,

$$\begin{aligned} 1 &\leq \left[\sum \int \left(\frac{\partial \ln \psi}{\partial \xi}\right)^2 \psi \, dx\right]^{1/2} \left[\sum \int \psi(\hat{\xi} - \xi)^2 \, dx\right]^{1/2}, \\ &= \left[\sum \int \left(\frac{\partial \ln \psi}{\partial \xi}\right)^2 \psi \, dx\right]^{1/2} \sigma_{\hat{\xi}}. \qquad \text{QED.} \end{aligned}$$

[3] By independence,

$$\frac{\partial \ln \psi}{\partial \xi} = \sum_{i=1}^{n} \frac{\partial \ln \psi_i}{\partial \xi}.$$

When substituted in Equation 8 we get squared terms, which constitute the right side of Equation 9, and cross-product terms which we now show are 0:

$$\sigma_{\hat{\xi}}^2 \geq \frac{1}{\sum_{i=1}^{n} \sum_{\kappa_i=0}^{\infty} \int_{x_i} \left(\frac{\partial \psi_i}{\partial \xi}\right)^2 \psi_i \, dx_i}. \tag{9}$$

Siebert based both of his analyses on Equation 8, but he proceeded somewhat differently in the two papers. In 1968, he assumed that for all fibers i $\kappa_i = 1$ and x_{i1} is the number of pulses observed on fiber i during some fixed listening interval when signal (I, f) is presented. Assuming that the fibers are independent, a plausible assumption which at present is difficult to justify or reject, and assuming that the process on each fiber is Poisson, so that the distribution of IATs is exponential, it is easy to write an explicit formula for ψ which can then be substituted into Equation 9. This does not really tell us anything until we assume how the Poisson parameters depend on I and f and how they vary over fibers. Siebert made a number of assumptions which were motivated by his attempt to fit some physiological data, and from these he concluded $\Delta I = \sigma_{\hat{I}}$ satisfies Weber's Law,

$$\frac{\Delta I}{I} = \frac{A}{I} + B,$$

and that $\Delta f = \sigma_{\hat{f}}$ also satisfies Weber's Law,

$$\frac{\Delta f}{f} = \frac{1}{14}\left(\frac{A}{I} + B\right),$$

where the constants A and B are independent of I and f and are the same in the two equations. We do not pursue this further because, although it may be a reasonably satisfactory way to analyze intensity discrimination, it is surely inappropriate for frequency since the Poisson assumption completely ignores the frequency information available on single channels. Fundamentally, this model denies any role to temporal mechanisms.

To overcome this limitation, Siebert (1970) undertook a more complete analysis in which temporal mechanisms were incorporated. Obviously, a count of pulses on each channel will not reveal the periodicity, so the basic sample of information on each fiber i must be the times $x_{i1}, \ldots, x_{i\kappa_i}$ of the

---

$$\begin{aligned}
\sum_{\kappa} \int_{x} \frac{\partial \ln \psi_i}{\partial \xi} \frac{\partial \ln \psi_j}{\partial \xi} \psi \, dx &= \left(\sum_{\kappa_i} \int_{x_i} \frac{\partial \ln \psi_i}{\partial \xi} \psi_i \, dx_i\right)\left(\sum_{\kappa_j} \int_{x_j} \frac{\partial \ln \psi_j}{\partial \xi} \psi_j \, dx_j\right) \\
&= \left(\sum_{\kappa_i} \int_{x_i} \frac{\partial \psi_i}{\partial \xi} dx_i\right)\left(\sum_{\kappa_j} \int_{x_j} \frac{\partial \psi_j}{\partial \xi} dx_j\right) \\
&= \frac{\partial}{\partial \xi}\left(\sum_{\kappa_i} \int_{x_i} \psi_i \, dx_i\right) \frac{\partial}{\partial \xi}\left(\sum_{\kappa_j} \int_{x_i} \psi_j \, dx_j\right) \\
&= \left[\frac{\partial(1)}{\partial \xi}\right]^2 \\
&= 0. \qquad \text{QED.}
\end{aligned}$$

pulses observed during a listening interval δ. Observe that the sample size κ_i is not fixed, but rather is an RV that depends both on the underlying probability mechanism and on δ. Assuming that these processes are Poisson with a hazard function $\mu_i(x; I,f)$, then it is easy to show that the probability density of κ spikes occurring at the unordered times x_{ij}, $j = 1, \ldots, \kappa_i$, is given by

$$\psi_i(x_i, \kappa_i) = \prod_{j=1}^{\kappa_i} \mu_i(x_{ij}; I,f) \exp\left[-\int_0^\delta \mu_i(x; I,f)\, dx\right] \Big/ \kappa_i!.$$

Assuming independent fibers and an efficient estimator, we obtain from Equation 9

$$(\Delta f)^2 = \frac{1}{\sum_{i=1}^{n} \int_0^\delta \frac{1}{\mu_i} \left(\frac{\partial \mu_i}{\partial f}\right)^2 dx}. \tag{10}$$

To evaluate Equation 10, Siebert substituted in Equation 1 for μ_i and he assumed forms for the functions G and H which accord well with current physiological knowledge. From this, he deduced

$$\frac{1}{(\Delta f)^2} \cong 3 \times 10^6 \frac{\delta}{f^2} + 1.5 \times 10^6 \delta^3 \ln A. \tag{11}$$

He then concluded that the first of these terms corresponds to the contribution of place information and the second to periodicity information. The argument is that if one assumes a spike train in which there is no periodicity information (achieved by deleting the cosine term fiom Eq. 1), then only the first term of Equation 11 arises; whereas, if one assumes that all of the spike trains are identical and so there is no place information (make G a constant function in Eq. 1), then only the second term arises. Choosing typical values of $\delta = 0.1$ sec, $f = 1000$ Hz, and $A = 300$, the terms of $1/\Delta f^2$ have values of about 0.3 sec^2 and 10^4 sec^2. Clearly, the periodicity information is far superior to place; however, the observed data of Δf approximately equal to 1 Hz are of about the same equality as the place information. And so Siebert (1970, p. 727) concludes:

> 1) the brain does *not* make full, efficient use of the periodicity information coded in the auditory nerve firing patterns, and
>
> 2) there *is* adequate information in the place pattern alone to account for behavior, if it is used efficiently.

If one conjectures some inefficiency in the brain, as seems plausible, the periodicity mechanisms are favored. In the following sections we show that a reasonable, but surely inefficient, periodicity decision scheme seems to yield a satisfactory prediction of both ΔI and Δf as a function of I and f.

4. HYPOTHETICAL CENTRAL REPRESENTATION AND DECISION MECHANISM

For the remainder of the chapter we use the renewal process model, with the simple probability laws given in Equations 2 through 7, rather than the Poisson model. What is complicated about this model is how intensity and frequency determine which fibers are active and how they affect the geometric probability p in an active fiber. Although complicated at the periphery, perhaps by the time the information is consolidated at a later decision stage it is simplified. We shall postulate that when it reaches the decision center the information continues to be in the form of Equations 2 through 7, but that it is simpler in two major respects.

First, we suppose that all active channels—we do not use the word 'fibers' because our assumptions are now hypothetical and functional, not necessarily anatomical—are statistically identical. This seems a relatively minor idealization in which we replace a set of different channels by the same number of identical average ones. There is no basic difficulty in dropping this assumption, but to do so would add a lot of extra baggage which, at this point, does not seem really useful. We shall continue to suppose, as is true at the periphery, that the number of active channels is a function both of intensity and frequency. In Section 7 we will be led to assumptions about that dependency.

Our second idealization is much more serious. At the periphery, pulse rates change by only a factor of from 2 to 10 over a relatively narrow intensity range for any given fiber. We have no very clear idea how this information is combined in order to give information about the full dynamic range. We shall, in any event, suppose that the combination is such that at the central mechanism p is a strictly increasing function of I. Put another way, by Equations 2, 3, and 5,

$$E(\mathrm{IAT}) = \frac{1}{pf} \tag{12}$$

is a decreasing function of I. In Section 7 we discuss more fully the form of this function which obviously must compress the physical range of 10^{10} into something manageable as firing rates. This assumption means that estimates of p, which by Equation 12 is determined by pulse rate for f fixed, provide estimates of I.

If one tentatively accepts Equations 2 through 7 together with these two idealizations as an adequate description of the form in which the information exists when decisions are made, our next task is to consider the nature of the decision rules employed. Our first assumption will be that the decision rule is sensitive only to the pulse trains and completely ignores all place information. Of course, it takes into account the dependence of the number of active channels on intensity and frequency, for that will alter the sample sizes of IATs on which to base decisions, but the decision procedure will completely ignore *which* channels are active. This is a radical assumption, and if we are unable to account for the psychophysical data, it is surely one to be reconsidered.

Now consider possible decision rules for intensity. According to what we have assumed, intensity is reflected solely by the pulse rate, and so any rule must involve some estimate of that rate. For example, if the subject is to decide whether a more or less intense signal has been presented, the simplest rule—so simple that some feel it should not be called a decision rule—consists of just comparing the estimate of the rate with a criterion, much as in the theory of signal detectability. More complex rules involve some form of increased sampling whenever the estimate is near the criterion, but we do not explore these here (except for a sequential rule that arises naturally in our analysis of simple reaction time in Sec. 10). Obviously, a sequential rule introduces more variance into the decision time than does a criterion one, and so it will be important to decide which rule is correct if we are to give a correct account of response times. With the simple criterion rule, the only remaining theoretical question is how the pulse rate is estimated. There are two extreme ways to do this. One is to fix a time interval and to count the total number of pulses that arrive during that time on all of the active channels; the estimated rate is the mean count per channel. Another is to fix a count per channel and to time how long it takes for that number of pulses to arrive on each channel; the estimated rate is the reciprocal of the mean time per IAT per channel. Using the former rule, one counts pulses; using the latter, one times IATs. For this reason, models based on the former are called *counting* ones (McGill, 1967), and those based on the latter are called either *timing* (Luce & Green, 1972) or *clocking* models (Uttal & Krissoff, 1965).

One might guess that these two estimation schemes would make little difference in behavior but, as we show below, that guess is wrong. Moreover, it appears that both rules are available to subjects. To establish these points, consider an absolute identification experiment in which one of two tones of different intensities is presented on each trial and the subject is to identify which was presented. Suppose further that each signal remains on until he responds (so that, from a theoretical point of view, we do not have to worry about running out of signal), but let us put the subject under pressure to respond rapidly as well as accurately. The time pressure can be effected by a deadline with fines for late responses, and accuracy can be affected by a payoff matrix. We analyze this experiment using both the counting and timing rules.

First consider the mean response time. It is composed of two parts, the mean decision delay and the mean of all other delays. Let the latter be denoted by $\bar{r}$. It is evident that, according to the counting rule based on a fixed listening interval δ, the decision time is independent both of the signal and the response. Since $\bar{r}$ is also, we predict that the MRT $= \bar{r} + \delta$ is the same in all four cells of the signal-response matrix. It is equally evident that according to the timing rule, the more intense signal should have faster response times than the less intense one because of the difference in pulse rates. In fact, let us suppose that κ IATs are collected per channel and that J channels are active. Then the distribution of decision times is the slowest of J samples from the distribution of the sum of κ independent IATs. If for signal $i = 1$ or 2 we write $M_i = E(\text{IAT}_i)$ and $V_i = V(\text{IAT}_i)$, it is easy to see that

$$\text{MRT}_i = \bar{r} + h(J, \kappa, V_i^{1/2}/M_i)M_i, \quad i = 1, 2, \tag{13}$$

where $h(J, \kappa, \sigma)$ is the mean decision time when $E(\text{IAT}) = 1$ and $V(\text{IAT}) = \sigma^2$. We do not need to derive the form of $h(J, \kappa, \sigma)$ here, although we will do so in the next section. Suppose that $M_1 > M_2$, then eliminating $h(J, \kappa, \sigma)$ from the equation (we assume that the differences in intensity are so small that we can neglect the differences in J and σ),

$$\text{MRT}_1 = \left(\frac{M_1}{M_2}\right)\text{MRT}_2 - \bar{r}\left(\frac{M_1}{M_2} - 1\right). \tag{14}$$

This we can test by varying the deadline and plotting the two signal MRTs against one another, checking for linearity and the value of the slope.

A second prediction is obtained from the two models by considering the ROC curves [plot of $P(1 \mid s_1)$ against $P(1 \mid s_2)$] obtained with a fixed deadline. To carry out this analysis we assume that the expected value M and variance V of the IAT distribution are related so that both V and M^3/V are strictly increasing functions of M. This is true, for example, in the exponential case where $V = M^2$ and for Equations 2 through 7 provided[4] $p \leq \frac{1}{3\theta}[1 - (3\theta)^{1/2}]$.

Consider the timing rule first. Assuming that $J\kappa$ is large, the central limit theorem implies that the sum of the IATs is approximately normally distributed with mean $J\kappa M$ and variance $J\kappa V$. Using the probability cutoff of c, the corresponding z score is

$$z = \frac{c - J\kappa M}{(J\kappa V)^{1/2}}.$$

---

[4] Observe, $M = E(\text{IAT}) = \frac{1}{pf}$ and $V = \frac{q + p^2\theta}{p^2 f^2} = M^2(q + p^2\theta)$ and so $\frac{M^3}{V} = \frac{1}{p(q + p^2\theta)f}$. For f fixed, M increases as p decreases. As is easily shown by computing its maximum, $p(q + p^2\theta)$ decreases with p for $p \leq \frac{1}{3\theta}[1 - (1 - 3\theta)^{1/2}]$ and V increases everywhere. For $\theta = 0.2$, this bound is 0.61, which covers the known data.

Considering two signals with mean IATs of $M_1 > M_2$ (i.e., the second signal is more intense), the ROC curve (in z scores) is obtained by eliminating c:

$$z_2 = \left(\frac{V_1}{V_2}\right)^{1/2} z_1 + \frac{(M_1 - M_2)}{V_2^{1/2}} (J\kappa)^{1/2}. \tag{15}$$

Since V is a strictly increasing function of M, the slope of this ROC curve is > 1. In the exponential case, it is M_1/M_2.

To work out the prediction for the counting model, we note that in a renewal process, the number $\mathbf{N}$ of pulses counted in a time δ per channel is, asymptotically as $J\delta \longrightarrow \infty$, given by

$$E(\mathbf{N}) = J\delta/M, \qquad V(\mathbf{N}) = J\delta V/M^3$$

(Parzen, 1962, p. 180). So the z score is of the form

$$z = \frac{c + E(\mathbf{N})}{V(\mathbf{N})^{1/2}} = \frac{(c + J\delta/M)M^{3/2}}{(J\delta V)^{1/2}},$$

whence the ROC curve is

$$z_2 = \left[\frac{V_1}{V_2}\left(\frac{M_2}{M_1}\right)^3\right]^{1/2} z_1 + (J\delta)^{1/2}\left(1 - \frac{M_2}{M_1}\right)\left(\frac{M_2}{V_2}\right)^{1/2}. \tag{16}$$

Since M^3/V is an increasing function of M, the slope is < 1. In the exponential case it is $(M_2/M_1)^{1/2}$.

Green and Luce (1973; also see Luce, 1972) ran the above experiment using a 1000 Hz signal in noise and noise alone. When the deadline was varied, the MRTs were virtually identical except for the very long deadlines (see the next section), and with the deadline fixed and the payoffs varied, the ROC curve was linear with slopes of 0.92, 0.90, and 0.69 for three subjects. Thus, the counting rule is more appropriate for this experiment than the timing one. When, however, the experiment was slightly altered so that the deadline applied only to those trials on which the more intense signal was presented, then both the MRT and ROC curves were approximately linear and the pairs of MRT and ROC slopes for three subjects were: (1.34, 1.30), (1.48, 1.47), and (1.38, 1.37). Obviously, the timing rule is now more appropriate than the counting one. Moreover, the close agreement of the values suggests $V = M^2$, in which case the exponential distribution is approximately correct.

Obviously, considerable care is going to have to be taken both experimentally and theoretically to make sure whether a timing or counting model is appropriate in any given context. Our experience to date with these models suggests that the timing ones are generally more plausible except in situations when it is distinctly to the subject's advantage to employ counting behavior. Such an advantage obtains when we impose a uniform deadline on all responses, as just discussed, and when we use very brief signals whose presence is well marked, as is typical of many psychophysical designs.

5. SPEED-ACCURACY TRADE OFF

The contrast between the counting and timing decision rules is nicely illustrated by the different trade off they predict between speed and accuracy. Speed is, of course, measured by MRT. Accuracy can be measured in several ways. Provided that the data are fairly linear on an ROC plot in z scores, one of these is d', i.e., the z_2 coordinate of the ROC curve corresponding to $z_1 = 0$. We see from Equation 16 that for the counting model

$$d' = A\delta^{1/2}, \tag{17}$$

where

$$A = J^{1/2}\left(1 - \frac{M_2}{M_1}\right)\left(\frac{M_2}{V_2}\right)^{1/2}. \tag{18}$$

As the listening interval δ is increased by manipulating the deadlines, both d' and $\text{MRT} = \bar{r} + \delta$ should increase with the trade off being of the form

$$d' = \begin{cases} A(\text{MRT} - \bar{r})^{1/2}, & \text{MRT} \geq \bar{r}, \\ 0, & \text{MRT} < \bar{r}. \end{cases} \tag{19}$$

In like manner, Equation 15 yields for the timing model

$$d' = \frac{M_1}{M_2^{1/2}} A\kappa^{1/2}, \tag{20}$$

where A is given in Equation 18, and the expression for MRT was derived in Equation 13. Again, increasing the deadline should increase κ and, hence, both d' and MRT. To derive the exact form of the trade off, we must see how $h(J, \kappa, \sigma)$ depends on κ. Assuming a train of pulses with mean 1 and standard deviation σ for the IATs, and letting ψ_κ denote the distribution of the $(\kappa + 1)$st pulse, we have by definition

$$h(J, \kappa, \sigma) = J \int_0^\infty x\psi_\kappa(x) \left[\int_0^x \psi_\kappa(y)\, dy\right]^{J-1} dx.$$

Assuming that ψ_κ is approximately normal, which by the central limit theorem it certainly is when κ is large, it is easy to see that

$$h(J, \kappa, \sigma) \cong \kappa + 1 + (\kappa + 1)^{1/2}\sigma H(J),$$

where $H(J)$ is the mean of the largest of J normally distributed RVs with mean 0 and variance 1. Selected values are $H(2) = 0.56$, $H(10) = 1.54$, $H(100) = 2.51$, and $H(1000) = 3.24$. Substituting in Equation 13,

$$\text{MRT}_i = [\bar{r} + M_i + V_i^{1/2}H(J)] \\ + M_i\{\kappa + [(\kappa + 1)^{1/2} - 1](V_i^{1/2}/M_i)H(J)\}, \quad (21)$$

where we have written the right side so that the first term on the right corresponds to the irreducible minimum MRT when $\kappa = 0$. Since $\kappa = 0$ implies no information ($d' = 0$), the first term also describes the intercept in an equation relating speed and accuracy. This suggests introducing the variable

$$T_i = \text{MRT}_i - \bar{r} - M_i - V_i^{1/2}H(J),$$

solving for κ in Equation 21, and substituting that in Equation 20, which yields

$$(d')^2 = \begin{cases} \dfrac{M_1^2}{M_2 M_i} A^2 \left\{T_i + \dfrac{M_i B_i}{2}\left((B_i + 2) - \left[(B_i + 2)^2 + \dfrac{4T_i}{M_i}\right]^{1/2}\right)\right\}, & T_i \geq 0, \\ 0, & T_i < 0 \end{cases} \quad (22)$$

where

$$B_i = V_i^{1/2}H(J)/M_i.$$

Comparison of Equations 19 and 22 reveals three differences in the two models. First, in the counting model the function starts at $\bar{r}$, whereas in the timing model it begins later, at $\bar{r} + M_i + V_i^{1/2}H(J)$. Second, in the counting model there is only a single function since $\text{MRT}_1 = \text{MRT}_2$, whereas in the timing model there are two distinct functions corresponding to $i = 1, 2$. And third, in the timing model the initial growth of the function corresponding to the weaker signal ($i = 1$) is $(M_1/M_2)^{1/2}$ times that of the counting model and that of the stronger signal is that much again, or M_1/M_2, times that of the counting model. This means that a plot of d' versus mean reaction time for the timing model should begin later than that of the counting model, but it should grow considerably more rapidly.

Because we collected data for a complete ROC curve at only one deadline, the values of d' at other deadlines were inferred by passing a line with the one estimated ROC slope through the single observed point at each deadline. According to both the counting and the timing models (Eqs. 15 and 16) the ROC slope should be independent of the deadline (δ or κ), so this method of estimation does not favor either model.

The data are shown in Figure 3. The first panel includes the data of three observers run with the deadline applied on all trials. The trade off appears to be substantially the same for all three, and the value of $\bar{r}$ is about 150 msec. The growth of the function is more nearly linear than the ½ power of $\text{MRT} - \bar{r}$, as was predicted; we return to this discrepancy shortly. The other three panels show another three observers run with the deadline applied only to the signal, i.e., to $i = 2$. Again, they are similar to each other: the inter-

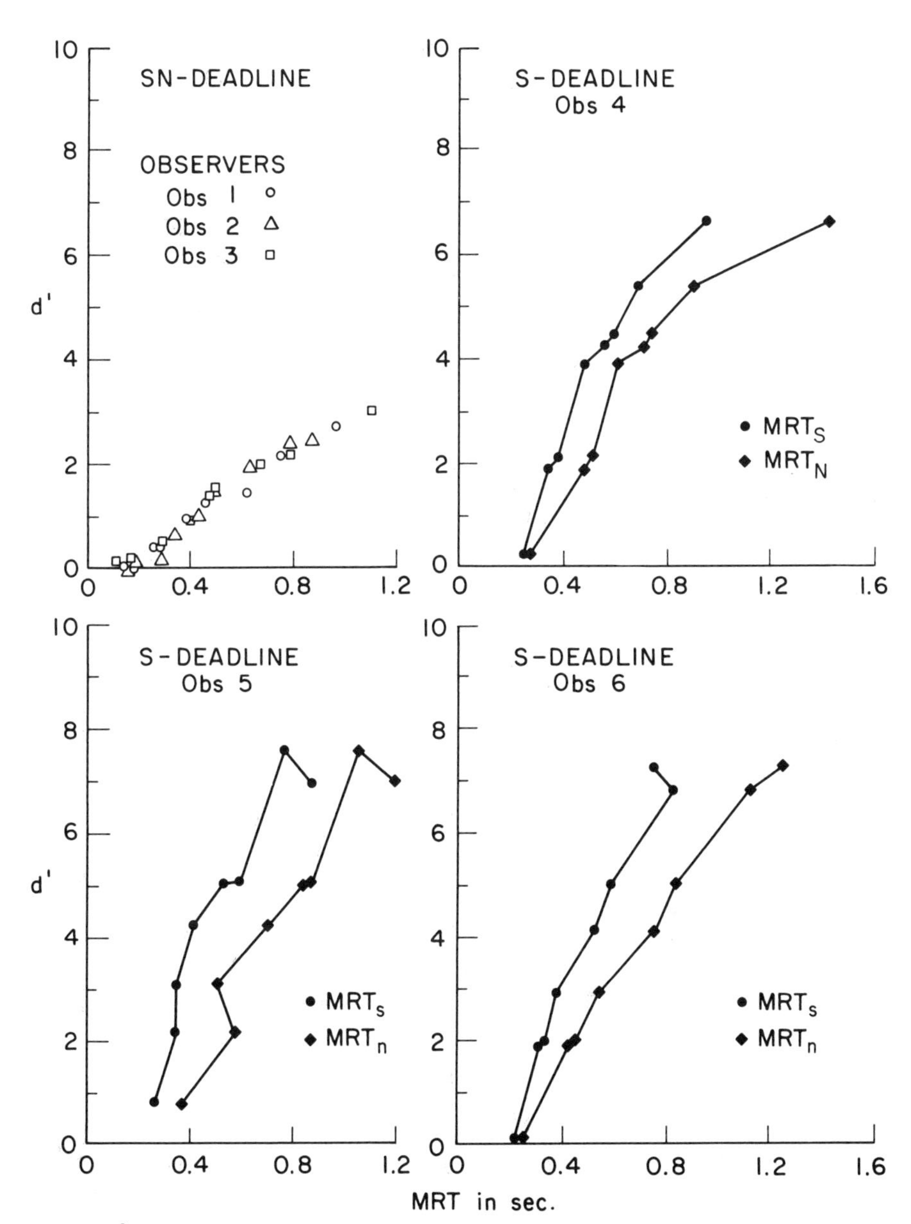

FIGURE 3.
Plots of d' versus MRT. The upper left panel combines the data for three observers when the deadline applies to all trials. The other three panels, one per observer, plot d' versus both MRTs and MRT_n when the deadline applies only to signal trials. (Green & Luce, 1973, Fig. 11.)

cepts of the functions are well over 200 msec and all of the functions rise distinctly more rapidly than in the first panel. Qualitatively, at least, these results are in excellent accord with the predictions of, respectively, the counting and timing models.

A direct numerical comparison is not possible for several reasons. First, different observers were used in the two experiments. Second, the predicted speed-accuracy trade off for the timing model requires estimates of M_i, V_i, and J, which we do not have. And third, the predicted square-root form for the speed-accuracy trade of the counting model was not substantiated by the data. There are at least two possible reasons for this discrepancy. First, estimates of d' near 0 are very unstable, and it is conceivable that the small d's in the region of 150 to 200 msec are in fact all really zero, in which case the square-root growth is not a bad approximation. Such an explanation is, however, inconsistent with our conclusion that the intercepts for the two experiments are different. Second, for long durations it is clearly to the subject's advantage to switch from a counting to a timing mode: he will thereby increase his accuracy. There is some suggestion of such a change in that, at the longest deadlines, the difference between MRT_1 and MRT_2 is not zero. For example, at 2000 msec it is 56, 65, and 9 msec for the three subjects. This suggests that at least two of the subjects were mixing the counting and the timing modes at the long deadlines. The degree of departure is, however, less than one would expect if they were trying to optimize their accuracy payoffs.

A number of authors, using various techniques to manipulate or classify response times, have examined the speed-accuracy trade off. Taylor, Lindsay, and Forbes (1967) predicted that d'^2 should be approximately linear with MRT and they confirmed it in data of Schouten and Bekker (1967). Other closely related studies are those by Fitts (1966), Lappin and Disch (1972), Pachella and Fisher (1972), Pachella, Fisher, and Karsh (1968), Pachella and Pew (1968), and Pew (1969).

In the case of readily detectable signals, Ollman (1966) and Yellott (1967) proposed a quite different, two-state model to account for the speed-accuracy trade off. They assumed that the subject chooses on each trial either to respond to the signal onset, in which case he is fast and inaccurate (chance level), or to wait until the signal is positively identified, in which case he is slow and accurate. By varying the probability of waiting, a trade off is effected. This is known as the *fast-guess* model. Without formalizing the model and working through the algebra, the following relation can be derived. Letting c (for correct) denote the 11 and 22 cells and e (for error) the 12 and 21 cells in the stimulus-response matrix, then

$$P_c\text{MRT}_c - P_e\text{MRT}_e = A(P_c - P_e),$$

where A is a constant and

$$P_c = (\tfrac{1}{2})[P(1 \mid 1) + P(2 \mid 2)],$$
$$P_e = (\tfrac{1}{2})[P(2 \mid 1) + P(1 \mid 2)].$$

We evaluate the same quantity for our two models, again omitting the algebra. For the counting model,

$$P_c\text{MRT}_c - P_e\text{MRT}_e = (\bar{r} + \delta)(P_c - P_e).$$

Since $P_c - P_e$ increases with δ, we see that an accelerated function is predicted rather than a linear one. For the timing model,

$$P_c\text{MRT}_c - P_e\text{MRT}_e = \bar{r}(P_c - P_e) + h(J, \kappa, \sigma)\{M_1[P(1 \mid 1) - \tfrac{1}{2}] + M_2[P(2 \mid 2) - \tfrac{1}{2}]\}.$$

The first term is simply the linear prediction on the fast-guess model, and the second one increases with κ and so with $P_c - P_e$, again yielding an accelerated function. Observe, however, that unlike the counting model, the degree of acceleration is a function of signal strength, M_i. For intense signals (small M_i), it is negligible and so we should obtain the linear relation; for weak ones, however, it should be convex.

The data for both weak and intense signals in the experiment with a deadline imposed on all trials (both $i = 1$ and 2) are shown in Figure 4, and the data for weak signals in the experiment with a deadline only on the more intense signal (deadline for $i = 2$ only) are shown in Figure 5. Obviously, the fast-guess model is wrong for weak signals since, for both types of deadline, the curves are decidedly convex. Qualitatively, both sets of data for weak signals agree with both the counting and timing models. The only data we have for intense signals come from the experiment where, for weak signals, we concluded earlier that the counting model held. The strong signal data appear linear and hence are consistent with only the timing model. It appears, therefore, that with increasing signal strength there is a tendency to switch into the timing mode. Presumably this is the more natural mode of behavior, and there is little advantage to using counting when the maximum number of IATs that can be collected per channel has a high probability of being within the deadline for either signal because M_i is so small.

Our tentative conclusion is that subjects use the timing rule except when it is decidedly disadvantageous to do so, in which case they shift to the counting one. Examples of experiments where the timing rule is disadvantageous are those involving weak signals that are either of short duration or to which fast responses (and so a short sample of the signal) are encouraged. As we shall see below, several other types of experiments appear to be accounted for by timing rules.

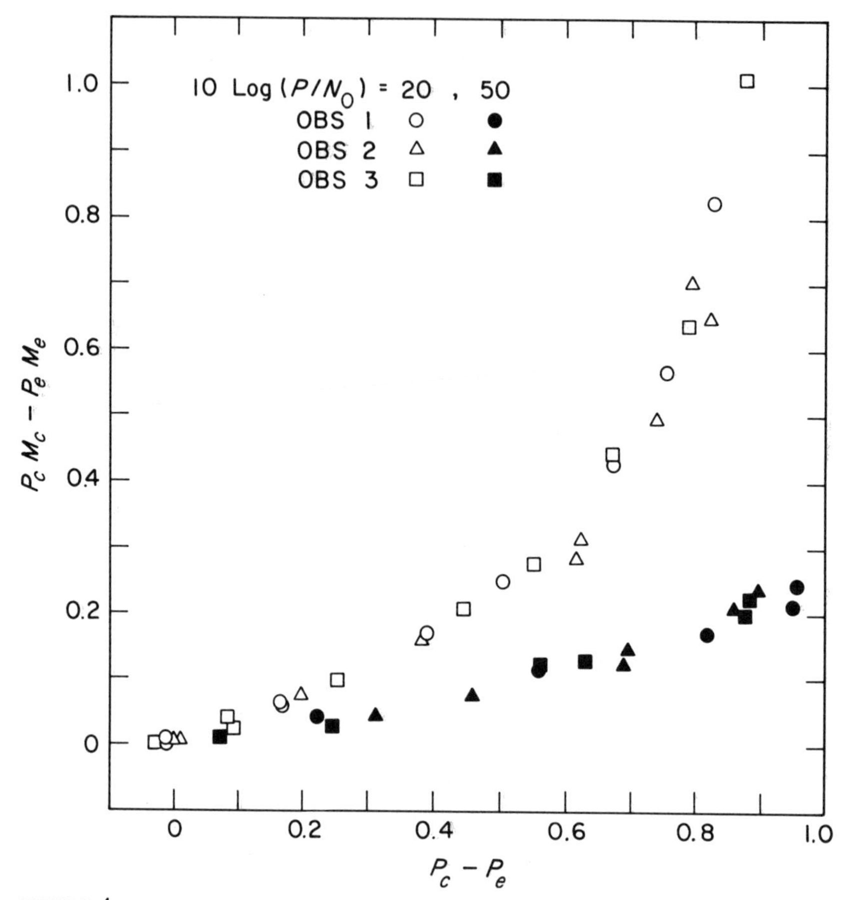

FIGURE 4.
Fast-guess analysis of speed-accuracy data when the deadline applies to all trials. The open points are for weak signals; the solid ones, for strong signals. (Green & Luce, 1973, Fig. 2.)

6. DISCRIMINATION FUNCTION FOR INTENSITY

One ultimate test of a theory of neural coding is its ability to account for psychophysical data on the limits of discriminability for small changes in both the frequency and the intensity of a sinusoidal signal. As we show here and in Section 8, the theory does indeed make such predictions. Although one does not anticipate any difficulty in comparing these predictions with the empirical data—since surely after a century of study these basic relations are

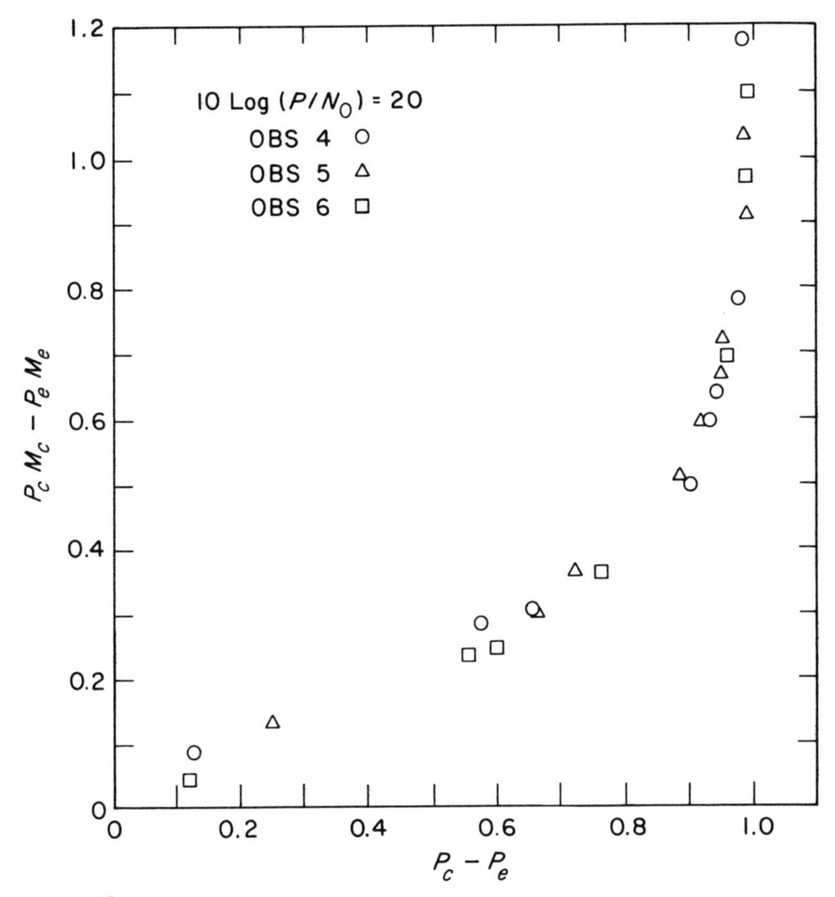

FIGURE 5.
Fast-guess analysis of speed-accuracy data (weak signals) when the deadline applies to signal trials. (Green & Luce, 1973, Fig. 5.)

well established and generally accepted—nothing could be further from the truth.

It has long been known that a variety of intensive stimuli obey Weber's in the form

$$\frac{\Delta I}{I} = \frac{A}{I} + B, \tag{23}$$

where I denotes a reference intensity, ΔI the deviation from I that is necessary to achieve a fixed level of discriminability, and A and B are constants. Hawkins and Stevens (1950) showed that the signal energy to noise power density needed for a sinusoid to be just detectable in wideband noise is remarkably constant over a range of about 90 dB. Miller (1947) showed that the just detectable increment in a wideband noise is some constant fraction of the background noise once the background is slightly above the absolute threshold. On the other hand, the detection of an intensity increment in a sine wave has never exhibited this relation over any reasonably large range of intensity. Rather, if one plots $\Delta I/I$ versus I, the function begins at some large value and generally decreases as the intensity of the background I is increased (Campbell & Lasky, 1967; Dimmick & Olsen, 1941; Green, 1967; McGill & Goldberg, 1968; Reisz, 1928; Viemeister, 1972).

McGill and Goldberg (1968) were the first to discuss carefully this deviation from Weber's law—what they call the near miss to Weber's law—and they pointed out that it is well approximated for large I by an equation of the form

$$\frac{\Delta I}{I} = BI^{c-1}, \tag{24}$$

where B and c are constants independent of I. The data are fit with c about 0.90. The explanation they offer is that the small deviation from Weber's law arises because of the nonlinear dependence of the number of neural spikes in a counting model as a function of intensity.

Another interpretation of these data has recently been suggested by Viemeister (1972), who seems to have shown empirically that the reason for the deviation from Weber's law is wholly different from a neural explanation. He noted that as intensity is increased, the inherent nonlinearity of the ear introduces amplitude distortion contributions at the various harmonics of the signal. At low intensities, the energy at each harmonic is considerably smaller than at the next lower harmonic. In fact, at sufficiently low intensities, all harmonics higher than the fundamental are inaudible. However, as the intensity of the fundamental is increased, the rate of growth is faster the higher the harmonic. For example, if the distortion follows a square law, then the amplitude of the second harmonic grows as the square of the amplitude of the primary. And so a 1 dB change in the primary produces a 2 dB change in the second harmonic. Thus, at the point at which the second harmonic becomes audible, it is better to use it in making discriminations because the change in dB is larger there than at the fundamental. Similar arguments apply to the other harmonics—an n dB change occurs at the nth harmonic. Thus, as the stimulus is increased in intensity, if the subject moves from harmonic to harmonic as each becomes audible, he will clearly do better than Weber's law even if that law applies to discrimination of the primary.

Viemeister analyzed existing data on amplitude distortion to see if this argument is viable, and it is. So he performed the following direct test of it. If this is what the subjects are doing and if one can make it impossible for them to do so, then they should exhibit Weber's law. He used high-pass masking noise (with its boundary above the fundamental but below the first harmonic) to obscure the harmonics, and found Weber's law to be almost exactly satisfied. Despite the fact that his experiment has not been replicated, we find the basic logic compelling. Nonetheless, we work out the predictions of the present theory. A more complete discussion, a counting alternative emphasizing and using data other than Reisz' is given by Luce and Green (1974).

We now compute how ΔI depends on I and f in the timing model. Usually ΔI is defined to be that stimulus increment producing 75 percent correct detections in a two-alternative forced-choice design, and this is very close to $d'_I = 1$. From Equation 15 we know that d'_I is given by

$$d'_I = (J_K)^{1/2}\left(\frac{M_I - M_{I+\Delta I}}{V_I^{1/2}}\right).$$

From Equations 2 through 7,

$$M = E(\mathrm{IAT}) = 1/pf$$

and

$$\begin{aligned} V = V(\mathrm{IAT}) &= (q + p^2\theta^2)/p^2f^2 \\ &\cong q/p^2f^2 \\ &= qM^2, \end{aligned}$$

where $q = 1 - p$ and $\theta \cong 0.2$ is a measure of the smearing of the modes (see Eq. 6). For $p \leq \frac{1}{2}$, the approximation is obviously good. Substituting and setting $d' = 1$ yields

$$\frac{\Delta p}{p} = \frac{1}{(J_K/q)^{1/2} - 1}. \tag{25}$$

To calculate ΔI, we must first determine how p depends on I. If we approximate the IAT distribution by an exponential (parameter μ) displaced by the amount $1/f$ from the origin, then equating means it is easy to see that

$$\frac{1}{1-p} = \frac{\mu}{f}.$$

As we show empirically in Section 10, reaction-time data strongly suggest that μ is a power function of intensity—at least for low intensities. Letting p_0 be the probability corresponding to the threshold intensity I_0 at frequency f, we may therefore rewrite the above equation and the empirical hypothesis as

$$\frac{p}{1-p} = \frac{p_0}{1-p_0}\left(\frac{I}{I_0}\right)^{\gamma}. \tag{26}$$

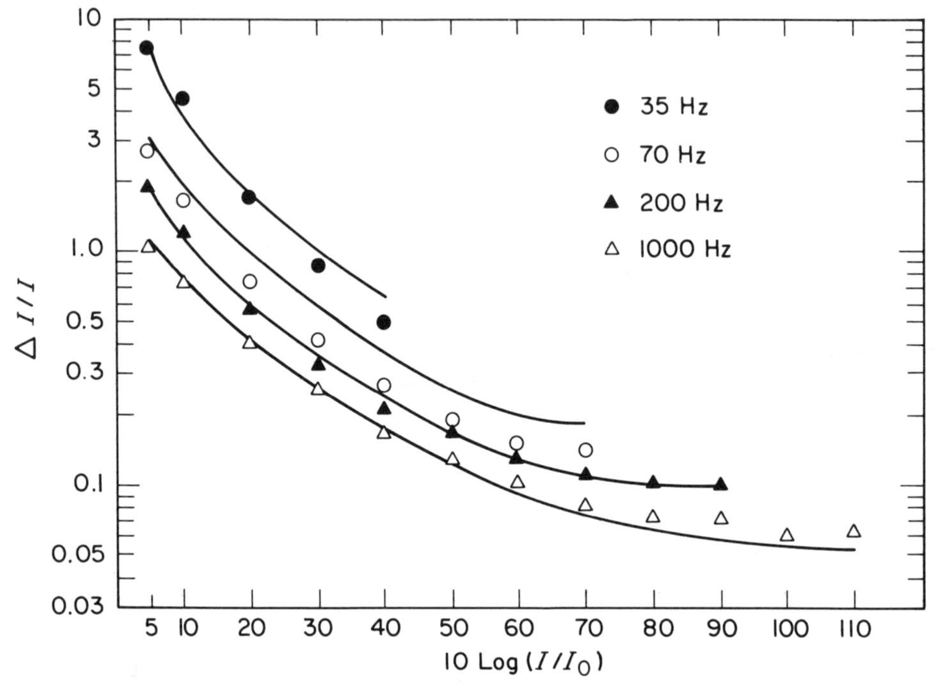

FIGURE 6.
$\Delta I/I$ versus I with f as a parameter. The data are taken from Reisz (1928). The theoretical curves are Equations 27 through 30 with the following parameter values: $\gamma = 0.20$, $\eta = 0.10$, $A_0 = 0.60$ sec, $\kappa_0 = 1$, $\kappa_m = 25$, $J_0 = 17.10$, $B = 1.246$.

It then follows readily that

$$\frac{\Delta I}{I} = \left[\frac{1}{1 - 1/(qJ\kappa)^{1/2}}\right]^{1/\gamma} - 1. \tag{27}$$

Thus, we now need to see how $qJ\kappa$ depends on I and f. For the present we simply write down the three equations that we use in our calculations; their partial justifications are provided in the next section.

$$p = \frac{(I/I_0)^\gamma}{A_0 f - 1 + (I/I_0)^\gamma}, \quad q = 1 - p = \frac{A_0 f - 1}{A_0 f - 1 + (I/I_0)^\gamma}. \tag{28}$$

$$J = J_0(\log_{10} f - B)(I/I_0)^\eta. \tag{29}$$

$$\kappa = \begin{cases} (\kappa_0 + 1)A_0 p f - 1, & \kappa \leq \kappa_m, \\ \kappa_m, & \text{otherwise.} \end{cases} \tag{30}$$

Using the parameter values shown in the caption of Figure 6 and substituting Equations 28 through 30 into Equation 27 yields the family of curves shown in Figure 6. The data points are those of Reisz. We discuss some of the choices of the parameter values as we 'derive' Equations 28 through 30.

7. THE DEPENDENCE OF p, J, AND κ ON I AND f

To arrive at the preceding three expressions, we make three rather speculative hypotheses which have very incomplete experimental support.

1. *The neural criterion of activity in a channel, and hence the criterion for a sensory threshold, is characterized by* $E(\mathrm{IAT}) = A_0$, *where* A_0 *is a constant that has to be estimated.*

In whatever way the nervous system decides whether or not a channel is active, it seems plausible that the decision is based solely on sample of activity from that channel. Since the only information available from a renewal process is contained in the IATs, the criterion must be based on them and certainly one of the simplest is merely to use the mean IAT. If this were true, then except for a possible small effect due to changes in J with I and f (Eq. 29), the MRT at threshold should be independent of f. The small amount of data that we have collected seem to support this; however, a more extensive study is needed.

Using the fact (Eq. 12) that $E(\mathrm{IAT}) = 1/pf$ and the hypothesis that at threshold $E(\mathrm{IAT}) = A_0$, Equation 28 follows immediately from Equation 26.

The only real difficulty we have had in choosing acceptable parameter values is with A_0. According to Equation 21,

$$\mathrm{MRT} \geq \bar{r} + A_0(1 + \kappa_0).$$

Using $\kappa_0 = 1$ and $A_0 = 600$ msec, as in Figure 6, and (from Sec. 5 or from simple RT data with intense signals) $\bar{r} = 150$ msec, we see that MRT is not smaller than 1350 msec—in fact, it is considerably larger since the inequality is crude. Our RT data at threshold suggest a value of about 1000 msec. Of course, our data were collected some 45 years after Reisz' and used a very different discrimination technique, so we cannot be certain whether or not there is a real difficulty.

Our second hypothesis is:[5]

[5] *Note added in proof.* Since writing this paper, our views on modeling discrimination data have changed in important ways. First, Reisz' data resulted from an unusual experimental procedure, one not particularly well modeled by the present considerations, and unlike all later intensity discrimination data they exhibit a strong dependency on frequency. Second, the later data all used relatively brief duration signals, suggesting that a counting model is appropriate. Luce and Green (1974b) fit the obvious one, using hypothesis 1 and constant values for δ and J; they had no need, therefore, for hypotheses 2 and 3. Third, and perhaps most important, a detailed study of magnitude estimates reported by Green and Luce (1974) suggest that the situation is actually somewhat different from, and much more interesting than, the conjectures that follow. See footnote 6 and also Luce and Green (1974a).

2. *When a subject is asked to magnitude estimate a signal, he collects κ* IATs *on each channel, forms the reciprocal of the total time on each channel, adds these numbers over all active channels, and then multiplies that by a constant and emits the result as his magnitude estimate.*

This hypothesis is a special case of the more basic hypothesis that the subject interprets magnitude-estimation instructions as a request to estimate and report the neural pulse rate. If so, there are a number of ways in which it might be estimated. For example, we assumed earlier (Luce & Green, 1972) an exponential IAT distribution and that the times were summed over channels prior to forming the reciprocal. In that model, magnitude estimates are reciprocals of a RV with a gamma distribution of order $J\kappa$. Because $J\kappa$ is necessarily large, the resulting distribution should have a stable mean and variance. Data we have collected make us doubt whether the variance really is very stable. This could arise if κ were small, 2 or 3, and if the reciprocal were formed on each channel individually before summing. For example, with $\kappa = 2$ the theoretical variance of the reciprocal of a second-order, gamma-distributed variable does not exist. The trouble arises, of course, because the exponential places so much density at 0 which the reciprocal maps into ∞. These observations have led us to think, first, that maybe only a very small number of IATs are collected on each channel; second, that the reciprocal is computed for each channel separately; and third, that for theoretical ME calculations the exponential distribution is a very poor approximation to Equation 7. Some indirect support to the idea that only a few IATs are collected per channel is provided by the studies of Stevens (1966) and Stevens and Hall (1966), in which it was shown that auditory magnitude estimates reach their maximum value when the signal duration is about 150 msec.

In working out the magnitude-estimation distribution using the IAT model of Equations 2 through 7 and this hypothesis, we make the simplifying assumption that $\mathbf{X}$ can be neglected. This introduces some error in $E(\text{ME})$ and, were we to compute it, a rather sizable error in $V(\text{ME})$, especially when p is large. With this approximation, we compute the expectation of $f/\mathbf{I}$ where $\mathbf{I}$ is an integer-valued random variable whose distribution is the convolution of κ geometric distributions, each with its origin at 1 rather than 0. As the calculations are routine, we exhibit only the mean (over J channels) for the first three values of κ:

$$E(\text{ME}) = Jf\left(\frac{p}{1-p}\right)\begin{cases} \ln\dfrac{1}{p}, & \kappa = 1, \\ 1 - \left(\dfrac{p}{1-p}\right)\ln\dfrac{1}{p}, & \kappa = 2, \\ \dfrac{1}{2} - \left(\dfrac{p}{1-p}\right)\left[1 - \left(\dfrac{p}{1-p}\right)\ln\dfrac{1}{p}\right], & \kappa = 3. \end{cases} \tag{31}$$

Since empirically $E(\mathrm{ME})$ is approximately a power function of I, with an exponent of about 0.3 in the case of loudness, since the dominant term appears to be $Jfp/(1 - p)$, and since $p/(1 - p)$ is a power function (Eq. 26) we are led to conjecture that

$$J \propto (I/I_0)^{\eta},$$

where $\gamma + \eta = 0.3$, a constraint we maintain in our choice of constants.

The further assumption that the effect of intensity and frequency contribute multiplicatively is pure speculation, and the choice of a logarithmic dependence on f in Equation 29 is entirely empirical.

Our final hypothesis is:

> 3. *The number of* IATs *collected on each channel is adjusted so that the total time consumed is approximately a constant up to some limit,* κ_m, *determined by the size of a buffer store for* IATs.

It seems plausible that the nervous system either holds κ constant independent of frequency or makes some sort of adjustment to maintain approximately constant times. The model does not fit the data if κ is kept constant, so we were led to hypothesis 3. Of course, if the intensity levels were randomized, one suspects that a constant κ would result.

Letting κ_0 denote the threshold value of κ, it follows that, since the number of pulse is one more than the number of IATs, for $\kappa \leq \kappa_m$,

$$(\kappa_0 + 1)A_0 = (\kappa + 1)E(\mathrm{IAT}) = (\kappa + 1)/pf,$$

from which Equation 30 is an immediate consequence.

8. DISCRIMINATION FUNCTION FOR FREQUENCY

The data on the discrimination of small changes in frequency are, if anything, less satisfactory than those for intensity. Three major papers (Harris, 1952; Nordmark, 1968; Shower & Biddulph, 1931) summarize how Δf varies with f and I (but only for 30 dB SPL in Harris and 45 dB SPL in Nordmark). The results at comparable SPL and low frequencies differ by almost an order of magnitude. It is still not known what differences in method are crucial to these differences in performance. There is one optimistic note in that, although the values of Δf differ considerably, their ratios at different frequencies are nearly the same. Figure 7 presents these data.

At high frequencies the agreement is better, but Henning (1966) has challenged all of these high-frequency measurements as being seriously influenced by artifacts. He demonstrated that above 4000 Hz, the discrimination of a

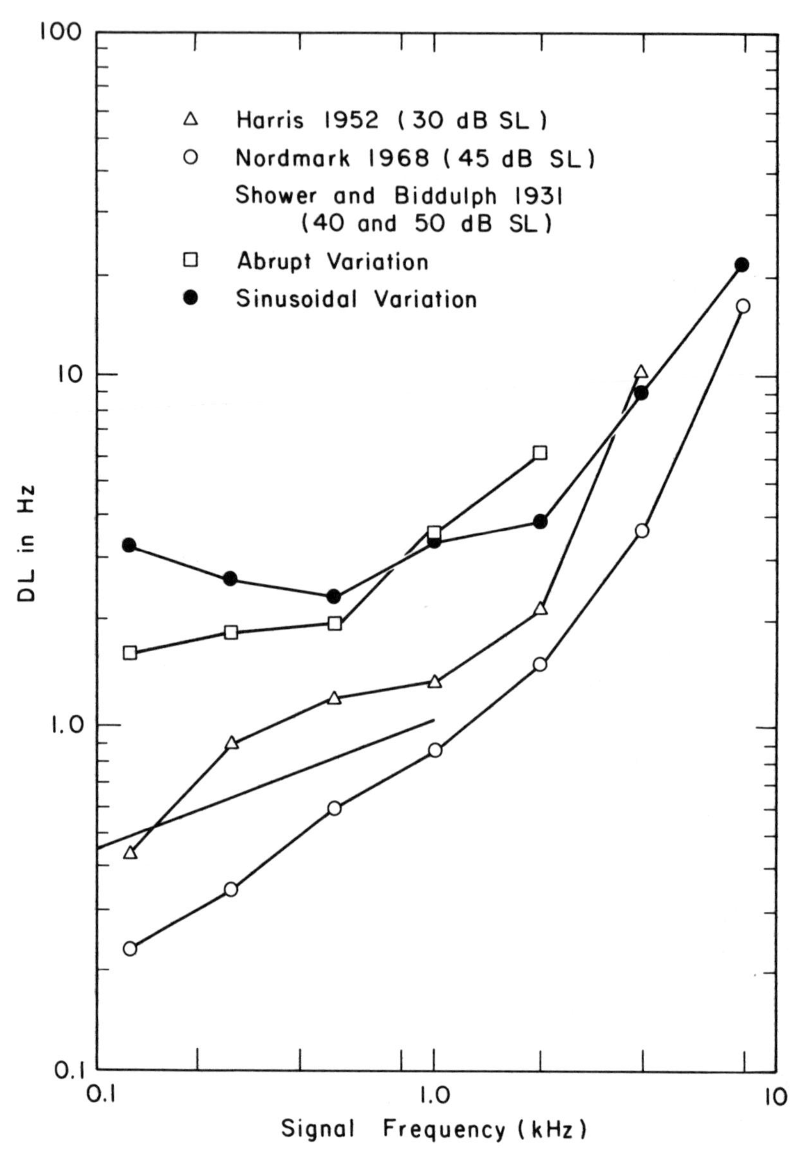

FIGURE 7.
Δf versus f. The data are from the sources shown. The theoretical curve is Equation 37 with the parameters given in the caption of Figure 5, $\theta = 0.20$, and $I/I_0 = 45$ dB.

frequency change is probably based largely on a change in loudness. The evidence is that when the intensity of the two tones is randomized, the accuracy of frequency discrimination deteriorates markedly. For example, at 8000 Hz the jnd is 16 Hz when the intensities are fixed and 100 Hz when they are randomized. No comparable change was observed at low frequencies, but one would not expect any since a change in frequency at low frequencies does not alter the amplitude of the signal appreciably.

The only study of Δf over the whole (I, f) plane that we know of is Shower and Biddulph's. Since this study is the earliest, it exhibited the poorest discrimination, and the technique of signal variation employed is very difficult to model, we remain very unsure how Δf in its usual sense depends on I.

Turning to theory, we first recall that Wever (1949) argued from various sorts of data that place mechanisms probably account for the discrimination of frequencies above approximately 2000 Hz and periodicity mechanisms probably are involved for lower frequencies. Such a break is almost certainly enforced by the inability of a nerve to fire faster than about every ½ msec. Turning to our low-frequency model given in Equation 1, we see that if **I** were known, then

$$\mathbf{T} = \frac{\mathbf{IAT}}{\mathbf{I}} = \frac{1}{f} + \frac{\mathbf{X}}{\mathbf{I}} \tag{32}$$

estimates the period, $1/f$, of the input signal. It is equally clear that given a reasonably large sample of IATs—one from each active channel should do—the variability due to **X** is so small that **I** can be determined correctly for each IAT. (This would not be true if the modes of the IAT distribution overlapped.) If we let $\overline{\mathbf{T}}$ be the average value of **T** over J_K IATs, then it is easy to show (Luce & Green, 1974) that

$$E(\overline{\mathbf{T}}) = 1/f,$$

$$V(\overline{\mathbf{T}}) = \left(\frac{\theta}{f}\right)^2 \frac{p}{q} \frac{\phi(q)}{J_K},$$

where

$$\phi(q) = \sum_{i=1}^{\infty} q^i/i^2 = -\int_0^q \frac{\ln(1-x)}{x}\, dx.$$

Defining d'_f as the difference in means divided by their (approximately common) standard deviation,

$$d'_f = \left(1 - \frac{f}{f + \Delta f}\right) \frac{1}{\theta} \left(\frac{J_K}{p\phi(q)/q}\right)^{1/2}.$$

Solving for Δf with $d'_f = 1$ yields

$$\Delta f = \frac{f\theta[\phi(q)/q]^{1/2}}{(J\kappa/p)^{1/2} - \theta[\phi(q)/q]^{1/2}} \cong \frac{f\theta[\phi(q)/q]^{1/2}}{(J\kappa/p)^{1/2}}, \tag{33}$$

where the approximation is justified because $\theta \cong 0.2$ and $[\phi(q)/q]^{1/2}$ has $\pi/6^{1/2}$ as its maximum value. Substituting p from Equation 28, J from Equation 29, and κ from Equation 30 in Equation 33, and using the constants listed in the caption of Figure 6 together with $\theta = 0.20$ yields the theoretical curve shown in Figure 7.

It is worth noting that the portion of the theoretical curve for low intensities decreases with increasing intensity; however, after κ reaches κ_m frequency discrimination for large f is predicted to deteriorate slowly with intensity. There are recurrent indications of this in Henning's (1967, Fig. 3) work, but it has never been carefully studied.

We conclude that the same set of constants is adequate to account for frequency as well as intensity discrimination.

9. OTHER SOURCES OF VARIABILITY

Up to this section, we have attempted to account for behavior in terms of the sum of three sources of variability: that due to the stimulus itself, that arising from its transduction into neural impulses, and that occurring during the transmission of these neural impulses up to the point at which the decision is made. Specific assumptions about the nature of these kinds of variability allowed us to predict a variety of classical psychophysical data. In particular, most of the data for simple discrimination and their corresponding response latencies can be analyzed from this point of view. Whether or not *all* simple discriminative behavior can be dealt with so simply is problematical, but many theorists, including us, are optimistic.

We would be remiss, however, to leave the impression that this is the only variability that ever arises or, indeed, that this approach can be extended to more complicated psychophysical tasks without introducing some fundamentally new assumptions. Although a full discussion of this topic is not appropriate here, we will review briefly some work both because it is timely and relevant and because it suggests that variability in applying the decision rule itself is necessary to explain some psychophysical behavior.

Consider an absolute identification, or categorization, experiment in which one of N signals is presented on each trial and the subject attempts to identify which it is. Assuming that the process is stationary and independent from trial to trial, then the basic data can be presented in an $N \times N$ matrix of probabilities where the ij entry is the relative frequency of the jth response when the ith signal is presented. In most such studies the signals are chosen from some ordered continuum, such as intensity level of a pure tone, and the responses are the first N integers assigned to the signals so that their natural order corresponds to the order of the sensory continuum.

One question is how to measure accuracy so we can study how accuracy depends on the number, spacing, and range of the signals. The most obvious measure, percentage of correct responses, has the drawback that as N increases it decreases without in any way taking into account how many responses were almost correct—it is insensitive to the magnitude of the errors. More satisfactory measures use more than just the main diagonal of the data matrix. Perhaps the most famous of these is the information measure; below we will describe a d'-type of measure.

A remarkable, but ubiquitous, finding of this area is that the information transmitted grows with N up to about $N = 7$, at which point it reaches an asymptote or, perhaps, a maximum (Miller, 1956). Note that we have stated this entirely in terms of the number of signals without qualification as to their spacing or range. To a good first approximation, such qualification is immaterial: various signal spacings and ranges have been studied and the results are surprisingly insensitive to those choices, except when the range is small, say, less than 20 dB in auditory intensity.

Recently, Durlach and his collaborators (Braida & Durlach, 1972; Durlach & Braida, 1969; Pynn, Braida, & Durlach, 1972) have embarked on a series of experimental and theoretical investigations of this and related paradigms. Although their data-analysis procedure is somewhat unconventional—a modified version of Thurstone's successive intervals rather than, for example, information theory—their main findings are consistent with earlier studies. Their basic measure is obtained by computing over all responses (with estimated probabilities not too near 0 or 1) the values of d' for each pair of adjacent signals. The sum of these values over all adjacent pairs of signals is their overall measure of accuracy. Among other things, they find that if N is held fixed and the signal range is increased, this measure reaches an asymptote.

To account for this at a theoretical level they assume that the variance which enters into the computation of d' for each pair of adjacent stimuli has two independent sources. The one is the usual variability in the representation of the signal, which is what we have discussed in the bulk of the chapter. The other source they interpret as due to fallible memory; e.g., it might result from variability in the location of category boundaries from trial to trial. Their crucial new assumption is not that imperfect memory introduces variability, but that this variance is proportional to the entire square of the range, measured in dB, of the signals being used. Why this should be is not obvious.

To make their theory quite explicit, we list its three principle features. First, signals are represented as normally distributed random variables with constant variance (the latter is not true in a pulse model). Second, the mean of the signal random variable is assumed to be a logarithmic function of signal intensity (this also is not true in our pulse model, where a power function growth was postulated). Third, in computing d' the effective variance consists of the sum of the constant variance of the signal representation and that of the category boundaries, which is assumed to be proportional to $(\log I_N/I_1)^2$, where I_1 and I_N are the intensities of the least and most intense signals, respectively.

According to this theory, for fixed N the total accuracy asymptotes with increasing range because the increase in variance in the category boundaries more than offsets the increase in discriminability in the signal representation due to increased signal separation. Indeed, these postulates account for the bulk of their data, but some notable discrepancies do exist. Without going into them in detail, they seem to result from, first, difficulties in the logarithmic assumption and, second, from what they interpret as 'edge' effects that arise from the end signals. The regularity and, in some experiments, the size of the discrepancies invite alternative models, but so far none has been suggested. We have attempted to construct a pulse model of the type discussed in this chapter, but we are not yet satisfied with it. Our concern has been to find some way to account for the enlarged variance without being forced to assume that it grows directly with range.[6]

The simple reaction-time situation, in which the subject attempts to detect the onset of a signal as rapidly as possible, is a second case when the natural decision rule introduces considerable additional variability. As this topic is important in its own right, we devote the final section to it.

10. SIMPLE REACTION TIMES TO WEAK SIGNALS

It has long been believed that the distribution of reaction times contains information about the decisions being carried out when the subject senses the

---

[6] *Note added in proof.* Our work on the variability of magnitude estimates, reported in Green and Luce (1974), has led to a hypothesis that seems to be consistent with these results. We postulate a band of attention, about 20 dB wide for auditory intensity, which can be located anywhere. Signals falling in the band give rise to samples of IATs that are about an order of magnitude larger than the samples for signals outside the band. Thus, as the range of signals increases, the probability of a small sample and hence increased variability also increases, possibly accounting for the apparent effect of range in degrading absolute identification below that predicted from two-signal data. Of course, detailed work will be needed to verify that this is in fact the explanation. Luce and Green (1974a) also pointed out that such an attention mechanism may underlie the sequential effects observed in both magnitude estimation and absolute identification experiments.

reaction signal, but until comparatively recently little has come of attempts to extract that information. One difficulty is that the temporal reflections of the decision process are almost certainly badly obscured by the ripples resulting from other delays in the overall system unless, of course, the decision process is slow relative to these other delays. Almost paradoxically, reducing the signal intensity serves to amplify the decision process. For a clear and audible signal, the pulse rates are high, decisions rapid, and what we observe—a fairly peaked density whose mean is perhaps 150–200 msec and whose standard deviation is about 10 percent of that—is mostly the result of delays other than those introduced by the decision process. By contrast, for a weak or, because of masking noise, barely audible signal, the pulse rates are low, decisions slow, and what we observe—a broad distribution whose mean is as large as 1000 msec and standard deviation is of the same magnitude—is some mix of the decision latency and other, fixed and variable, residual latencies. Observe that such an increase in the standard deviation is to be expected under the hypothesis that the basic intensity information is encoded in a near-Poisson pulse train since the mean and standard deviation are equal in an exponential distribution.

This general view is probably not very controversial; however, the exact realization of these assumptions differs considerably from one theory to another. All of our own models have explicitly assumed that the decision process is statistically independent of the residual one. In particular, we assume that changes in the signal intensity affect the IATs and so the decision process, but they do not affect the residual process. This is controversial. By independence, the distribution of reaction times is the convolution of the distributions of the decision and residual processes. Other closely related views are the classical one of an irreducible reaction time and Donders' (1868) method of subtraction, which in recent times has been refurbished and exploited effectively by Sternberg (1969). These are weaker models because they assume only additivity, not independence, of the two (or more) stages.

In all such models, no matter what the detailed assumptions are, the effect of the residual latency is simply to obscure the decision process. One way or another, one attempts to evade this noise. A key idea, first pointed out in this context by Christie and Luce (1956) and recently exploited by us, is that the classical transforms—e.g., Laplace and Fourier—convert a convolution into a product. Let capital letters stand for the Fourier transform of the lower case density, i.e.,

$$F(\omega) = \int_{-\infty}^{\infty} e^{-i\omega t} f(t)\, dt, \quad i = (-1)^{1/2}.$$

If

$$f(t) = \begin{cases} \int_0^\infty \ell(t - x) r(x)\, dx, & t \geq 0, \\ 0, & t < 0, \end{cases} \tag{34}$$

then

$$F(\omega) = L(\omega)R(\omega), \tag{35}$$

where ℓ is the density of the decision process and r that of the residual one. Given a theory of the decision process, we can derive explicit expressions for ℓ or L which, however, will depend on several parameters that must be estimated from the data. One such parameter is the pulse rate. Two approaches suggest themselves, and there may be others.

Assume for the moment that we have estimates of the parameters. In the one approach we form the histogram that approximates f, smooth it, and then obtain its transform $\hat{F}$. Substituting the estimated parameters into the decision theory yields an estimate of L, $\hat{L}$. And so by Equation 35 we estimate R by $\hat{R} = \hat{F}/\hat{L}$, and then by taking the inverse transform, we obtain $\hat{r}$. In other words, given the data and estimates of the parameters of the theory, we determine what the residual density must be in order for the theory to yield the observed data. Our interest is not in r per se, but rather in whether what we compute is a possible density function. In Green and Luce (1971) we actually carried this out. The theory studied there, which postulated that the occurrence of pulses is treated as evidence that the signal is present, led to a distribution function with two free parameters. They were estimated by a method described below. When we solved for $\hat{r}$ we found a function that had a mean of about 300 msec and that was appreciably negative in the region from about 400 to 500 msec. From data on intense signals we know that this mean is from 100 to 150 msec too large and, of course, the negative region is inconsistent with it being a density function. So we concluded that that theory of the decision process is wrong, and we were led to the IAT theories discussed earlier.

It should be noted that this method of analysis, and any other involving transforms of empirical data, is beset with an inherent difficulty. Considerable effort must be expended to avoid the so-called Gibbs phenomenon—high-frequency oscillations resulting from the discrete jumps that are inherent in histograms—but without losing all resolution. We used a running average on the histogram to suppress some of the discontinuities, but obviously that introduces some temporal smear. Further theoretical work is needed on how best to effect the compromise before this can become a practical technique of analysis.

A second approach is to use very intense signals to estimate $\hat{R}$ (assuming that the decision time is negligible) and then to compare the fit between $\hat{L} = \hat{F}/\hat{R}$ and L. This raises the question, to which we do not know the answer, of how one best evaluates the quality of fit between transformed data and the transform of a theoretical distribution, i.e., between $\hat{L}$ and L. So far as we know, mathematical statisticians have not formulated an answer to this.

Aside from this (apparently) unresolved question of goodness of fit in the transformed domain, two related problems remain. First, we wish to formulate a plausible decision rule for detecting a change in the intensity parameter of a Poisson process and from that to derive the form of the theoretical distribution ℓ (or its transform L). Second, with that in hand, we need to arrive at plausible ways to estimate the parameters of the model, especially the noise and signal Poisson parameters. In practice, we only have incomplete information about ℓ, but just enough to permit crude estimates of the parameters.

As before, we shall assume the simplest decision rule, namely, that an estimate of pulse rate is compared with a criterion, and the subject responds whenever he has evidence that the pulse rate has increased, which suggests that the reaction signal has been presented. Since the subject is under time pressure, we assume that he uses the smallest sample available, namely, one IAT. But unlike the other models we have examined, the number of IATs actually observed before he initiates a response is a random variable, and this fact makes the analysis considerably more difficult. The key result is that (under certain conditions) the tails of the decision distributions (i.e., for $t > \tau'$) are approximately exponential, with time constants that are simple functions of the Poisson parameters of the noise and signal-plus-noise pulse trains.

To fit such a model to data, we must estimate these Poisson parameters. In practice, we have tried only one way and it is not completely satisfactory. Suppose, as seems plausible, that the residual latencies are bounded in the sense that for some $\tau > 0$, $r(t) = 0$ for $t \geq \tau$. If we substitute this together with the exponential character of the decision process into Equation 34, we see that $f(t) \sim e^{-\mu t} \int_0^\tau e^{\mu x} r(x)\, dx, \quad t \geq \tau + \tau'$. So the tails of the observed distributions should exhibit the same exponential character as the decision distribution when the residual times are bounded. Obviously, the same argument is approximately valid if the residual times are not actually bounded, provided that large times are exceedingly rare.

The data in Figure 8 show that the tails of the response-time distributions in a simple reaction-time design (with response terminated signals and exponentially distributed foreperiods) are very nearly exponential. So we may use these data to estimate the Poisson parameters for noise, ν, and for signal plus noise, μ. The ratio of these estimated parameters as a function of signal intensity in dB is shown in Figure 9, and we see that to a good approximation it is a power function of intensity. We used this fact in deriving Equation 26. Luce and Green (1972) provide an argument, based on the assumption of processing on multiple channels, to show that the exponent estimated from Figure 9 is consistent with that estimated from ME data; we do not reproduce that argument here.

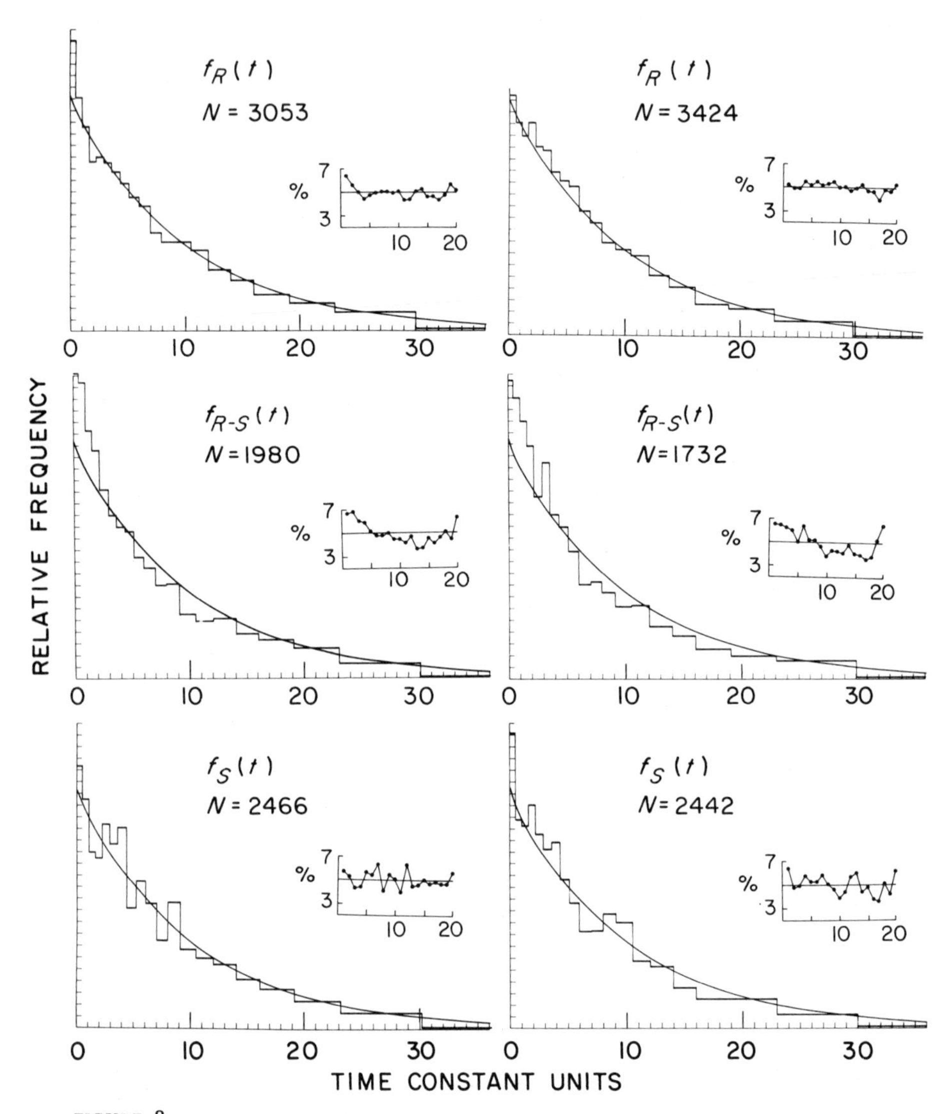

FIGURE 8.
Comparison of tails of distributions of time and false alarm to weak signal with best-fitting exponentials (Luce & Green, 1970, Figs. 6, 7).

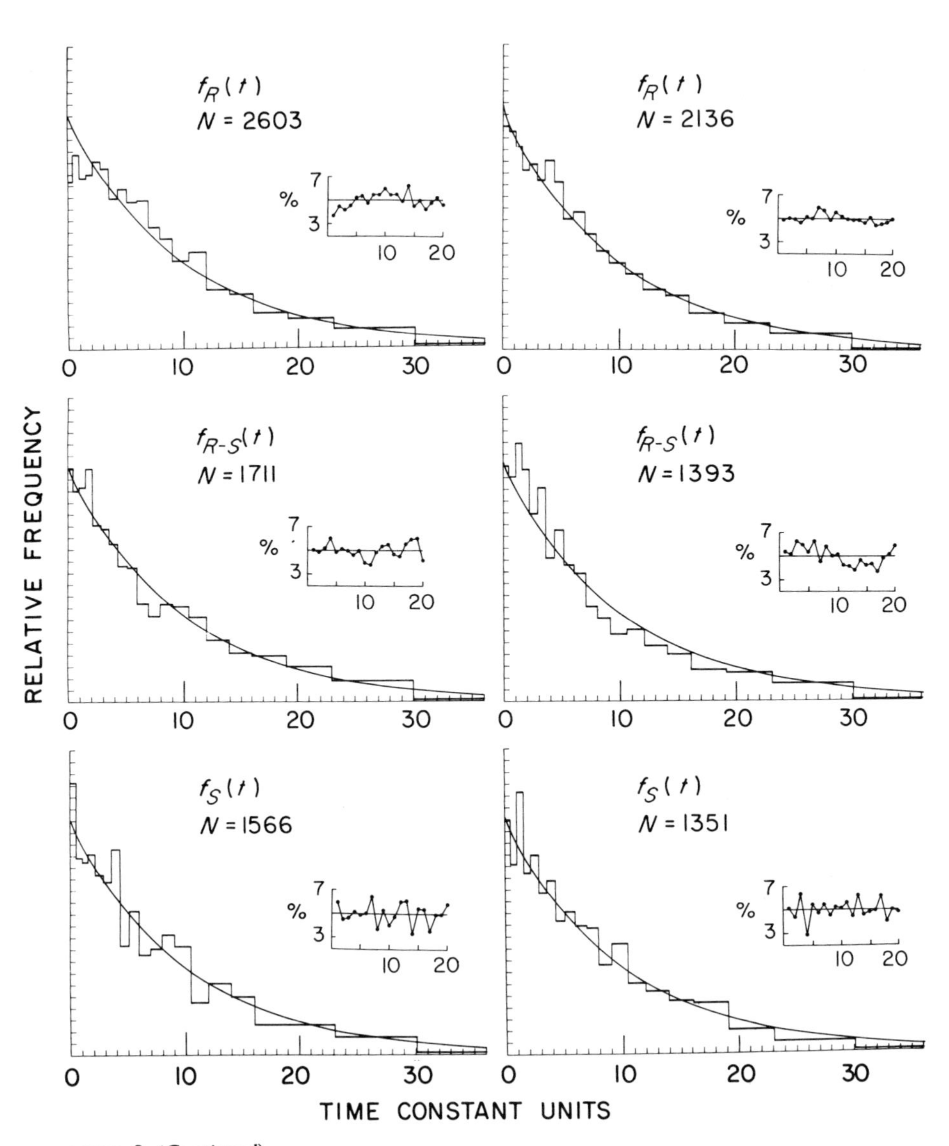

FIGURE 8. (*Continued*)

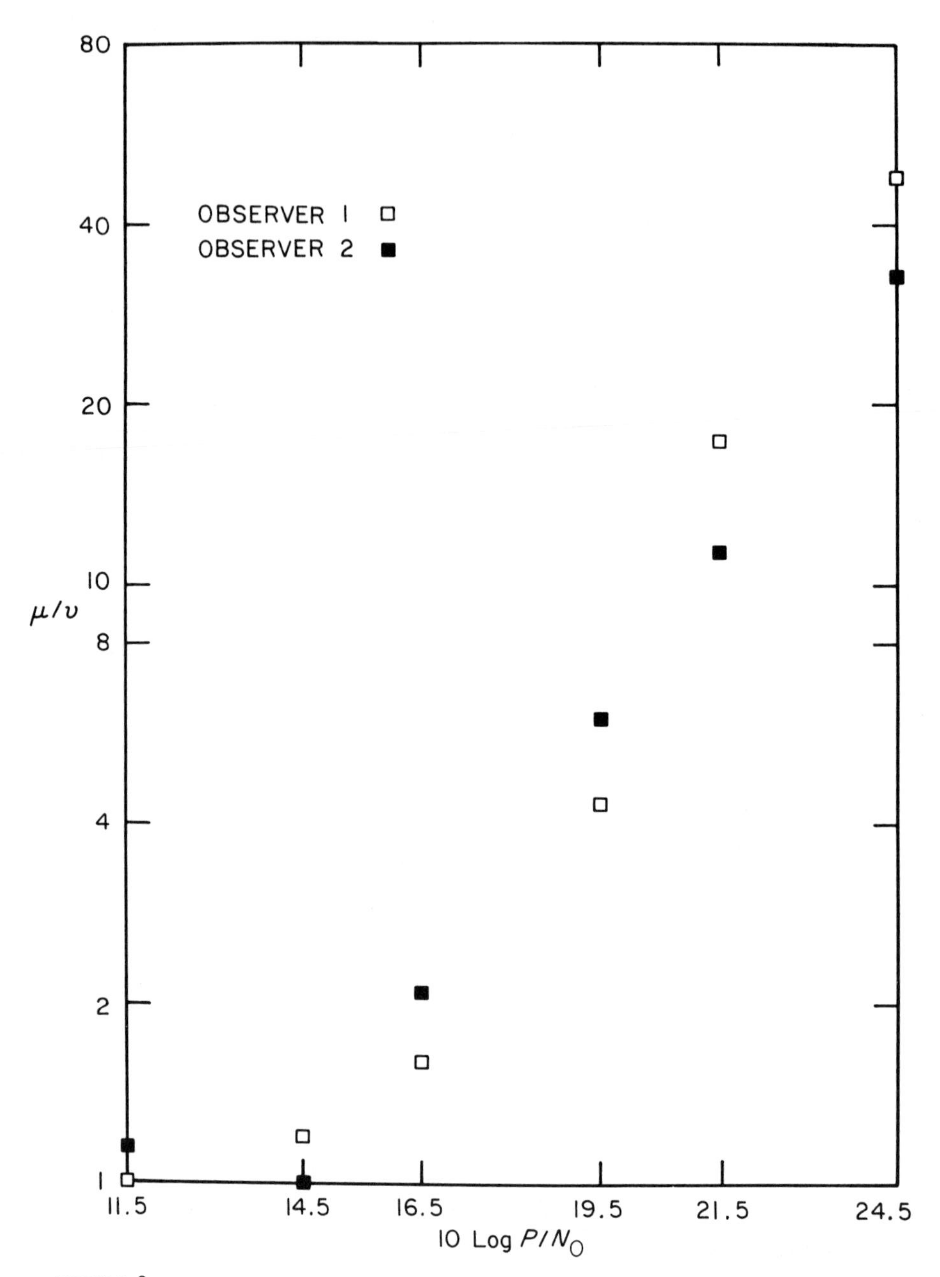

FIGURE 9.
Ratio of Poisson parameters, estimated from tails of simple RT distributions, versus signal power to noise density in decibels (Luce & Green, 1970, Fig. 9).

ACKNOWLEDGMENTS

This work was supported in part by a grant from the National Science Foundation to the University of California, San Diego, and in part by a grant from the Alfred P. Sloan Foundation to the Institute for Advanced Study, Princeton, N.J.

REFERENCES

Braida, L. D., & Durlach, N. I. Intensity perception. II. Resolution in one-interval paradigms. *J. Acoustical Society America*, 1972, **51,** 483–502.

Campbell, R. A., & Lasky, E. Z. Masker level and sinusoidal-signal detection. *J. Acoustical Society America*, 1967, **42,** 972–976.

Chocholle, R. Variations des temps de réaction auditifs en fonction de l'intensité à diverses fréquences. *L'Année Psychologique*, 1940, **41,** 5–124.

Christie, L. S., & Luce, R. D. Decision structure and time relations in simple choice behavior. *Bulletin of Mathematical Biophysics*, 1956, **18,** 89–112. Reprinted in R. D. Luce, R. R. Bush, and E. Galanter (Eds.), *Readings in mathematical psychology*, I. New York: Wiley, 1963.

Cramér, H. *Mathematical methods of statistics.* Princeton: Princeton University Press, 1946.

Dimmick, F. L., & Olson, R. M. The intensive difference limen in audition. *J. Acoustical Society America*, 1941, **12,** 517–525.

Donders, F. C. Over de snelheid van psychische processen. Onderzoekingen gedaan in het Physiologisch Laboratorium der Utrechtsche Hoogeschool, 1868–1869, Tweede reeks, II, 92–120. Translated by W. G. Koster, *Acta Psychologica*, 1969, **30,** 412–431.

Durlach, N. I., & Braida, L. D. Intensity perception. I. Preliminary theory of intensity resolution. *J. Acoustical Society America*, 1969, **46,** 372–383.

Fechner, G. T. *Elemente der Psychophysik.* Leipzig: Breitkopf und Härtel, 1860. English translation of Vol. 1 by Adler, H. E. (Howes, D. H., & Boring, E. G., Eds.). New York: Holt, Rinehart and Winston.

Fitts, P. M. Cognitive aspects of information processing: III. Set for speed versus accuracy. *J. Experimental Psychology*, 1966, **71,** 849–857.

Galambos, R., & Davis, H. The response of single auditory-nerve fibers to acoustic stimulation. *J. Neurophysiology*, 1943, **6,** 39–57.

Green, D. M. Additivity of masking. *J. Acoustical Society America*, 1967, **41,** 1517–1525.

Green, D. M., & Luce, R. D. Detection of auditory signals presented at random times: III. *Perception & Psychophysics*, 1971, **9,** 257–268.

Green, D. M., & Luce, R. D. Speed-accuracy trade-off in auditory detection. In S. Kornblum (Ed.), *Attention and performance*. Vol. 4. New York: Academic Press, 1973.

Green, D. M., & Luce, R. D. Variability of magnitude estimates: A timing theory analysis. *Perceptional Psychophysics*, 1974, in press.

Green, D. M., & Swets, J. A. *Signal detection theory and psychophysics.* New York: Wiley, 1966.

Harris, J. D. Pitch discrimination. *J. Acoustical Society America*, 1952, **24,** 750–755.

Hawkins, J. E., Jr., & Stevens, S. S. The masking of pure tones and of speech by white noise. *J. Acoustical Society America*, 1950, **22,** 6–13.

Hecht, S., Shlaer, S., & Pirenne, M. H. Energy, quanta, and vision. *J. General Physiology*, 1942, **25,** 819–840.

Henning, G. B. Frequency discrimination of random-amplitude tones. *J. Acoustical Society America*, 1966, **39,** 336–339.

Henning, G. B. Frequency discrimination in noise. *J. Acoustical Society America*, 1967, **41,** 774–777.

Kiang, N. Y-S. *Discharge patterns of single fibers in the cat's auditory nerve*. Cambridge: MIT Press, 1965.

Kiang, N. Y-S. A survey of recent developments in the study of auditory physiology. *Annals of Otology, Rhinology and Laryngology*. 1968, **77,** 656–676.

Lappin, J. S., & Disch, K. The latency operating characteristic: I. Effects of stimulus probability on choice reaction time. *J. Experimental Psychology*, 1972, **92,** 419–427.

Luce, R. D. *Individual choice behavior*. New York: Wiley, 1959.

Luce, R. D. What sort of measurement is psychophysical measurement? *American Psychologist*, 1972, **27,** 96–106.

Luce, R. D., & Green, D. M. Detection of auditory signals presented at random times, II. *Perception and Psychophysics*, 1970, **7,** 1–14.

Luce, R. D., & Green, D. M. A neural timing theory for response times and the psychophysics of intensity. *Psychological Review*, 1972, **79,** 14–57.

Luce, R. D., & Green, D. M. The response ratio hypothesis for magnitude estimation. *J. Mathematical Psychology*, 1974, in press. (a)

Luce, R. D., & Green, D. M. Neural coding and psychophysical discrimination data. *J. Acoustical Society America*, 1974, submitted. (b)

McGill, W. J. Loudness and reaction time: A guided tour of the listener's private world. *Proceedings of the* XVI *international congress of pyschology*, 1960. Amsterdam: North-Holland, 1960.

McGill, W. J. Neural counting mechanisms and energy detection in audition. *J. Mathematical Psychology*, 1967, **4,** 351–376.

McGill, W. J., & Goldberg, J. P. Pure-tone intensity discrimination and energy detection. *J. Acoustical Society America*, 1968, **44,** 576–581.

Miller, G. A. Sensitivity to changes in the intensity of white noise and its relation to masking and loudness. *J. Acoustical Society America*, 1947, **19,** 609–619.

Miller, G. A. The magical number seven, plus or minus two: Some limits on our capacity for processing information. *Psychological Review*, 1956, **63,** 81–97.

Nordmark, J. O. Mechanisms of frequency discrimination. *J. Acoustical Society America*, 1968, **44,** 1533–1540.

Ollman, R. Fast guesses in choice reaction times. *Psychonomic Science*, 1966, **6,** 155–156.

Pachella, R. G., & Fisher, D. Hick's law and the speed-accuracy trade-off in absolute judgment. *J. Experimental Psychology*, 1972, **92,** 378–384.

Pachella, R. G., Fisher, D. F., & Karsh, R. Absolute judgments in speeded tasks: Quantification of the trade-off between speed and accuracy. *Psychonomic Science*, 1968, **12,** 225–226.

Pachella, R. G., & Pew, R. W. Speed-accuracy trade-off in reaction time: Effect of discrete criterion times. *J. Experimental Psychology*, 1968, **76,** 19–24.

Parzen, E. *Stochastic processes.* San Francisco: Holden-Day, 1962.

Pew, R. W. The speed-accuracy operating characteristic. *Acta Psychologica*, 1969, **30,** 16–26.

Pynn, C. T., Braida, L. D., & Durlach, N. I. Intensity perception. III. Resolution in small range identification. *J. Acoustical Society America*, 1972, **51,** 559–566.

Reisz, R. R. Differential sensitivity of the ear for pure tones. *Physical Review*, 1928, **31,** 867–875.

Rose, J. E., Brugge, J. F., Anderson, D. J., & Hind, J. E. Phase-locked response to low-frequency tones in single auditory nerve fibers of the squirrel monkey. *J. Neurophysiology*, 1967, **30,** 769–793.

Schouten, J. F., & Bekker, J. A. M. Reaction time and accuracy. *Acta Psychologica*, 1967, **27,** 143–153.

Shower, E. G., & Biddulph, R. Differential pitch sensitivity of the ear. *J. Acoustical Society America*, 1931, **3,** 275–287.

Siebert, W. M. Stimulus transformations in the peripheral auditory system. In P. A. Kolers and M. Eden (Eds.), *Recognizing patterns.* Cambridge, Mass.: MIT Press, 1968.

Siebert, W. M. Frequency discrimination in the auditory system: Place or periodicity mechanisms? *Proceedings of the IEEE*, 1970, **58,** 723–730.

Sternberg, S. The discovery of processing stages: Extensions of Donder's method. *Acta Psychologica*, 1969, **30,** 276–315.

Stevens, J. C., & Hall, J. W. Brightness and loudness as functions of stimulus duration. *Perception and Psychophysics*, 1966, **1,** 319–327.

Stevens, S. S. On the psychophysical law. *Psychological Review*, 1957, **64,** 153–181.

Stevens, S. S. Duration, luminance and the brightness exponent. *Perception and Psychophysics*, 1966, **1,** 96–100.

Stevens, S. S. Issues in psychophysical measurement. *Psychological Review*, 1971, **78,** 426–450.

Tanner, W. P., Jr., & Swets, J. A. A decision-making theory of visual detection. *Psychological Review*, 1954, **61,** 401–409.

Taylor, M. M., Lindsay, P. H., & Forbes, S. M. Quantification of shared capacity processing in auditory and visual discrimination. *Acta Psychologica*, 1967, **27,** 223–229.

Thurstone, L. L. A law of comparative judgment. *Psychological Review*, 1927, **34,** 273–286.

Uttal, W. R., & Krissoff, M. Response of the somesthetic system to patterned trains of electrical stimuli. In D. R. Kenshalo (Ed.), *The skin senses.* Springfield, Ill.: Charles C Thomas, 1965.

Viemeister, N. F. Intensity discrimination of pulsed sinusoids: the effects of filtered noise. *J. Acoustical Society America*, 1972, **51,** 1265–1269.

Wever, E. G. *Theory of hearing.* New York: Wiley, 1949. Reprinted, New York: Dover, 1970.

Yellott, J. Correction for guessing in choice reaction times. *Psychonomic Science*, 1967, **8,** 321–322.

The Holographic Hypothesis of Memory Structure in Brain Function and Perception

Karl H. Pribram *Marc Nuwer*
STANFORD UNIVERSITY

Robert J. Baron
UNIVERSITY OF IOWA

WHY A HOLOGRAPHIC HYPOTHESIS?

Introduction

Recently a growing number of theorists have invoked the principles of holography to explain one or another aspect of brain function. Historically the ideas can be traced to problems posed during neurogenesis when the activity of relatively remote circuits of the developing nervous system must become integrated to account for such simple behaviors as swimming. Among others, the principle of chemical 'resonances' that 'tune' these circuits has had a long and influential life (see, e.g., Loeb, 1907; Weiss, 1939). More specifically, however, Goldscheider (1906) and Horton (1925) proposed that the establishment of tuned resonances in the form of interference patterns in the adult brain could account for a variety of perceptual phenomena. More recently, Lashley (1942) spelled out a mechanism of neural interference patterns to explain stimulus equivalence and Beurle (1956) developed a mathematically rigorous formulation of the origin of such patterns of plane wave interferences in neural tissue. But it was not until the advent of holography with its powerful damage-resistant image storage and reconstructive capabilities that the promise of an interference pattern mechanism of brain function became fully appreciated. As the properties of physical holograms became known (see Stroke, 1966; Goodman, 1968; Collier, Burckhardt, & Lin, 1971), a number of physical and computer scientists saw the relevance of holography to the problems of brain function, memory, and perception (e.g., van Heerden, 1963; Julesz & Pennington, 1965; Westlake, 1968; Baron, 1970; Cavanagh, 1972).

The purpose of the present chapter is threefold: (a) to summarize the neurological evidence that makes holographic processing, storage (temporary or permanent), and image reconstruction a plausible analogy to students of brain function; (b) to present a mathematical network model for the holographic process as a demonstration that optical systems are not necessary for its realization; and (c) to examine the evidence sustaining or negating the neurological assumptions involved in this and alternate realizations.

The Anatomical Problem

One of the best established, yet puzzling, facts about brain mechanisms and memory is that large destructions within a neural system do not seriously impair its function. Various controlled experiments have been performed to investigate this puzzle. Lashley (1950) showed that 80 percent or more of the visual cortex of a rat could be damaged without loss of the ability to correctly respond to patterns; Galambos, Norton, and Frommer (1967) have severed as much as 98 percent of the optic tracts of cats with similar negative results; and Chow (1968) has combined the two experiments into one simultaneous assault, again with little effect on visual recognition behavior. In man, of course, hemianopia and other large scotomata also fail to impair the recognition mechanism. Even small punctal irritative lesions peppered throughout the cortical mantle of monkeys and shown to disrupt its electrical activity leave response to visual patterns intact (Kraft, Obrist, & Pribram, 1960).

These findings have been interpreted by everyone to indicate that the neural elements necessary to the recognition and recall processes must be distributed throughout the brain systems involved. The questions that arise are (a) how is the distribution effected, (b) how does recognition occur, and (c) how are associated events recalled by the network?

An answer that is often given is to consider the input systems of the neuraxis to be composed of large numbers of randomly connected neural elements (Rosenblatt, 1962) and to show either by computer simulation or by mathematical analysis that in a random network of neurons, replication and distribution of signals can occur. Unfortunately for this explanation, the anatomical facts are largely otherwise. In the visual system, for instance, the retina and cortex are connected by a system of fibers that run to a great extent in parallel. Only two modifications of this parallelity occur.

1. The optic tracts and radiations that carry signals between the retina and cortex constitute a sheaf within which the retinal events converge to some extent onto the lateral geniculate nucleus of the thalamus from where they diverge to the cortex. The final effect of this parallel network is that each fiber in the system connects ten retinal outputs to about 5,000 cortical receiving cells.
2. In the process of termination of the fibers at various locations in the pathway, an effective overlap develops (to about 5° of visual angle) between neighboring branches of the conducting fibers.

Equally striking and perhaps more important than these exceptions, however, is the interpolation at every cell station of a sheet of horizontally connected neurons in a plane perpendicular to the parallel fiber system. These horizontal cells are characterized by short or absent axons but spreading dendrites. It has been shown in the retina (Werblin & Dowling, 1969) and to some extent also in the cortex (Creutzfeldt, 1961), that such spreading dendritic networks may not generate nerve impulses; in fact, they usually may not even depolarize. Their activity is characterized by hyperpolarization that tends to organize the functions of the system by inhibitory rather than excitatory processes. In the retina, for instance, no nerve impulses are generated prior to the ganglion cells from which the optic nerve fibers originate. Thus, practically all of the complexity manifest in the optic nerve is a reflection of the organizing properties of depolarizing and hyperpolarizing events, not of interactions among nerve impulses.

Some Neurophysiological Considerations

Two mechanisms are therefore available to account for the distribution of signals within the neural system. One relies on the convergence and divergence of nerve impulses onto and from a neuronal pool. The other relies on the presence of lateral (mostly inhibitory) interactions taking place in sheets of horizontal dendritic networks situated at every cell station perpendicular to the essentially parallel system of input fibers. Let us explore the possible role of both of these mechanisms in explaining the results of the lesion studies.

Evidence is supplied by experiments in which conditions of anesthesia are used that suppress the functions of small nerve fibers thus leaving intact and clearly discernible the connectivity by way of major nerve impulse pathways. These experiments have shown that localized retinal stimulation evokes a receptive field at the cortex over an area no greater than a few degrees in diameter (e.g., Talbot & Marshall, 1941). Yet, the data that must be explained indicate that some 80 percent or more of the visual cortex including the foveal region can be extirpated without marked impairment of the recognition of a previously learned visual pattern. Thus, whatever the mechanisms, distribution of input cannot be due to the major pathways but must involve the fine fibered connectivity in the visual system, either via the divergence of nerve impulses and/or via the interactions taking place in the horizontal cell dendritic networks.

Both are probably to some extent responsible. It must be remembered that nerve impulses occurring in the fine fibers tend to decrement in amplitude and speed of conduction thus becoming slow graded potentials. Further, these graded slow potentials or minispikes usually occur in the same anatomical location as the horizontal dendritic inhibitory hyperpolarizations and thus interact with them. In fact, the resulting micro-organization of junctional neural activity (synaptic and ephaptic) could be regarded as a simple summation of graded excitatory (depolarizing) and inhibitory (hyperpolarizing) slow potential processes.

These structural arrangements of slow potentials are especially evident in sheets of neural tissue such as in the retina and cortex. The cerebral cortex, for instance, may be thought of as consisting of columnar units that can be considered more or less independent basic computational elements, each of which is capable of performing a similar computation (Mountcastle, 1957; Hubel & Wiesel, 1968). Inputs to the basic computational elements are processed in a direction essentially perpendicular to the sheet of the cortex, and therefore cortical processing occurs in stages, each stage transforming the activation pattern of the cells in one of the cortical layers to the cells of another cortical layer. Analyses by Kabrisky (1966) and by Werner (1970) show that processing by one basic computational element remains essentially within that element, and therefore the cortex can be considered to consist of a large number of essentially similar parallel processing elements. Furthermore, the processing done by any one of the basic computational elements is itself a parallel process (see, for example, Spinelli, 1970), each layer transforming the pattern of activity that arrived from the previous layer by the process of temporal and spatial summation, the summation of slow hyper- and depolarizations in the dendritic microstructure of the cortex. Analyses by Ratliff (1965) and Rodieck (1965) have shown that processing (at least at the sensory level) that occurs through successive stages in such a layered neural network can be described by linear equations. Each computational element is thus capable of transforming its inputs through a succession of stages, and each stage produces a linear transformation of the pattern of activity at the previous stage.

HOLOGRAPHY

Optical Computing

The problem that thus confronts us is essentially this: how can the relationships between neural activity become distributed and stored (temporarily or more permanently) by a neural network in which such patterns are transmitted and transformed through several successive stages in which processing is an essentially linear parallel process?

Fortunately for neurophysiology, physicists have been concerned with such systems for a long time: optical devices are parallel transmission systems, and during the past 25 years their processing characteristics have been studied intensively. One property of optical processing initially called it to our attention. As we shall see, records can be produced in which the input becomes distributed throughout the storage medium. This makes the record resistant to damage, and, in fact, loss of all but a small portion does not destroy its image reconstructing potential.

Most of us are familiar with the image-generating aspects of optical systems. A camera records on photographic film a copy of the light intensities reflected from the objects within the camera's field of view. Each point on the film stores the intensity (the square of the amplitude) of the light that arrives from a corresponding point in the field of view, and thus the film's record 'looks like' the visual field. What have been studied more recently are the properties of records made when a film does not lie in an image plane of an optical system. When a piece of film is exposed to coherent light that is reflected and scattered by objects in the visual field, there is no ordinary image produced on the film. In fact, the film becomes so blurred that there is no resemblance whatever between the pattern that is stored on the film and the visual field itself. However, when properly illuminated, the film reconstructs the wavefronts of light that were present when the exposure was made. As a result, if an observer looks toward the film, it appears as if the entire visual scene were present behind it. The reconstructed image appears exactly as it did during the exposure, complete in every detail and in three dimensions! The light waves from each point of the visual field had interacted to produce an interference pattern at the film, and it is this interference pattern that was stored throughout the film. Interference patterns give rise to the remarkable characteristics of optical information storage as we shall show.

Even before the practical demonstration of the use of interference patterns in the reconstruction of images, Gabor (1948) had mathematically proposed a way of producing images from photographic records. Gabor began with the intent to increase the resolution of electron microphotographs. He suggested that a coherent background wave and the waves refracted by the tissue could produce interference patterns that would store both amplitude and spatial phase information. Then, in a second step, these stored patterns could be used to reconstruct an image of the original tissue. Gabor christened his film record a *hologram* because it contains all necessary information to reconstruct the whole (hol-) image. The use of this term for that type of photographic record has since become common.

As the art and science of holography developed, it became clear that a variety of methods described by a number of mathematical procedures could result in holograms. This chapter briefly describes two elementary types—the Fresnel and the Fourier—and provides a network realization of the Fourier holographic process. Other types of holograms have been found useful: they go by names such as Reflection, Volume, Phase, Color, Pulsed-laser, Incoherent, and Digital holograms. All are basically similar to the elementary Fourier and Fresnel types but have special properties that make them especially useful in one or another application. The following exposition is therefore meant to provide only a guideline to holographic processing by presenting the requirements necessary for a holographic hypothesis of memory storage in brain function.

Lensless Optical Holography

The problem faced when trying to store a wavefront of light on photographic film is that film does not store the amplitude or phase distributions (patterns) of light, but instead it records only the intensity (amplitude squared) distribution. The image that is stored on the film is a static representation of the dynamic wavefront of light that arrived during the exposure process, but the stored image has no phase information. As a result, it is impossible to recreate the dynamic pattern of light from which the image was made. Holography offers a way to overcome this problem by recording on film the interference pattern formed by two different wavefronts of light. As we shall show, if the intensity pattern formed by the interference (superposition) of two different wavefronts of light is recorded on the film, sufficient information is retained to enable a reconstruction of either wavefront when only the other one is present at a later time.

Let us suppose than an object, say O_A, is placed near a piece of film and illuminated by a coherent source of light. The wavefront produced at the film by light that is scattered from the object will be denoted by A. The ideal situation would result if film could store the wavefront A. However, when the film is exposed and developed, the image recorded by the film is proportional to the intensity pattern $|A|^2 = AA^*$, not the desired pattern A. As we mentioned earlier, the wavefront A cannot be reconstructed from AA^* alone. Several different holographic techniques have been studied that enable a wavefront of light to be stored, and we shall discuss four of them now.

One technique that has been widely discussed in the literature consists not only of illuminating the object O_A with a plane coherent wavefront of light, but at the same time reflecting some of the coherent light by a plane mirror directly toward the film. The second wavefront of light is called a reference wave and will be denoted by R. When the two different arrival patterns A and R interact, the superposition (algebraic sum of electric and magnetic components) given by $A + R$ is formed. Because A and R are generated by the same light source, they have the same frequency. A stable interference pattern is formed, and the film records the intensity of the interference pattern given by $(A + R)(A + R)^*$. If the film is transparent, A and R continue to propagate independently, and we call these output portions of the wavefronts the output patterns or departure patterns.

Now suppose the film is developed and replaced in the optical system. Also suppose that the object O_A is removed so that only the reference wave R is allowed to illuminate the film. The departure pattern is given by the product of the incoming wavefront attenuated at each point by the transmission coefficient of the film at that point. The departure pattern is therefore given by $R(A + R)(A + R)^*$, which can be expanded mathematically to

$$R(|A|^2 + |R|^2) + A|R|^2 + A^*R \cdot R.$$

The first term of the expansion describes the reference wave R attenuated by an amount $|A|^2 + |R|^2$. The second term describes a reconstructed copy of the desired wavefront A attenuated by an amount $|R|^2$. This wavefront has all the properties of the original wavefront present during the exposure process. As a result, a person looking toward the film would 'see' the object O_A. Because O_A is not present, the reconstructed image is called a 'ghost image,' and since the reconstructed wavefront is an exact copy of A, the ghost image appears in three dimensions and has all other properties that could be seen during the exposure. The last term describes noise, which is introduced into the system by the holographic process. The film is called a hologram and the wavefront A is said to be stored. (Note that if A could have been stored directly, the noise term would not have been produced and the reconstructed image could have been formed directly from the stored image.)

Now let us suppose that the mirror rather than the object had been removed during the reconstruction process. The departure pattern would have been given by $A(A + R)(A + R)^* = A(|A|^2 + |R|^2) + R|A|^2 + R^*A \cdot A$. In this case, the reference wave is reconstructed (the second term), the wavefront A attenuated (the first term), and again noise is produced. There is thus a natural symmetry between the two wavefronts in the system.

Recognition Holography

The second type of holography discussed is a slight modification of the first type. Rather than using a plane mirror to produce the reference wave, let us use a spherical mirror that focuses the reflected light onto a point P in front of (on the output side of) the film. When the hologram produced by this system is developed and replaced in the system, two different departure patterns result depending on how the hologram is illuminated. If the object is removed and the hologram is illuminated only by light reflected by the mirror, the wavefront A is reconstructed. Both the focused wavefront and a noise term are also produced. If the object is used to illuminate the hologram and the mirror removed, the focused wavefront is reconstructed, and a bright spot of light is focused at the point P. A light detector placed at P could be used to detect the bright spot, and since a bright spot is produced only when object O_A is present in the system, the detector can be used to 'recognize' the presence of the object O_A. This optical system can be used for the recognition of three-dimensional objects.

Association Holography

The third lensless system described is similar to the two previous systems. This time, however, the mirror is replaced by a second object, say O_B. See Figure 1. When objects O_A and O_B are illuminated by a coherent light source, two wavefronts A and B are produced. If a film is exposed to the interference pattern produced by A and B and given by $A + B$, the film will record the static pattern $(A + B)(A + B)^*$.

Now let us see what happens if the film is developed and placed exactly where it was during the exposure. Assume that object O_A is removed. If O_B is illuminated, the arrival wavefront that reaches the film is B, and the static distribution of transmission coefficients on the film is given by $(A + B)(A + B)^*$. The departure wavefront is given by $B(A + B)(A + B)^* = B(|A|^2 + |B|^2) + A|B|^2 + A^*B \cdot B$. The first term shows that the wavefront B is transmitted, the second term shows that the wavefront A is also reconstructed, and the third term shows that noise is produced.

If the objects do not cause light to be focused on the film, the intensity distributions $|A|^2$ and $|B|^2$ are nearly uniform across the film even over small (millimeter) distances. However, the interference pattern $(A + B)$ gives rise to an intensity pattern that varies considerably over small distances, and the resulting stored pattern $(A + B)(A + B)^*$ resembles a very complex diffraction grating. It is precisely for this reason that the departure pattern is a reconstruction of the wavefront of A that is not present during the reconstruction process.

Figures 1 and 2 illustrate the holographic association process. Figure 2 shows the exposure process. Light is reflected from two objects, a wide block O_A in the background, and a tall block O_B in the foreground. Both objects are illuminated by a single coherent monochromatic plane wave generated by a laser. After exposure, the film is developed and replaced exactly where it was during the exposure process. Now, however, the object O_A is 'seen' so that the film is exposed to light which is reflected only from the tall block O_B. Because the departure wavefront A is reconstructed by the hologram, the object is 'seen' by an observer even though it is no longer present. Again the reconstructed wavefront is an exact copy of the departure pattern that was present during the exposure, so the object appears in three dimensions and has all other properties that could be seen during the exposure.

The three lensless holographic processes described above are among many similar techniques used for producing holograms. Holograms can be produced by exposure to objects that are either near the film (near-field holograms) or far away from the film (far-field holograms). The optical far-field transforms used above are known as Fresnel transforms. However, both the near-field and far-field lensless holograms—i.e., those produced with scattered wavefronts as reference waves (e.g., the third system)—are usually referred to as Fresnel holograms. Holograms produced with infinitely far-field transforms are called Fourier holograms.

One optical system of the Fourier type, using a plane coherent reference wave produced by lenses, is of special interest to the neurotheorist because a direct analogy can be drawn between it and a layered neural network. Because it is this analogy that we wish to pursue, the mathematical properties of the system are described in somewhat greater detail.

THE FOURIER HOLOGRAPHIC PROCESS

The Lens System

The system of interest consists of two spherical lenses arranged so that the second focal plane of the first lens is coincident with the first focal plane of the second lens. This is shown in Figure 3. The three focal planes of the two lenses are also of interest to this analysis and will be called the *input plane*, *transform* or *memory plane*, and *output plane* of the system.

It is well known that when a photographic image is placed in one focal plane of a spherical lens and illuminated by a plane, coherent light wave, the Fourier transform of the image is produced in the other focal plane of the lens. (For a nice proof of this result, see Preston, 1965.) For the optical sys-

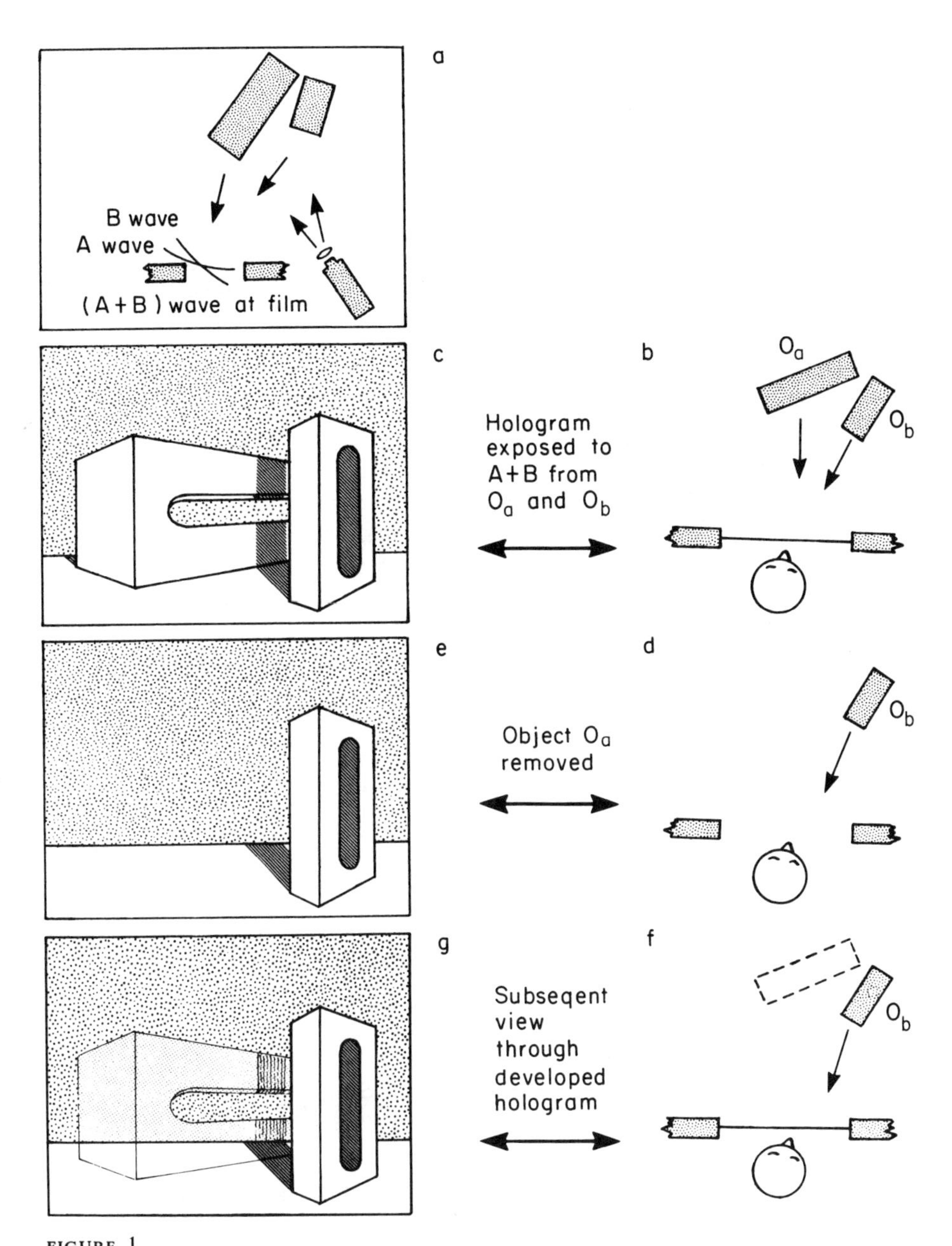

FIGURE 1.
A hypothetical holographic experiment. The apparatus for a Fresnel hologram is shown in (a). The view as the film is being exposed is shown in (b) and (c); the view after the film and one object are removed in (d) and (e); and the view after the developed hologram is returned, in (f) and (g), demonstrating the ghost image of the missing object. (Adapted from Collier & Pennington, 1966.)

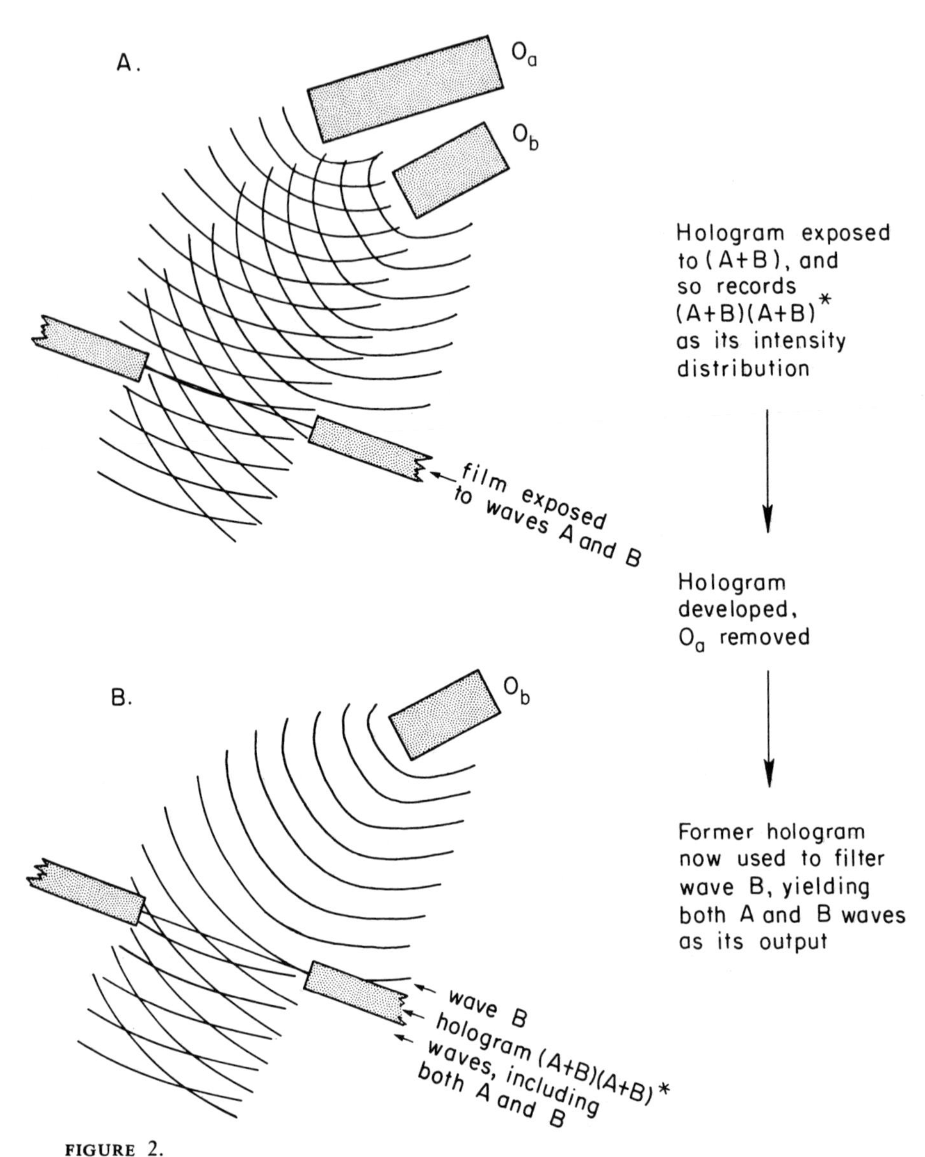

FIGURE 2.
Diagram of waves in creation (A) and reconstruction (B) processes. Note how in (B) the filtering of B through hologram $(A + B)(A + B)^*$ results in waves including both A and B.

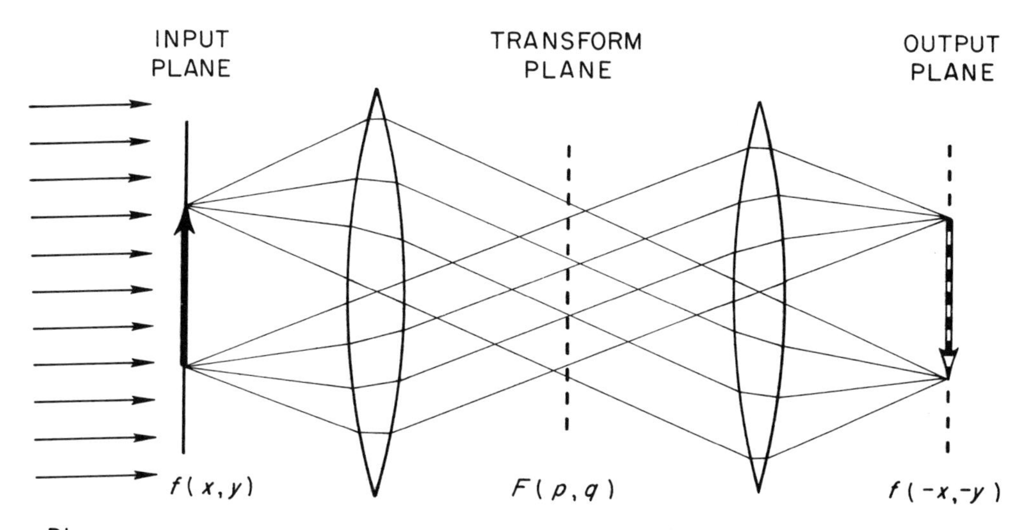

FIGURE 3.
Fourier transforming property of a lens. Note how the second lens effects a second 'reverse' transform, causing an image in the output plane like the original object in the input plane.

tem pictured in Figure 3, this transformed image occurs in the transform plane. The transformed image also lies in the first focal plane of the second spherical lens and therefore the Fourier transform of the transformed image appears in the third focal plane or output plane of the system. An elementary theorem of Fourier analysis tells us that the resulting output image of the system is precisely the input image, only it appears upside down and backwards.

This process may be stated more precisely as follows. In an optical system, wavefront patterns are of two distinctly different types: *static* and *dynamic*. The static patterns are the stored photographic images that are described in terms of their transmission coefficients for light as a function of position on the photographic film. We will let (x, y) represent the geometric coordinates of position and $f(x, y)$ represent the transmission coefficient at the point (x, y). The function $f(x, y)$ on a piece of film represents a static storage pattern for the optical system. The dynamic or active patterns are the wavefronts of light that are processed and transformed by the optical systems. The specific transformations that occur in an optical system depend on the components (lenses, prisms, etc.), their placement, and the properties of the light waves used. For the optical system being described, Fourier transformation results because of the choice of spherical lenses, their arrangement, and the use of coherent light. Let $F(p, q)$ be the Fourier transform of $f(x, y)$. Then for the optical system of Figure 3, $F(p, q)$ is imaged in the transform plane. The variables p and q represent the position of the image in the transform plane. If we denote by $\longrightarrow$ the process of Fourier transformation, then we write

$$f(x, y) \longrightarrow F(p, q)$$

to represent the first stage of processing, and

$$F(p, q) \longrightarrow f(-x, -y)$$

to represent the second stage.

Now consider the process that occurs at the photographic film itself. A light wave of uniform intensity, say A, illuminates the image $f(x, y)$. The intensities of the departure wavefront are attenuated by the presence of the input film. In fact, the intensity at point (x, y) is proportional to the transmission coefficient of the film at that point. The static storage pattern is converted to a dynamic processing pattern with amplitude distribution proportional to the spatial distribution of transmission coefficients of the image. This conversion is multiplicative. Thus, if the amplitude of the arrival pattern at point (x, y) is A, then the amplitude of the departure pattern in front of the image at point (x, y) is $Af(x, y)$.

Now let us suppose that an actual photographic image of the transform $F(p, q)$ could be placed in the transform plane of the optical system. Also suppose that the transform image could be illuminated by a plane coherent wavefront. (A uniform illumination across the transform plane can easily be produced by placing a point source of light in the center of the input plane.) The departure pattern produced by the film would have an intensity distribution $F(p, q)$ and the image $f(-x, -y)$ would be produced in the output plane. We find that both images $f(x, y)$ and $F(p, q)$ contain exactly the same information, the only difference is in the way in which information is coded. In fact, a film record of the transformed image, if made with a plane reference wave, is a Fourier hologram of the input image and from now on we refer to these transform images as holograms. They are the 'memories' of the optical system.

As described earlier, the static images are intensity distributions whereas the dynamic images are amplitude and phase distributions. Static distributions can be represented by positive real functions whose value is less than unity (light is not produced and the phase of transmitted light is not changed by a piece of film), whereas dynamic patterns are represented by arbitrary complex (in the mathematical sense) quantities. In general, the Fourier transform of a positive real image is a complex quantity and therefore cannot be stored directly as a static pattern. To produce a hologram of the image $f(x, y)$, we cannot simply expose a piece of film in the transform plane. However, by taking the superposition of the desired wavefront and a plane reference wave, a hologram is produced in which the desired information is not lost. This is illustrated in Figures 4 and 5. The details of this process are similar to the lensless case, which was described earlier using a plane reference wave, and may be pursued in detail elsewhere by the interested reader (Stroke, 1966; Tippett, Berkowitz, Clapp, Koester, & Vanderburgh, 1965).

A second theorem from Fourier theory is important to the discussion that follows. When the Fourier transform $F(p, q)$ of an image $f(x, y)$ is multiplied by the complex conjugate of the Fourier transform $G(p, q)^*$ of a second image $g(x, y)$, and the Fourier transform of the product taken, the result is the cross correlation of the two initial images. This is important because cross correlation is a measure of the similarity of the original two images. A measure of similarity is precisely what is required for recognition.

For our optical system, if the hologram $(G(p, q) + R(p, q))(G(p, q) + R(p, q))^*$ is placed in the transform plane while the image $f(x, y)$ is placed in the input plane, the departure pattern that results just after the hologram is the product of the two functions $F(p, q)$ and the hologram. The result that follows from the above theorem is that the cross correlation of the functions $f(x, y)$ and $g(x, y)$ is produced in the output plane. See Figure 5. If the two images are similar, a bright spot appears in the output plane and the brightness of the spot indicates how similar the two images are. The system instantaneously cross correlates two spatial patterns. In fact, this technique has been applied successfully to the instantaneous recognition of human faces! (In general, a thresholding light detector is placed in the output plane to determine whether or not a given input image should be 'recognized.') The reader should note that the hologram is formed from the interference pattern between the desired transform $G(p, q)$ and a plane coherent wavefront $R(p, q)$. The result is that both the cross correlation and convolution functions of the two images are formed in the output plane. Figure 6 illustrates the formation of these functions by the optical system, and Figure 7 gives a geometric interpretation of the cross correlation and convolution functions. Also note the similarity between this system and recognition holography described earlier.

There is another property of the Fourier transform of an image that is of interest to the neurophysiologist. Each point of the transform indicates the presence of specific spatial frequencies that are present in the input image. If, for example, the input image is a simple sinusoidally varying intensity pattern of spatial frequency p, then the Fourier transform at position $F(p,0)$—or $F(0, p)$ depending on the orientation of the image—would have intensity proportional to the brightness of the image, and the rest of the transform would have zero intensity. By analyzing the Fourier transform of an input image, one can determine its exact spatial components.

The analysis of the optical system is similar to the analyses both of conventional optical holography in which Fourier or Fresnel holograms are pro-

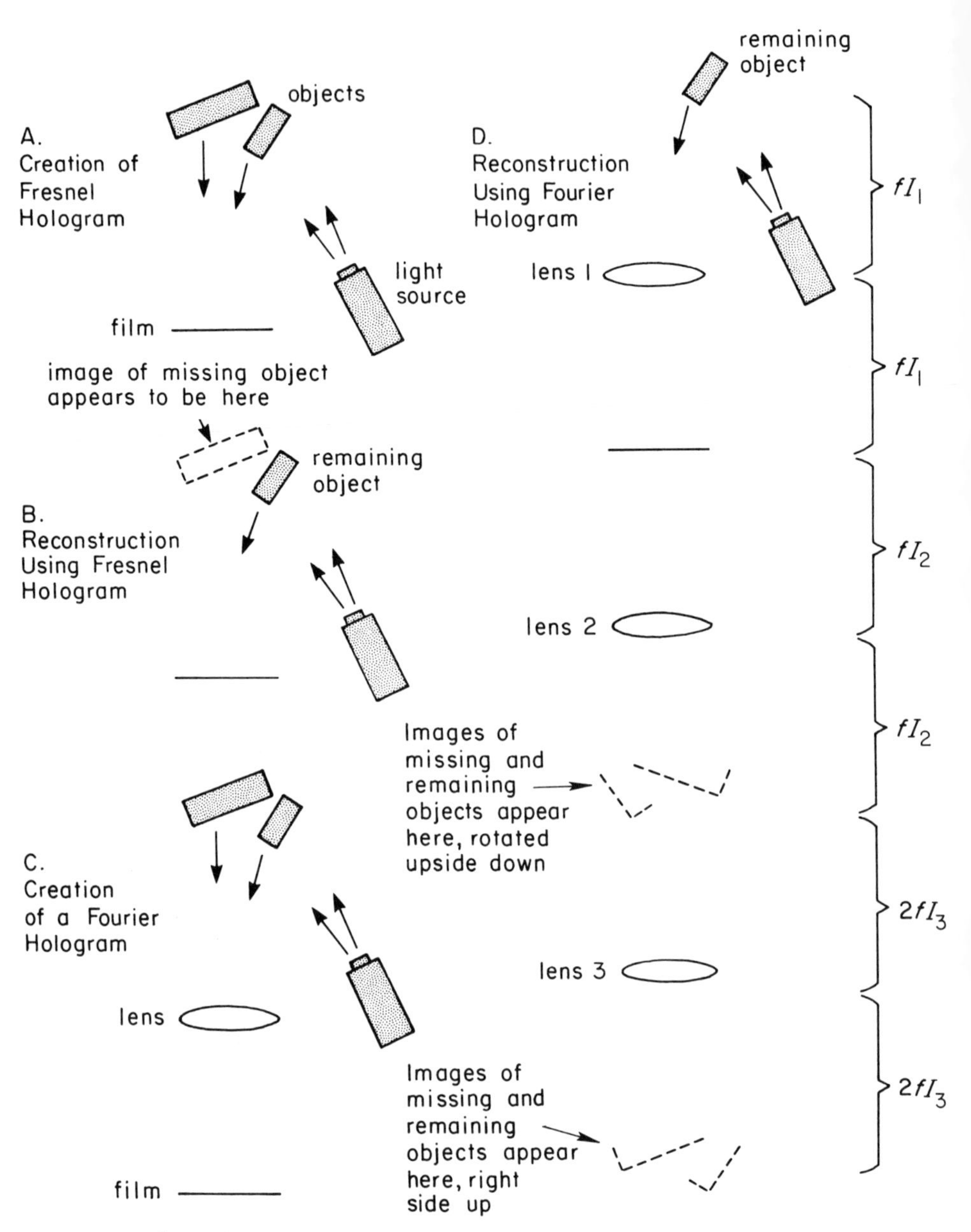

FIGURE 4.
Comparison of apparatus and images in creation and reconstruction using Fresnel- and Fourier-type holograms. Note that the third lens in (D) is optional, and simply inverts the first reconstructed image.

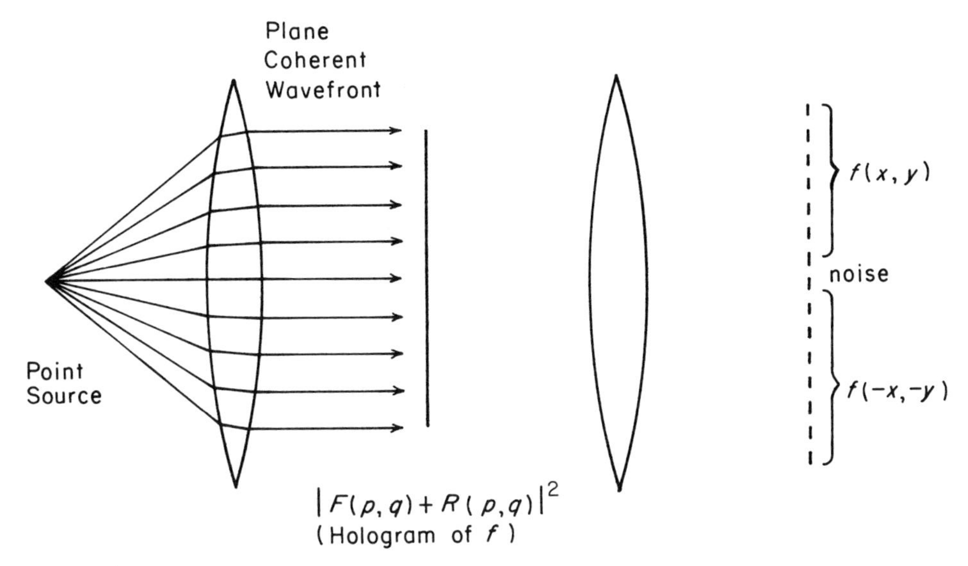

FIGURE 5.
Reconstruction of $f(x, y)$ using Fourier holography. Here the plane wave $R(p, q)$ acts as the second, reference wave in capturing $F(p, q)$ on the hologram. $R(p, q)$ is the Fourier plane image of an *off-centered* point source. Reconstruction is effected here through use of an *on-center* point source. This trick separates $f(x, y)$ from $f(-x, -y)$ in the output plane.

duced and of van Heerden's (1963) method of information storage in solids. There are also analogies between cross correlation in the optical system described here and in recognition (and in fact association) techniques both in van Heerden's system and in conventional holography. The reader is directed for the details to the references cited.

A Mathematical Network Model

Enough of the formal attributes of the holographic optical systems that we have found to have the essential properties for storage, recognition, and recall. What needs to be done to make holography into a useful metaphor for students of brain function in memory and perception is to see whether the lens system, or even an optical system, is necessary to the accomplishment of the holographic process.

We found earlier that there are both static (storage) and dynamic (processing) patterns in the optical system. We found that optical processing occurs because of the geometry and components of an optical system. We found how static and dynamic information patterns interact, and finally we found that a system in which there are two stages of linear information processing

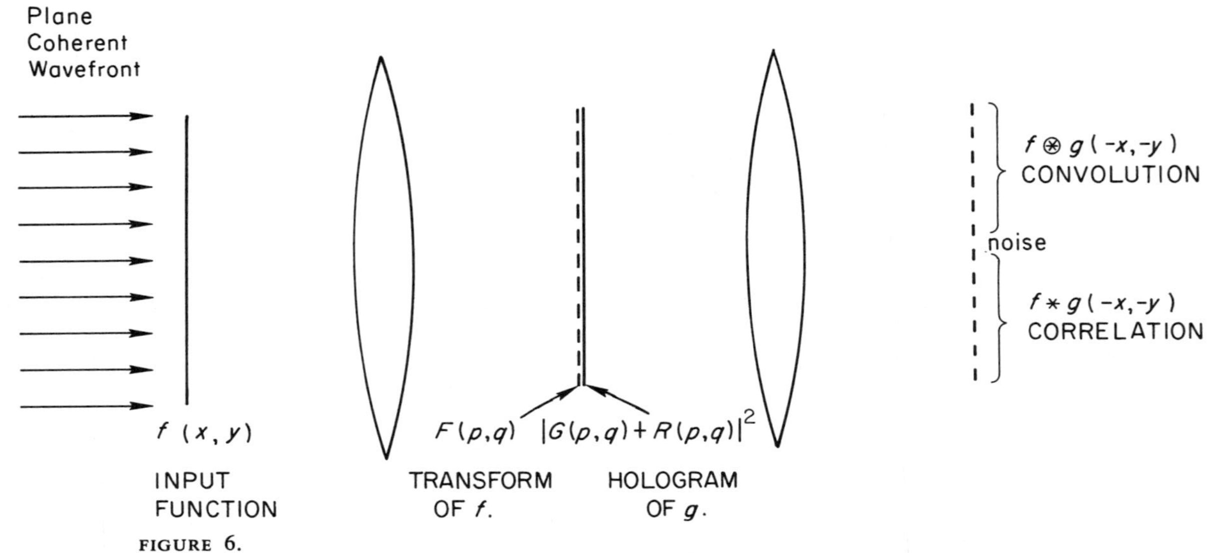

FIGURE 6.
Results from illuminating a Fourier hologram with a wave, $F(p, q)$, which is not the same as the waves which created the hologram, $G(p, q)$ and $R(p, q)$. Convolution and correlation functions are formed in the output plane.

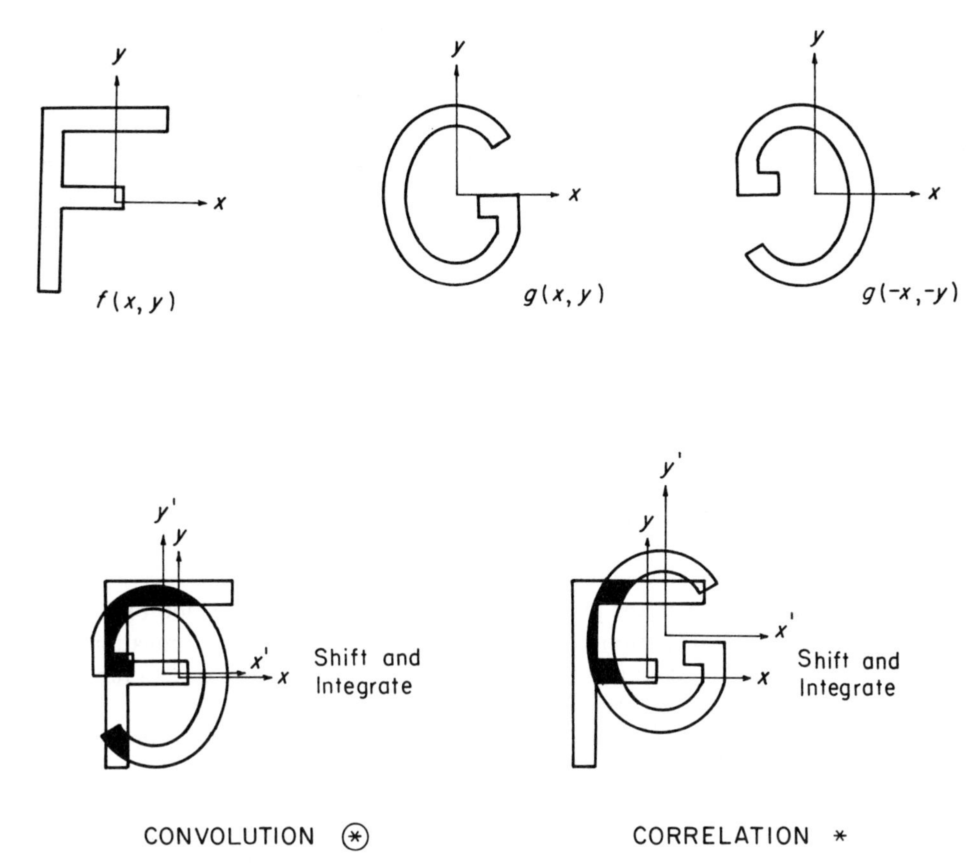

FIGURE 7.
Geometric interpretations of the convolution and correlation functions. Note how the convolution involves a rotated version of one function, while the correlation does not.

is capable of storing, recognizing, and recalling visual information. We now draw an analogy between each of these quantities in optical systems and corresponding quantities in computer simulated systems.

There are both static (storage) and dynamic (processing) patterns in our network model. The static patterns are sensitivity values of the junctional contacts between two cells. (For convenience we call our mathematical processing elements 'cells' and the loci of changes of sensitivity values 'junctions.') The junctions are assumed to be distributed throughout the volume of the assembly of cells. The contribution that a unit (not to be confused with a processing unit) of transmissivity has toward the rate of activation of the postjunctional cell may differ depending on the geometric position of its arrival. We therefore define the *local sensitivity* at point (x, y, z) of a cell to be a measure of the effective contribution toward activation that is produced at position (x, y, z) from prejunctional cells. Static patterns are described by giving for each cell (p, q) of the network the local sensitivity function $k_{pq}(x, y, z)$. Local sensitivity values are a function of the junctional microstructure of contacts among the cells.

In our model, dynamic or processing patterns are the patterns of activation in collections of processing cells. We assume for simplicity that the activation of a cell depends only on the average value of local sensitivity throughout its junctional microstructure. (This assumption is made here for expository purposes only. In a more comprehensive model by Baron, 1970, this assumption is not made.) We define the *sensitivity* of unit (p, q), denoted by K_{pq}, to be the volume integral over the microstructure of the local sensitivity function. That is,

$$K_{pq} = \iiint k_{pq}(x, y, z)\, dx\, dy\, dz.$$

These sensitivity values correspond to the transmission coefficients of the optical systems. Memories, the static stored patterns, are preserved in these sensitivity values as we now show.

The dynamic patterns process information from one collection of cells to another. Dynamic patterns are described by giving the rate of activation and inhibition of the junctional contacts of each cell in the collection as a function of time. Dynamic patterns must be real-valued functions of time.

There is also a direct analogy between the transformations of the optical system as specified by its geometry and components, and the transformations of the network model. In the network model, patterns of excitation are transformed as they propagate from one layer of cells to another because of the *coupling coefficients* between cells. If, for example, one cell excites another at a high rate, the coupling between the two cells is high and positive. If the one cell tends to inhibit another, the coupling is negative. Coupling coefficients are functions of pairs of cells and must be specified for every connected pair of cells of a network.

Static and dynamic patterns interact in the network model in much the same way that they do in optical systems. In an optical system, light travels in a straight line and the 'coupling' is determined geometrically by the precise distance the light has to travel from its origin to its destination. In the optical system described above, the irradiating light wave is coherent, and therefore the phase of light transmitted from each point is the same. If the distance a light ray travels is a multiple of the wavelength, that ray adds; if the distance is an odd multiple of half the wavelength, the ray subtracts. These correspond to maximum excitatory and inhibitory coupling in the network model, respectively. See Figure 8. Because the distance that the light has to travel varies systematically across the transform plane, the Fourier transform of the input pattern is formed in the transform plane. As noted, in the network

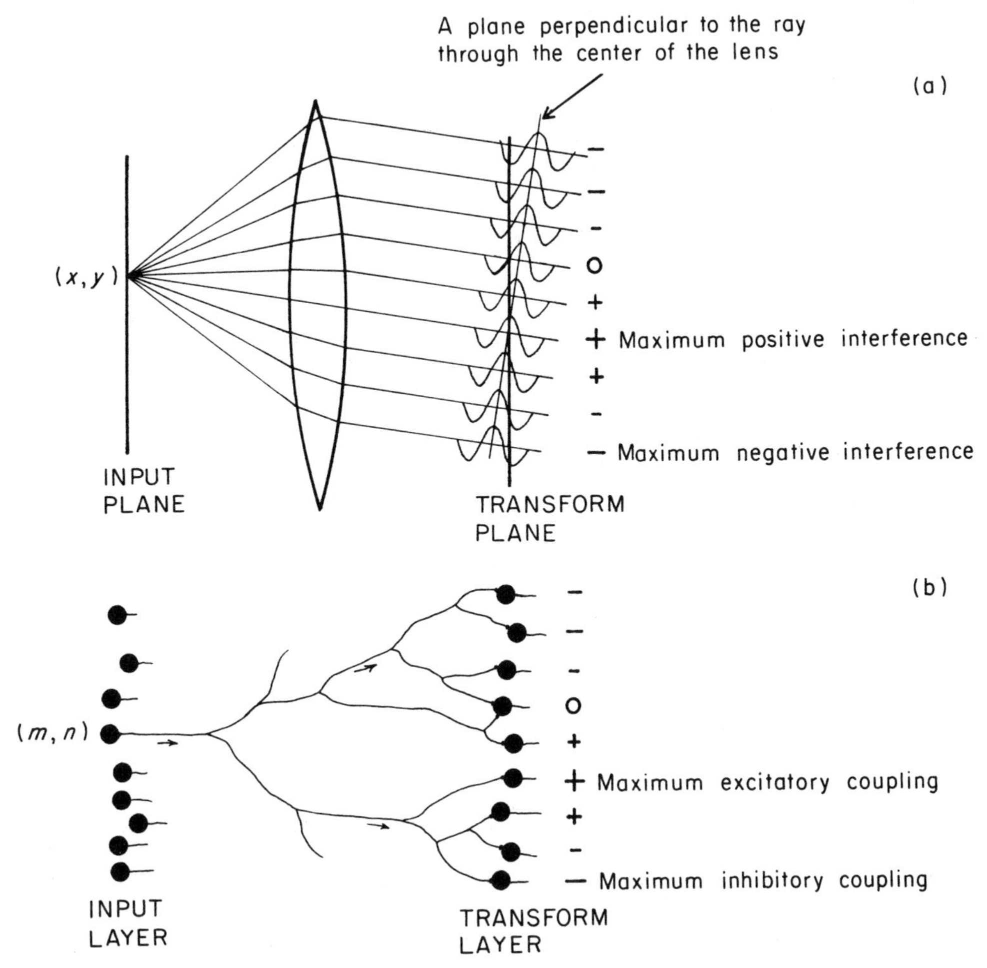

FIGURE 8.
Correspondence between optical and neural systems. The couplings in the neural system may be considered with respect to some positive baseline. Thus an 'inhibitory' coupling is in reality just the decrease in excitation below baseline level, and still could be excitatory in physiological terms.

model 'coupling' is determined by the amount of activity occurring in pairs of cells of an ensemble.

In the optical system, if the superposition of all light waves has amplitude A at point (x, y), and the film at point (x, y) has transmission coefficient $f(x, y)$, then the output pattern has amplitude $Af(x, y)$. In the network, if the spatial summation of all activity results in a net excitatory quantity A at cell (p, q) in the network, the contribution of activity to the next cell is determined by A and by the sensitivity K_{pq} of the receiving cell. The quantity A is not determined by coherence properties and it is at this point that a geometric analogy with the optical system fails. If the interactions between static and dynamic processes are multiplicative (which we will later propose) the analogy is direct, and the contribution from all prejunctional cells toward the activation of postjunctional cell (p, q) is given by AK_{pq}.

The resulting properties, storage, recognition, and recall depend on specific architecture, coupling coefficients, and sensitivity values, and are now to be studied in detail.

Achieving the Fourier Transformation

Consider the three-layered network illustrated in Figure 9a. We will call the cells in the three layers the *input cells*, *transform* or *memory cells*, and *output cells*, respectively. The activation patterns in these three collections of cells will correspond directly to the dynamic patterns which occur at the input, transform, and output planes of the optical system of Figure 3.

In order to make an analogy between the optical system and the network of Figure 3, we will need to choose a standard labeling scheme for the cells of the three collections. We will begin by assuming that cells are labeled with a two-dimensional system of variables. Thus a particular cell in a collection may be the (4, 6)th cell, or in general, we consider the (m, n)th cell. The activity of the (m, n)th cell at time t will be given by $f(m, n)(t)$. This corresponds to the dynamic input image $f(x, y)$ of the optical system.

For each pair of connected cells we must specify the coupling coefficient between that pair of cells. For the (m, n)th cell in the input layer and the (p, q)th cell in the transform layer we will designate by c_{pq}^{mn} the coupling coefficient. (Superscripts refer to prejunctional cells, and subscripts refer to postjunctional cells.) We assume that the coupling coefficients do not depend on the geometric position (x, y, z) in the network. We must also designate a local sensitivity function $k_{pq}(x, y, z)$ for each cell (p, q) of the transform layer. (For the analysis that follows, the only cells for which the local sensitivity values are of interest are the cells of the transform or memory layer. We will therefore assume for simplicity that the local sensitivity values of all other cells have value 1 and may therefore be omitted from the discussion.)

FIGURE 9.
In part a a neural network is described schematically. The network is analogous to the optical diagram of Figure 3. Part b represents a more realistic diagram of microstructure of synaptic domains in cortex. The ensemble of overlapping circles represented the junctions between branches of input axons and cortical dendrites. (Redrawn after Scheibel & Scheibel in Pribram, 1971.)

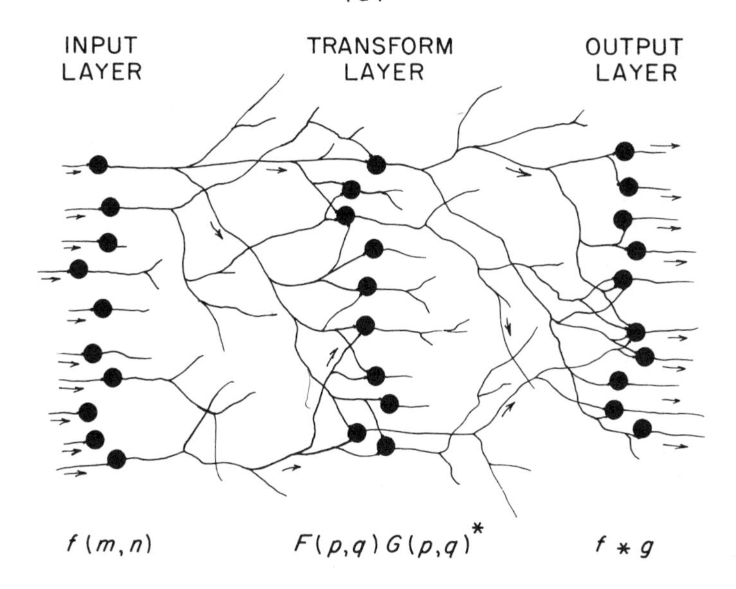
(a)
INPUT
LAYER
TRANSFORM
LAYER
OUTPUT
LAYER
f (m, n)
F (p,q) G (p,q)*
f * g

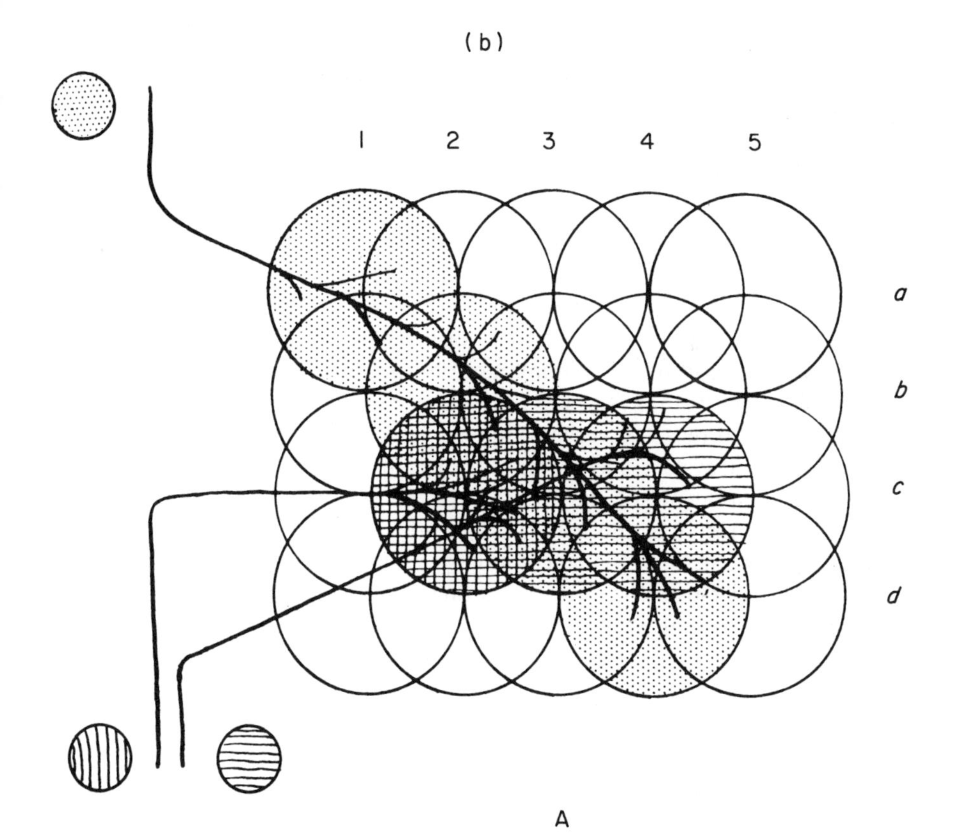
(b)
1
2
3
4
5
a
b
c
d
A

This network model corresponds to one basic computational element: one collection of cells that are closely coupled together. Our assumption is more precisely that every cell of the input layer makes connections to every cell of the transform layer, and every cell of the transform layer makes connections to every cell of the output layer. We will designate by $f(m, n)(t)$, $F(p, q)(t)$, and $f'(m', n')(t)$ the dynamic activity patterns in the input, transform, and output layers, respectively at time t. Since we designate by c_{pq}^{mn} the coupling coefficient between the (m, n)th input cell and the (p, q)th transform cell, then the net amount of excitation less the amount of inhibition provided by all input cells at any point in the transform layer of the network is given by

$$\sum_{(m,n)} c_{pq}^{mn} f(m, n)(t). \tag{1}$$

The rate of activation of the (p, q)th transform cell is given by

$$F(p, q)(t) = \iiint k_{pq}(x, y, z) \sum_{(m,n)} c_{pq}^{mn} f(m, n)(t)\, dx\, dy\, dz, \tag{2}$$

where integration would be over the volume containing all prejunctional connections of the transform unit. (In each case, if the rate of activation is negative, we assume the cell is inhibited and will not transmit at all.) Because the coupling coefficients do not depend on position, the integration can be performed and Equation 2 reduces to

$$F(p, q) = K_{pq} \sum_{(m,n)} c_{pq}^{mn} f(m, n)(t). \tag{3}$$

(In a somewhat more complicated model presented by Baron, 1970, the effective coupling between prejunctional and postjunctional cells during storage and recall depends systematically on time and geometric position (x, y, z) in the network. The result is that the local sensitivity values cannot be averaged. The addition of this 'timing mechanism' enables the network to store and recall patterns that vary as a function of time. Thus, for example, storage recognition and recall of verbal (auditory) information is possible.)

In the optical system, specific transformations resulted because of the geometry and components of the specific system. Under the analogy, transformations that result in the network depend directly on the coupling coefficients, c_{pq}^{mn}. We assume that the coupling coefficients are chosen so that the network Fourier-transforms the input image. To see how this might be done, we need only to look at the specific equations for Fourier transformation in optical systems. For the optical system, the Fourier transform is given by

$$F(p, q) = \iint \exp\left((px + qy)2\pi i\right) f(x, y)\, dx\, dy. \tag{4}$$

If the system is discrete rather than continuous, Equation 4 becomes

$$F(p, q) = \sum_{x=1}^{m} \sum_{y=1}^{m} M^{-1} \exp\left((px + qy)2\pi i/M\right) f(x, y), \tag{5}$$

where M^2 is the number in discrete points of the image.

For our network model, if we could choose the coupling coefficients in the analagous way, that is,

$$c_{pq}^{mn} = M^{-1} \exp\left((mp + nq)2\pi i/M\right), \tag{6}$$

the network would Fourier-transform the input pattern. However, we are not at liberty to use mathematically complex quantities. (Our early assumption was that dynamic patterns are nonnegative real functions.) In order to build a network model that resembles a biological neural network that will preserve the Fourier transform, we must preserve negative and complex quantities. In all optical systems described, negative and complex quantities were preserved by storing the intensity distribution of the superposition of the desired signal with a second (reference) light wave. Because coherent light was used as a source of illumination, a stable interference pattern resulted whose intensity distribution enabled recall (reconstruction) of the stored pattern. One possible mechanism for a neural network to encode negative and complex quantities is to allow independent neurons to convey the positive and negative, real and imaginary components of the desired signal. This approach was used by Baron (1970). Another possible method is to have the cells activated at a background rate (which represents an actual value of 0), and to assume that inhibitory effects reduce a cell's activity to below background firing. These slow rates would represent negative quantities. The real and complex components of the transform would still have to be preserved by independent cells. Several researchers have also considered coherent neural activity in direct analogy with the optical system. In each case an internal signal is used to insure coherence. See for example Westlake (1968), Swigert (1967), and Barrett (1969).

The Neural Hologram

We note that the assumed neural processing that underlies this formalism is relatively simple. Chemical transmitters are released at synaptic junctions between presynaptic and postsynaptic cells. These chemical transmitters diffuse across the synaptic cleft and modify the resting potential of the postsynaptic cell. The local modifications to the resting potential near the synaptic junctions propagate to some extent away from the junctions. Excitatory transmitters cause depolarizations; inhibitory transmitters cause hyperpolarizations in the dendritic microstructure of the postsynaptic cell. Excitatory and inhibitory contributions sum by spatial summation, and the result is a small contribution to the activation of that cell. The contribution to depolarization from any junction is determined by the local sensitivity at that junction that depends in turn on its membrane properties (see below). The local fluctuations, slow potentials, or minispikes interacting with the junctional microstructure, propagate toward the soma and cause it to depolarize at a rate given by Equations 2 or 3. The result is a linear transformation of the input depolarization pattern $f(m, n)(t)$ to the transform pattern $F(p, q)(t)$ through one stage of neural processing.

In the optical system, the exposure of an undeveloped piece of film to an optical pattern sensitizes the film so that development at a later time causes points that are exposed to bright light (light with a large amplitude) to become black upon development, and points that are exposed to dim light (or no light at all) to remain transparent. If one then makes a 'positive' image of the film, the result is that the transmission coefficients of points that received light having a high intensity will become high, and the transmission coefficients of points that received no light at all will become low. This is the optical hologram or 'memory trace' of the optical system. By analogy, we suggest that in the model during the 'exposure' period the local sensitivity values become altered in regions that receive a large net amount of activation and become altered in the opposite direction in regions that received a small net amount of activation. Some neurons store the 'real' part of the transform, other neurons store the 'imaginary' part. In particular, we propose that the local sensitivity values for the 'real' half of the population of junctions become proportional to the net (excitatory less inhibitory) amount of the activation that arrived during the exposure period, while for the 'imaginary' half we propose the opposite, i.e., deactivation (inhibition less excitation) occurs. That is,

$$K_{pq} \text{ is set proportional to } \sum_{(m,n)} (c_{pq}^{mn})^* f(m, n)(t), \tag{7}$$

where $f(m, n)(t)$ is the pattern to be stored. Because the net amount of arriving activation has a spatial distribution (Eq. 1) that is precisely the conjugate Fourier transform of the dynamic input pattern, the resultant distribution of local sensitivity values corresponds to one term of the optical Fourier hologram and it is for this reason that we call these 'memory traces' a neural hologram.

In short, a *neural hologram* is the pattern of sensitivity values that correspond to one element of an optical Fourier hologram and it is a function of the junctional microstructure of the memory units. The sensitivity values of a neural hologram preserve the conjugate Fourier transform of the patterns of excitation and inhibition that are initiated by the input.

Recognition and Recall

We have shown that Fourier transformation can be performed by a single stage of incoherent neural processing, at least in principle. We now extend our analogy between the optical system shown in Figure 3 and the three-layered network shown in Figure 9a. We assume that the transformation that occurs between the memory layer and the output layer is also a Fourier transformation. That is, the dynamic activation pattern of the output layer is the Fourier transformation of the dynamic pattern of the memory layer. Once this happens, there are predictable patterns of activity in the output cells, and this activity takes on two distinctly different forms.

For an arbitrary input pattern (i.e., the encoded form of the sensory stimulus), the output pattern is the cross correlation of the input pattern with the pattern represented by the local sensitivity values. In this case, the system gives a strong signal when the input pattern is similar to the stored pattern. This is the recognition process. In the optical system of Figure 3, the recognition signal appears as a bright spot in the output plane of the system. By analogy, the recognition signal for a memory node is the rapid firing of a small group of cells in the output layer. The sensitivity pattern has a 'focusing' effect on the surrounding activity, and when the input pattern is similar to the stored pattern, the recognition information is gathered at a group of cells. The activity in such cells could easily be monitored by other networks and used for selecting the memory nodes from which to later recall information. This process is completely analogous to recognition holography as described earlier, and to the cross-correlation process of the lens system.

By contrast, the output pattern has a completely different nature when the cells of the transform layer are excited in a uniform way. If the memory cells are uniformly excited, that is, by a single cell whose coupling coefficients are the same to all memory cells, the uniform activation is modified by the sensitivity values of the memory cells and the result is that each memory cell fires at a rate that is proportional to its sensitivity value. In this case, their pattern of depolarization is a reactivation of the stored pattern. It is, in fact, the conjugate Fourier transform of the pattern that is stored. The output pattern is in this case a copy of the original pattern of departure activity that was input when the local sensitivity values were established. The output pattern is a recalled copy of the stored information.

Alternate Models

The Fourier process is not the only process our formalism can usefully describe. Equation 3 gives the pattern of activity of the units in the memory layer in terms of the input pattern $f(m, n)(t)$, the coupling coefficients c_{pq}^{mn}, and the sensitivity values k_{pq}. The coupling coefficients that enabled the network to Fourier-transform the dynamic patterns of activity were an arbitrary choice, and in principle any choice of coupling coefficients could be made. In order for the network to be able to store information, the input transformation should not lose information. It must have an inverse. If the sensitivity values of the memory units preserve the transformed pattern and the transformation between the memory units and the output units produces the inverse transformation, then a uniform excitation of the memory cells will cause the stored pattern to be recalled.

However, if the network is to adequately recognize input information, the recognition signal should be much stronger when the stored pattern arrives than when an arbitrary different pattern arrives. It is well known that for a linear system (Turin, 1960), the best possible recognition signal is achieved when the transfer function of the system is the complex conjugate of the Fourier transform of the pattern to be recognized divided by the frequency spectrum of the noise. This is precisely the result achieved by the holographic model presented in this chapter. (We have assumed for simplicity that there is no systematic noise in the network. If noise is present in the system, the sensitivity values must be divided by the spectrum of the noise.) Thus, although other linear transformations can support the storage and recall processes, they are not as ideally suited for recognition as the Fourier processes.

A very recent neural holographic model was proposed by Cavanagh (1972) in direct analogy with the associative holographic system described earlier. Cavanagh proposes that two independent collections of input units contact the memory units, and that two independent collections of output units carry the departure patterns from each computational element of the system. The initial input patterns, say a and b are converted to the transformed patterns A and B by the first layer of processing, and the intensity pattern $(A + B)(A + B)^*$ is stored by the sensitivity values of the memory units. (The patterns A and B correspond to the arrival patterns A and B in the associative holography system described earlier.) The coupling between the memory units and the output units is assumed to have two properties: if no information is stored in the memory units, the departure patterns in the two collections of cells correspond to the arrival patterns. If, however, consolidation had occurred when arrival patterns A and B were present, then at a later time, input pattern a alone will cause the departure pattern b to be reconstructed, and vice versa. Thus, if a is present in the input, a 'ghost image' of b is automatically generated, and this reconstruction process is instantaneous. No search procedure or external control is necessary.

In contrast to our model, Cavanagh has demonstrated explicitly the two collections of input units that correspond to the two wavefronts of light used in optical holography. Also in contrast to our model, Cavanagh proposes that intensity values are stored rather than the real and complex quantities of the transformed image. The essential point is that many models can be suggested, and each one captures one or more of the essential features of optical holography. At present, each model must be considered as a suitable alternative to the other, and until more direct experimental evidence is at hand, no model can be used to preclude another.

Many models fall into the class of neural holographic storage models, and it is this general class of models we are trying to describe here.

NEURAL PROCESSES

Storage Properties

Network models of the holographic processes have accomplished several aims:

1. they have shown that optical systems are not required for holographic transformations to be realized; and
2. by keeping in mind some simple characteristics of nervous tissue, they have spelled out requirements for (a) temporary or permanent modification of any one layer of neurons by way of changes in the static patterns of sensitivity values within a junctional microstructure, and (b) processing between layers of neurons by specification of the coupling coefficients between them.

Let us begin by citing evidence consonant with the requirements for neural modification (storage, temporary, or permanent) demanded by the model. The model suggests that the arrival patterns to a unit—a neuron and its dendrites—produce a microstructure of slow potentials (depolarizations and hyperpolarizations) in the form of an interference pattern on that neuron's dendritic and somatic membranes. Eccles, Ito, and Szentagothai (1967) have described the mechanism of production of such interference patterns in the cerebellar cortex. At the regions of greatest constructive interference, changes are produced in the membrane's sensitivity to excitation and inhibition. Whenever new arrival patterns similar to the original occur, perturbations of these sensitive areas are produced. Thus the likelihood of depolarization and conduction in that cell are greater than when nonsimilar inputs arrive. It is precisely this, and *only* this, property that we have shown to be necessary for neural holography to be possible.

There is good reason to believe that a similar process occurs at the cerebral cortex. Benevento, Creutzfeldt, and Kuhnt (1973) have suggested on the basis of intracellular recordings that all input to the cortex results in excitatory (depolarizing) processes while the effects of horizontal interactions are essentially inhibitory (hyperpolarizing). Extracellular recordings, testing the effects of double simultaneous visual stimulation in our laboratory, are most readily interpreted in the same way (Phelps, 1972). These results give some justification to the emphasis given in our strictly holographic model to a single process by which depolarization and hyperpolarization affect changes in sensitivity values. Neither hyper- nor depolarization per se are therefore considered agents for membrane modification.

That this effect is restricted in locus is not unlikely: a rough inverse square law for the effects of slow graded potentials has been assumed by model builders since Beurle (1956) and has sufficient backing of evidence (e.g., Phelps, 1972) to be taken seriously.

The holographic storage hypotheses require that the modified synaptic sensitivity be proportional either to the input signal strength (our model) or to the square of the input signal strength (Cavanagh). This could be accomplished in the following way. Permanent or some reversible semipermanent change to a membrane would be proportional to the square of the voltage difference between neighboring input patterns of electrical activity, since perturbation in the postsynaptic domain is a function of the differences in distribution of hyper- and depolarizations produced by the arrival of input patterns of presynaptic potentials, resulting in voltage differences *parallel* to the postsynaptic membrane. A testable physical description of such a mechanism has been developed by Richard Gauthier for the special, though not unusual, case of two synapses from different axons forming adjacent junctions into a dendritic or somatic membrane. The interactions between excitatory and inhibitory processes can be conceived to occur somewhat like this: with no synaptic input to the membrane, there is a resting electric potential across the membrane, with the voltage gradient or electric field lines perpendicular to the membrane surface (Fig. 10a). Suppose that in neighboring terminals the input to the presynaptic terminals causes the postsynaptic junctions to become locally depolarized and hyperpolarized, respectively (Fig. 10b). The effect is to produce a pair of horizontally oriented electric dipoles at the surfaces of the cell membrane, which superimpose their electric fields on the vertical fields already present (Julesz, 1971; Barrett, 1969; Pribram, 1971). The net effect is to produce significant electric fields or voltage gradient components that are parallel to the surface of the membrane. We propose that these transient horizontal components of electric field trigger structural (e.g., conformational) changes in the membrane that outlast these horizontal fields. The induced structural changes in the membrane which in themselves may be reversible could then set off further biochemical processes leading to long-lasting ion permeability changes.

When either synapse is activated again, these structural or permeability changes can cause the effects (i.e., postsynaptic potentials) of one synaptic input to diffract and mimic the effects of the other, as if the latter were present. Thus the activation of one synapse produces the effects of activating both. The contribution of any such pair of synaptic inputs is small, but when many

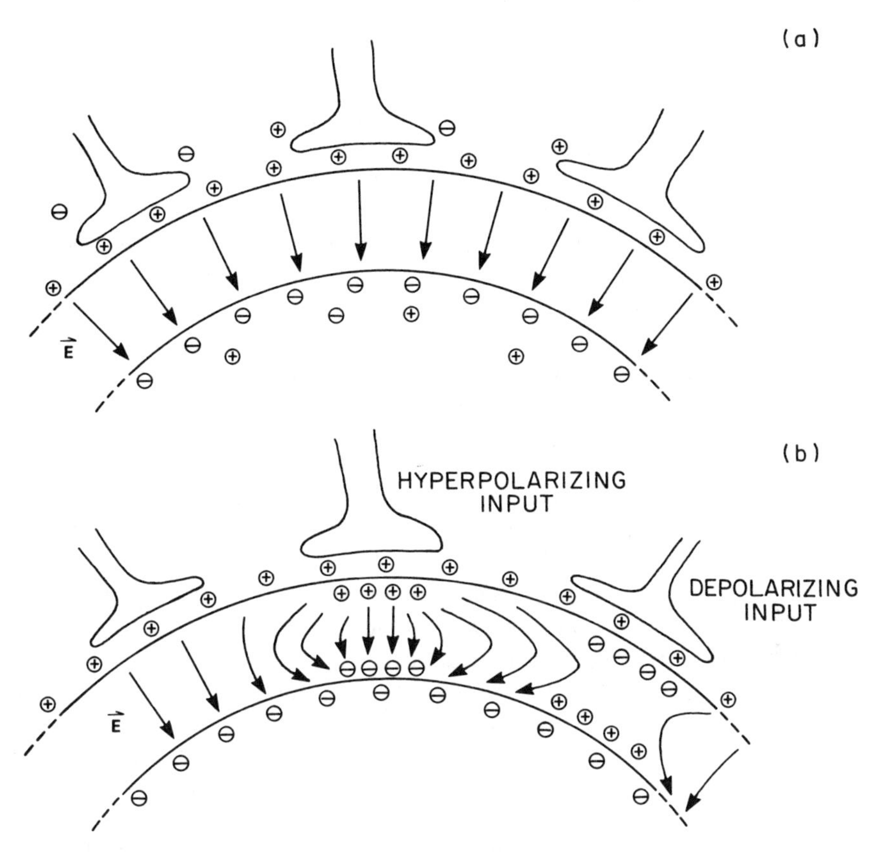

FIGURE 10.
Membrane electrical fields. The upper figure (a) shows the field when undisturbed. The lower figure (b) shows the field when inputs arrive, demonstrating the consequent horizontal components of the field within the membrane. (From Richard Gauthier, personal communication, 1972.)

identical effects throughout the microstructure of a synaptic domain are summed, the physiology of the network would be significantly affected. While this example concerns two adjacent synapses, it is important to keep in mind that the process also applies to the more remote effects that sets of synapses within a domain can have on each other through the interactions of the slow potential activity induced in the membrane of the postsynaptic cell.

These membrane changes should be detectable with present technology. Short- and long-term changes in sensitivity of a local area of membrane to excitation can be tested by means of intracellular recordings using large invertebrate neurons. In such preparations, cells whose membranes are repeatedly exposed to pairs of input patterns should come to produce equivalent output patterns even when only one input is received. In more complex neural aggregates such a demonstration has already been achieved (Chow, 1964; Dewson, Chow, & Engle, 1964).

The long-lasting modifications of membrane structure are most likely to occur postsynaptically. Although at a somewhat grosser level, evidence has recently been produced that in fact such postsynaptic changes do occur.

Recent evidence from electron micrography comparing the cortex of rats raised in deprived and enriched environments clearly shows the importance of the postsynaptic membrane in modifications of neural structure by experience. These studies show that though the number of synapses per unit volume of the cortex is greater in the deprived rats, a characteristic thickening of the postsynaptic membrane occurs in many of the synapses of the enriched rats (Rosenzweig, Bennett, & Diamond, 1972). Thus a homogeneous network becomes modified into a patterned one.

Processing Properties

Given plausibility for temporarily or more permanently storing the static pattern required by the models, what about the coupling parameters that provide their processing determinants? Let us start with the receptors of the visual system, the rods and cones of the retina—upon which a true visual image is produced by the optical lens system of the eyeball. Each rod and cone can be thought of as an omatidium—a relatively discrete independent intensity transducer of a small part of the total retinal image mosaic.

Now for a moment we turn to the various interactions that occur in the deeper layers of the retina as inferred from the output of the ganglion cell layer that has been studied so extensively by making extracellular microelectrode recordings from the optic nerve. Two major quantitative descriptions are of special interest here. The first details the mathematical description of the receptive field characteristics of ganglion cell responses. It is well known (Kuffler, 1953) that the ganglion cell responses are of two types—'on' cells characterized by increased firing of the central portion of the receptive field to onset of illumination, and 'off' cells characterized by increased firing to the offset of illumination. Both 'on' and 'off' cells show a roughly concentric arrangement of their receptive fields: an inhibitory or excitatory penumbra surrounds the center. These surrounds have been shown due to lateral inhibition—hyperpolarizations that are produced in the horizontal and amacrine networks that lie perpendicular to the input-output fibers of the retina. Mathematical descriptions of the relationships obtained in these receptive field configurations are reviewed by Ratliff (1965) and Rodieck (1965). Most of these descriptions involve a convolution of luminance change of the retinal input with inferred inhibitory characteristics of the network to compose the observed ganglion-cell receptive field properties. In short, a set of convolutional integrals has been found to adequately describe the transformations that occur between the retinal receptor mosaic and the ganglion cell output from the eye to the brain.

The second quantitative description comes from Enroth-Cugell and Robson (1966). In their account they have demonstrated in the range of ganglion-cell receptive fields a variety of relationships between center and surround, a finding also emphasized by Spinelli (1966). Enroth-Cugell, Robson, etc., then showed that they could explain this variety on the assumption of an opponent mechanism—separate excitatory (depolarizing) and inhibitory (hyperpolarizing) retinal processes, each process displaying an essentially Gaussian distribution.

Where do these separate excitatory and inhibitory processes take place? Dowling and Boycott (1965) have shown that, prior to the ganglion cell layer, few if any retinal neurons generate nerve impulses. All interactions are performed by way of slow potentials. These are of two opposing types—depolarizing excitations and hyperpolarizing inhibitory effects. Intracellular recordings (Svaetichin, 1967) have suggested that the excitatory potentials are generated along the input transmission paths (bipolar cells) of the retina while the inhibitory potentials are due to the horizontal layers (amacrine and horizontal cells) that cross the transmission channels. This 'lateral' inhibitory process has been studied extensively and made the basis for quantitative descriptions of sensory interaction by Hartline (see Ratliff, 1965) and by von Bekesy (1959) in their treatment of the Mach band phenomenon. The equations they invoke are similar to those used by Rodieck in his description of ganglion-cell receptive fields.

The gist of these experimental analyses is that the retinal mosaic becomes decomposed into an opponent process by depolarizing and hyperpolarizing slow potentials and transforms into more or less concentric receptive fields in which center and surround are of opposite sign. Sets of convolutional integrals fully describe this transformation.

The next cell station in the visual pathway is the lateral geniculate nucleus of the thalamus. The receptive field characteristics of the output from neurons of this nucleus are in some respects similar to the more or less concentric organization obtained at the ganglion cell level. Now, however, the concentric organization is more symmetrical, the surround usually has more clear-cut boundaries and is somewhat more extensive (e.g., Spinelli & Pribram, 1967). Furthermore, a second penumbra of the same sign as the center can be shown to be present though its intensity (number of nerve impulses generated) is not nearly so great as that of the center. Occasionally, a third penumbra, again of opposite sign, can be made out beyond the second (Hammond, 1972).

Again, a transformation has occurred between the output of the retina and the output of the lateral geniculate nucleus. Each geniculate cell acts as a peephole 'viewing' a part of the retinal image mosaic. This is due to the fact that each geniculate cell has converging upon it some 10,000 ganglion cell fibers. This receptive field peephole of each geniculate cell is made of concentric rings of opposing sign, whose amplitudes fall off sharply with distance from the center of the field. In these ways the transformation accomplished is like very near-field optics.

Pollen, Lee, and Taylor (1971), though supportive of the suggestion that the visual mechanism as a whole may function in a Fourier-like manner, emphasize that the geniculate output is essentially topographic and punctate, is not frequency specific, and does not show translational invariance—i.e., every illuminated point within the receptive field does not produce the same effect. Further, the opponent properties noted at the retinal level of organization are maintained and enhanced at the cost of overall translational invariance. Yet a step toward a discrete transform domain has been taken since the output of an individual element of the retinal mosaic—a rod or cone receptor —is the origin of the signal transformed at the lateral geniculate level.

When the output of lateral geniculate cells reaches the cerebral cortex, further transformations take place. One set of cortical cells, christened 'simple' by their discoverers (Hubel & Wiesel, 1968), has been suggested to be characterized by a receptive field organization composed by a literally linelike arrangement of the outputs of lateral geniculate cells. This proposal is supported by the fact that the simple-cell receptive field is accompanied by side bands of opposite sign and occasionally by a second side band of the same sign as the central field. Hubel and Wiesel proposed that these simple cells thus serve as line detectors in the first stage of a hierarchical arrangement of pattern detectors. Pollen et al. (1971) have countered this proposal on the basis that the output from simple cells varies with contrast luminance as well as orientation and that the receptive field is too narrow to show translational invariance. They argue, therefore, that an ensemble of simple cells would be needed to detect orientation. They suggest that such an ensemble would act much as the strip integrator used by astronomers to cull data from a wide area with instruments of limited topographic capacity (as is found to be the case in lateral geniculate cells). Whether in fact strip integration occurs, the linelike arrangement could be conceived as a preparatory step in Fresnel, Fourier or other frequency-type processing—now a slit rather than a peephole 'views' the retinal mosaic.

But it is not necessary to view simple cells only as way stations in a hierarchy—these cortical units clearly have functions in their own right. A series of ingenious studies by Henry and Bishop (1971) have confirmed that these simple cells are exquisitely tuned to the *edges* (luminance contrast) of lines in the visual receptive field independent of line width. Some are tuned to the leading, some to the trailing edges. Responses are of two types, excitatory and inhibitory, and very often show opponent properties: i.e., when the edge is moved in one direction across the receptive field the effect (e.g., excitation) is the converse of that (e.g., inhibition) produced when the edge is moved in the opposite direction. These investigators have shown that this effect is binocularly activated. Only when the excitation zones are in phase is an output signal generated. This occurs exclusively when the image on the two retinas superimposes—i.e., when 'objects' are in focus. Thus, simple cells act as gates that let pass only binocularly fused information.

Another class of cortical cells has generated great interest. These cells were christened 'complex' by their discoverers, Hubel and Wiesel, and thought by them (as well as by Pollen) to be the next step in the pattern recognition hierarchy. Some doubt has been raised (Hoffman & Stone, 1971) because of their relatively short latency of response as to whether all complex cells receive their input from simple cells. Whether their input comes directly from the geniculate or by way of simple cell processing, however, the output from complex cells of the visual cortex displays transformations of the retinal input, characteristic of the holographic domain.

A series of elegant experiments by Fergus Campbell and his group have suggested that these complex cortical cells are spatial-frequency sensitive elements. Initially, Campbell showed that the response of the potential evoked in man and cat by repeated flashed exposure to a variety of gratings of certain spacing (spatial frequency), adapted not only to that fundamental frequency but also to any component harmonics present. He concluded therefore that the visual system must be encoding spatial frequency (perhaps in Fourier terms) rather than the intensity values of the grating. He further showed that when a square wave grating was used adaptation was limited to the fundamental and its third harmonic as would be predicted by Fourier theory. Finally, he found neural units in the cat's cortex that behaved as did the gross potential recordings.

Pollen (1973) has evidence that suggests that these spatial-frequency sensitive units are Hubel and Wiesel's complex cells, although both his work and that of Maffei and Fiorentini (1973) have found that simple cells have the properties of spatial frequency filters, in that they are broadly sensitive to a selective band of spatial frequencies. In addition, the latter investigators have found that the simple cells can transmit contrast and spatial phase information in terms of two different parameters of their response: contrast is coded in terms of impulses per second and spatial phase in terms of firing pattern.

The receptive field of complex cells is characterized by the broad extent (when compared with simple cells) over which a line of relatively indeterminate length but a certain orientation will elicit a response. Pollen demonstrated that the output of complex cells was not invariant to orientation alone—number of lines and their spacing appeared also to influence response. He concluded, therefore, as had Fergus Campbell, that these cells were spatial frequency sensitive and that the spatial frequency domain was fully achieved at this level of visual processing. Additional corroborating evidence has recently been presented from the Pavlov Institute of Physiology in Leningrad by Glezer, Ivanoff, and Tscherbach (1973), who relate their findings on complex (and hypercomplex) receptive fields as Fourier analyzers to the dendritic neurostructure of the visual cortex much as we have done here.

There is, however, still another set of problems that must be disposed of before the conception of a spatial frequency transformation of the retinal image by cortex can be accepted. These problems deal with the tuning characteristics of each spatial frequency sensitive element and the extent of retinal image which this element transforms. The evidence to date suggests that each simple cell is rather broadly tuned, but that the receptive fields of complex cells are narrower in their tuning characteristics. Richards and Spitzberg (1972), among others, have suggested the pattern recognition mechanism be considered analogous to that which obtains for color where stepwise recombination of opponent processes sharpens broadly tuned receptor characteristics into a magnificent tool for subtle color perception. We have already detailed the evidence that shows opponent processes to exist at various levels of the visual pattern transformation mechanism. What remains to be done is to show quantitatively how by combinations of opponent processes, sharper tuning characteristics of the spatial frequency mechanism can be achieved.

The problem is not much different for obtaining the greater visual angle over which the spatial frequency mechanism must operate. Some combinational process must occur—the question is, where and how. Evidence pertaining to this point is presented toward the end of this chapter, but first let us take leave from the transform process and look again into the distribution mechanism without which the brain lesion effects cannot be explained.

A Limiting Specification

We thus have evidence that neural transformations occur in the visual system that could, given appropriate storage, result in Fresnel-like (simple cell) and Fourier-like (complex cell) holograms. But all problems are not yet out of the way. Perhaps the most critical current question that is posed by the holographic hypothesis of memory storage is the question of the extent to which input becomes distributed at any one stage of processing. As indicated in an earlier part of this chapter, visual inputs even to the complex cell level of the cortex still represent only a few degrees of visual space. Obviously, input does not become distributed over the whole brain in one pass, if ever. What are the limits on distribution that would yet allow one to use the hologram as a model? This question may not have a single answer and probably depends on the coding and control mechanisms available to the organism for this purpose at any given moment. There is considerable evidence from verbal learning experiments that rehearsal accomplishes internal distribution of the events rehearsed (Voss, 1969; Trabasso & Bower, 1968). Neisser (1967) points out that retinal translation should destroy the congruence necessary for recognition were a simple point-to-point template involved—and it does not. Moyer (1970) on the other hand, in a series of experiments, has shown that recognition at a nonexposed retinal locus is impaired when a complex unfamiliar pattern is presented tachistoscopically once to a restricted retinal locus. Even a single repetition of the exposure with no change of locus will, however, significantly enhance recognition at a distant locus. Rehearsal is obviously a potent source of distribution of information.

Another way of approaching this question is to ask just how much replication and how much distribution is demanded by the holographic hypothesis. Pollen's data suggest that in the visual cortex only small regions of visual space become encoded. However, Hubel and Wiesel (1968) describe considerable overlap of receptive fields within a single penetration for cells at the same orientation and preferred slit width. How this overlap becomes usefully integrated is an experimental question under present laboratory study (Pribram, 1974) and is discussed to some extent in the final section of the chapter.

The evidence cited above thus gives strong support to the concept that local regions of the cortex are responsible for storing the memories of experience. Does this contradict the results of the ablation studies cited earlier? No, for it is quite possible as already noted, that the proposed system, in response to rehearsal, stores multiple copies of the same experience in remote regions of the cortex, and that each of these records is a complete description of the given event. Evidence that in fact such multiple copies occur has been obtained: small macroelectrodes were implanted over the visual cortex of monkeys and electrical activity recorded in a discrimination experiment. In randomly distributed locations over the visual cortex, localized electrical activity was reliably found to be related to either the stimulus or the response (or reinforcement) events in the experiment (Pribram, Spinelli, & Kamback, 1967).

Let us suppose that the strictly holographic transformations are a local phenomenon, and that integration of information across the cortex is done either by a hierarchical summing onto the next level of connectivity, or by a parallel processing mechanism via subcortical connections. These alternatives are being explored experimentally at present (see Pribram, 1974, for review). Since the hierarchical alternative is almost universally espoused, it needs little explanation. Let us for a moment therefore consider the advantages of parallel processing mechanisms that make the experimental investigation worthwhile.

Relation to Control Processes

An essential distinction between a hierarchical serial process and a parallel process is that the latter allows control functions to be exercised before or at the transform or memory plane rather than after transformations and storage have been accomplished. In their operations, control functions can influence several biological memory processes that are independent of the particular storage model, yet each one can crucially affect the way in which the memory store operates. They are the following.

1. Permanence: Are the memory traces permanent (they never change once they are made), or are they temporary (the same memory location can store a different pattern at later times)?

2. Modifiability: Are the memory traces adaptive—do they change slowly after the initial consolidation process? If so, when are the modifications allowed to occur?

3. Consolidation: Do the memory traces become permanent upon a single presentation of the input pattern, or do they slowly become established upon repeated presentations of the same input information? If the latter is true, what are the control processes to insure that only the proper memory traces are allowed to consolidate at any given time?

4. Interference: If more than a single presentation of the input pattern is necessary, how sensitive is the nascent memory trace to interference?

5. Decay: Are the memory traces actually 'permanent' after they are established, or do they slowly degrade as a function of time? If they degrade, what is the rate of degradation? How does degradation affect recognition and recall?

Several memory *control functions* can operate on these storage functions. These control functions are not particular to the holographic model and in fact are not part of the holographic model. They are the following.

1. Start-stop: When will a particular memory location begin the 'exposure' process, and when will it finish? This start-stop process may be considered for a particular memory location, or it may be considered for all the memory locations in a memory store.

2. Selection: For a memory store, the question may be restated: why will one memory location store the current input information rather than another? The memory may be stored initially on the basis of innate competences of the neural tissue involved and later on the basis of changes in competence produced by experience. Or, memory locations may be arbitrarily (pseudorandomly) selected initially and then input deliberately channeled to compatible sites on the basis of temporal or conceptual contiguity.

3. Recognition threshold: An input is judged as 'recognized' when it evokes associated memories. The implication is that whenever memories are called

forth from an associative (content addressable) store, then the effects of an earlier similar input must indeed already be in the memory store. How similar must similar be to become recognized by the memory? The simulation of the holographic process developed here proposes that a cross correlation take place in parallel with all stored memory. Then if any region of any cross correlation exceeds some arbitrary threshold, the input is considered recognized. How is that threshold established and altered?

4. Recall: If storage is truly associative, recall occurs to the extent that an input evokes the effects of earlier associated inputs. How are the limits placed on such evocation? A control mechanism must be present that is able to decide which of the memory locations that have been addressed are to be used in further processing.

The above questions and comments are typical concerning the relationship between memory and its control functions and some of these have been touched on by Spinelli (1970) in the discussion of "Occam." What is important to understand is that the general holographic hypothesis is strictly a model of the interaction between storage and input processing and does not address these decisional questions. However, the holographic hypothesis does propose that these decisional properties involving storage, recognition, and recall occur by way of a content addressable parallel-processing mechanism. We therefore consider briefly the evidence that, in fact, the organization and processing of memory occurs in a content addressable parallel-processing system.

There are a growing number of experiments that are designed to determine whether or not the recognition of a stimulus and the retrieval of related (associated) information are sequential or parallel processes. Results presented by Sternberg (1969) and more recently by Atkinson and Juola (1974) suggest in fact that both sequential and parallel processes are involved. It is generally believed true that there is an initial sequence of processes that encode the stimulus information into the form used by the memory stores. There next appears to be an initial addressing of memory that results in a 'familiarity index' for the stimulus involved. This is the recognition signal or correlation value produced by an associative memory store. The reaction time studies indicate that this is a parallel process—that the encoded stimulus pattern is presented simultaneously to all memory locations (nodes or basic computational elements) of the memory stores.

On the other hand, the experiments indicate that once the familiarity of a stimulus item is obtained, additional information can only be recalled by a secondary sequential search of memory. This secondary search process locates specific items from among those which are related to the stimulus. The sequential search appears to be restricted to those memory locations that contain information associated with the stimulus—memory locations that recognized the encoded stimulus pattern. Furthermore, these experimental results imply that individual memory locations may be accessed independently and their stored information recalled.

SUMMARY

Our studies of holographic processes have detailed possible mechanisms for the distributed memory required by the results of experiments on brain function. Several issues became clarified. Holography depends on two separable functions: (a) storage of interference patterns or their equivalents; (b) patterns created by superposition or other Fresnel- or Fourier-like input processing. A network model of one limiting case of holography—the Fourier Hologram—has been accomplished using elements and junctional characteristics plausibly like those in neural networks. Other models have also been touched upon. In addition, evidence has been adduced that, in fact, holographic storage and spatial frequency—Fourier-like—processing occurs in the visual system.

Thus the advantages of a holographic memory as a model for brain function in perception can be fruitfully pursued with vigor. Aside from the property of distributed storage, holographic memories show large capacities, parallel processing, and content addressability for rapid recognition, associative storage for perceptual completion and for associative recall. The holographic hypothesis serves therefore not only as guide to neurophysiological experiment, but also as a possible explanatory tool in understanding the mechanisms involved in behaviorally derived problems in the study of memory and perception.

ACKNOWLEDGMENTS

We wish to thank Richard Gauthier, Erich Sutter, Arthur Lange, and Charles Stromeyer for their assistance and critical comments which enlivened considerably the writing of this paper.

REFERENCES

Atkinson, R. C., & Juola, J. F. Search and decision processes in recognition memory. In D. H. Krantz, R. C. Atkinson, R. D. Luce, and P. Suppes (Eds.), *Contemporary developments in mathematical psychology*. Vol. 1. San Francisco: Freeman, 1974.

Baron, R. J. A model for cortical memory. *J. of Mathematical Psychology*, 1970, **7,** 37–59.

Barrett, T. W. The cortex as interferometer: The transmission of amplitude, frequency and phase in cortical structures. *Neuropsychologia*, 1969, **7,** 135–148.

Benevento, L. A., Creutzfeldt, O. C., & Kuhnt, U. Significance of intracortical inhibition in the visual cortex: Data and model. *Nature*, 1973, in press.

Beurle, R. L. Properties of a mass of cells capable of regenerating pulses. *Philosophical Transactions of the Royal Society of London, Ser. B*, 1956, **240,** 55–94.

Cavanagh, J. P. Holographic processes realizable in the neural realm: Prediction of short-term memory and performance. Unpublished doctoral dissertation, Carnegie-Mellon University, 1972.

Chow, K. L. Bioelectrical activity of isolated cortex, III. Conditioned electrographic responses in chronically isolated cortex. *Neuropsychologia*, 1964, **2,** 175–187.

Chow, K. L. Visual discrimination after ablation of optic tract and visual cortex in cats. *Brain Res.*, 1968, **9,** 363–366.

Collier, R. J., Burckhardt, C. B., & Lin, L. H. *Optical holography.* New York: Academic Press, 1971.

Collier, R. J., & Pennington, K. S. Ghost imaging by holograms formed in the near field. *App. Phys. Lett.*, 1966, **8,** 44.

Creutzfeldt, O. D. General physiology of cortical neurons and neuronal information in the visual system. In M. B. A. Brazier (Ed.), *Brain and behavior.* Washington, D.C.: American Institute of Biological Sciences, 1961.

Dewson, J. H., III, Chow, K. L., & Engle, J., Jr. Bioelectrical activity of isolated cortex, II. Steady potentials and induced surface-negative cortical responses. *Neuropsychologia*, 1964, **2,** 167–174.

Dowling, J. E., & Boycott, B. B. Neural connections of the retina: Fine structure of the inner plexiform layer. *Quant. Biol.*, 1965, **30,** 393–402.

Eccles, J. C., Ito, M., & Szentagothai, J. *The cerebellum as a neuronal machine.* New York: Springer, 1967.

Enroth-Cugell, C., & Robson, J. G. The contrast selectivity of retinal ganglion cells of the cat. *J. Physiol.*, 1966, **187,** 517–552.

Gabor, D. A new microscopic principle. *Nature*, 1948, **161,** 777–778.

Galambos, R., Norton, T. T., & Frommer, C. P. Optic tract lesions sparing pattern vision in cats. *Experimental Neurology*, 1967, **18,** 8–25.

Glezer, V. D., Ivanoff, V. A., & Tscherbach, T. A. Investigation of complex and hypercomplex receptive fields of visual cortex of the cat as spatial frequency filters. *Vision Res.*, 1973, **13,** 1875–1904.

Goldscheider, A. Über die materiellen Veränderungen bei der Assoziationsbildung. *Neurol. Zentralblatt*, 1906, **25,** 146.

Goodman, J. W. *Introduction to Fourier optics.* San Francisco: McGraw-Hill, 1968.

Hammond, P. Spatial organization of receptive fields of LGN neurons. *J. Physiol.*, 1972, **222,** 53–54.

Henry, G. H., & Bishop, P. O. Simple cells of the striate cortex. In W. D. Neff (Ed.), *Contributions to sensory physiology.* New York: Academic Press, 1971.

Hoffman, K. P., & Stone, J. Conduction velocity of afferents to cat visual cortex: A correlation with cortical receptive field properties. *Brain Res.*, 1971, **32,** 460–466.

Horton, L. H. *Dissertation on the dream problem.* Philadelphia: Cartesian Research Society of Philadelphia, 1925.

Hubel, D. H., & Wiesel, T. N. Receptive fields and functional architecture of monkey striate cortex. *J. Physiol.*, 1968, **195,** 215–243.

Julesz, B. *Foundations of cyclopean perception.* Chicago: University of Chicago Press, 1971.

Julesz, B., & Pennington, K. S. Equidistributed information mapping: An analogy to holograms and memory. *J. Opt. Soc. Am.*, 1965, **55,** 604.

Kabrisky, M. *A proposed model for visual information processing in the human brain.* Urbana: University of Illinois Press, 1966.

Kraft, M. S., Obrist, W. D., & Pribram, K. H. The effect of irritative lesions of the striate cortex on learning of visual discrimination in monkeys. *J. Comp. Physiol. Psychol.*, 1960, **53,** 17–22.

Kuffler, S. W. Discharge patterns and functional organization of mammalian retina. *J. Neurophysiol.*, 1953, **16,** 37–69.

Lashley, K. S. The problem of cerebral organization in vision. In, *Biological symposia.* Vol. 7. *Visual mechanisms.* Lancaster, Pa.: Jacques Cattell Press, 1942.

Lashley, K. S. In search of the engram. In, *Society for experimental biology (Great Britain): Physiological mechanisms in animal behavior.* New York: Academic Press, 1950.

Loeb, J. *Comparative physiology of the brain and comparative psychology.* Science Series. New York: Putman, 1907.

Maffei, L., & Fiorentini, A. The visual cortex as a spatial frequency analyzer. *Vision Res.*, 1973, **13,** 1255–1267.

Mountcastle, V. B. Modality and topographic properties of single neurons of cat's somatic sensory cortex. *J. Neurophysiol.*, 1957, **20,** 408–434.

Moyer, R. S. On the possibility of localizing visual memory. Unpublished doctoral dissertation, Stanford University, 1970.

Neisser, U. *Cognitive psychology.* New York: Appleton-Century-Crofts, 1967.

Phelps, R. W. Inhibitory interactions in the visual cortex of the cat. Unpublished doctoral dissertation, Stanford University, 1972.

Pollen, D. A. Striate cortex and the reconstruction of visual space. In, *The neurosciences study program*, III. Cambridge, Mass.: MIT Press, 1973.

Pollen, D. A., Lee, J. R., & Taylor, J. H. How does the striate cortex begin the reconstruction of the visual world? *Science*, 1971, **173,** 74–77.

Preston, K., Jr. Use of the Fourier transformable properties of lenses for signal spectrum analysis. In J. T. Tippett, D. A. Berkowitz, L. C. Clapp, C. J. Koester, and A. Vanderburgh (Eds.), *Optical and electro-optical information processing.* Cambridge, Mass.: MIT Press, 1965.

Pribram, K. H. *Languages of the brain.* Englewood Cliffs, N.J.: Prentice-Hall, 1971.

Pribram, K. H. Why is it that sensing so much we can do so little? In, *The neurosciences study program*, III. Cambridge, Mass.: MIT Press, 1974.

Pribram, K. H., Spinelli, D. N., & Kamback, M. C. Electrocortical correlates of stimulus response and reinforcement. *Science*, 1967, **157,** 94–96.

Ratliff, F. *Mach bands: Quantitative studies in neural networks in the retina.* San Francisco: Holden-Day, 1965.

Richards, W., & Spitzberg, R. Spatial frequency channels: Many or few? *J. Opt. Soc. Am.*, 1972, **62,** 1394.

Rodieck, R. W. Quantitative analysis of cat retinal ganglion cell response to visual stimuli. *Vision Res.*, 1965, **5,** 583–601.

Rosenblatt, F. *Principles of neurodynamics: Perceptrons and the theory of brain mechanism*. Washington, D.C.: Spartan Books, 1962.

Rosenzweig, M., Bennett, E., & Diamond, M. Brain changes in response to experience. *Scientific American*, 1972, **2**, 22.

Spinelli, D. N. Visual receptive fields in the cat's retina: Complication. *Science*, 1966, **152**, 1768–1769.

Spinelli, D. N. Occam, a content addressable memory model for the brain. In K. H. Pribram and D. Broadbent (Eds.), *The biology of memory*. New York: Academic Press, 1970.

Spinelli, D. N., & Pribram, K. H. Changes in visual recovery function and unit activity produced by frontal and temporal cortex stimulation. *Electroenceph. Clin. Neurophysiol.*, 1967, **22**, 143–149.

Sternberg, S. Memory-scanning: Mental processes revealed by reaction-time experiments. *American Scientist*, 1969, **57**, 421–457.

Stroke, G. W. *An introduction to coherent optics and holography*. New York: Academic Press, 1966.

Svaetichin, G. Horizontal and amacrine cells of retina-properties and mechanisms of their control upon bipolar and ganglion cells. *Act. Cient.*, 1967, **18**, 254.

Swigert, C. J. Computational properties of a nerve and nerve net model. Unpublished doctoral dissertation, University of California, Berkeley, 1967.

Talbot, S. A., & Marshall, U. H. Physiological studies on neural mechanisms of visual localization and discrimination. *Amer. J. Ophthal.*, 1941, **24**, 1255–1264.

Tippett, J. T., Berkowitz, D. A., Clapp, L. C., Koester, C. J., & Vanderburgh, A. (Eds.) *Optical and electro-optical information processing*. Cambridge, Mass.: MIT Press, 1965.

Trabasso, T., & Bower, G. H. *Attention in learning theory and research*. New York: Wiley, 1968.

Turin, G. L. An introduction to matched filters. *IRE Transactions on Information Theory*, 1960, **6**, 311–329.

van Heerden, P. J. A new method of storing and retrieving information. *Applied Optics*, 1963, **2**, 387–392.

von Bekesy, G. Synchronism of neural discharges and their demultiplication in pitch perception on the skin and in hearing. *J. of Acoustical Society of America*, 1959, **31**, 338–349.

Voss, J. F. Associative learning and thought: The nature of an association and its relation to thought. In J. F. Voss (Ed.), *Approaches to thought*. Columbus, Ohio: Meredith, 1969.

Weiss, P. *Principles of development*. New York: Holt, 1939.

Werblin, F. S., & Dowling, J. E. Organization of the retina of the mud puppy, *Necturus maculosus*, II. Intracellular recording. *J. Neurophysiol.*, 1969, **32**, 339–355.

Werner, G. The topology of the body representation in the somatic afferent pathway. In, *The neurosciences study program*, II. New York: Rockefeller University Press, 1970.

Westlake, P. R. Towards a theory of brain functioning: A detailed investigation of the possibilities of neural holographic processes. Unpublished doctoral dissertation, University of California, Los Angeles, 1968.

Index
(VOLUME II)